HANDBOOK OF ARTIFICIAL INTELLIGENCE IN BIOMEDICAL ENGINEERING

Biomedical Engineering: Techniques and Applications

HANDBOOK OF ARTIFICIAL INTELLIGENCE IN BIOMEDICAL ENGINEERING

Edited by

Saravanan Krishnan, PhD

Ramesh Kesavan, PhD

B. Surendiran, PhD

G. S. Mahalakshmi, PhD

First edition published 2021

Apple Academic Press Inc.
1265 Goldenrod Circle, NE,
Palm Bay, FL 32905 USA
4164 Lakeshore Road, Burlington,
ON, L7L 1A4 Canada

CRC Press
6000 Broken Sound Parkway NW,
Suite 300, Boca Raton, FL 33487-2742 USA
2 Park Square, Milton Park,
Abingdon, Oxon, OX14 4RN UK

First issued in paperback 2021

Apple Academic Press exclusively co-publishes with CRC Press, an imprint of Taylor & Francis Group, LLC

Library and Archives Canada Cataloguing in Publication

Title: Handbook of artificial intelligence in biomedical engineering / edited by Saravanan Krishnan, PhD, Ramesh Kesavan, PhD, B. Surendiran, PhD., G. S. Mahalakshmi, PhD.

Names: Saravanan, Krishnan, 1982- editor. | Kesavan, Ramesh, editor. | Surendiran, B., editor. | Mahalakshmi, G. S., editor.

Series: Biomedical engineering series (Boca Raton, Fla.)

Description: Series statement: Biomedical engineering: techniques and applications | Includes bibliographical references and index.

Identifiers: Canadiana (print) 20200316303 | Canadiana (ebook) 20200316737 | ISBN 9781771889209 (hardcover) | ISBN 9781003045564 (ebook)

Subjects: LCSH: Artificial intelligence—Medical applications. | LCSH: Biomedical engineering.

Classification: LCC R859.7.A78 H36 2021 | DDC 610.28563—dc23

Library of Congress Cataloging-in-Publication Data

Names: Saravanan, Krishnan, 1982- editor. | Kesavan, Ramesh, editor. | Surendiran, B., editor. | Mahalakshmi, G. S., editor.

Title: Handbook of artificial intelligence in biomedical engineering / edited by Saravanan Krishnan, Ramesh Kesavan, B. Surendiran, G. S. Mahalakshmi.

Other titles: Biomedical engineering (Apple Academic Press)

Description: Palm Bay, FL : Apple Academic Press, [2021] | Series: Biomedical engineering: techniques and applications | Includes bibliographical references and index. | Summary: "Handbook of Artificial Intelligence in Biomedical Engineering focuses on recent AI technologies and applications that provide some very promising solutions and enhanced technology in the biomedical field. Recent advancements in computational techniques, such as machine learning, Internet of Things (IoT), and big data, accelerate the deployment of biomedical devices in various healthcare applications. This volume explores how artificial intelligence (AI) can be applied to these expert systems by mimicking the human expert's knowledge in order to predict and monitor the health status in real time. The accuracy of the AI systems is drastically increasing by using machine learning, digitized medical data acquisition, wireless medical data communication, and computing infrastructure AI approaches, helping to solve complex issues in the biomedical industry and playing a vital role in future healthcare applications. The volume takes a multidisciplinary perspective of employing these new applications in biomedical engineering, exploring the combination of engineering principles with biological knowledge that contributes to the development of revolutionary and life-saving concepts. Topics include: Security and privacy issues in biomedical AI systems and potential solutions Healthcare applications using biomedical AI systems Machine learning in biomedical engineering Live patient monitoring systems Semantic annotation of healthcare data This book presents a broad exploration of biomedical systems using artificial intelligence techniques with detailed coverage of the applications, techniques, algorithms, platforms, and tools in biomedical AI systems. This book will benefit researchers, medical and industry practitioners, academicians, and students"-- Provided by publisher.

Identifiers: LCCN 2020038313 (print) | LCCN 2020038314 (ebook) | ISBN 9781771889209 (hardcover) | ISBN 9781003045564 (ebook)

Subjects: MESH: Artificial Intelligence | Biomedical Engineering--methods | Medical Informatics Applications

Classification: LCC R855.3 (print) | LCC R855.3 (ebook) | NLM W 26.55.A7 | DDC 610.285--dc23

LC record available at https://lccn.loc.gov/2020038313

LC ebook record available at https://lccn.loc.gov/2020038314

ISBN: 978-1-77188-920-9 (hbk)
ISBN: 978-1-77463-761-6 (pbk)
ISBN: 978-1-00304-556-4 (ebk)

ABOUT THE BOOK SERIES: BIOMEDICAL ENGINEERING: TECHNIQUES AND APPLICATIONS

This new book series aims to cover important research issues and concepts of the biomedical engineering progress in alignment with the latest technologies and applications. The books in the series include chapters on the recent research developments in the field of biomedical engineering. The series explores various real-time/offline medical applications that directly or indirectly rely on medical and information technology. Books in the series include case studies in the fields of medical science, i.e., biomedical engineering, medical information security, interdisciplinary tools along with modern tools, and technologies used.

Coverage & Approach

- In-depth information about biomedical engineering along with applications.
- Technical approaches in solving real-time health problems
- Practical solutions through case studies in biomedical data
- Health and medical data collection, monitoring, and security

The editors welcome book chapters and book proposals on all topics in the biomedical engineering and associated domains, including Big Data, IoT, ML, and emerging trends and research opportunities.

Book Series Editors:
Raghvendra Kumar, PhD
Associate Professor, Computer Science & Engineering Department,
GIET University, India
Email: raghvendraagrawal7@gmail.com

Vijender Kumar Solanki, PhD
Associate Professor, Department of CSE, CMR Institute of Technology (Autonomous), Hyderabad, India
Email: spesinfo@yahoo.com

Noor Zaman, PhD
School of Computing and Information Technology, Taylor's University, Selangor, Malaysia
Email: noorzaman650@hotmail.com

Brojo Kishore Mishra, PhD
Professor, Department of CSE, School of Engineering, GIET University, Gunupur, Osidha, India
Email: bkmishra@giet.edu

FORTHCOMING BOOKS IN THE SERIES

The Congruence of IoT in Biomedical Engineering: An Emerging Field of Research in the Arena of Modern Technology
Editors: Sushree Bibhuprada B. Priyadarshini, Rohit Sharma, Devendra Kumar Sharma, and Korhan Cengiz

Handbook of Artificial Intelligence in Biomedical Engineering
Editors: Saravanan Krishnan, Ramesh Kesavan, and B. Surendiran

Handbook of Deep Learning in Biomedical Engineering and Health Informatics
Editors: E. Golden Julie, S. M. Jai Sakthi, and Harold Y. Robinson

Biomedical Devices for Different Health Applications
Editors: Garima Srivastava and Manju Khari

Handbook of Research on Emerging Paradigms for Biomedical and Rehabilitation Engineering
Editors: Manuel Cardona and Cecilia García Cena

High-Performance Medical Image Processing
Editors: Sanjay Saxena and Sudip Paul

ABOUT THE EDITORS

Saravanan Krishnan, PhD, is Senior Assistant Professor in the Department of Computer Science & Engineering at Anna University, Regional Campus, Tirunelveli, Tamilnadu, India. He has 14 years of experience in academia and the IT industry and has published papers in 14 international conferences and 24 international journals. He has also written six book chapters and has edited three books with international publishers. He has conducted four research projects and two consultancy projects with the total worth of Rs.70 Lakhs. He is an active researcher and academician, and he is reviewer for many reputed journals. He also received an outstanding reviewer certificate from Elsevier, Inc. He is a Mentor of Change, for Atal Tinkering Lab of NITI Aayog, and has professional membership with several organizations. He previously worked at Cognizant Technology Solutions, Pvt Ltd. as software associate. He completed his ME (Software Engineering) in 2007 and earned his PhD in 2015.

Ramesh Kesavan, PhD, is Assistant Professor in the Department of Computer Applications, Anna University Regional Campus, Tirunelveli, India. His areas of research include cloud computing, big data analytics, data mining, and machine learning. He earned his PhD degree in Computer Science from Anna University, Chennai, India.

B. Surendiran, PhD, is Associate Dean (Academic) and Assistant Professor in the Department of Computer Science and Engineering at the National Institute of Technology, Puducherry, Karaikal, India. His research interests include medical imaging, machine learning, dimensionality reduction, and intrusion detection. He has published over 20 papers in international journals and has several conference publications to his credit. He is an active reviewer for various SCI and Scopus journals. He earned his PhD at the National Institute of Technology, Tiruchirappalli, India.

G. S. Mahalakshmi, PhD, is Associate Professor in the Computer Science and Engineering department at College of Engineering, Guindy, Anna University, Chennai, INDIA. She has vast research experience and published 180 papers in reputed journals and international conferences. She is also deputy director for Centre for Entrepreneurship Development, Anna University. She is active reviewer for various SCI, Scopus Journals. Her research interests include machine learning, artificial intelligence, text mining, and natural language processing.

CONTENTS

CONTRIBUTORS

S. Anto
School of Computer Science and Engineering, Vellore Institute of Technology, Vellore, India

Manivanna Boopathi Arumugam
Instrumentation & Chemicals Division, Bahrain Training Institute, Kingdom of Bahrain

Sarojini Balakrishanan
Department of Computer Science, Avinashilingam Institute for Home Science and Higher Education for Women, Coimbatore 641043, India

B. Bhavya
Deloitte Consulting India Private Limited, Bengaluru, Karnataka

Bichitrananda Behera
Department of Computer Science, Pondicherry University, Karaikal, India

J. V. Bibal Benifa
Department of Computer Science and Engineering, Indian Institute of Information Technology, Kottayam, India

Deya Chatterjee
Department of Computer Science and Engineering, SRM Institute of Science and Technology, Kattankulathur, Chennai 603203, India

Zafer Comert
Department of Software Engineering, Samsun University, Turkey

D. Renuka Devi
Department of Computer Science, IDE, University of Madras, Chennai 600005, Tamil Nadu, India

S. Siamala Devi
Department of Computer Science and Engineering, Sri Krishna College of Technology, Coimbatore, India

J. Satya Eswari
Department of Biotechnology, National Institute of Technology Raipur, Raipur, Chhattisgarh 492010, India

N. Gopikarani
Department of Computer Science and Engineering, PSG College of Technology, Coimbatore, Tamil Nadu

S. Shymala Gowri
Department of Computer Science and Engineering, PSG College of Technology, Coimbatore, Tamil Nadu

Vinit Kumar Gunjan
Department of Computer Science & Engineering, CMRIT, Hyderabad, India

Praveen Kumar Gupta
Department of Biotechnology, R. V. College of Engineering, Bangalore, India

Lingaiya Hiremat
Department of Biotechnology, R. V. College of Engineering, Bangalore, India

Dennis Hsu
Department of Computer Science, San Jose State University, San Jose, CA, USA

K. R. Jothi
School of Computer Science and Engineering, Vellore Institute of Technology, Vellore, India

Rhutu Kallur
Department of Electronics and Communication, R. V. College of Engineering, Bangalore, India

K. V. N. Kavitha
School of Electronics Engineering, Vellore Institute of Technology, Vellore, India

T. Ananth Kumar
Department of Computer Science and Engineering, IFET college of Engineering, Tamil Nadu, India

G. Kumaravelan
Department of Computer Science, Pondicherry University, Karaikal, India

R. Lokeshkumar
School of Computer Science and Engineering, Vellore Institute of Technology, Vellore, India

P. Mahalakshmi
Department of Electronics and Communication Engineering, Anna University Regional Campus, Tirunelveli, Tamil Nadu, India

M. Manonmani
Department of Computer Science, Avinashilingam Institute for Home Science and Higher Education for Women, Coimbatore 641043, India

S. Shyni Carmel Mary
Department of Computer Science, IDE, University of Madras, Cheapuk, Chennai 600 005, Tamil Nadu, India

G. Venifa Mini
Department of Computer Science and Engineering, Noorul Islam Centre for Higher Education, Kumaracoil, India

Diane Moh
College of Pharmacy, Touro University, Vallejo, CA, USA

Melody Moh
Department of Computer Science, San Jose State University, San Jose, CA, USA

Teng-Sheng Moh
Department of Computer Science, San Jose State University, San Jose, CA, USA

Saurabh Mukherjee
Banasthali Vidyapith Banasthali, Rajasthan, India

Srilakshmi Mutyala
Stratalycs Technologies Pvt. Ltd., Bangalore, India

A. Aafreen Nawresh
Department of Computer Science, Institute of Distance Education, University of Madras, Chennai, India
E-mail: anawresh@gmail.com

G. Niranjana
Department of Computer Science and Engineering, SRM Institute of Science and Technology

K. Padmavathi
Department of Computer Science, PSG College of Arts and Science, Coimbatore 641014, Tamil Nadu, India

Rashmi Pathak
Siddhant College of Engineering, Sudumbre, Pune, Maharashtra, India

P. Sivananaintha Perumal
Department of Computer Science and Engineering, Manonmaniam Sundaranar University, Tirunelveli, India

Puja Sahay Prasad
Department of Computer Science & Engineering, GCET, Hyderabad, India

N. Hema Priya
Department of Information Technology, PSG College of Technology, Coimbatore, Tamil Nadu

S. Suja Priyadharsini
Department of Electronics and Communication Engineering, Anna University Regional Campus, Tirunelveli, Tamil Nadu, India

Sindhu Rajendran
Department of Electronics and Communication, R. V. College of Engineering, Bangalore, India

R. S. Rajesh
Department of Computer Science and Engineering, Manonmaniam Sundaranar University, Tirunelveli, India

T. V. K. Hanumantha Rao
Department of Electronics and Communication Engineering, National Institute of Technology, Warangal, Telangana 506004, India, E-mail: tvkhrao75@nitw.ac.in

A. S. Saranya
Department of Computer Science, PSG College of Arts and Science, Coimbatore 641014, Tamil Nadu, India

S. Sasikala
Department of Computer Science, IDE, University of Madras, Cheapuk, Chennai 600 005, Tamil Nadu, India

S. Arunmozhi Selvi
Department of Computer Science and Engineering, Manonmaniam Sundaranar University, Tirunelveli, India

A. Sharmila
School of Electrical Engineering, Vellore Institute of Technology, Vellore, India

Christa I. L. Sharon
Department of Information Science and Engineering, Dayananda Sagar College of Engineering, Bangalore, Karnataka, India

Vidhya Shree
Department of Electronics and Instrumentation, R. V. College of Engineering, Bangalore, India

Rajendran Sindhu
Department of Electronics and Communication, R.V. College of Engineering, Bangalore 560059, India

Pradeep Singh
Department of Computer Science and Engineering, National Institute of Technology Raipur, Raipur, Chhattisgarh 492010, India

P. Srividya
Department of Electronics and Communication, R.V. College of Engineering, Bangalore 560059, India

J. Stalin
Department of Computer Science and Engineering, Manonmaniam Sundaranar University, Tirunelveli, India

V. Suma
Department of Information Science and Engineering, Dayananda Sagar College of Engineering, Bangalore, Karnataka, India

P. Sundareswaran
Department of Computer Science and Engineering, Manonmaniam Sundaranar University, Tirunelveli, India

Meghamadhuri Vakil
Department of Electronics and Communication, R. V. College of Engineering, Bangalore, India

Santhosh Kumar Veeramalla
Department of Electronics and Communication Engineering, National Institute of Technology, Warangal, Telangana 506004, India

Subha Velappan
Department of Computer Science & Engineering, Manonmaniam Sundaranar University, Tirunelveli, India

Hao-Yu Yang
CuraCloud Corporation, Seattle, WA, USA

ABBREVIATIONS

ABC	artificial bee colony
ACHE	adaptive energy efficient cluster head estimation
ACO	ant colony optimization
ADEs	adverse drug events
ADHD	attention deficit hyperactivity disorder
AI	artificial intelligence
ANNs	artificial neural networks
API	application program interface
apriori TID	apriori algorithm for transaction database
ARM	augmented reality microscope
AUC	area under curve
BANs	body area networks
BCOA	binary cuckoo optimization algorithm
BIDMC	Beth Israel Deaconess Medical Center
BP	backpropagation
BPN	backpropagation network
CAD	computer-aided diagnosis
CADs	coronary artery diseases
CART	classification and regression tree
CBR	case-based reasoning
CCA	clear channel assessment
CDSS	clinical decision support systems
CFCNs	cascade of completely convolutional neural systems
CKD	chronic kidney disease
CM	classification measure
CNN	convolutional neural arrange
CNN	convolutional neural network
CRFs	conditional random fields
CT	computed tomography
CTG	cardiotocography
CV	cross-validation
CVS	cross-validation score
DAG	directed acyclic graph
DBN	deep belief network

DILI	drug-induced liver damage
DIP	digital image processing
DME	diabetic macular edema
DNN	deep neural network
DP	data preprocessing
DPW	distributional profile of a word
DPWC	distributional profile of multiple word categories
DRE	digital rectal examination
DT	decision tree
DTC	decision tree classifier
DTFs	directed transfer functions
DWT	discrete wavelet transform
EC	ensemble classifier
ECG	electrocardiography/electrocardiogram
EEG	electro-encephalography
EHG	electrohysterography
EHR	electronic health record
ELM	extreme learning machine
EMG	electromyography
EOG	electrooculogram
FCN	fully convolutional network
FE	feature extraction
FER	facial expression recognition
FFNNs	feedforward neural networks
FHR	fetal heart rate
fMRI	functional MRI
FN	false negative
FP	false positive
FP	frequent pattern
FS	feature selection
FS	Fisher score
GA	genetic algorithm
GAN	generative adversarial network
GA-SA	genetic algorithm-simulated annealing
GC	Granger causality
GC	grouping and choosing
GD	Gaussian distribution
GE	General Electric
GLCM	gray level co-occurrence matrices
GLMC	gray-level co-occurrence lattice

GNB	Gaussian naïve Bayes
HBA	heartbeat analysis
HBC	human body communication
HEDIS	Healthcare Effectiveness Data and Information Set
ID3	iterative dichotomiser-3
IG-OBFA	OBFA melded with Information Gain
IMFs	intrinsic mode functions
IR	information retrieval
KDE	kernel density estimation
KNN	*k*-Nearest neighbor
LDA	linear discriminant analysis
LGR	logistic regression
LNMF	local non-negative matrix factorization
LOS	line of sight
LOSO	leave-one-subject-out
LS	least square
LSI	latent semantic indexing
LSSVM	late slightest squares bolster vector machine
LSTM	long short-term memory
LYNA	lymph node assistant
MAC	medium access control
MC-CNN	multi-edit convolution neural organize
MEG	magnetoencephalography
ML	machine learning
MLP	multilayer perceptron
MMK	medical monitor kit
MRF	Markov random field
MRI	magnetic resonance imaging
MTRs	medical transportation robots
MVAR	multivariate autoregressive
NB	naïve Bayes
NB	narrow band
NEE	neural edge enhancer
NFs	network filters
NLP	natural language processing
NMF	non-negative matrix factorization
NN	neural network
OBFA	Opposition-Based Firefly Algorithm
OBL	opposition-based learning
OWC	optical wireless communication

OWL	web ontology language
PA	passive–aggressive
PCA	principal component analysis
PDC	partial directed coherence
PET	positron emission tomography
PF	particle filter
PID	PIMA Indians Diabetes
POS	Part of Speech
PPDM	privacy-preserving data mining
PPN	perceptron
PSA	prostate-specific antigen
PSG	polysomnography
PWM	position weight matrices
QSTR	quantity structure toxicity relationship
RADBAS	radial basis function
RBF	radial basis function
RBM	restricted Boltzmann machine
RBNN	radial basis function neural network
RC	Rocchio classifier
RDD	resilient distributed dataset
RDF	resource description framework
RElim	recursive elimination
ReLU	rectified linear unit
REM	rapid eye movement
RF	random forest
RFC	random forest classifier
RFSs	random finite sets
RNN	recurrent neural network
RNN	repetitive or recursive neural arrange
ROI	region of interest
SA	simulated annealing
SGD	stochastic gradient descent
SGN	stochastic gradient descent
SI	swarm intelligence
SLFN	single hidden layer feed forward network
SOM	self-organizing map
SVM	support vector machine
TEEN	threshold sensitive energy efficient network
TF	term frequency
TN	true negative

TP	true positive
TRIBAS	triangular basis function
TS	training set
UMLS	Unified Medical Language System
UWB	ultra-wide band
VAE	variationalautoencoder
VLC	visible light communication
VSM	vector space method
VSVM	vicinal back vector machine
WAC	weighted associative classifier
WBAN	wireless body area network
WDBC	Wisconsin Diagnostic Breast Cancer
WSNs	wireless sensor networks

PREFACE

Biomedical engineering is a multidisciplinary field that applies engineering principles and materials to medicine and healthcare. The combination of engineering principles with biological knowledge has contributed to the development of revolutionary and life-saving concepts. Artificial intelligence (AI) is an area of computer science and engineering that provides intelligence to machines. AI in biomedical engineering uses machine-learning algorithms and software to analyze complicated medical data and perform automatic diagnosis.

With the recent rapid advancement in digitized medical data acquisition, wireless medical data communication, and computing infrastructure, AI has drastically changed medical practice. AI has wide applications in the field of biomedical engineering, namely, health applications, wearable devices, medical image processing, telemedicine, and surgical robots.

Biomedical engineering applications are associated with many domains, such as Big Data, IoT, machine learning, and AI. Many technologies, modern tools, and methodologies have been adopted in this field. Information technology solutions empower biomedical engineering and healthcare. AI contributes many research advancements in medical applications.

This book is focussed on recent AI technologies and applications that contribute to the biomedical field. This edited book explores the applications and the research solutions. The book is organized into 22 chapters as follow:

Chapter 1 explores an expert system based on fuzzy logic and ant colony optimization (ACO) for automatic diagnosis and clinical decision-making. It also proposes an expert system based on hybrid genetic algorithm-simulated annealing and support vector machine (GASA-SVM) for disease diagnosis. Finally, it suggests a decision-support system based on Fisher score-extreme learning machine-simulated annealing.

Chapter 2 describes the data acquisition prospects in biomedical systems and the different approaches that can be followed in the knowledge representation. Further, it describes the design issues and feature-selection techniques that can be adapted to achieve an optimum learning model. It also addresses the design and validation challenges faced while adapting machine learning techniques in the biomedical engineering specialization.

Chapter 3 describes the uses of AI and their related techniques in the biomedical and healthcare. This chapter also explore the field of biomedical and informatics within the branches of AI.

Chapter 4 gives an insight into the diagnosis of heart disease using hybrid genetic algorithm. It describes the basis of genetic algorithm and its limitations. A hybrid genetic algorithm is proposed by combining genetic algorithm and classification techniques. It also brief classical decision algorithms, includes CHAID algorithm, CART algorithm, ID3 algorithm, C4.5 algorithm, and C5.0. Finally, it describes a novel hybrid genetic algorithm devised by combining image processing technique.

Chapter 5 gives a comprehensive description of healthcare applications. Furthermore, a number of applications such as breast mass lesion, pharmaceutical, and rehabilitation robotics are included. Finally, it provides a summary of healthcare applications of biomedical AI system.

Chapter 6 introduces the current status of AI in healthcare. Motivated by the widespread applications of AI techniques in biomedical field, it discusses the current applications and its issues in detail. It also reviews the real-time application of AI.

Chapter 7 gives a clear insight into basic biomedical research, translational research, and clinical practice. It also gives a detailed survey on biomedical imaging capturing techniques like X-ray, computed tomography (CT) scan, positron emission tomography (PET), magnetic resonance imaging (MRI), and ultrasound. It also illustrates AI-based biomedical imaging components developed by various companies. Finally, it concludes with the challenges in the field of biomedical imaging.

Chapter 8 deals with the prediction of heart disease. A comparative study is done with KNN, decision tree, support vector machine, random forest, and multilayer perceptron algorithms for heart disease detection. It also concludes that a random forest algorithm gives a better prediction ratio.

Chapter 9 is a review on an AI-based patient monitoring system. It discusses the key biomedical challenges and their wide range of AI-enabled medical applications. The chapter also discusses different segmentation approaches and their advantages and disadvantages. It also gives a short insight on prognostic models and the usage of prognostic scores to quantify severity or intensity of diseases.

Chapter 10 deals with the management of heterogeneous healthcare data. It describes semantic annotation process based data handling and methods for incorporating semantic models in an AI expert system for the prediction of chronic diseases. The main objective envisaged in this chapter is to propose a semantic annotation model for identifying patients suffering from chronic kidney disease (CKD). This chapter stresses the need for achieving semantic annotation by surpassing the implementation challenges by using ontology.

Chapter 11 describes a drug side effect analysis system based on reviews from the social media. It details the techniques to extract information from social media website Twitter using sentiment analysis and machine learning. Then, it describes the procedure to process and handle large datasets using distributed computing through Apache Spark. This is followed by an experimental results analysis of the best and most efficient ways to correctly extract the frequency of adverse drug events using the techniques previously described. Afterward, a detailed pharmaceutical analysis is provided for the results, with insight from a domain expert.

Chapter 12 focuses on deep learning applications in brain image segmentation. It starts off with a brief introduction of brain imaging modalities, image segmentation, followed by the essential image processing procedures. Then it gives an introduction to the basic building blocks of CNNs and lays the foundation for modern CNN architectural designs. Next, it reviews the state-of-the-art deep learning models for brain segmentation and draw comparisons with traditional machine learning methods. Finally, this chapter discusses the current state, challenges of clinical integration, and future trends of deep learning in brain segmentation.

Chapter 13 deals with security and privacy issues in biomedical AI systems. It outlines various ranges of security and privacy threats to biomedical AI systems such as linkage attacks, inference attacks, adversarial examples, and so on. Similarly, solutions to such problems have been discussed, such as conventional techniques like auditing, etc., and newer advancements in research like differential privacy and federated learning.

Chapter 14 proposes a real-time patient monitoring system. It describes a visible-light-based wireless technology called Li-Fi for real-time monitoring. It proposes a Li-Fi Monitoring System framework (LiMoS) and also gives a detailed note on the components of the system. It also describes the procedure for experimental setup. Finally, it concludes with a result analysis to justify the performance of the system.

Chapter 15 involves a comparative analysis of facial expression recognition techniques using the classic machine learning algorithms—*k*-nearest neighbor (KNN), support vector machine (SVM), ensemble classifiers, and the most advanced deep learning technique using convolutional neural networks.

Chapter 16 offers insights into the development of an intelligent system for the early diagnosis of premature birth by correctly identifying true/false labor pains. Raw term preterm electrohysterography (TPEHG) signals from PhysioNet were analyzed in this work. The performance of classifiers such as the SVM, ELM, KNN, ANN, radial basis function neural network, and random forest is individually evaluated in terms of classifying EHG signals.

Chapter 17 briefly presents the feature selection techniques to improve the performance of SVM classifier. Cardiotography (CTG) data is examined to diagnose fetal hypoxia for fetal state anticipation. In this chapter, three efficient feature selection techniques based on evolutionary methodologies such as firefly algorithm (FA), opposition-based firefly algorithm (OBFA), and opposition-based firefly algorithm melded with information gain (IG-OBFA) are presented in detail.

Chapter 18 investigates the deployment of the state-of-the-art ML algorithms like decision tree, *k*-nearest neighborhood, Rocchio, ridge, passive–aggressive, multinomial naïve Bayes, Bernoulli naïve Bayes, support vector machine, and artificial neural network classifiers such as perceptron, random gradient descent, backpropagation neural network in automatic classification of biomedical text documents on benchmark datasets like BioCreative Corpus III(BC3), Farm Ads, and TREC 2006 genetics Track.

Chapter 19 gives a detailed note on body area networks (BAN) and the properties of wearable devices. To extend the life of wearable sensors, it proposes a novel protocol to find the best possible cluster heads for a single round of operations in reactive body area networks.

In Chapter 20, a practical solution is proposed through a novel algorithm by segmenting the tumor part and healthy part from MR image that works based on image segmentation and self-organizing neural networks. The segmentation algorithm identifies the tumor regions and further boundary parameters will be gathered from the segmented images to feed into the neural network system.

Chapter 21 explores the fundamentals and application of machine learning in biomedical domain. It interprets various aspects of AI and machine learning (ML)-enabled technologies, prodigies, and its independent applications. Finally, it reviews the research developments and challenges in biomedical engineering.

Chapter 22 suggests a new strategy to identify brain sources with their corresponding locations and amplitudes depending on a particle filter. Modeling of the time series is used to detect movement and time dependence among the brain sources. Finally, the Granger causality techniques have been applied to assess directional causal flow across the sources. It provides a framework to test the analytical pipeline on real EEG information.

—Saravanan Krishnan
Ramesh Kesawan
B. Surendiran
G. S. Mahalakshmi

CHAPTER 1

DESIGN OF MEDICAL EXPERT SYSTEMS USING MACHINE LEARNING TECHNIQUES

S. ANTO[1*], S. SIAMALA DEVI[2], K. R. JOTHI[3], and R. LOKESHKUMAR[4]

[1,3,4]School of Computer Science and Engineering, Vellore Institute of Technology, Vellore, India.

[2]Department of Computer Science and Engineering, Sri Krishna College of Technology, Coimbatore, India

[*]Corresponding author. E-mail: anto.s@vit.ac.in

ABSTRACT

The number of qualified doctors in India is around 7 per 10,000 people, which leads to the various challenges faced by the medical field such as inadequacy of physicians, the slower rate of diagnosis, and unavailability of medical assistance to common people in time. In this scenario, as clinical decision demands the utmost accuracy of diagnosis, it is a tedious and challenging task for physicians. These issues can be addressed by an automated system that helps the clinicians in disease diagnosis. Attaining the maximum accuracy is an open challenge in this scenario. The scope for improving the accuracy of diagnosis has motivated many researchers to propose medical expert systems using various machine learning and optimization techniques. This leads to proposing four different medical expert system designs as follows.

As a first step, an expert system based on fuzzy logic and ant colony optimization (ACO) is used. This system uses the ACO algorithm to generate fuzzy classification rules from the training patterns. The stochastic behavior of the ACO algorithm encourages the ants to find more accurate rules. Second, an expert system based on a hybrid genetic algorithm–simulated annealing and support vector machine is proposed for the disease diagnosis. The hybrid GASA is used for selecting the most significant feature subset as well as to optimize the kernel parameters of the SVM classifier. As a next step, a decision support system based on Fisher score-extreme learning machine (ELM)-simulated annealing is proposed. The ELM-based learning machine uses a single-hidden layer feedforward neural network. Finally, an expert system based on least square–support vector machine–simulated annealing (LS-SVM-SA) is proposed. FS is used for the selection of the most significant features. To improve the performance of the system, LS-SVM with radial basis function is used for classification and the SA is used for the optimization of the kernel parameters of the LS-SVM.

1.1 INTRODUCTION

1.1.1 ARTIFICIAL INTELLIGENCE

Artificial intelligence (AI) is a replication of human intelligence by computer systems. It is an interdisciplinary field that embraces a number of sciences, professions, and specialized areas of technology. To be precise, AI will not replace people but will augment their capabilities. AI helps to devise an intelligent agent in which, an agent observes its environment and takes actions to maximize the probability of success. AI encompasses an extensive range of areas in decision-making process. One of the subfields of AI is an automated diagnosis. It is related to the development of algorithms and techniques to confirm the behavior of a system. Thus, the developed algorithm should be capable enough to discover its cause, whenever something goes wrong.

1.1.2 EXPERT SYSTEMS

An expert system is "a computer program that represents and reasons with knowledge of some specialist subject with a view to solving problems or giving advice" (Jackson, 1999).

It consists of a knowledge source and a mechanism that solves problems and returns a response based on the information provided by the query. Direct input from domain experts and evidence from literature are the sources of knowledge to the expert systems. To solve expert-level problems, efficient access to a substantial domain knowledge base and a reasoning mechanism to apply the knowledge to the problems is mandatory.

Knowledge acquisition, the process of transforming human knowledge to machine-usable form is considered a bottleneck (Feigenbaum, 1977) as it demands more time and labor. Further, maintaining the knowledge base is also a challenging task (Coenen and Bench-Capon, 1992; Watson et al., 1992). Techniques such as case-based reasoning (CBR) (Watson and Marir, 1994) and machine learning (ML) methods based on data are used for inference as they avoid the knowledge acquisition problem. In CBR, the knowledge consists of preceding cases that include the problem, solution, and the outcome stored in the case library. To obtain a solution for a new case, it is needed to identify a case that resembles the problem in the case library and adopt the proposed solution from the retrieved case. Similar to CBR, ML-based expert systems avoid the bottleneck of knowledge acquisition as knowledge is directly obtained from data. Recommendations are generated by nonlinear forms of knowledge and easily updated by simply adding new cases.

1.1.2.1 MEDICAL EXPERT SYSTEM

A decision support system is computer software that attempts to act like a human being. During the past few years, medical expert systems for the diagnosis of different diseases have received more attention (Kourou et al., 2015; Fan et al. 2011). Knowledge discovery in patient's data and machine learning are used to design such expert systems. These decision support systems can play a major role in assisting the physicians, while making complex clinical decisions, thereby, can improve the accuracy of diagnosis. Such systems have higher optimization potential and reduced financial costs. Pattern recognition and data mining are the techniques used in these expert systems that allow retrieval of meaningful information from large scale medical data.

1.1.3 MACHINE LEARNING

ML, a subfield of computer science and statistics, is a scientific discipline that deals with the design and study of algorithms to learn from data and to make autonomous decisions. It has strong ties to data mining (Mannila and Heikki, 1996), AI, and optimization. It does not explicitly program computers to acquire knowledge but emphases on the development of computer programs that grow and change by teaching themselves when exposed to new data. Further, it focuses more on exploratory data analysis (Friedman, 1998) and on the improvement of machine projects that develop and change when given new information. Knowledge representation and generalization are the core of ML.

To be precise, the algorithms do not involve programmed instructions but build a model based on inputs and make predictions or decisions on their own. It comprises of a set of methods that automatically detect patterns in data, use the apparent patterns to predict the data, and performs better for exact decision making. It is employed in applications like spam filtering, optical character recognition (Wernick et al., 2010) search engines, and computer vision, where designing explicit rule-based algorithms is not feasible.

1.1.4 DATASET FOR PERFORMANCE EVALUATION

The datasets that are considered for analyzing the performance of the proposed systems are given below. These data sets are taken from UCI ML Repository.

- **Breast Cancer Wisconsin (Original) Dataset**
  Wisconsin Diagnostic Breast Cancer (WDBC; Original) is one of the standard datasets considered for Breast Cancer diagnosis and it has 699 instances, out of which 458 are benign and 241 are malignant with 11 attributes including the class attribute.
- **PIMA Indians Diabetes (PID) Dataset**
  The PID dataset for diabetes is used in which the patients are females above 21 years of age having Pima Indian heritage. There are 768 instances with nine attributes including the "class" variable. The attributes are numeric-valued. A total of 500 instances belong to Class "0" and 268 instances belong to Class "1."
- **Breast Cancer Wisconsin (Diagnostic) Dataset**
  WDBC includes 569 instances, out of which 357 are benign and 212 are malignant with 32 attributes including ID, diagnosis, and 30 real-valued input features.
- **Hepatitis Dataset**
  The Hepatitis domain consists of mostly Boolean or numeric-valued attribute types which have 155 instances with 20 attributes including the "class" attribute. The "BILIRUBIN" attribute is continuously-valued. A total of 32 instances belong to "DIE" class and 123 instances belong to "LIVE" class.
- **Cardiac Arrhythmia Dataset**
  The Cardiac Arrhythmia dataset records the presence/absence of cardiac arrhythmia and classifies it

in one of the 16 groups. The database has 452 instances with 279 attributes, out of which 206 are linear valued and the rest are nominal.

1.1.5 PERFORMANCE METRICS

To evaluate the performance of the medical expert system model, the following performance metrics are used and the results are given in Table 1.3

(i) Confusion Matrix: The confusion matrix holds both the actual and predicted instances classified by the classifier system. A confusion matrix for a classification problem with two classes is of size 2×2 as shown in Table 1.1.

TABLE 1.1 Confusion Matrix

Predicted	Actual	
	Positive	**Negative**
Positive	TP (true positive)	FP (false positive)
Negative	FN (false negative)	TN (true negative)

TN is the correct predictions of an instance as negative.
FN is the incorrect predictions of an instance as positive.
FP is the incorrect predictions of an instance as negative.
TP is the correct predictions of an instance as positive.

(ii) Cross-Validation (CV): CV is a widely used statistical method to evaluate the classifier's performances by splitting a data set into two sets as training and testing. In CV, the training and the testing sets must cross over in successive rounds, and in this way, each record has a chance of being validated against.

(iii) Classification Accuracy: Classification accuracy is the most commonly used measure for determining the performance of classifiers. It is the rate of a number of correct predictions made by a model over a data set as shown in Equation (1.1)

$$\text{Accuracy} = \frac{\text{TP}+\text{TN}}{\text{TP}+\text{TN}+\text{FP}+\text{FN}} \quad (1.1)$$

(iv) Sensitivity and Specificity: Sensitivity is the true positive rate, and specificity is the true negative rate. They are defined in (1.2) and (1.3)

$$\text{Sensitivity} = \frac{\text{TP}}{\text{TP}+\text{FN}} \quad (1.2)$$

$$\text{Specificity} = \frac{\text{TN}}{\text{FP}+\text{TN}} \quad (1.3)$$

1.2 MEDICAL EXPERT SYSTEM BASED ON FUZZY CLASSIFIER WITH ANT COLONY OPTIMIZATION (ACO)

A medical decision support system based on fuzzy logic and ACO is proposed for the diagnosis of various disease datasets like breast cancer (original), breast cancer (diagnostic), diabetes, heart disease (Niranjana and Anto, 2014), and hepatitis. The datasets are accessed from the UCI ML repository. A set of fuzzy rules is extracted from the patient's dataset using fuzzy logic and ACO (Anto and Chandramathi, 2015). ACO algorithm optimizes these extracted fuzzy rules and generates an optimized set of rules. The fuzzy inference system uses these optimized rules to perform the classification of the test data. A 10-fold cross-validation procedure is used to evaluate the performance of the system in terms of the classification accuracy.

1.2.1 FEATURE SELECTION

The selection of the most significant features is done by using correlation-based feature selection. It assigns higher scores to feature subsets that are highly correlated to the class labels but uncorrelated to each other.

The merit "*S*" of an attribute set is given by Equation (1.4)

$$S = \frac{k.\,\mathrm{rcf}}{\sqrt{\left(k + k(k-1)\,\mathrm{rff}\right)}} \tag{1.4}$$

where,

k is the number of attributes in the set *S*;

rcf models the correlation of the attributes to the class label;

rff is the inter-correlation between attributes.

It selects a constant value of the 10 most significant features of the given medical dataset.

1.2.2 FUZZY-ACO CLASSIFIER

The proposed system uses the ACO algorithm to generate fuzzy classification rules from training patterns of the dataset. The artificial ants make candidate fuzzy rules gradually in search space. The stochastic behavior of the ACO algorithm encourages the ants to find more accurate rules.

These optimized rules are used by the fuzzy inference engine to perform decision making on testing patters as shown in Figure 1.1.

1.2.3 MEMBERSHIP VALUE ASSIGNMENT BY NORMALIZATION OF DATASET

Normalization of dataset between 0.0 and 1.0 is done using the min–max normalization method as shown in Equation (1.5)

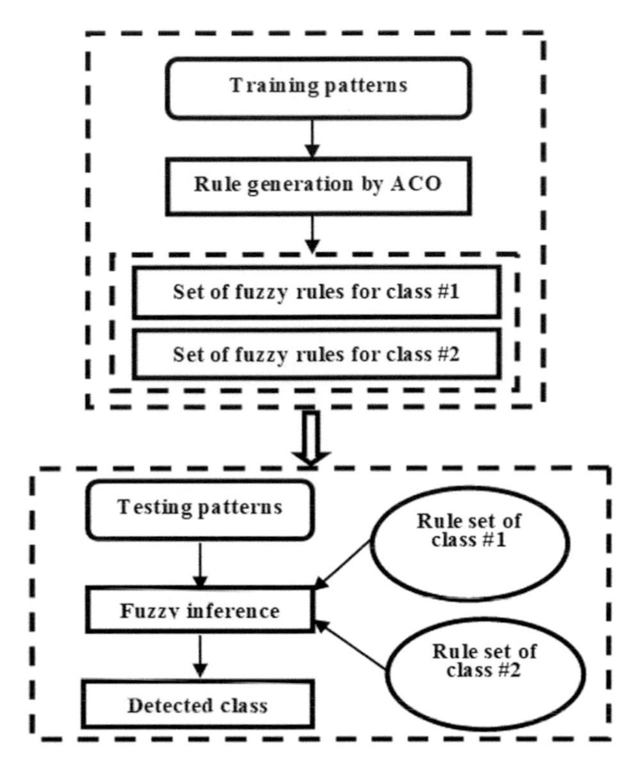

FIGURE 1.1 Stages of the proposed system.

$$\text{Normalize}\ (X) = \frac{X - X_{min}}{X_{max} - X_{min}} \quad (1.5)$$

where

(X) is the membership function;

X is the linguistic value;

X_{min} is the least linguistic value;

X_{max} is the maximum linguistic value.

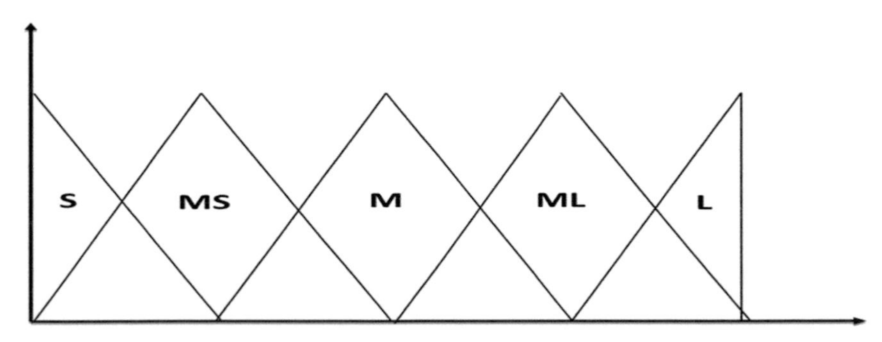

FIGURE 1.2 Antecedent fuzzy sets.

The domain of each attribute is homogeneously partitioned into symmetric triangular fuzzy sets. The membership function of each linguistic value is determined from the domain as shown in Figure 1.2. The antecedent fuzzy sets that are used are small (S), medium small (MS), medium (M), medium large (ML), and large (L).

1.2.4 RULE GENERATION

In the training stage, a set of fuzzy rules is generated using the training patterns. The fuzzy rules are of the form:

Rule R_j: If x_1 is A_{j1} and … and x_n is A_{jn}, then Class C_j with CF = CF_j

where "R_j" is the label of *j*th fuzzy IF–THEN rule, $A_{j1} \ldots A_{jn}$ are the antecedent fuzzy sets in the unit interval [0, 1], "C_j" is the consequent class and CF_j is the grade of certainty of the fuzzy IF–THEN rule.

The rule learning process is done separately for each class. The list of discovered rules is initially empty and the training samples consist of all the training samples. The first ant constructs the rule "R_j" randomly by adding one term at a time. The ants modify the rule "R_j" according to the maximum change parameter.

1.2.5 RULE MODIFICATION

During initial iteration ($t = 0$), the ant creates a rule and in subsequent iterations ($t \geq 1$), the ants modify the rule. In an iteration ($t \geq 1$), the maximum number of times each ant modifies the rule is decided by maximum possible change value. Each ant chooses $term_{ij}$ to modify based on the probability given in Equation (1.6)

$$P_{ij} = \frac{\tau ij(t).\eta ij}{\sum_i^a \sum_j^{bi} \tau_{ij}(t), \eta_{ij}, \forall i \in I} \tag{1.6}$$

where

$\tau_{ij}(t)$ is the pheromone currently available on the path between attribute "i" and antecedent fuzzy set "j";

η_{ij} is the heuristic value for $term_{ij}$;

"a" is the total number of attributes;

"b_i" is the total number of an antecedent fuzzy set for $attribute_i$;

"I" is the set of attributes that are not yet used by the ant.

Corresponding to the quality of the modified rules, a pheromone is assigned to each trail of the ant. The ants choose the trail with high pheromone density.

1.2.6 HEURISTIC INFORMATION

The ants modify the rule using heuristic information and the amount of pheromone. The model uses a set of two-dimensional matrices as heuristic information for each class. The rows represent the attributes and the columns representing the fuzzy values. These matrices help the ants to choose more accurate rules.

1.2.7 PHEROMONE UPDATE RULE

Pheromone update is carried out only when there is an improvement in the quality of the rule before modification. It is carried out using the following equations:

$$\Delta_Q = Q_i^{\text{After modification}} - Q_i^{\text{Before modification}} \quad (1.7)$$

$$\tau_{ij}(t+1) = \tau_{ij}(t) + \tau_{ij}(t)(\Delta_i^Q . C) \quad (1.8)$$

where "Δ_i^Q" shows the difference between the quality of rule after and before modification and "C" is the parameter to regulate the influence of "Δ_i^Q" to increase the pheromone.

1.2.8 FUZZY INFERENCE

The fuzzy-ACO (Anto and Chandramathi, 2015) system generates and optimizes a set of fuzzy rules to classify the test data. To compute the certainty grade of each fuzzy IF-THEN rule, the following steps are followed:

1. The compatibility of each training pattern $x_p=(x_{p1},x_{p2},\ldots x_{pn})$ is determined with the fuzzy IF-THEN rule "R_j" using the following product operation:

$$\mu_j(X_p) = \mu_{j1}(X_{p1})\, x \ldots .\ \mu_{jn}(X_{pm}),\ p = 1, 2, 3, m \quad (1.9)$$

where $\mu_{ji}\,(x_{pi})$ = membership function of ith attribute of pth pattern,

where "m" is the total number of patterns.

2. The relative sum of compatibility grades of the training pattern is computed with each fuzzy IF–THEN rule "R_j."

$$\beta_{classh}(R_j) = \sum_{xp \in classh} \frac{\mu_j(x_p)}{N_{classh}}, h = 1,2,\ldots,c \quad (1.10)$$

where

$\beta_{classh}\,(R_j)$ is the sum of the compatibility grades of the training patterns in class;

R_j is the fuzzy rule;

N_{classh} is the number of training patterns.

3. The grade of certainty CF_j is determined as follows:

$$CF_j = \frac{\beta_{class}\hat{h}(R_j) - \bar{\beta}}{\sum_{h=1}^{c} \beta_{classh}(R_j)} \quad (1.11)$$

where

$$\bar{\beta} = \frac{\sum_{h \neq \hat{h}_j} \beta_{classh}(R_j)}{(C-1)} \quad (1.12)$$

The certainty grade for any combination of antecedent fuzzy sets can be specified. Combinations of antecedent fuzzy sets for generating a rule set with high classification ability are to be generated by the fuzzy classification system. When a rule set is given, an input pattern is classified by a single rule as given below:

$$\mu_j(x_p).CF_j = \max\{\mu_j(x_p).CF_j \mid R_j\} \quad (1.13)$$

The winner rule has the maximum product of the compatibility and certainty grade CF_j.

1.3 MEDICAL EXPERT SYSTEM BASED ON SVM AND HYBRID GENETIC ALGORITHM (GA)-SIMULATED ANNEALING (SA) OPTIMIZATION

1.3.1 FEATURE SELECTION USING GA AND SA

Feature selection is an optimization problem, which is based on the principle of picking a subset of attributes that are most significant in deciding the class label. It reduces the dimension of the data. When the input to an algorithm is too large to be processed and is suspected to be extremely redundant, then the input data is represented with a reduced set of features. The selection of the most significant subset of features from a dataset is an optimization problem. In this system, the feature selection is done using hybrid GA–SA local search mechanism.

1.3.2 OPTIMIZATION USING HYBRID GA–SA

This hybrid GA–SA optimization technique is used for feature selection and SVM parameter optimization. The performance of an SVM classifier depends mainly on the values of the kernel function parameter, Gamma (γ), and penalty function parameter (*C*). Finding the best values of these two parameters to achieve a maximum classification accuracy of the SVM classifier is an optimization problem. A hybrid GA–SA algorithm is used to solve this problem and find the optimal values of "*C*" and "γ."

1.3.2.1 STEPS OF GA

1. Randomly generate an initial source population with "*n*" chromosomes.
2. Calculate the fitness function *f*(*x*) of all chromosomes in the source population using $\min f(x) = 100^*(x(1)^2 - x(2))^2 + (1 - x(1))^2$

2. Create an empty successor population and then repeat the following steps until "n" chromosomes have been created.
3. Using the fitness value, select two chromosomes "x_1" and "x_2" from the source population.
   - Apply crossover to "x_1" and "x_2" to obtain a child chromosome "n."
   - Apply mutation to "n," to produce a dissimilar new offspring.
   - Place the new offspring in a new population.
   - Replace the source population with the successor population.
4. If the termination criterion is satisfied, stop the algorithm. Otherwise, go to step 2.

1.3.2.2 ACCEPTANCE FUNCTION

The calculation chooses the best answer to avoid local optimums. Initially, the neighbor solution is checked to see whether it improves the current solution. If so, the current solution is accepted. Else, the following couple of factors are considered.

- How greatly worse the neighboring solution is?
- How high the current temperature of our system is?

At high temperatures, the system is more likely to accept worse solutions.

1.3.2.3 STEPS OF SA

Table 1.2 lists the terminologies used in the SA algorithm.

TABLE 1.2 Terminology of the SA Algorithm

Terminology	Explanation
X	Design vector
f_c	System energy (i.e., objective function value)
T	Temperature
Δ	The difference in system energy between two configuration vectors

Step 1: Choose a random "X_i," select the initial temperature "t_1" and specify the cooling schedule.

Step 2: Evaluate $f_c(X_i)$ using a simulation model.

Step 3: Perturb "X_i" to get a neighboring design vector (X_{i+1}).

Step 4: Evaluate $f_c(X_{i+1})$ using a simulation model.

Step 5: If $f_c(X_{i+1})<f_c(X_i)$, X_{i+1} is the new current solution.

Step 6: If $f_c(X_{i+1})>f_c(X_i)$, then accept X_{i+1} as the new current solution with a probability using $\exp(-\Delta/t)$ where $\Delta=f_c(X_{i+1})-f_c(X_i)$

Step 7: Reduce the system temperature according to the cooling schedule.

Step 8: Terminate the algorithm.

1.3.2.4 NEIGHBORHOOD SEARCH (HYBRID GA–SA)

In the hybrid SA algorithm, the best-obtained solution in each GA generation is transferred to SA to improve the quality of solution through neighborhood search to produce a solution close to the current solution in the search space, by randomly choosing one gene in a chromosome, removing it from its original position and inserting it at another random position in the same chromosome. According to this criterion, even when the value of the next solution is worse, the solution can be accepted based on the current temperature to avoid algorithm to stick in a local optimum. In the cooling phase, the new temperature is determined by the decrement function *t*.

1.3.3 CLASSIFICATION USING SVM

The main objective of SVM in classification is to separate data into two different classes with maximum margin. Here, SVM is applied for classification and for optimizing SVM parameters GA– SA is applied. The overall flow of the proposed system including feature selection, classification, and optimization is shown in Figure 1.3.

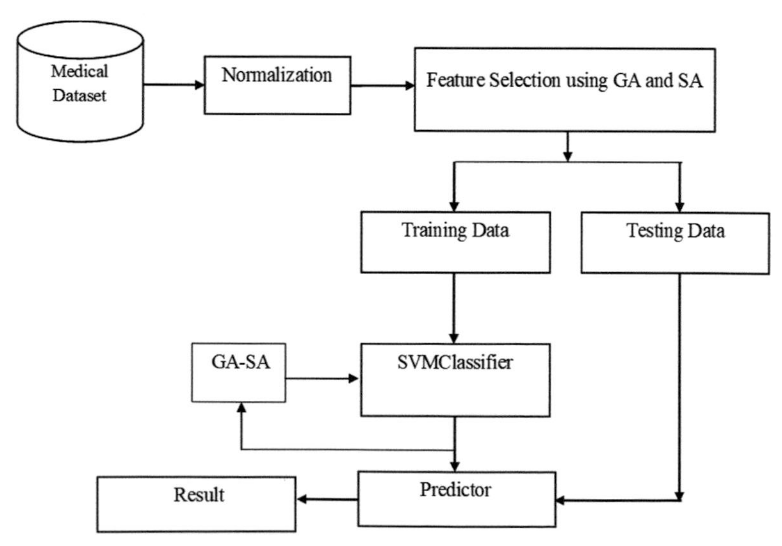

FIGURE 1.3 Flow diagram of the proposed GASA-SVM model.

Given the training sample of instance-label pairs (x_i, y_i), $i = 1, \ldots, l$, $X_i \in R^n, y_i, \in \{1,-1\}$, SVMs require the solution of the following (primal) problem (Keerthi and Lin, 2003).

$$\min_{w,b,\varepsilon} \frac{1}{2} W^T W + C \sum_{i=1}^{l} \varepsilon_i \quad (1.14)$$

Subject to $y_i \left(W^T Z_i + b\right) {}^3 1 - \varepsilon_i, \varepsilon_i {}^3 0, i = 1, \ldots, l.$

where the training vector "x_i" is mapped onto a high dimension space by mapping function φ as $z_i = \varphi(x_i)$. $C > 0$ is the penalty parameter of the error term.

Usually, Equation (1.14) is resolved by sorting out the following dual problem:

$$\min_{\alpha} F(\alpha) \frac{1}{2} \alpha^T Q_\alpha - e^T \alpha \quad (1.15)$$

Subject to $0 \le \alpha_i \le C, \ i = 1, \ldots l$

$$\gamma^T \alpha = 0 \quad (1.16)$$

where "e" is the vector of all 1's and "Q" is a positive semidefinite matrix.

The (i, j)th element of "Q" is given by

$$Q_{i,j} \equiv y_i y_j K\left(x_i x_j\right) \quad (1.17)$$

The kernel function is

$$K\left(x_i, x_j\right) \equiv \varphi^T \left(x_i\right) \varphi\left(x_i\right) \quad (1.18)$$

$\{\alpha_i\}_{i=1}^{i}$ is Lagrange multiplier and W= $\sum_{i=1}^{l} \alpha_i y_i \varphi, (x_i)$ is the weight vector.

The classification decision function is given as

$$\operatorname{sgn}\left(W^T \Phi(x) + b\right) = \operatorname{sgn}\left(\sum_{i=1}^{l} \alpha_i y_i K(x_i, x) + b\right) \quad (1.19)$$

The kernel function $K(x_i, x_j)$ has manifold forms. In this work, the Gaussian kernel function is shown in Equation (1.20) or (1.21) is used:

$$K(x, x_i) = \exp\left(-\gamma \|x - x_i\|^2\right) \quad (1.20)$$

$$K\left(x, x_i\right) = \exp\left(-\frac{1}{\sigma^2} \|x - x_i\|^2\right) \quad (1.21)$$

Both Equations (1.20) and (1.21), which are in the same context, can transform parameters "γ" and "σ^2." The Gaussian kernel parameter "c" is determined by $\gamma = \frac{1}{\sigma^2}$.

The parameters of SVMs with Gaussian radial basis function (RBF) kernel refer to the pair—the error penalty parameter "C" and the Gaussian kernel parameter "γ," usually depicted as (C, γ).

1.4 MEDICAL EXPERT SYSTEM BASED ON ELM AND SA

The key problem in a neural network (NN) is determining the number of hidden nodes that affect accuracy. To overcome this problem, the proposed system uses ELM on single hidden layer feedforward network (SLFN), in which the hidden nodes are randomly selected and the optimal number of hidden nodes is determined by SA. The performance of an ELM is mainly decided by the number nodes present in the hidden layer. This parameter (number of nodes) is optimized by the SA.

1.4.1 FEEDFORWARD NEURAL NETWORKS (FFNNS)

FFNNs is the most widely used models in classification problems. It has a single hidden layer feed-forward network with inputs "x_i" and output "O_j" (Figure 1.4). Each arrow symbolizes a parameter in the network. It is an artificial neural network (ANN; Vinotha et al., 2017), where a directed cycle is not formed in the connections between the units. The network consists of three layers namely, the input layer, the hidden layer, and the output layer.

- The input layer consists of the inputs to the network.
- The hidden layer consists of neurons or hidden units placed in parallel. Each neuron performs a weighted summation of the inputs and then passes a nonlinear activation function called the neuron function.
- The output layer.

Standard FFNNs with hidden nodes have universal approximation and separation capabilities.

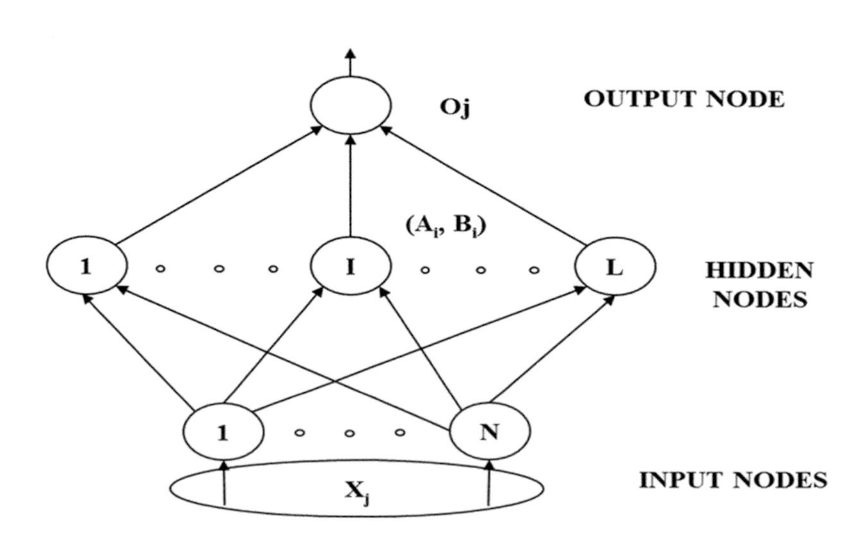

FIGURE 1.4 Single hidden layer feedforward network.

1.4.2 FEATURE SELECTION

The most significant features in the dataset are to be selected. The dimensionality of the dataset has a higher influence on the accuracy of the system. Hence, it is necessary to consider the aspects that reduce the complexity of the system (Refaeilzadeh, 2007). Fisher score (FS) is the most suitable method used to select the best features of all the four medical datasets.

1.4.2.1 FISHER SCORE

FS selects the most relevant feature "*m*" from the given set of features. The datasets consist of (x_i, y_i) for "*N*" instances, where "x_i" is the input vector with "*p*" features and "y_i" is the class label. To find the most discriminating features, two basic steps are involved:

1. Calculation of feature score for all features.
2. Selection of top "*m*" features based on the score.

FS is computed using equation (1.22)

$$F(x_j) = \sum_{k=1}^{c} \frac{n_k \left(\mu_k^j - \mu_j\right)^2}{\left(\sigma_j\right)^2} \quad j = 1, 2 \ldots p \qquad (1.22)$$

where

- n_k is the number of instances in each class *k*;
- μ_j is the *j*th feature means of the whole dataset;
- σ^j is the *j*th feature standard deviation (SD) of the whole dataset.

SD is computed as shown in Equation (1.23)

$$\left(\sigma_j\right)^2 = \sum_{k=1}^{G} n_k \left(\sigma_k^j\right)^2 \qquad (1.23)$$

1.4.2.2 NORMALIZATION

Scaling is done to evade the dominance of attributes with greater numerical values over the smaller values by computing the linear transformation of numerical values within a range. The values of the selected features from the dataset are normalized from 0 to 1 as shown in Equation (1.24).

$$X_{norm} = \frac{X - X_{min}}{X_{max} - X_{min}} (\text{upperbound} - \text{lowerbound}) \qquad (1.24)$$

where "*X*" is the original data, "X_{max}" is the maximum value of *X*, "X_{min}" is the minimum value in *X* and "X_{norm}" is the normalized value within the given upper and lower bound.

1.4.3 CLASSIFICATION USING EXTREME LEARNING MACHINE (ELM)

ELM, a learning algorithm, is a SLFNs proposed by Huang et al. (2004). The hidden node parameters may be randomly chosen and fixed, followed by analytically determining the output weights. The parameters of the hidden nodes are not only independent of the target functions or the training datasets but also of each other. The parameters of ELM are analytically determined instead of being tuned. Once the weights of the SLFNs are randomly assigned, then SLFNs is considered as a linear system. The output weights are analytically obtained by a generalized inverse operation of the hidden layer output matrices.

In contrast to the conventional learning methods that see the training data before generating the hidden neuron parameters, ELM randomly generates the hidden neuron parameters even before seeing the training data. ELMs have fast learning speed, easy to implement, and involve minimal human intervention. They seem to be a feasible alternate for large-scale computing and ML.

1.4.3.1 ELM ALGORITHM

The formal ELM algorithm is given below;

Given:

- A training set of input/output values:
- $(x_i, t_i) \in R^n x R^m$, for i = 1, 2, ..., *N*. (1.25)
- An activation function:
- $G(a_i, X_j, b_i) = g(b_i \| X_j - a_i \|)$ (1.26)
- The number of hidden nodes L.

Step 1: By using continuous sampling distribution, assign hidden nodes by randomly generating parameters (a_i, b_i), for i=1, 2,...,N

Step 2: Compute the hidden layer output matrix "H."

Step 3: Compute the output weight "$\tilde{\beta}$," by using the relation

$$\tilde{\beta} = H^{\#}T \tag{1.27}$$

1.4.4 OPTIMIZATION USING SIMULATED ANNEALING

The number of hidden nodes (L) has a high impact on the performance of the ELM-based classification system. Computing the optimal value for "L" is a demanding task. Here, SA is employed in computing the optimal value for "L," to improve the performance of ELM. It is one of the most popular optimization techniques used for finding solutions for optimization problems.

It is a local heuristic search algorithm that employs a nongreedy method for finding optimal solutions that usually do not settle in local maxima. The strength of SA is that it does not get caught at local maxima. It searches for the best solution by generating a random initial solution and exploring the area nearby. If a neighboring solution is better than the current one, the algorithm moves to it. It is a kind of Monte Carlo method used for examining the state and frozen state of the *n*-body systems.

In the hierarchical SA search, classification measure (CM) is used as the basic optimization parameter. The classification accuracy of ELM is considered as the CM. Initially, the SA and the ELM parameters are initialized. Then, the neighbors of the ELM parameters are selected. The neighbors are tuned using the SA optimization search. The output of the first stage is given as the input of the next stage. This decides whether the parameter is acceptable or not. If it is not acceptable, the parameter is tuned further.

A maximum of "k_{max}" iterations are performed until maximum accuracy is achieved. CallNeighbor( ) function finds the ensuing values of "*N*." CallRandom() function generates a random value from "0" and "1." The basic SA process is shown in Figure 1.5. To divide the set into training and testing sets, the *k*-fold cross-validation procedure is applied to the selected feature set.

```
I←I0; A←A (I);
Ibest←I; Abest←A; //Initial Iteration, Accuracy.
k←0; kmax←Constant Value; //Initial "best" solution
MaxAccuracy←A constant Value //Evaluation count
whilek <kmax and A < = MaxAccuracy
{ //while time left & not good enough
Inew←Neighbor (I) //Select number of Neurons
Anew←A (Inew) //Compute it's Accuracy
If exp (Anew−A) > Random () then
I←Inew; A←Anew   // next Iteration
ifAnew>Abestthen   //Is this a new best?
Ibest←Inew; Abest←Anew  // Best Accuracy
k←k+ 1  //One more evaluation done
}
returnIbest, Abest //Return the best Accuracy
```

FIGURE 1.5 Simulated annealing search algorithm.

The cross-validation technique returns a CM for "*k*" classifiers built by the ELM algorithm. Each fold of the dataset is optimized using the hyperparameter search strategy. The procedure for cross-validation is as follows:

- *Division of datasets:* The medical datasets are divided into training and testing sets. The *k*-non overlapping equal-sized subsets are formed from the given dataset "D_i," where $i = 1, 2,...,k$.
- *Classifier training:* "$k-1$" folds are trained using the classifier algorithm and the remaining one fold is tested on the trained classifier. Each classifier output generates accuracy for the predicted sets. The class performance is analyzed using performance parameters.
- *CM:* The CM is obtained by Equation (1.28)

$$CM = \frac{\text{Number of True Records Predicted}}{\text{Number of Total Records}} \quad (1.28)$$

CM is calculated for every sequential increase in the number of hidden nodes (L).

- *Optimization parameters:* To find the optimal number of neurons (L), the SA optimization technique is used. The CM is calculated by increasing the number of neurons from 1 to 200. The value of the parameter "L" for which maximum CM is obtained is chosen as the best value.

1.4.5 FISHER SCORE-EXTREME LEARNING MACHINE-SIMULATED ANNEALING (FS-ELM-SA)

In the proposed ELM-based learning machine that uses SA for optimization, the SLFN has "L" hidden nodes. It can be approximated by the given "N" pairs of input/output values namely, $(x_i, t_j) \in R^n x\ R^m$ with zero errors. The overall flow and working of the proposed system are depicted in Figure 1.6

$$\sum_{i=1}^{p} \beta_i G(a_i, X_j, b_i) = t_j, \text{ for } j = 1, 2,\ldots,L \quad (1.29)$$

where (a_i, b_i) is the parameter associated with "ith" hidden node, and "β_i" is the output weight that links the "ith" hidden node to the output node. In this work, a nonlinear activation function called RBF as shown in Equation (1.30) is used:

$$G(a_i, X_j, b_i) = g(b_i \|X_j - a_i\|) \quad (1.30)$$

Hence, Equation (1.30) can be rewritten as,

$$H\beta = T \quad (1.31)$$

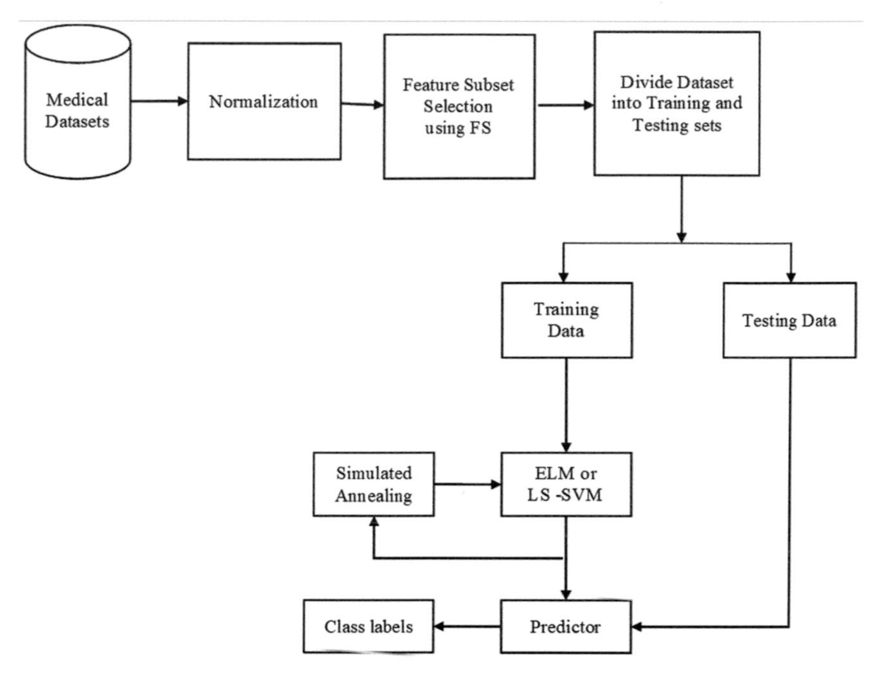

FIGURE 1.6 Flow diagram of the proposed FS-ELM-LSSVM-SA.

where

$$H = \begin{bmatrix} G(a_1, X_1, b_1) G(a_2, X_1, b_2) \ldots\ldots\ldots G(a_p, X_1, b_p) \\ G(a_1, X_2, b_1) G(a_2, X_2, b_2) \ldots\ldots\ldots G(a_p, X_2, b_p) \\ \cdot \\ \cdot \\ \cdot \\ G(a_1, X_N, b_1) G(a_2, X_N, b_2) \ldots\ldots\ldots G(a_p, X_N, b_p) \end{bmatrix} \tag{1.32}$$

$$\beta = \left[\beta_1^T, \beta_2^T, \beta_3^T \ldots . \beta_p^T\right]^T \tag{1.33}$$

$$T = \left[t_1^T, t_2^T, t_3^T \ldots t_N^T\right]^T \tag{1.34}$$

$$\tilde{\beta} = H^{\#} T \tag{1.35}$$

"$\tilde{\beta}$" is used as the estimated value of "β," where "H#" is the Moore–Penrose generalized inverse of the hidden layer output matrix "H" (Serre, 2002).

1.5 MEDICAL EXPERT SYSTEM BASED ON LS-SVM AND SA

To improve the classification performance of the expert system further, LS-SVM with RBF kernel is used for classification and the SA is used for the optimization of the kernel parameters of the LS-SVM. The performance of the SVM classifier is highly influenced by the kernel function parameter gamma (γ) and penalty function parameter (C). The critical parameters "C" and "γ" are optimized using SA to get the best combination of kernel parameters, leading to the highest classification accuracy.

1.5.1 FEATURE SELECTION USING FISHER SCORE

The FS algorithm is used for many supervised learning systems to determine the most relevant and discriminative features for classification (Yilmaz, 2013). Based on the prominence of the attributes in the dataset, it generates a score for each attribute and vital features are selected based on the scores. It uses discriminative methods and generative statistical models to perform feature selection.

1.5.2 CLASSIFICATION USING LEAST SQUARE SUPPORT VECTOR MACHINE (LS-SVM)

The main objective of SVM in classification is to separate data into two different classes with maximum margin. A higher computational load due to the quadratic programming problem is a challenge in SVM. To balance this, Suykens and Vandewalle (1999) have proposed LS-SVM. LS-SVM uses linear equations instead of quadratic programming of SVM.

A linear SVM is a binary classifier used to separate data between two classes, $y_i \in \{1,-1\}$. Separating hyperplanes are formed using Equation (1.36) and inequality for both classes is found using Equation (1.37).

$$D(x) = (w * x) + w_0 \quad (1.36)$$

$$y_i \left[(w * x_i) + w_0 \right] \geq 1,\ i = 1, \ldots, n. \quad (1.37)$$

where "x_i" is the input vector, "m" the number of features and "y_i" is the class label. Support vectors are the margin values that are formed when the equality of Equation (1.37) holds. Classification of data is done using these support vectors.

1.5.3 OPTIMIZATION USING SIMULATED ANNEALING

The performance of the LS-SVM classifier (Aishwarya and Anto, 2014) is also subjective to the values of 'C' and 'γ'. As finding the best values of these parameters is monotonous, optimization techniques are used along with LS-SVM. SA is one of the most popular optimization techniques used for finding a solution to the optimization problem. It is a local heuristic search algorithm that uses a greedy method for finding an optimal solution.

```
I←I0; A←A(I)    //Initial Iteration, Accuracy
Ibest←I; Abest←A  //Initial "best" solution
k←0; kmax←Constant Value ;
MaxScore←A constant Value   //evaluation count
while k <kmax and A ≤ MaxScore
{
        Inew←Neighbour(I)  //Select next C and Gamma Value
        Anew←A(Inew)  //Compute its Accuracy
        if exp(Anew −A) > Random() then
        I←Inew; A←Anew
        if Anew>Abest then   //Is this a new best?
        Ibest←Inew; Abest←Anew   //Save 'new neighbour' to 'best found'
        k←k + 1  //One more evaluation done
}
return Ibest, Abest  //Return the best solution found
```

FIGURE 1.7 Simulated annealing search.

In least squares-support vector machine (LS-SVM), simulated annealing (SA) delivers the best values for "C" and "γ" by trying random variations of the current solution (local). SA is a kind of Monte Carlo method used for finding the global minimum of a nonlinear error function. Figure 1.7 shows the basic SA search algorithm.

Iteration "I_0" is initialized and continues to the maximum of "k_{max}" steps. Let "k_{max}" be the maximum number of iterations done until maximum accuracy is achieved. The function Neighbor () finds the next values of "C" and "γ." The function Random () generates a random value between a range "0" and "1."

Cross-validation score (CVS) is the basic parameter that should be maximized in the hierarchical SA search optimization technique. CVS gives the classification accuracy of the LS-SVM classifier. To commence with, the SA and SVM parameters are initialized. The neighbors of the SVM parameters can be selected and tuned using the SA optimization search. This helps in deciding whether the parameter is acceptable or more tuning is required.

Cross-validation procedure involves the following steps.

1. *Division of datasets:* The diabetes dataset is divided into training and testing sets. The "k" nonoverlapping

equal-sized subsets are formed from the given dataset "D_i" where $i = 1, 2,\ldots,k$.

2. *Classifier training:* The "k–1" folds are trained using the classifier and the remaining one fold is tested using the trained classifier. The output of each classifier generates accuracy for the predicted sets. The classifier performance is analyzed using performance metrics.
3. *Calculation of CVS:* It is calculated for each combination of "C" and "γ" values. CVS is obtained from Equation (1.38)

$$CVS = \frac{\#\text{Records Predicted True}}{\#\text{Total Records}} \cdot \quad (1.38)$$

4. *Optimization parameters:* Find an optimal solution for the kernel parameters "C" and "γ" is tedious in case of any SVM.

Hence, the SA optimization technique is used. Here, the kernel values are varied between 2^{-5} to 2^{15} for "C" and 2^{-15} to 2^{5} for "γ."

1.5.4 FS-LSSVM-SA

The proposed FS-LSSVM-SA (Aishwarya and Anto, 2014) involves the following steps namely, feature selection using FS, classification using LSSVM, and optimization using SA.

The FS is computed using Equation (1.39).

$$F(x^j) = \frac{\sum_{k=1}^{c} n_k \left(\mu_k^j - \mu^j\right)^2}{(\sigma^j)^2}, j = 1,2,\ldots,p. \quad (1.39)$$

where "n_k" is the number of instances in class "k," "μ_j" is the mean of the whole dataset for jth feature, "σ_j" is the SD of the whole dataset for jth feature. The SD is given by

$$(\sigma^j)^2 = \sum_{k=1}^{c} n_k \left(\sigma_k^j\right)^2. \quad (1.40)$$

The values of the selected features from the datasets are normalized between the range "0" and "1"as shown below

$$X_{\text{normalized}} = \frac{X - X_{\min}}{X_{\max} - X_{\min}} (\text{Upper Bound} - \text{Lower Bound}). \quad (1.41)$$

where "X" is the original data, "X_{max}" is the maximum value of X, "X_{min}" is the minimum value of "X" and "$X_{normalized}$" is the normalized value within the given upper and lower bounds.

The following inequality holds for the margins (support vectors) of the hyper-planes:

$$\frac{yk \times D(xk)}{\|w\|} \geq \Gamma, k = 1,\ldots,n. \quad (1.42)$$

Margin (Γ) is inversely proportional to "w," thus minimizing "w" and maximizing the margin.

Equation (1.43) is written to reduce the number of solutions for the norm of "w."

$$\Gamma \times \|w\| = 1. \tag{1.43}$$

Minimizing "w,"

$$\frac{1}{2}\|w\|^2. \tag{1.44}$$

Slack variables "ξ_i" is added to Equation (1.38) and expression (1.44)

$$y_i\left[\left(wx_i\right) + w_0\right] \geq 1 - \xi_i. \tag{1.45}$$

$$C\sum_{i=1}^{n} \xi_i + \frac{1}{2}\|w\|^2. \tag{1.46}$$

SVMs are used to classify linear data. In SVM, it is difficult to achieve better classification for nonlinear data. To overcome this problem, SVM uses kernel functions. Input datasets are distributed in nonlinear dimensional space. These are converted into high dimensional linear feature space by using kernels. RBF is used for such mapping of medical dataset as given in Equation (1.47)

$$\text{RBF kernels: } K(x, x') = \exp\left(\frac{-\|x - x'\|^2}{\sigma^2}\right). \tag{1.47}$$

The major difference between SVM and LSSVM is that, LSSVM uses linear equation whereas SVM uses quadratic optimization problem. Equations (1.45) and (1.48) are minimized as

$$y_i\left[\left(wx_i\right) + w_0\right] = 1 - \xi_i, i = 1, \ldots, n \tag{1.48}$$

$$\frac{1}{2}\|w\|^2 + \frac{C}{2}\sum_{i=1}^{n} \xi_i^{\;2} \tag{1.49}$$

Based on Equations (1.45) and (1.46), the dual problem can be built as shown in Equation (1.50)

$$(w, b, \alpha, \xi) = \frac{1}{2}\|w\|^2 + \frac{C}{2}\sum_{i=1}^{n} \xi_i^{\;2} - \sum_{i=1}^{n} \alpha_i \left\{y_i\left[\left(wx_i\right) + w_0\right] - 1 + \xi_i\right\} \tag{1.50}$$

Lagrange Multiplier "α_i" can be either positive or negative for LSSVM, whereas it should be positive for SVM. LS-SVM can be expressed as

$$f(x) = \text{sign}\sum_{i=1}^{N} y_i \alpha_i K\left(x, x'\right) + b. \tag{1.51}$$

The k-fold cross-validation procedure is applied to the selected feature set to divide the set into training and testing sets. Cross-validation technique returns the CVS for "k" classifiers which are built by the LS-SVM algorithm.

1.6 CONCLUSION

It has been observed that people in the medical field face lots of challenges in the timely delivery of diagnosis and medication. This situation has led to delay or nonavailability of medical assistance to patients especially suffering from chronic diseases. The introduction of an automated computing system to assist the physicians can drastically improve the reach of a medical facility to the common public. Recent advances in the field of AI have created an opportunity for the researchers to investigate several intelligent systems to assist the physicians in disease diagnosis. These systems are designed using knowledge discovery in patient's data and ML. The major components of such a system include feature selection, classification, and optimization. Several research works in the field of the medical expert system aim at improving the performance of these components.

The core objective is to improve the diagnosis accuracy of the medical decision support system using several ML techniques. With this objective, four different decision support system designs were proposed. The metrics that are considered to evaluate the performance of the proposed medical expert system model are discussed in Section 1.1.5.

The performance of the proposed systems is evaluated based on classification accuracy, sensitivity, and specificity by constructing a 2 × 2 matrix named confusion matrix using the predicted values of the classifier. The prediction accuracy, also known as classification accuracy, is calculated from the values of the constructed confusion matrix. The performance of a classification system is evaluated by keeping all the instances of the database as a test set, that is, the splitting of the dataset is done using k-fold cross-validation. The dataset was divided into 10 equal partitions and 1 partition out of 10 is kept as a test set and the remaining instances are used for training the classifier.

The datasets that are used for analyzing the performance of the proposed systems are discussed in Section 1.1.4.

The performance of the proposed fuzzy-ACO system for all the datasets is shown in terms of maximum accuracy. The comparison of the performance of the proposed system with other existing systems such as GA–SVM (Tan et al., 2009), GA and grid algorithm (Martens et al., 2007), MLPNN (Nauck and Kruse, 1999) and AR2+NN (Dorigo and Blum, 2005), SVM (Sartakhti et al., 2012), ANN (Pradhan et al., 2011, Vinotha et al., 2017), RIPPER, C4.5 and 1NN (Martens et al., 2007), Naïve Bayes, and KNN (Sartakhti et al., 2012). It is found that the fuzzy–ACO system performs better when compared to the existing methodologies for all the datasets.

As a next step, a clinical decision support system based on SVM and hybrid ^^GA–SA is used for diagnosis. The SVM with Gaussian RBF kernel performs the classification process. The hybrid GA–SA is used for two purposes, one is to select the most significant feature subset of the dataset, and the other is to optimize the kernel parameters of SVM. While the existing RST based model offered an accuracy of 85.46%, the proposed GASA-SVM yields the maximum accuracy of 93.6% for breast cancer (diagnostic) dataset. For the diabetes dataset, SVM offers the least accuracy of 74% while the proposed GASA-SVM yields the maximum accuracy of 91.2%. On the hepatitis dataset, the proposed GASA-SVM gives the maximum accuracy of 87%. SVM-Gaussian kernel model yields the minimum accuracy of 76.1% while the proposed GASA-SVM yields the maximum accuracy of 89.3% for cardiac arrhythmia dataset.

In GA, an average of 30 generations is taken. The best fitness value is found to be 0.1481 and the mean fitness values for the four datasets are also calculated as 0.17 for PID, 0.19 for breast cancer, 0.18 for hepatitis, and 0.18 for cardiac arrhythmia. For SA, the initial temperature is set as 6.455, and the final temperature as 0.333. The temperature of SA is gradually reduced from the initial value to the final in 50 cycles. GA receives the best chromosome with the help of SA. The comparisons of the performance of the proposed system in terms of the classification accuracy with the existing systems along with systems such as grid algorithm (Huang et al., 2006), RST (Azar et al., 2014), decision tree (Zangooei et al., 2014), BP (Orkcu and Bal 2011), SVM NSGA-II (Zangooei et al., 2014), LDA-ANFIS (Dogantekin et al., 2010), PSO, and SVM (Sartakhti et al., 2012).

Subsequently, a medical expert system based on ELM and SA is proposed. Classification is performed using ELM while optimization of ELM parameter is carried out by SA heuristic. The performance of the proposed model is compared with several existing works. The RST based system offers an accuracy of 85.46% while the proposed ELM-SA yields the maximum accuracy of 94.39% for breast cancer (diagnostic) dataset. The SVM based system offers the least accuracy of 77.73% while the proposed GASA-SVM yields the maximum accuracy of 96.45% for the diabetes dataset. For the hepatitis dataset, the Naïve Bayes system yields a minimum accuracy of 82.05% while the proposed GASA-SVM yields the maximum accuracy of 81.08%. For the cardiac arrhythmia dataset, KNN-HITON yields the minimum accuracy of 65.3% while the proposed GASA-SVM yields a maximum accuracy of 76.09%.

Experimental results show the highest accuracy on the best folds, the average accuracy over 10-folds, sensitivity, and specificity of the proposed system for the four medical datasets. The classification accuracy of the proposed system is compared with the existing systems such as RST (Azar et al., 2014), CART (Ster and Dobnikar, 1996), GRID algorithm (Chen et al., 2012), GA-based approach, MKS-SSVM (Purnami et al., 2009), MABC fuzzy (Fayssal and Chikh, 2013), VFI5-GA (Yilmaz, 2013), RF-CBFS (Ozcift 2011), and AIRS-FWP (Polat and Gunes, 2009).

Finally, a medical decision support system based on LSSVM and SA heuristic for the disease diagnosis is proposed. FS method is used to select the most significant features from the given feature set. LS-SVM with RBF is used for classification and the SA for optimization of the kernel parameters of the LS-SVM. For breast cancer dataset the existing RST-based system offered the least accuracy of 85.46% while the proposed LSSVM-SA yields the maximum accuracy of 97.54%. The SVM based approach offers the minimum accuracy of 77.73% while the proposed LSSVM-SA yields the maximum accuracy of 99.29% for the diabetes dataset. For the hepatitis dataset, the existing GRNN model offers the least accuracy of 80% while the proposed LSSVM-SA yields the maximum accuracy of 90.26%. The VFI5-GA system offers a minimum accuracy of 68% while the proposed LSSVM-SA yields the maximum accuracy of 77.96% for the cardiac arrhythmia dataset.

To conclude, it is observed that the medical expert systems proposed in this chapter applied over breast cancer, PID, hepatitis, and cardiac arrhythmia dataset produced improved classification accuracy when compared with other existing systems as shown in Table 1.3. The proposed system based on LSSVM-SA produced the highest accuracy over breast cancer, PID, and hepatitis dataset. Moreover, the proposed system based on GASA-SVM gives maximum accuracy over the cardiac arrhythmia dataset. Since clinical decision making requires the utmost accuracy of diagnosis, medical expert systems design with the highest classification accuracy can help the physicians to carry out an accurate diagnosis of diseases.

TABLE 1.3 Accuracies of the Proposed Systems and the Existing Systems

Breast Cancer Dataset		Diabetes Dataset		Hepatitis Dataset		Cardiac Arrhythmia Dataset	
Methods	**Accuracy (%)**	**Methods**	**Accuracy (%)**	**Methods**	**Accuracy (%)**	**Methods**	**Accuracy (%)**
RIPPER	79.75	SVM	74	SVM	74	VFI5-GA	68
C4.5	79.34	GA–SVM	82.98	KNN	75	PRUNING APPROACH	61.4
1NN	80.37	ANN	73.4	C4.5	83.6	KNN-HITON	65.3
SVM ACO	81.93	LDA-ANFIS	84.61	NAIVE BAYES	82.05	KDFW-KNN	70.66
RST	85.46	SVM NSGA-II	86.13	KNN	83.45	AIRS-FWP	76.2
GRID ALGORITHM	90.78	ANN	73.4	GA–SVM	86.12	SVM-GAUSSIAN KERNEL	76.1
CART	93.5	SVM	77.73	PSO	82.66	HLVQ	76.92
DECISION TREE	92.81	GA-SVM	71.64	SVM	84.67	NEWFM	81.32
BP	93.1	GA BASED APPROACH	82.98	GRNN	80	RF-CBFS	76.3
PROPOSED FUZZY-ACO	87.19	MKS-SSVM	93.2	PROPOSED FUZZY-ACO	84.95	PROPOSED FUZZY-ACO	75.7
PROPOSED GASA-SVM	93.6	MABC FUZZY	84.21	PROPOSED GASA-SVM	87	PROPOSED GASA-SVM	89.3
PROPOSED FS-ELM-SA	94.39	PROPOSED FUZZY-ACO	85.38	PROPOSED FS-ELM-SA	81.08	PROPOSED FS-ELM-SA	76.09
PROPOSED LSSVM-SA	97.54	PROPOSED GASA-SVM	91.2	PROPOSED LSSVM_SA	90.26	PROPOSED LSSVM_SA	77.96
		PROPOSED FS-ELM-SA	96.45				
		PROPOSED LSSVM-SA	99.29				

KEYWORDS

- **medical expert systems**
- **machine learning**
- **classifier optimization**
- **clinical decision making**

REFERENCES

Aishwarya S, Anto S, 2014, 'A medical decision support system based on genetic algorithm and least square support vector machine for diabetes disease diagnosis' International Journal of Engineering Sciences & Research Technology, vol. 3, no. 4, pp. 4042–4046.

Anto S , Chandramathi S, 2015, 'An expert system for breast cancer diagnosis using fuzzy classifier with ant colony optimization,' Australian Journal of Basic and Applied Sciences, vol. 9, no. 13, pp. 172–177.

Azar, AT, Elshazly, HI, Hassanien, AE & Elkorany AM, 2014, 'A random forest classifier for lymph diseases,' Computer Methods and Programs in Biomedicine, vol. 113, no. 2, pp. 465–473.

Chen, HL, Yang, B, Wang, G, Wang, S. J, Liu, J & Liu, DY, 2012, 'Support vector machine based diagnostic system for breast cancer using swarm intelligence,' Journal of Medical Systems, vol. 36, no. 4, pp. 2505–2519.

Coenen, F & Bench-Capon, TJM, 1992, 'Maintenance and maintainability in regulation based systems,' ICL Technical Journal, vol. 5, pp. 76–84.

Dogantekin, E, Dogantekin, A, Avci, D & Avci, L, 2010, 'An intelligent diagnosis system for diabetes on linear discriminant analysis and adaptive network based fuzzy inference system: LDA-ANFIS,' Digital Signal Processing, vol. 20, no. 4, pp. 1248–1255.

Dorigo, M & Blum, C, 2005, 'Ant colony optimization theory: A survey,' Theoretical Computer Science,' vol. 344, no. 2, pp. 243–278.

Fan, CY, Chang, PC, Lin, JJ & Hsieh, JC, 2011, 'A hybrid model combining case-based reasoning and fuzzy decision tree for medical data classification,' Applied Soft Computing, vol. 11, no. 1, pp. 632–644.

Feigenbaum, EA, 1977, 'The art of artificial intelligence. 1. Themes and case studies of knowledge engineering (No. STAN-CS-77-621), Stanford University CA, Department of Computer Science.

Friedman, JH, 1998, 'Data mining and statistics: What's the connection?,' Computing Science and Statistics, vol. 29, no. 1, pp. 3–9.

Huang, GB, Zhu, QY & Siew, CK, 2004, 'Extreme learning machine: A new learning scheme of feedforward neural networks,' Proceedings IEEE International Joint Conference on Neural Networks, pp. 985–990.

Jackson P, 1999, 'Introduction to Expert Systems,' 3rd Edition, Addison Wesley, Reading, MA, USA.

Kourou, K, Exarchos, TP, Exarchos, KP, Karamouzis, MV, & Fotiadis, DI, 2015, 'Machine learning applications in cancer prognosis and prediction,' Computational and Structural Biotechnology Journal, vol. 13, pp. 8–17.

Mannila, H, 1996, 'Data mining: Machine learning, statistics, and databases,' in SSDBM, pp.2–9.

Martens, D, De Backer, M, Haesen, R, Vanthienen, J, Snoeck, M & Baesens, B, 2007, ' Classification with ant colony optimization,' IEEE Transactions on Evolutionary Computation, vol. 11, no. 5, pp. 651–665.

Nauck, D & Kruse, R, 1997, 'A neuro-fuzzy method to learn fuzzy classification rules from data,' Fuzzy Sets and Systems, vol. 89, no. 3, pp. 277–281.

Niranjana Devi Y, & Anto S, 2014, 'An evolutionary-fuzzy expert system for the diagnosis of coronary artery disease,' International Journal of Advanced Research in Computer Engineering & Technology, vol. 3, no. 4, pp. 1478–1484.

Orkcu, HH & Bal, H, 2011, 'Comparing performances of backpropagation and genetic algorithms in the data classification,' Expert Systems with Applications, vol. 38, no. 4, pp. 3703–3709.

Polat, K & Güneş, S, 2009, 'A new feature selection method on classification of medical datasets: Kernel F-score feature selection,' Expert Systems with Applications, vol. 36, no. 7, pp. 10367–10373.

Pradhan, M & Sahu, RK, 2011, 'Predict the onset of diabetes disease using artificial neural network (ANN),' International Journal of Computer Science & Emerging Technologies, vol. 2, no. 2, pp. 2044–6004.

Purnami, SW, Embong, A, Zain, J.M & Rahayu, SP, 2009, 'A new smooth support vector machine and its applications in diabetes disease diagnosis,' Journal of Computer Science, vol. 5, no. 12, 1003.

Refaeilzadeh, P, Tang, L & Liu, H, 2007, 'On comparison of feature selection algorithms,' Proceedings of Association for the Advancement of Artificial Intelligence," pp. 35–39.

Sartakhti, JS, Zangooei, MH & Mozafari, K, 2012 'Hepatitis disease diagnosis using a novel hybrid method based on support vector machine and simulated annealing (SVM-SA),' Computer Methods and Programs in Biomedicine, vol.108, no. 2, pp. 570–579.

Ster, B & Dobnikar, A, 1996, 'Neural networks in medical diagnosis: Comparison with other methods,' In Proceedings of the International Conference EANN, pp. 427–430.

Suykens, JA & Vandewalle, J, 1999, 'Least squares support vector machine classifiers,' Neural Processing Letters, vol. 9, no. 3, pp. 293–300.

Tan, KC, Teoh, EJ, Yu, Q & Goh, KC, 2009, 'A hybrid evolutionary algorithm for attribute selection in data mining,' Expert Systems with Applications, vol. 36, no. 4, pp. 8616–8630.

Vinotha PG, Uthra V, Dr Anto S, 2017, ' Medoid Based Approach for Missing Values in the Data Sets Using AANN Classifier,' International Journal of Advanced Research in Computer Science and Software Engineering, vol. 7, no. 3, pp. 51–55.

Watson, I, Basden, A & Brandon, P, 1992, 'The client-centred approach: Expert system maintenance,' Expert Systems, vol. 9, no. 4, pp. 189–196.

Watson, I & Marir, F, 1994, 'Case-based reasoning: A review,' The Knowledge Engineering Review, vol. 9, no. 04, pp. 327–354.

Wernick, MN, Yang, Y, Brankov, JG, Yourganov, G & Strother, SC, 2010, 'Machine learning in medical imaging,' IEEE Signal Processing Magazine, vol. 27, no. 4, pp. 25–31.

Yilmaz, E, 2013, 'An expert system based on Fisher score and LS-SVM for cardiac arrhythmia diagnosis,' Computational and Mathematical Methods in Medicine, vol. 5, pp. 1–6.

Zangooei, MH, Habibi, J & Alizadehsani, R, 2014, 'Disease diagnosis with a hybrid method SVR using NSGA-II,' Neurocomputing, vol. 136, pp. 14–29.

CHAPTER 2

FROM DESIGN ISSUES TO VALIDATION: MACHINE LEARNING IN BIOMEDICAL ENGINEERING

CHRISTA I. L. SHARON* and V. SUMA

Department of Information Science and Engineering,
Dayananda Sagar College of Engineering, Bangalore, Karnataka, India

**Corresponding author. E-mail: sharonchrista-ise@dayanandasagar.edu*

ABSTRACT

This chapter is all about the design and validation challenges that are faced when adapting machine learning techniques in biomedical engineering specialization. The issues concerning the acquisition of data from biomedical sources to concerns in the design and validation of the machine learning models are wrapped up and presented. The chapter describes the data acquisition prospects in the biomedical systems and the different approaches that can be followed in the knowledge representation. Further, the design issues and feature selection techniques that can be followed to achieve an optimum learning model are presented. The different ways in which the outcome of the biomedical based machine learning models can be validated is also depicted further.

2.1 INTRODUCTION

With the integration of software systems in day to day life, by 2020, 44 trillion GB of data will be accumulated, of which a huge share is from the biomedical sector. The computing system has crossed the era of limited, slower, and expensive memory storage, therefore, storage is not a hurdle anymore. The advancements in technology-enabled storing huge amounts of data. Data in the repository can be utilized for analytics that gives insights on common trends, anomalies patterns, and relationships. Researchers and scientists all over the world are using advanced technology and algorithms in predicting and detecting diseases. Even though the proposal to integrate AI and biological data was spurred in the early 1970s, the approach becomes popular in the 1980s. Even then it had limited results because of the unavailability of the apt technology. Watson's system, a machine learning-based research initiative by IBM and similar machine learning algorithms produces results that are accurate than doctors themselves. Precision medicine using artificial intelligence techniques is an emerging field. For any of the technology to work in the desired manner, the availability of data is essential.

Data in the right format enables one to apply the available advanced technologies to produce desired results. When data analytics enables the researcher to identify the patterns and trends, data analysis is the process that acts as a support system to perform data analytics. Data analysis cleans and transforms the data into the required format. The process is important especially in biomedical data because of the complicated and diverse nature of the medical records. Computerizing the test results of different kinds and the hand-written medical notes are pretty challenging. Organizing and automating the same is pretty challenging. The computerized data that is available has types that range from binary type, image type to temporal and fuzzy types.

While developing an AI-based system for biomedical data, the application is very important. The accuracy of the models will be based on the initialization parameters, data input type, and the required output (Villanueva et al., 2019). These are considered the design issues that need to be addressed. Further, analysis and the factors to be considered in the analysis, the methods of validating and evaluating the output is application specific.

The chapter gives an introduction and details on the design issues that have to be considered in modeling an artificial intelligence-based biomedical system. Design issues cover the approach to be considered in identifying the model and its objectives. An approach to identify the right data source followed by a different approach to extract data from medical records is put forth. Data collection from available datasheets and processing is presented. Since missing data is a major concern and it affects the accuracy of the models, the approaches in filling out the missing data are presented further. This is followed by parameters that can be considered in identifying the number of data points as well as requirements in the training and testing of the data. Further, the factors that have to be considered in designing the model as well as what factors have to be considered are presented in the next part of the chapter as analyses of the data model.

Approaches for the validation and evaluation of the data models are presented along with the issues that have to be considered for which the different aspects of the system that has to be considered are presented further. The analysis to be followed to identify the appropriateness of the model will give an insight into the factors that have to be considered in a model before adapting a training set as well as the model itself. The practical considerations as well as the appropriateness of the algorithm based on the performance evaluation and the different ways to perform the same will conclude the chapter.

2.2 DATA AND KNOWLEDGE ACQUISITION

An AI-based system will have two parts, a knowledge base that is fed with data acquired from different sources along with the inputs from domain experts and an expert system that derives inferences from the knowledge base. The development of a knowledge base in a required domain is a major issue since these knowledge bases require inputs from experts and domain associated databases. To develop the knowledge base, the data of biomedical signals like bioelectric signals, blood oxygenation, respiration, etc. are acquired by different biomedical instruments like EEG, ECG, etc. These data need to be stored for the proper performance of the AI models.

The different ways in which a knowledge base can be populated is presented further:

The raw biomedical signals collected from different biomedical instruments are stored in different forms presented in Section 2.3.1. These data are stored in structured storage systems called databases. On the other hand, the relationships between different variables of the databases are maintained in the knowledge base. To identify the same requires inputs from the domain experts. The initial step is to populate the database and then establish the relationships.

There are different approaches in which the knowledge base can be populated. Rule-based knowledge base is one of the typical and primary approaches that is adapted. In rule-based approach, the rules are derived according to the criteria that are based on the inputs from the emergency room details. Figure 2.1 specifies a sample ER data and the rules based on the specifies sample ER data are presented further in Figure 2.2.

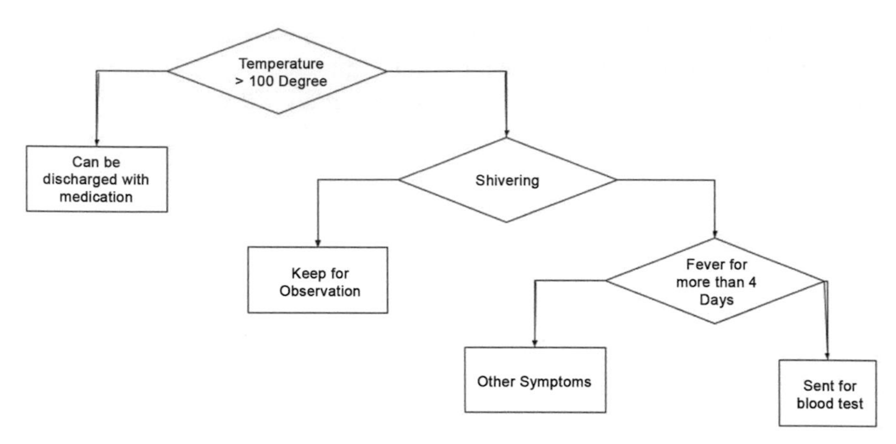

FIGURE 2.1 ER data and the rules.

If Fever is Greater than 100 Degree AND Shivering Present AND Fever present for more than 4 Days

Proceed to Blood Test

FIGURE 2.2 ER-based rules.

Once the rules are developed, the models are implemented. Further, the same is evaluated based on the data. The rules can be modified or changed if required as per the inputs from the evaluation data.

Another approach is to extract knowledge from the experts. Eliciting information from the experts requires sound knowledge of the requirements and the type of knowledge that is required. The question that is raised to the experts should as well map very much with the decision-making process. The actual decision-making process should be very well-reflected. Another important factor is to streamline the data recording process where the factors that are being recorded should be represented in the same way as it is followed in the medical field. The outcome of the knowledge elicitation is to lead to a state where knowledge state should lead to decision making state.

When multiple experts are concerned about knowledge elicitation, the factors that should be taken care of are different; experts will follow different strategies to diagnose and treat some diseases.

Furthermore, the rules specified by different experts can contradict and it can be inconsistent. Therefore, the inconsistencies should be removed to get the most accurate and optimum model. Therefore, contradictory information if present in the knowledge base as well as the rules that will lead to inconsistent conclusions should be removed (Parmar, 2018).

One of the major concerns of the knowledge base and its related model is that the models cannot be complete. All the different cases of symptoms cannot be included and there are boundaries in the expert system that should be accepted.

The boundary can be overcome to some extend by updating the information in the knowledge base continuously according to the field changes and expansion of knowledge.

An alternative to the expert elicitation is to learn through examples. Rote learning, advice taking, and induction are the three different types of learning through examples. Rote learning is nothing but storing the information that is presented to the system that is similar to the human-based learning. At this level, knowledge organization is required. In the advice taking system, the steps involved are based on deducing and understanding the consequences of the actions and updating the knowledge base according to the inputs on the actions. The evaluation enables a better understanding of the knowledge base. Induction is based on example-based learning that involves supervised and unsupervised learning methods, representing simple and complex concepts, etc. In all the cases, representation of the primary knowledge and the associated rules is the biggest challenge. Each of the training data points must be mapped as the rule (Li, 2018).

Knowledge acquisition requires information that is associated with the knowledge base and that can be acquired by a better understanding of the meta-data. The meta-data is the data that deals with domain knowledge. The data represents the depth of the understanding of the knowledge base and the associated rules and its correctness. Meta-data further represents the process and steps that are taken by the AI-based models to reason and reach a decision.

2.3 KNOWLEDGE REPRESENTATION

Biomedical data takes different forms and has to be derived from different forms of medical records. The different types of data encountered are patient history, physical exams, laboratory tests, pathology reports, imaging reports, and different tests. The conversion of the medical data to electronic data is important since models developed using erroneous data will result in faulty AI models. The different data forms that can be considered while converting the biomedical data to electronic data are specified in Section 2.3.1. The conversion of medical data to electronic data is the first step that will be discussed in Section 2.3.2. The ways in which knowledge is represented in the AI-based systems play a major role in the decision support system. There are various approaches to it. Section 2.3.3 onward gives the details of the different approaches in knowledge representation.

2.3.1 DATA FORMS

Different forms of data that can be derived from medical records are presented further.

- Binary/bipolar data with two possible responses that can be assumed values 0/1 or –1/1
- Categorical data with values that can be ranked worst to best or so. These types of data need to be numerically coded in the ordered form.
- Integer and continuous data include variables like temperature with an inherent order. The main consideration in such data is precision and accuracy.
- Fuzzy data can be used in the representation of information that is associated with imprecision.
- Temporal data represents the findings related to the biomedical data over a period of time and time-series data is a category of temporal data that records and presents the patterns of readings associated with different biological functions.
- Image data: digitized images are the outcome of different medical imaging technologies that are used in the diagnosis of different diseases. Images are represented as pixels that determine its resolution. Image data is analyzed using pixel values.

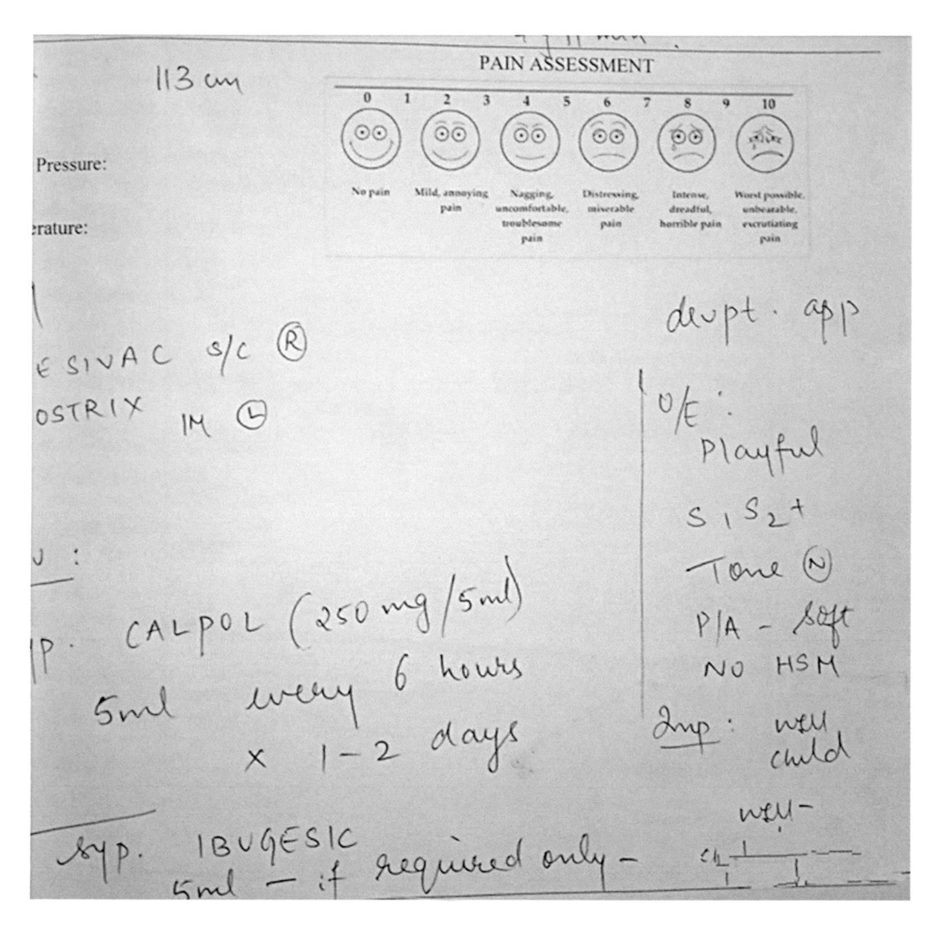

FIGURE 2.3 Patient record.

2.3.2 MEDICAL DATA TO ELECTRONIC DATA

Figure 2.3 shows a sample patient record from which data need to be extracted. Typically patients will have different volumes of medical records. The extraction of medical information from the handwritten records is challenging since it will be illegible with multiple interpretations (Sarjapu, 2016). Therefore, the format should be made in such a way that the information relevant to the current decision-making process is specified. A sample data collection format associated with a model for cardiac diseases is presented in Figure 2.4 (Hudson and Cohen, 2000).

ID Number	***(Integer)***
Age	***(Integer)***
Sex	***(Binary)***
Hx of bypass	***(Binary)***
Hx of MI	***(Binary)***
Presence of symptoms (Incfreased, Decreased, Stable)	***(Categorical)***
Dyspnea	***(Binary)***
Orthopnea	***(Binary)***
PND	***(Binary)***
Duration of Symptoms	***(Continuous)***
Physical Findings	
Resting Heart Rate	***(Continuous)***
Edema	***(Binary)***
Rales	***(Binary)***
Gallup	***(Binary)***
Mitroregurgitation	***(Binary)***
Functional Impairment (NYHA)	***(Categorical)***
LV Ejection Fraction	***(Continuous)***
Echo	***(Subjective)***
ETT Data	***(Binary)***
Resting Heart Rate	***(Continuous)***
Resting Blood Pressure	***(Continuous)***
Time of Maximum ST Depression	***(Continuous)***
Time of Angina	***(Continuous)***
Total Exercise Time	***(Continuous)***
Heart Rate at End of Test	***(Continuous)***
Blood Pressure at End of Test	***(Continuous)***
Reason for Stopping ETT	***(Subjective)***
Holter Data	***(Subjective)***
Electrolytes	
Na	***(Continuous)***
K	***(Continuous)***
Mg	***(Continuous)***
BUN	***(Continuous)***
Cr	***(Continuous)***
Drugs	
Digitalis	***(Binary)***
Diuretic	***(Binary)***
ACE Inhibitor	***(Binary)***
Vasodilators	***(Binary)***
Anti-arrhythmic	***(Binary)***
URI/Viral Syndrome	***(Binary)***

FIGURE 2.4 ER data converted to electronic data.

2.3.3 DIFFERENT REPRESENTATIONS

The data once derived from the biomedical records need to be processed and represented in the proper format based on the type of information that is available (Friedman, 2018). These different ways of representing the data will aid in building the decision support systems of the AI models. There are different approaches that can be adapted in representing the knowledge based on the requirements.

2.3.3.1 DATABASE

Databases provide primary physical storage at a logical level of the data organization that makes storage of the records more efficient, that is, database is used in the logical organization of data. There are different database organizations based on mathematical relations that are being used at different levels to reduce the complexity of data retrieval like hierarchical structure, relational structure, object-oriented structure, etc. Different database product and service providers provide databases that are based on relational database structure since it has a strong mathematical basis. Data in the database is stored as tables with each characteristic of the biomedical signal stored as an attribute. The records of each patient will be stored as a row in the table. The rows are termed as tuples and are typically unordered. Furthermore, attributes are also unordered. Typically, the entities represented as a table will be related to each other where relationships are specified. Further, top-level databases have fields that are referred to databases at the lower levels when the databases follow hierarchical structure (Liu, 2018).

In general, database follows a relational database structure. Figure 2.5 shows a sample patient relation (Healthcare, 2019). The advantage of a database over traditional file structure is that database provides data consistency and integrity. It further avoids data inconsistency. The relational databases fail to integrate the knowledge of the domain to the raw data. To achieve the same, an advanced database architecture can be adapted, which is the object-oriented database. The features of object-oriented modeling like a combination of procedures and data, hierarchy, and inheritance is adapted to integrate the knowledge base to the raw database. Further, object-oriented databases are highly adaptable and can be used in the inclusion of temporal data.

Hospital Patient ID	Patient Age	DRG	Attending Physician ID	Admit Date	Discharge Date	Discharge Disposition
1234509	42	465	1389	6-05-06	6-10-06	Home
3445676	75	110	3409	5-16-06	6-10-06	SNF
5678932	22	322	6704	6-07-06	6-08-06	Home
7890111	6	201	3422	6-10-06	6-11-06	Transfer

FIGURE 2.5 Sample patient relation (Reprinted from Khan, 2010).

The database enables efficient addition, deletion, and retrieval of data from the database. It is accomplished by the querying language. The query language is based on the relational algebra. Logical constructs like AND, OR, and NOT are used in the language for data manipulation. Further, different constructs like SELECT, WHERE, COUNT, AVERAGE, MAX, etc. are used in deriving the data in the desired form (Agrawal, 2018).

Based on the location, a database can be centralized or distributed. The database if present in a single location then it is termed centralized and if it is spread across multiple locations it will be distributed. The adaptation of the different types of databases is mainly based on the requirements and other factors that concern the client.

2.3.3.2 DATABASE AND KNOWLEDGE-BASED SYSTEMS FOR TEMPORAL DATA

The focus of storing temporal data is to identify the occurrence of the event and the time at which the event occurred. These data are termed as point events. Further, the point events are classified as simple, complex, and interval events where complex events are a combination of simple and interval events and interval events are events that happened for the interval of time. An example for a simple event is the ECG reading. Interval events can be stress, blood pressure, etc. Complex reading can be chest pain where heartbeat reading and BP are also considered. The representation of the same in the database requires the time also to be captured along with the event. In knowledge-based systems, the temporal variable whereever relevant should be taken and recorded in a usable format.

2.3.3.3 FRAMES

A frame is another approach for knowledge representation that includes general information related to the event. The same can be modified and adjusted to fit different situations. This further mimics the knowledge representation in humans where the preliminary knowledge on the situation if encountered will be used to perform the actions and accordingly the knowledge of the humans will be updated (Priyadarshini and Hemanth, 2018). The same is achieved by adopting the object-oriented property called inheritance. Each frame associated with a user will inherit the general information in the mainframe. Further, user-specified details are filled in the associated fields and subframes. Triggers can be further used to follow the data-driven approach in the frames. The triggers will be invoked in case of a change in the frame field values. Figure 2.6 shows a general frame structure and the inherited user-specific frame.

ER Monitoring Template
Specialization:
Complaint:
Physical Characteristics (Name, Height, Weight, Age)
Types(Blood Pressure, ECG,...)

Patient Name: John Doe
Specialization Cardiology
Complaint: Chest Pain, Breathlessness
Physical Characteristics (Height: 150 cm, Weight: 70 kg, Age: 50)
Types: BP Reading
ECG Reading:

FIGURE 2.6 Sample general frame structure and inherited user specific frame.

2.3.3.4 PRODUCTION RULES

The other approach for the representation of the knowledge base is termed as production rules. It is one of the earliest, flexible, and most important knowledge representation approach. A production rule-based approach can be categorized into two parts. The first part will identify the different situations that can be encountered in a domain. The second part will identify actions that can be adopted in different situations. The situation, in general, is termed as a condition and that is followed by actions.

The system works in such a way that the occurrence of a condition will trigger the action. There are cases where multiple conditions occur. In that case, the conditions will be conjugated and the actions will be performed only if the conditions specified are all true. To confirm the occurrence of the conditions, the system works in such a way that question are directed to the patient/user. Once the inputs for the cases for conditions are obtained, the model will give the specified outcome (Panyam, 2018).

Even with all the associated rules and specifications, processing of the language is very challenging. Since the expert systems are domain-specific the language used will only be the subset of the natural language. Further, to overcome the same there are systems that allow some conditions to be matched and give responses accordingly. The rule-based systems should be data driven, where the user will be allowed to enter the information and this information can be matched with the conditions.

The rules or combinations of rules are represented logically with AND/OR Tree. The tree presents a logical grouping of the conditions and the actions. The leaf node of the tree represents the actions and the root and internal nodes represent facts or conditions. Figure 2.7 presents a general structure of AND/OR Tree

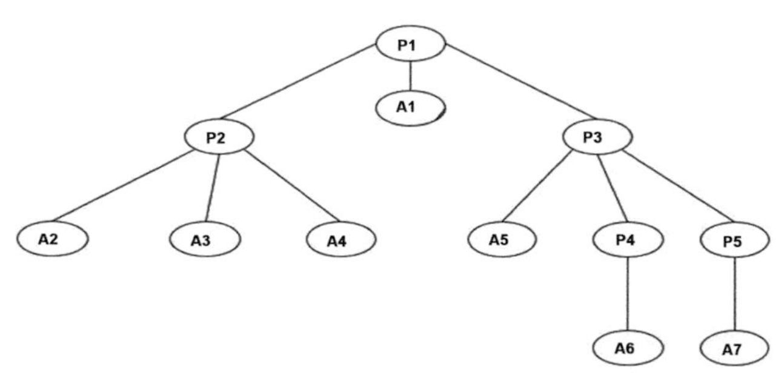

FIGURE 2.7 AND/OR Tree Pi is the production rule and AI is the action. The edges from same nodes represent AND operation.

based on the production rules. The situations where values of multiple conditions are considered the order of the conditions are analyzed to see if it matters. In that case, the order of the conditions is set. Further, in case of conflict in the condition, the priority in which the conditions should be considered is set.

The main drawback of the production rule is that the model is limited to the factors presented in the production rules. The scenario in which other conditions are present will not be considered in such a system. The rule search strategies are limited only to forward chaining and backward chaining. In certain rule-based systems, the rules allow standard conditions and the inclusion of AND, OR NOT, COUNT, etc. along with the certainty factor will provide more accurate results.

Each production rule is associated with a certainty factor that is a value between zero and one. The AND nodes in the AND/OR Tree will have the certainty factor that is calculated based on the outcome of the situation and is carried forward. 1—certainty factor is considered for the evaluation of the outcome (Yu, 2015). This can be considered mainly in cases where ad hoc approach is considered with uncertain information. Even then there are factors that make the production of rule-based system inefficient. Primarily, it is due to the restriction to the formal specification and domain. Also, it can be used in keeping track of the rule searching order. Further, it mimics human reasoning in the standard conditions with a uniform structure.

2.4 FEATURE SELECTION

The primary factor that influences the design and development of an AI model is the input variables. The process of identification of these input variables is the preliminary step for the design of an AI model and the same is called feature extraction. The process is uniform for supervised and unsupervised learning models.

The first step of any feature extraction phase is to identify the types of variables that can be used to develop the models that will give optimized results. The feature selection process starts with the feature extraction process and it depends on the type of model and type of variable. The biomedical systems have continuous variables as well as time series variables; the models developed should be able to accept such inputs.

One aspect of feature extraction is to identify all the possible variables that will contribute positively to the development of AI-based models. Even though only a limited number of variables are allowed to be included as input variables, an adequate amount of data under each variable should be available for the training of the models. In the learning process, the weight of the unimportant variables will approach zero. An example of the features that can be considered while modeling using brain signals is (Chen et al., 2016).

- WM stage is labels.
- Frequency is a continuous variable.
- Feature weight is a discrete variable.

All these variables will be enumerated to develop the model. Of the three variables presented in the example, feature weight is an integer, WM stage changes based on the electrode used so categorically and, frequency has continuous values. The parameter can be clinical and nonclinical. Again the consideration of parameters is purely based on the requirements of the model. Further, images and image-related data play an important role in medical diagnosis, therefore, feature extraction from medical images depends on the goal the model has to achieve. The recognition and classification of different patterns of the images are very much important for the accurate development of the models. Researchers have worked in this area and have put forth the best practices that can be adapted to classifying the images (Duda and Hart, 1973). Changes in gray levels (Vannier et al., 1992), areas with irregular borders (Wang, 2011), and changes from previous images of the same patient (Zhao and Peng, 2010), asymmetry are some of the features.

Further, moving on to the time series data, biomedical signals are recorded in different ways and they can be classified as signals with built-in patterns and without built-in patterns (Beam, 2018). Body signals with built-in patterns can be recorded based on the normal variations from its standards (Pyrkov, 2018), whereas signals without built-in patterns need the consultation of the medical experts to identify and extract the information that has to be recorded (Peng, 2018).

While considering the number of modeling of the system, the number of variables corresponds to the number of dimensions that would generalize the models result and it is practically not possible to visualize the models with more than three-dimension. Therefore, an increase in the number of variables will result in the increase in the dimensions that will make the training of the model more time-consuming. Therefore, feature selection is crucial in building AI-based models with better accuracy and reliability.

2.5 DESIGN ISSUES

Once the required data is stored in the proper format, AI models can be developed. To achieve the same, it is important to consider different issues related to designing the system. Availability of data, identification of the requirements of the output, and objectives plays a major role in the system design. Further, the major concerns are the specific models that can be adapted for a particular application.

The parameters that will aid in the best suitable model is and will be a factor that will have an impact on the designing of the AI-based models. The decision-making process involved in the AI-based models required appropriate information sources. The different approaches that can be considered are knowledge-based approach and data-driven approach. The choice of the approach depends mainly on the availability of the data. Expert opinions if required can also be integrated into case of knowledge-based approach (Sniecinski, 2018).

The input to the system will be the data extracted from the medical records that are encoded as per the model requirements. As mentioned in Section 2.3, data represented in the different formats will be manipulated for the modeling. Multiple responses like medicines prescribed, categorical data like symptoms, fuzzy data like a range of blood sugar are represented as raw data that need to be encoded in the most suitable way to get the most accurate outcome.

Further, one of the major features of biomedical data is the presence of the missing data. If the decision-making AI models have to rely on the charts then protocols have to be established to deal with the missing data. The different protocols that can be followed are to remove the cases where the data fields are missing. The integrity of the data is ensured in this particular case but the number of data points will reduce. Or wherever the data is missing, enter the minimum possible value or the maximum possible value. The average value can also be considered as the entry in the records. The usage of maximum, minimum, or average depends mainly on the significance. In case of the missing blood pressure value, an average of the previous and the next reading can be considered as the input of the missing field (Shah and Chircu, 2018).

Time series data as well as image data needs special consideration. The different categories of time-series data include the data reading from ECG, EEG, biosensors, etc. The monitoring of related AI models requires real-time analysis and decision-making systems. Therefore, the models should have the least trade-offs in terms of analysis time, execution time as well as interpretation time. When it comes to image data, different imaging techniques are in use for the proper diagnosis of the diseases. The image encoding and analysis is the primary concern along with the determination of anomalies (Hudson and Cohen, 2000).

Further, the choice of the learning algorithm and the choice of the features of the algorithm should be considered for the stability and convergence of the AI-based system. The integration and testing of multiple algorithms are very much required to ensure that the selected model that is developed for the application produces the best possible outcome than producing meaningless outcomes or simply fails to produce an outcome.

The different approaches that can be considered in the development of the learning models are supervised learning models and unsupervised learning models. To develop a well-rounded supervised learning model a very detailed labeled dataset is required. The model produced will be able to give results under two different categories, namely, classification and regression. The supervised learning algorithms can be used in the classification of malignant tumors or can be used in the prediction of heart failure. Whereas unsupervised learning algorithms on the other hand are build based on the unlabeled dataset. The models build with the dataset comes under clustering or self-organizing networks where it can be used to determine the type of malignancy in the tumors. These systems have little or zero control over the mapping of patterns to the clustering. Since the data is unlabeled, the data is highly sensitive to the presence of noisy data. Further, it cannot be determined with a fair accuracy if the solution is correct or not (Yu, 2018).

The results can be interpreted based on various factors, and the approach depends on the type of algorithm. In the case of the supervised learning algorithm, the process of developing the model is divided into the training phase, testing phase, and the validation phase. In the training phase, the most important parameters that have to be considered for modeling are determined. The different cases that have to be classified are also determined. An analysis of whether the classification cases that have to be considered is present in the dataset should be performed. Further, the model performance is measured based on the outcome of the models during the testing phase. The classified outcome by the models is compared with the already known outcome to determine the error (Yu, 2018).

In the case of the unsupervised learning models, the primary objectives are to identify the number of clusters that can be derived and the models need to be evaluated based on generally available knowledge. The establishment of different models enable in a more accurate differential diagnosis where the new data presented for the modeling will be almost similar to the original data.

To get the most out of the models, it is always beneficial to scale the data to the same range. Since different attributes will have a different range of data, the implication of the same will be different. Therefore, the data and the different attributes need to be scaled down to the same range. To achieve the same normalization of the data can be performed where all the attribute values can be narrowed down to values between 0 and −1 or +1.

A major limitation of the development of AI-based models is its dependence on the training dataset. The performance of the models mainly depends on the specification of the training model. The design issues presented in Section 5 are summarized in Table 2.1.

TABLE 2.1 Design Issues Associated with Building AI Models

Sr. No.	Design Issues
1	Information source/availability of data
2	Parameter selection
3	Encoding approach required for different types of data
4	Handling the missing data
5	Choosing the most appropriate learning algorithm
6	Choosing the required number of classification/cluster and its presence in the available data
7	Determining the most appropriate data preprocessing techniques

2.6 VALIDATION

Validation of AI-based decision support systems has different levels of problems. Primarily the outcome of the systems is based on the knowledge base and the parameters that should be considered. Further, the problems and the points that need to be considered in validating the outcome of a decision support system are discussed. Therefore, the approaches that can be adapted in the evaluation of performance as well as the evaluation of the model is presented further in this chapter.

2.6.1 INPUTS FROM THE DOMAIN EXPERT

The knowledge base and algorithms are not interrelated. But it can influence the outcome of the expert systems. The development of the knowledge base is based on the inputs from domain experts. The inputs from the domain expert guarantee neither completeness nor consistency of the knowledge base (Ozel, 2016). Upon the completion of the development of the knowledge base, the same should be verified by the domain expert/'s for its appropriateness.

2.6.2 VALIDATION OF THE DATA AND TRAINING ALGORITHM

Further, checking of the data can be performed by different approaches. The accuracy of the model based on the training data can be verified as per the reviews from databases and charts. Further, studies on the specific research area can be conducted. The data collected in each case can be considered in verifying the accuracy of the models. Further, how appropriate the training data is needed to be verified to determine if the most appropriate parameters are included. To achieve the same, statistical analysis of the training dataset needs to be performed. Possibilities of the accuracy of the data, scaling consistency of the dataset, standards to be followed are verified since the model will perform well only if the training dataset is most appropriate and optimized with the presence of all the required attributes. Further, in cases where the standards that are present for the correct classification, the cases where the standards are not followed need assumptions to be made. Thus, the model can produce inaccurate results. Such cases need to be considered when it comes to validation of the appropriateness of the training dataset (Wehbe, 2018).

The validation of the learning algorithm requires the developed model to be tested thoroughly. To ensure the same, the points to be considered are whether the selected algorithms are suitable for the input data and the model outcome is interpretable. Also, one of the major trade-offs of AI-based models is the training time. To get a well-rounded model, other approaches need to be considered and the performance of the developed model needs to be compared with the existing other models.

2.6.3 PERFORMANCE EVALUATION

The factors that can be considered in measuring the performance is the accuracy of the output generated by the model. The performance of the developed models can be verified by analyzing the outcome of the model using a test dataset. The same can be compared with the results obtained from other models for the same test dataset. Different error measures like mean absolute error, mean absolute percentage error, etc. can be considered in verifying the performance of the models. The values of the error measures should be one of the benchmarks in the evaluation of the models. The error measures considered in the verification of the model and the model with the minimum error will be considered the best performing model for the deployment (Emmert, 2016). When it comes to unsupervised models, apart from all the points mentioned, the applicability of other datasets in the model should be the parameter that has to be considered. To further narrow down the applicability of other datasets, it should have characteristics that are similar to the training dataset with the same parameters and accuracy.

2.7 CONCLUSION

In the process of the development of AI-based biomedical systems, the primary factor to be considered is the availability of the data and its representation. This chapter presents an overview of the different ways in which biomedical data can be represented and the associated knowledge can be acquired. Further, the different approaches in representing the knowledge are presented. The criteria that have to be considered in the selection of features and designing the AI models are presented. It should be noted that to realize an AI-based model, the concepts presented are not definite. Furthermore, the evaluation of a number of parameters is required. Different methodologies when combined can be beneficial also. The performance measure should be considered as the evaluation parameter of the model. The AI-based models have the capability to adapt to unseen scenarios even then the performance of the models in such cases depends mainly on the similarity of the training dataset.

KEYWORDS

- **data processing**
- **data acquisition**
- **biomedical data source**
- **feature selection**
- **data validation**

REFERENCES

Agrawal, S., Khan, A. and Kumar, K., International Business Machines Corp, 2018. Query modification in a database management system. U.S. Patent Application 15/424,769.

Ahmed, I.O., Ibraheem, B.A. and Mustafa, Z.A., 2018. Detection of eye melanoma using artificial neural network. Journal of Clinical Engineering, 43(1), pp. 22–28.

Beam, A.L. and Kohane, I.S., 2018. Big data and machine learning in health care. JAMA, 319(13), pp. 1317–1318.

Chen, C.M.A., Johannesen, J.K., Bi, J., Jiang, R., and Kenney, J.G. 2016. Machine learning identification of EEG features predicting working memory performance in schizophrenia and healthy adults. Neuropsychiatric Electrophysiology, 2(1), p. 3.

Codd, E.F., 1970. A relational model of data for large shared data banks. Communications of the ACM, 13(6), pp. 377–387.

Duda, R.O., and Hart, P.E. 1973. Pattern Classification and Scene Analysis. New York, NY, USA: John Wiley & Sons.

Emmert-Streib, F., Dehmer, M. and Yli-Harja, O., 2016. Against dataism and for data sharing of big biomedical and clinical data with research parasites. Frontiers in Genetics, 7, p. 154.

Friedman, C., 2018. Mobilizing Computable Biomedical Knowledge Conference October 18, 2017. Overview/Opening Remarks.

http://slideplayer.com/slide/6207174/20/images/28/Healthcare+example+of+relational+databases.jpg accessed on 06/05/2019

Hudson, D.L. and Cohen, M.E., 2000. Neural Networks and Artificial Intelligence for Biomedical Engineering. Institute of Electrical and Electronics Engineers. Piscataway, NJ, USA

Ju, Z., Wang, J. and Zhu, F., 2011. Named entity recognition from biomedical text using SVM. In 2011 5th International Conference on Bioinformatics and Biomedical Engineering (pp. 1–4). IEEE.

Khan, R.S. and Saber, M., 2010. Design of a hospital-based database system (A case study of BIRDEM). International Journal on Computer Science and Engineering, 2(8), pp. 2616–2621.

Li, B., Li, J., Lan, X., An, Y., Gao, W. and Jiang, Y., 2018. Experiences of building a medical data acquisition system based on two-level modeling. International Journal of Medical Informatics, 112, pp. 114–122.

Liu, Z.H., Lu, J., Gawlick, D., Helskyaho, H., Pogossiants, G. and Wu, Z., 2018. Multi-Model Database Management Systems—A Look Forward. In Heterogeneous Data Management, Polystores, and Analytics for Healthcare (pp. 16–29). Springer, Cham.

Ozel, T., Bártolo, P.J., Ceretti, E., Gay, J.D.C., Rodriguez, C.A. and Da Silva, J.V.L. eds., 2016. Biomedical Devices: Design, Prototyping, and Manufacturing. John Wiley & Sons, New York.

Panyam, N.C., Verspoor, K., Cohn, T. and Ramamohanarao, K., 2018. Exploiting graph kernels for high performance biomedical relation extraction. Journal of biomedical semantics, 9(1), p. 7.

Parmar, C., Barry, J.D., Hosny, A., Quackenbush, J. and Aerts, H.J., 2018b. Data analysis strategies in medical imaging. Clinical Cancer Research, 24(15), pp. 3492–3499.

Peng, L., Peng, M., Liao, B., Huang, G., Li, W. and Xie, D., 2018. The advances and challenges of deep learning application in biological big data processing. Current Bioinformatics, 13(4), pp. 352–359.

Priyadarshini, S.J. and Hemanth, D.J., 2018. Investigation and reduction methods of specific absorption rate for biomedical applications: A survey. International Journal of RF and Microwave Computer-Aided Engineering, 28(3), p. e21211.

Pyrkov, T.V., Slipensky, K., Barg, M., Kondrashin, A., Zhurov, B., Zenin, A., Pyatnitskiy, M., Menshikov, L., Markov, S. and Fedichev, P.O., 2018. Extracting biological age from biomedical data via deep learning: too much of a good thing?. Scientific Reports, 8(1), p. 5210.

Sarjapur, K., Suma, V., Christa, S. and Rao, J., 2016. Big data management system for personal privacy using SW and SDF. In Information Systems Design and

Intelligent Applications (pp. 757–763). New Delhi, India: Springer.

Shah, R. and Chircu, A., 2018. IOT and AI in healthcare: a systematic literature review. Issues in Information Systems, 19(3), pp. 33–41.

Sniecinski, I. and Seghatchian, J., 2018. Artificial intelligence: A joint narrative on potential use in pediatric stem and immune cell therapies and regenerative medicine. Transfusion and Apheresis Science, 57(3), pp. 422–424.

Vannier, M.W., Yates, R.E., and Whitestone, J.J. (eds.). 1992. Electronic Imaging of the Human Body. Wright Paterson Air Force Base, Ohio, USA: CSERIAC.

Villanueva, A.G., Cook-Deegan, R., Koenig, B.A., Deverka, P.A., Versalovic, E., McGuire, A.L. and Majumder, M.A., 2019. Characterizing the Biomedical Data-Sharing Landscape. The Journal of Law, Medicine & Ethics, 47(1), pp. 21–30.

Wehbe, Y., Al Zaabi, M. and Svetinovic, D., 2018, November. Blockchain AI Framework for Healthcare Records Management: Constrained Goal Model. In 2018 26th Telecommunications Forum (TELFOR) (pp. 420–425). IEEE.

Yu, H., Jung, J., Lee, D. and Kim, S., 2015, October. What-if Analysis in Biomedical Networks based on Production Rule System. In Proceedings of the ACM Ninth International Workshop on Data and Text Mining in Biomedical Informatics (pp. 28–28). ACM.

Yu, K.H., Beam, A.L. and Kohane, I.S., 2018. Artificial intelligence in healthcare. Nature Biomedical Engineering, 2(10), p. 719.

Zhao, Q., Peng, H., Hu, B., Liu, Q., Liu, L., Qi, Y. and Li, L., 2010, August. Improving individual identification in security check with an EEG based biometric solution. In International Conference on Brain Informatics (pp. 145–155). Springer, Berlin, Heidelberg.

CHAPTER 3

BIOMEDICAL ENGINEERING AND INFORMATICS USING ARTIFICIAL INTELLIGENCE

K. PADMAVATHI* and A. S. SARANYA

Department of Computer Science, PSG College of Arts and Science, Coimbatore 641014, Tamil Nadu, India

Corresponding author. E-mail: padmasakthivel@gmail.com

ABSTRACT

In recent decades, artificial intelligence (AI) is widely used in various fields of human life. One of the most promising areas of AI is medical imaging. Medical imaging provides an increasing number of features derived from different types of analysis, including AI, neural networks, fuzzy logic, etc. The selection of image processing features and AI technologies can be used as a medical diagnostics tools. AI is used to help and improve numerous aspects of the healthcare system. AI tools and techniques provide considerable insight to power predictive visual analysis. The recent growing applications of AI techniques in biomedical engineering and informatics used knowledge-based reasoning in disease classification, which is used to learn and discover novel biomedical knowledge for disease treatment. AI is used to detect disease in an earlier stage and guide diagnosis for early treatment with imaging technologies. AI applications are used and implemented in various biomedical fields for analyzing diseases like myocardial infarction, skin disorders, etc. The tools and techniques of AI are useful for solving many biomedical problems with the use of computer-related equipped hardware and software applications. This chapter provides a thorough overview of the ongoing evolution in the application of biomedical engineering and informatics using AI techniques and tools. It gives a deeper insight into the technological background of AI and the impacts of new and emerging technologies on biomedical engineering and informatics.

3.1 INTRODUCTION

Artificial intelligence (AI) is the human and computer interaction and development system that uses human intelligence to do various tasks like visual perception, speech recognition, language translations, robotics, and decision-making. AI and its related technologies offer real practical benefits and innovations in many research areas and in their applications. AI is a revolutionized technology that combines intelligent machines and software that work and react like human beings. AI and its applications are used in various fields of human life to solve complex problems in various areas like science, engineering, business, and medicine. Recent technological developments and related areas like biomedical engineering, medical informatics, and biomedicine use an innovative computer-based system for decision-making. AI is also used in various fields like biology, engineering, and medicine that give a great impact by using machine learning, neural networks (NNs), expert systems, fuzzy logic, and genetic algorithms.

In biomedical engineering, AI could be used to aid the doctors in making decisions without consulting the specialists directly. AI and related decision-support systems help to make clinical decisions for health professionals. They use medical data and knowledge domains in diagnosis to analyze patient's conditions as well as recommend suitable treatments for the patients. In addition, AI and related decision-support systems provide help to the patient and medical practitioner to improve the quality of medical decision-making, increase patient compliance, and minimize estrogenic disease and medical errors. This chapter presents a detailed approach to each application of AI in biomedical engineering such as diagnosis, medical imaging, waveform analysis, outcome prediction, and clinical pharmacology (Singh et al., 2014).

3.2 ARTIFICIAL INTELLIGENCE IN MEDICAL IMAGING AND DIAGNOSIS

Medical imaging is a collection of techniques that are used to create visual representations of a body interior for clinical analysis and medical intervention. Medical imaging plays an important role in medical diagnosis, treatment, and medical applications, which seeks to reveal internal structures hidden by the skin and bones for diagnosing and treating disease. In medicine, AI is used to identify diagnosis and give therapy recommendations. In medical diagnosis, artificial neural networks (ANNs) is used to get the result of the diagnosis. ANN provides an extraordinary level of achievement in the medical field. ANN has been applied to various areas in medicine like disease diagnosis, biochemical analysis, image analysis, etc. In recent years, medical image processing uses ANNs for analyzing medical images. The main components of medical image processing that heavily depend on ANNs are medical image object detection and recognition, medical image segmentation, and medical image preprocessing. The various AI imaging technologies help to examine various factors of the human body using radiography, MRI, nuclear medicine, ultrasound imaging, tomography, cardiograph, and so on (Smita et al., 2012).

3.2.1 COMMON ARTIFICIAL NEURAL NETWORKS IN MEDICAL IMAGE PROCESSING

In recent years, NNs algorithms and techniques are used in medical image processing because of their good performance in classification and function approximation. NN techniques are mostly used in image preprocessing (e.g., construction and restoration), segmentation, registration, and recognition. Table 3.1 shows the different types of NNs used in the medical field (Rajesh et al., 2016; Yasmin et al., 2013).

TABLE 3.1 Neural Network Used in the Medical Field

Neural Network	Preprocessing	Segmentation	Registration	Recognition
Hopfield NN	√	√	–	√
Radial basis function NN	–	–	√	√
Feedforward NN	√	√	–	√
Self-organizing feature NN	√	√	√	
Probabilistic NN	–	√	–	√
Fuzzy NN	√	√	–	√
Neural ensemble		√	√	√
Massive training NN	√	–	–	√

Image segmentation is an indispensable process in outlining boundaries of organs and tumors and the visualization of human tissues during clinical analysis. Segmentation of medical image processing is very important for clinical data analysis, diagnosis, and applications, leading to the requirement of robust, reliable, and adaptive segmentation techniques. Image segmentation and edge detection often follow image registration and can serve as an additional preprocessing step in multistep medical imaging applications (Pratyush et al., 2014). The following subsections describe applications of ANNs where segmentation or edge detection is the primary goal.

3.2.2 USES OF ARTIFICIAL NEURAL NETWORK IN MEDICAL IMAGE PROCESSING

Medical image processing is a collection of techniques, used for disease diagnosis in the medical field. Medical image processing techniques and applications meet these challenges and provide an enduring bridge in the field of medical imaging. It helps for quantitative image analysis using authoritative resources and sophisticated techniques, with a focus on medical applications. The common techniques used in medical image processing are

(i) preprocessing
(ii) pegmentation
(iii) andobject recognition

The preprocessing is used to improve image data by suppressing unwanted distortions and enhances or highlights image features for further processing. Medical image segmentation is a process, used to divide the image into meaningful regions using homogeneous properties. It performs operations on images to detect patterns and to retrieve information from it. Object recognition is used to recognize one or several prespecified or the learned objects or object classes in medical images.

(i) Preprocessing

In medical image processing, preprocessing is used to enhance the quality of the images when medical images have a poor noise-to-signal ratio. Image reconstruction and image restoration are the two categories that use neural networks to reconstruct and restore the medical images. In medical diagnosis, Hopfield NN and Kohonen NN are used for image reconstruction, and neural network filters (NFs) are used for image restoration.

Hopfield NN is a special type of feedback NN that is a fully interconnected network of artificial neurons, in this; each neuron is connected to each other. All artificial neurons have N inputs in which each input *i* have associated weight w_i. The weight w_i is computed and not changed.

All artificial neurons also have an output. All neurons are having both input and output, and all neurons are connected to each other in both directions using patterns. The inputs are received simultaneously by all the neurons, they output to each other continuously until a stable state is reached.

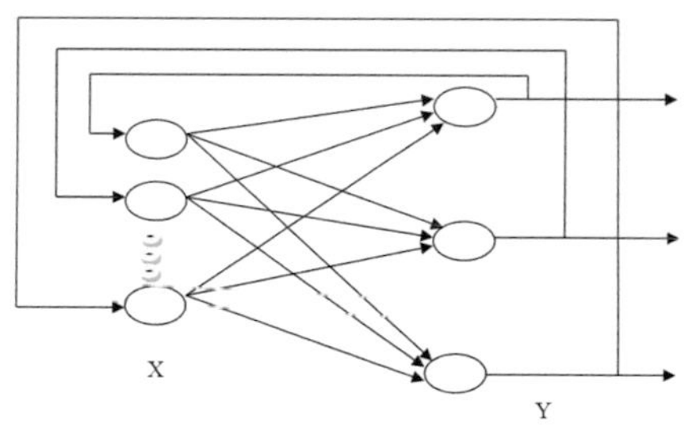

FIGURE 3.1 Hopfield neural network.

Medical image reconstruction is the process of reconstructing an image from a number of parameters acquired from sensors. The Hopfield NN (Figure 3.1) is used to reconstruct medical images that can always be conceptualized as an optimization problem making the convergence of the network to a stable position and at the same minimizing the energy functions. Electrical impedance tomography reconstruction on data that is noisy requires the solution to employ nonlinear inverse concepts. The problem is generally ill-conditioned and needs regularization based on prior knowledge or simplifying assumptions. Another method of ANN is self-organizing Kohonen NN that is the feedforward NN, which provides the most advantage for the problem of medical image reconstruction as compared to other methods.

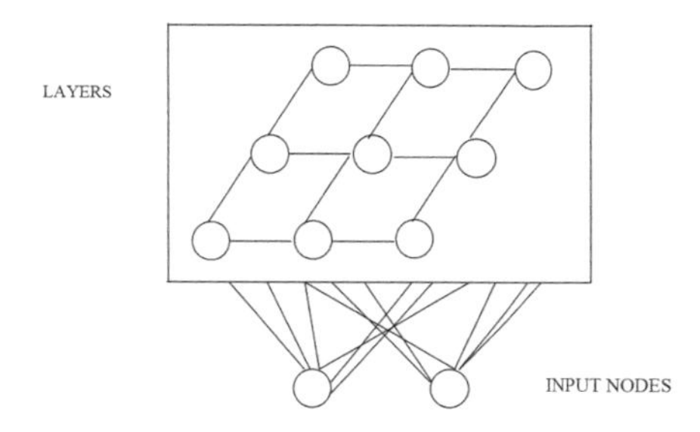

FIGURE 3.2 Kohonen network.

The Kohonen network (Figure 3.2) consists of an input layer, which distributes the inputs to each node in a second layer, the so-called competitive layer. In a competitive layer, each node acts as an output node. Each neuron in the competitive layer is connected to other neurons in its neighborhood and feedback is restricted to neighbors through these lateral connections. Neurons in the competitive layer have excitatory connections to immediate neighbors and inhibitory connections to more distant neurons. All neurons in the competitive layer receive a mixture of excitatory and inhibitory signals from the input layer neurons and other competitive layer neurons.

This method is used to compute the linear approximation associated with the inverse problem directly from the forward problem of finite element simulation. A high number of NN applications in medical image preprocessing are concentrated in medical reconstruction. In medical image restoration, neural NFs are used to remove noise from the image. Neural NFs use a neural edge enhancer (NEE), which is based on a modified multilayer NN to enhance edges that are desired. NEE is robust against noise, to enhance continuous edges found in images that are noisy.

(ii) Image segmentation

Image segmentation is used to extract the specific part from the medical images. In AI-based medical image segmentation, feedforward NNs are used to formulate image segments. Segmentation is the partitioning of an image into smaller parts that are coherent according to some criteria. In this classification task, segmentation is achieved by assigning labels to individual pixels.

In a feedforward network, information flows in one direction from the input layer to the final output layer via the intermediate hidden layers. The feedforward network uses the backpropagation (BP) supervised learning algorithm to dynamically alter the weight and bias values for each neuron in the network. A multilayer perceptron (MLP) is a special type of feedforward network employing three or more layers, with nonlinear transfer functions in the hidden layer neurons. MLPs are able to associate training patterns with outputs for nonlinearly separable data. In medical image processing, MLP performs segmentation directly on the pixel data or image features. Therefore, MLP performs segmentation in two different ways: (a) pixel-based segmentation and (b) feature-based segmentation.

(a) Pixel-based segmentation: Feedforward ANN (Figure 3.3) is used to segment the medical images using image pixels. BP supervised learning algorithm classifies and segments the medical images content based on

- texture or a combination of texture and local shape,
- connecting edge pixels,
- identification of surfaces, and
- clustering of pixels.

In medical applications, supervised classifiers are used to perform the desired segmentation.

(b) Feature-based segmentation: Feedforward ANNs classify and segment the medical images based on

- texture or a combination of texture and local shape,
- estimation of ranges,
- connecting edges and lines, and
- region growing.

Texture segregation is one of the most frequently performed segmentations by feature-based ANNs. In feedforward NNs-based segmentation, segmented images appear less noisy, and the classifier is also less sensitive to the selection of the training sets.

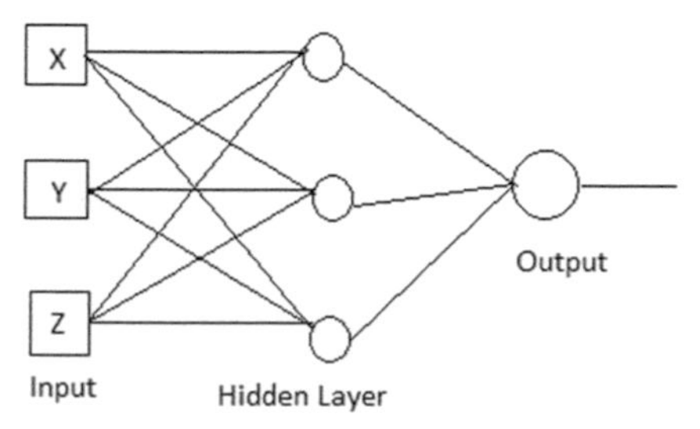

FIGURE 3.3 Multilayer feedforward network.

(iii) Object recognition

Object recognition consists of locating the positions and possibly orientations and scales of instances of objects in an image. The purpose of object recognition is to assign a class label to a detected object. In medical applications, ANNs are used to recognize and locate individual objects based on pixel data. Recurrent ANNs are used for object recognition in medical image processing. RNNs use supervised machine learning models for object recognition, made of artificial neurons with one or more feedback loops. The feedback loops are recurrent cycles over time or sequence. It is used to minimize the difference between the output and target pairs (i.e., the loss value) by optimizing the weights of the network for object recognition. The usage of recurrence in RNN uses averaging, which helps to give a more robust performance in object recognition.

3.3 WAVEFORM ANALYSIS

A waveform is the graphical representation of a signal in the form of a wave, derived by plotting a characteristics wave against time. The various inputs are used to create a waveform. Waveforms are used to represent different things in various fields like science, biochemistry, and medicine. In the medical field different kinds of waveforms are used to diagnosis the diseases. The most frequently used waveforms are

- ECG—electrocardiography,
- EEG—electroencephalography, and
- EMG—electromyography.

The waveforms are bioelectric signals, produced in human bodies by coordinated activity of a large group of body cells. ECG,, and EMG systems are used to measure cell activity of heart, brain, and muscle/nerve (Table 3.2). These waveforms are measured by bioelectric potentials on the surface of active tissue.

TABLE 3.2 Waveforms–Frequency Ranges

Bioelectric Signal	Frequency Range (μV)
ECG	50–5
EEG	2–100
EMG	20–5

ECG: ECG is the process of displaying the electrical activity of the heart over a period of time using electrodes placed on a human body (Figure 3.4). These electrodes detect the electrical changes on the membrane that arise from the heart muscle depolarizing during the heartbeat. The electrocardiogram is a diagnostic tool that is used to measure the electrical and muscular functions of the heart. The electrocardiogram is used to measure the rate and rhythm of the heartbeat and provide evidence of blood flow to the heart muscle.

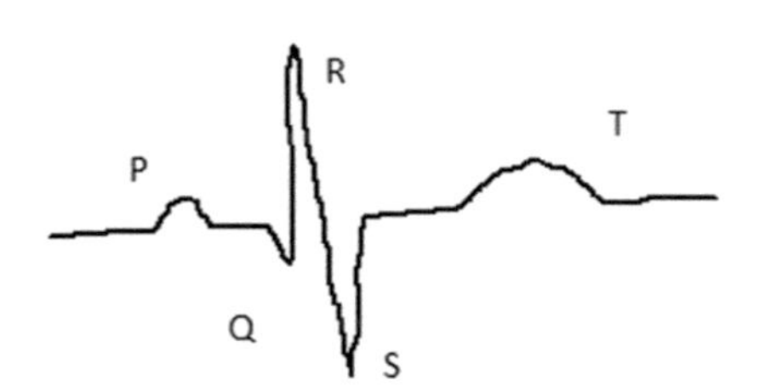

FIGURE 3.4 ECG waveform.

EEG: An electroencephalogram is a process that is used to detect electrical activity of the brain using small, metal discs (electrodes) attached to the scalp (Figure 3.5). Human brain cells communicate via electrical impulses and these pulses are active all the time, even when a human is asleep. This brain activity shows as wavy lines and it can be recorded as an EEG. EEG is used to measure changes in brain activity. It is one of the main diagnostic tools for assessing brain disorders like epilepsy, a seizure disorder. An EEG can also help in diagnosing and treating the following brain disorders:

- Brain tumor
- Brain damage from a head injury
  - o Brain dysfunction that can have a variety of causes (encephalopathy)
  - o Inflammation of the brain (encephalitis)
- Stroke
- Sleep disorders

FIGURE 3.5 EEG waveforms.

EMG: EMG is a method used to assess the health of muscles and the nerve cells of the body. EMG can identify nerve dysfunction, muscle dysfunction, or other problems with nerve-to-muscle signal transmission. In the human body, motor neurons are transmitted electrical signals between muscles. EMG uses electrodes to translate these signals into graphs, sounds, or any numerical values (Figure 3.6). The following nerve or muscle disorder symptoms can be identified by using EMG:

- muscular dystrophy or polymyositis myasthenia gravis,
- spinal cord disorders like carpal tunnel syndrome or peripheral neuropathies,
- amyotrophic lateral sclerosis or polio, and
- herniated disk in the spine.

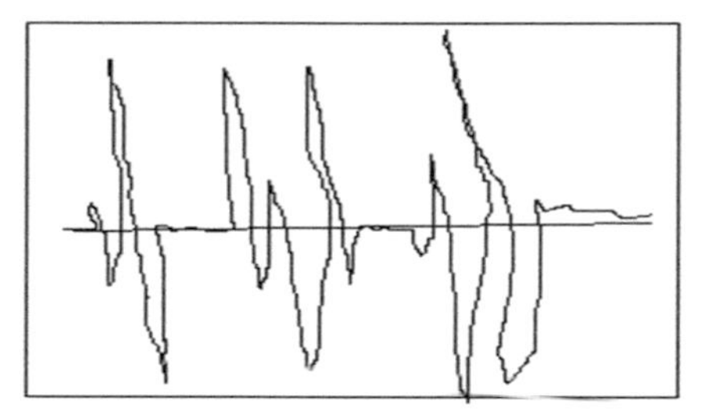

FIGURE 3.6 EMG waveform.

3.3.1 WAVEFORM ANALYSIS IN THE MEDICAL FIELD

3.3.1.1 DIAGNOSIS OF HEART DISEASE USING ANN AND ECG

The ECG is used to measure the bioelectrical activity of the heart. Variations in the signal amplitude and duration of the ECG are used to detect the cardiac abnormality. These variations are implemented in a computer-aided diagnosis system that can help in monitoring and diagnosis of cardiac health status. The information extraction can be done easily using ECG because of its nonlinear dynamic behavior. ANNs are effectively used for detecting morphological changes in nonlinear signals such as the ECG signal because ANNs use a pattern matching technique based on the nonlinear input–output mapping. A feedforward multilayer NN with error BP learning algorithm is used to investigate, monitor, recognize, and diagnose heart disease using ECG signals. The following steps are used to identify heart disease using feedforward multilayer NN with error BP learning algorithm (Sayad et al., 2014; Olaniyi et al., 2015) (Figure 3.7).

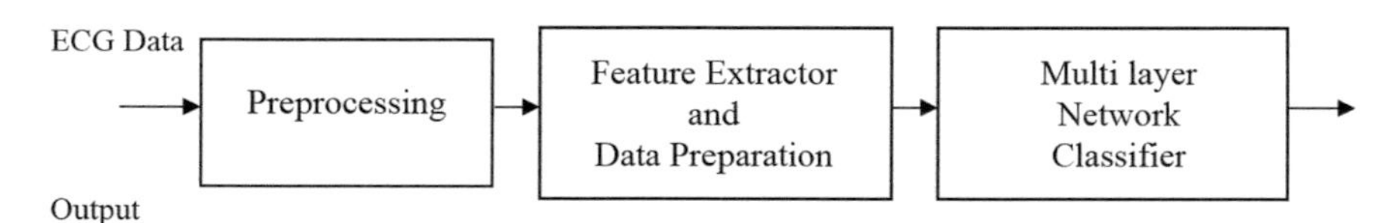

FIGURE 3.7 Common steps used in heart disease detection.

3.3.2 STEPS USED IN HEART DISEASE DETECTION

(i) *Preprocessing*: ECG signals may have corrupted various kinds of noises like baseline wander, alternating current power noise, muscular contraction noise, and electrode contact noise. To remove these noises, ECG signals are preprocessed using a filtering algorithm.

(ii) *QRS complex detection:* QRS complex detection is used to detect and recognize QRS complexes based on analyses of the slope, amplitude, and width. The recognition of the ECG onset and offset is necessary for computer-based analysis to assess the QRS complexes in ECG. The QRS onset is used to identify the beginning of the Q wave, or the R wave if no Q wave is present. The QRS offset is used to identify at the end of the S wave. The edges are detected as the point with zero slopes when there is a sudden change or by a minimum distance method.

(iii) *ST-segment analyzation:* The ST segment represents the period of the ECG after depolarization of the QRS complex and before depolarization of the T wave. Changes in the ST segment of the ECG indicate a deficiency in blood supply to the heart muscle. An ST segment analyzer is used to make measurements of the ST segment.

(iv) Image classification. In image classification, the various ECG characteristic and features are used.

Heart rate: It is the interval between two successive QRS complexes.

Change in heart rate: It is the difference between two successive heart rates.

QRS complex width: It is the duration between the QRS complex onset and offset.

Normalized source entropy of QRS complex: It is determined by the part of the signal containing in the QRS complex.

Normalized source entropy of ST wave: It is determined by the part of the signal containing in the ST segment.

Complexity parameter for the QRS complex: It is a Lempel and Ziv temporal complexity parameter that is determined by the part of the signal containing the QRS complex.

Complexity parameter for the ST wave: It is a Lempel and Ziv temporal complexity parameter that is determined by the part of the ECG signal containing the ST segment.

Spectral entropy: It is Shannon's spectral entropy that is determined by using the entire heartbeat

RT interval: It is the time between the occurrence of the R peak and T peak.

ST segment length: ST segment deviation and ST segment angle of deviation.

Using these characteristics and features as the input data set, the BP learning algorithm is implemented as classifiers. In the BP learning algorithm, the system weights are randomly assigned at the beginning and then progressively modified in the light of the desired outputs for a set of training inputs. The difference between the desired output and the actual output is calculated for every input, and the weights are altered in proportion to the error factor. The process is continued until the system error is reduced to an acceptable limit.

The modified weights correspond to the boundary between various classes, and to draw this boundary accurately, the ANN requires a large training data set that is evenly spread throughout the class domain. For quick and effective training, it is desirable to feed the data from each class in a routine sequence, so the correct message about the class boundaries is communicated to the ANN.

In the BP algorithm, the modifications are affected starting from the output layer and progress towards the input. The weight increments are calculated according to the formula listed in the following equations:

$$wij = \eta\, \delta j\, oi \quad (3.1)$$

$$\delta j = fj'(\text{net}j)\,(tj - oj) \quad \text{if unit } j \text{ is an output unit} \quad (3.2)$$

$$= fj'(\text{net}j) \sum k \delta k\, wjk \quad \text{if unit } j \text{ is and hidden unit} \quad (3.3)$$

where η denotes the learning factor (a constant); δj denotes error (the difference between the real output and teaching input) of unit j; tj denotes the teaching input of unit j; oi denotes the output of preceding unit i; i denotes the index of the predecessor to the current unit j, with link wij from i to j; j denotes the index of the current unit; and k denotes the index of successor to the current unit j with link wij from j to k.

The BPA requires both the activation function of the neuron and its (first) derivative to be of a finite magnitude and single valued. The input layer consists of nodes to accept data, and the subsequent layers process the data using the activation function. The output layer has three neurons, giving rise to an output domain of 16 possible classes.

3.3.1.2 DIAGNOSIS OF BRAIN TUMOR USING ANN AND EEG

ANN and EEG help to identify brain tumors with image processing techniques. Brain tumor classification is done by two stages: feature extraction and classification. In ANN, backpropagation network (BPN) classifier is used to evaluate the performance and classification accuracies. In brain tumor classification, the EEG signal is recorded and stored in the digital forms. The necessary features are extracted using principal component analysis (PCA). The steps of the PCA algorithm are as follows:

Step 1: Prepare the data:

- Center the data: Subtract the mean from each variable. This produces a data set whose mean is zero.
- Scale the data: If the variances of the variables in your data are significantly different, it's a good idea to scale the data to unit variance. This is achieved by dividing each variable by its standard deviation.

Step 2: Calculate the covariance/correlation matrix.

Step 3: Calculate the eigenvectors and the eigenvalues of the covariance matrix.

Step 4: Choose principal components: eigenvectors are ordered by eigenvalues from the highest to the lowest. The number of chosen eigenvectors will be the number of dimensions of the new data set.

Eigenvectors = (eig_1, eig_2,…, eig_n)

Step 5: Compute the new data set:

- Transpose eigenvectors : rows are eigenvectors
- Transpose the adjusted data (rows are variables and columns are individuals)
- New. Data = eigenvectors. Transposed * adjustedData. Transposed

In BPN classifier, two-layer feedforward network is used for classification. Feedforward network covers smaller regions for classification because it has a set of hidden layer and the output nodes.

The steps used in BPN are

- storing,
- sampling,
- finding similarity matching,
- by updating,
- repeating the four steps again, and
- spreads chosen by normalization.

3.3.1.3 USES OF ANN AND EMG IN DIAGNOSIS OF MUSCULAR DISORDERS

EMG is used to find the function of muscles and the nerves of the human body. EMG signals studies are used to help in the diagnosis of muscles and the nerve disorders like dystrophies and neuropathies. The classification of EMG diseases is done by various techniques like feedforward network, BPN, PCA, support vector machine (SVM). PCA is used for extracting desired information from relevant or irrelevant data sets because it uses a nonparametric method for information extraction. Classification of EMG diseases is a challenging and complex task.

The classification is done by BPN with feedforward network and SVM classifier.

3.4 OUTCOME PREDICTION USING ANN

The ability to predict the pathological stage of a patient with prostate cancer is important because it enables clinicians to better determine the optimal treatment and management strategies. This is to the patient's considerable benefit, as many of the therapeutic options can be associated with significant short- and long-term side effects. For example, radical prostatectomy, that is, the surgical removal of the prostate gland offers the best chance for curing the disease when prostate cancer is localized, and the accurate prediction of the pathological stage is fundamental to determining which patients would benefit most from this approach. Currently, clinicians use monograms to predict a prognostic clinical outcome for prostate cancer, and these are based on statistical methods such as logistic regression. However, cancer staging continues to present significant challenges to the clinical community (Koprowsk et al., 2012).

The prostate cancer staging monograms that are used to predict the pathological stage of the cancer are based on results from the clinical tests. Cancer prediction systems that consider various variables for the prediction of an outcome require computational intelligent methods for efficient prediction outcomes. Although computational intelligence approaches have been used to predict prostate cancer outcomes, very few models for predicting the pathological stage of prostate cancer exists. In essence, classification models based on computational intelligence are utilized for prediction tasks. A classification is a form of data analysis that extracts classifier models describing data classes, and uses these models to predict categorical labels (classes) or numeric values. When the classifier is used to predict a numeric value, as opposed to a class label, it is referred to as a predictor. Classification and numeric prediction are both types of prediction problems, and classification models are widely adopted to analyze patient data and extract a prediction model in the medical setting. Computational intelligence approaches, and in particular fuzzy-based approaches, are based on mathematical models that are specially developed for dealing with the uncertainty and imprecision that is typically found in the clinical data that are used for prognosis and the diagnosis of diseases in patients. These characteristics make these algorithms a suitable platform to base new strategies for diagnosing and staging prostate cancer. In prostate cancer, cancer staging prediction is a process for estimating the likelihood that the disease has spread before treatment is given to the patient. Cancer staging evaluation occurs before (i.e., at the prognosis stage) and after (i.e., at the diagnosis stage) the tumor is removed. There are three primary clinical-stage tests for prostate cancer:

Biopsy: It is used to detect the presence of cancer in the prostate and to evaluate the degree of cancer aggressiveness.

Digital Rectal Examination (DRE): It is a physical examination that can determine the existence of disease and possibly provide sufficient information to predict the stage of cancer.

A limitation of the PSA test is that abnormally high PSA levels may not necessarily indicate the presence of prostate cancer nor might normal PSA levels reflect the absence of prostate cancer. Pathological staging can be determined following surgery and the examination of the removed tumor tissue and is likely to be more accurate than clinical staging, as it allows direct insight into the extent and nature of the disease.

Prostate-Specific Antigen (PSA): The PSA test is a blood test that measures the level of PSA in the bloodstream. The PSA test is currently the best method for identifying an increased risk of localized prostate cancer. PSA values tend to rise with age, and the total PSA levels (ng/ml) recommended by the Prostate Cancer Risk Management Programme are as follows:

50–59 years: PSA $\geq$ 3.0,
60–69 years: PSA $\geq$ 4.0, and
70 and over: PSA > 5.0.

Abnormally high and raised PSA levels may indicate the presence of prostate cancer. The screening for prostate cancer can be reduced death from prostate cancer.

3.4.1 PRIMARY AND SECONDARY GLEASON PATTERNS

A tissue sample (biopsy) is used to detect the presence of cancer in the prostate and to evaluate its aggressiveness. The results from a prostate biopsy are usually provided in the form of the Gleason grade score. For each biopsy sample, pathologists examine the most common tumor pattern (primary Gleason pattern) and the second most common pattern (secondary Gleason pattern), with each pattern being given a grade of 3–5. These grades are then combined to create the Gleason score that is used to describe how abnormal the glandular architecture appears under a microscope.

For example, if the most common tumor pattern is grade 3, and the next most common tumor pattern is grade 4, the Gleason score is 3 + 4 or 7. A score of 6 is regarded as a low-risk disease, as it poses little danger of becoming aggressive, and a score of 3 + 4 = 7 indicates intermediate risk. Because the first number represents the majority of abnormal tissue in the biopsy sample, a 3 + 4 is considered less aggressive than a 4 + 3. Scores of 4 + 3 = 7 or 8–10 indicate that the glandular architecture is increasingly more abnormal and associated with a high-risk disease that is likely to be aggressive.

3.4.2 CLINICAL AND PATHOLOGICAL STAGES

The clinical stage is an estimate of the prostate cancer stage, and this is based on the results of the DRE. The pathological stage can be determined if a patient has had surgery and hence is based on the examination of the removed tissue. Pathological staging is likely to be more accurate than clinical staging because it can provide direct insight into the extent of the disease. At the clinical stage, there are four categories for describing the local extent of a prostate tumor (T1–T4). Clinical and pathological staging uses the same categories, except that the T1 category is not used for pathological staging.

- The stages T1 and T2 describe a cancer that is probably organ-confined.
- T3 describes a cancer that is beginning to spread outside the prostate.
- T4 describes a cancer that has likely begun to spread to nearby organs.

Category T1 is when the tumor cannot be felt during the DRE or be seen with imaging such as transrectal ultrasound. Category T1 has three subcategories:

- T1a: Cancer is found incidentally during a transurethral resection of the prostate (TURP), which will have been performed for the treatment of benign prostatic hyperplasia, and the cancer is present in no more than 5% of the tissue removed.
- T1b: Cancer is found during a TURP but is present in more than 5% of the tissue removed.
- T1c: Cancer is found in a needle biopsy that has been performed due to an elevated PSA level.

Category T2 is when the tumor can be felt during a DRE or seen with imaging but still appears to be confined to the prostate gland. Category T2 has three subcategories:

- T2a: Cancer is in one half or less of only one side (left or right) of the prostate.
- T2b: Cancer is in more than half of only one side (left or right) of the prostate.
- T2c: Cancer is on both sides, that is, left and right sides of the prostate.
- T3a: Cancer can extend outside the prostate.

- T3b: Cancer may spread to the seminal vesicles.

Category T4 cancer has grown into tissues next to the prostate like the urethral sphincter, the rectum, the bladder, and/or the wall of the pelvis. The TNM staging is the most widely used system for prostate cancer staging and aims to determine the extent of

- T-stage: primary tumor,
- N-stage: the absence or presence of regional lymph node involvement, and
- M-stage: the absence or presence of distant metastases.

Most medical facilities use the TNM system as an important method for cancer reporting. In prostate cancer staging prediction, classification can be done by various classification algorithms. They are

- classification using ANN,
- classifier classification using the naive Bayes classifier, and
- classification using the SVM classifier.

3.4.3 CLASSIFICATION USING THE ARTIFICIAL NEURAL NETWORK CLASSIFIER

In prostate cancer staging prediction, ANN is trained to recognize the patients who have organ-confined disease or extra-prostatic disease. The pattern recognition NN is used as a two-layer feedforward network, in which the first layer has a connection from the network input and is connected to the output layer that produces the network's output. A log-sigmoid transfer function is embedded in the hidden layer, and a softmax transfer function is embedded in the output layer.

A neuron has R number of inputs, where R is the number of elements in an input vector. Let an input vector X be a patient record Xi belonging to a class organ-confined disease or extra-prostatic disease. Each input Xi is weighted with an appropriate weight w. The sum of the weighted inputs and the bias forms the input to the transfer function f. Neurons can use a differentiable transfer function f to generate their output. The log-sigmoid function that generates outputs between 0 and 1 as the neuron's net input goes from negative to positive infinity is used. The Softmax neural transfer function is used to calculate a layer's output from its net input. Softmax functions convert a raw value into a posterior probability and this provides a measure of certainty. The maximum number of repetitions is set to $\epsilon = 200$ and to avoid over-fitting, training stops when the maximum number of repetitions is reached. The ANN is trained using the scaled conjugate gradient for fast supervised learning that is suitable for large-scale problems. The process of training the

ANN involves tuning the values of the weights and biases of the network to optimize network performance that is measured by the mean squared error network function.

3.4.4 CLASSIFICATION USING THE NAIVE BAYES CLASSIFIER

The naive Bayes classifier is designed for use when predictors within each class are independent of one another within each class. The naive Bayes classifies data in two steps. The first one is training and prediction. The training step uses the training data, which are patient cases and their corresponding pathological cancer stage (i.e., organ-confined disease or extra-prostatic disease), to estimate the parameters of a probability distribution, assuming predictors are conditionally independent given the class. In the prediction step, the classifier predicts any unseen test data and computes the posterior probability of that sample belonging to each class. It subsequently classifies the test data according to the largest posterior probability. The following naive Bayes description is used in the classification process.

Let $P(ci|X)$ be the posterior probability that a patient record Xi will belong to a class ci (class can be organ-confined disease or extra-prostatic disease), given the attributes of vector Xi. Let $P(ci)$ be the prior probability that a patient's record will fall in a given class regardless of the record's characteristics; and $P(X)$ is the prior probability of record X, and hence the probability of the attribute values of each record. The naive Bayes classifier predicts that a record Xi belongs to the class ci having the *highest posterior probability*, conditioned on Xi if and only if $P(ci|X) > P(cj|X)$ for $1 j \leq m, j \neq i$, maximizing $P(ci|X)$. The class ci for which $P(ci|X)$ is maximized is called the *maximum posteriori hypothesis.* The classifier predicts that the class label of record Xi is the class ci if and only if

$$P(X|ci)P(ci) > P(X|cj)P(cj)$$
$$\text{when } 1 \leq j \leq m, j \neq I \qquad (3.4)$$

The naive Bayes outcome is that each patient's record, which is represented as a vector Xi, is mapped to exactly one class ci, where ci = 1,…, n where n is the total number of classes, that is, $n = 2$. The naive Bayes classification function can be tuned on the basis of an assumption regarding the distribution of the data. The naive Bayes classifier used two functions for classification:

- Gaussian distribution (GD) and
- kernel density eestimation (KDE).

GD assumes that the variables are conditionally independent given the class label and thereby exhibit a multivariate normal distribution, whereas *KDE* does not assume a normal distribution, and hence, it is a nonparametric technique.

3.4.5 CLASSIFICATION USING THE SUPPORT VECTOR MACHINE CLASSIFIER

The SVM classification method uses nonlinear mapping to transform the original training data (i.e., the patient dataset) into a higher dimensional feature space. It determines the best separating hyperplane, which serves as a boundary separating the data from two classes. The best separating hyperplane for an SVM means the one with the largest margin between the two classes. The bigger the margin, the better the generalization error of the linear classifier is defined by the separating hyperplane. Support vectors are the points that reside on the canonical hyperplanes and are the elements of the training set that would change the position of the dividing hyperplane if removed. As with all supervised learning models, an SVM is initially trained on existing data records, after which the trained machine is used to classify (predict) new data. Various SVM kernel functions can be utilized to obtain satisfactory predictive accuracy.

The SVM finds the maximum marginal hyperplane and the support vectors using a Lagrangian formulation and solving the equation using the Karush–Kuhn–Tucker conditions. Once the SVM has been trained, the classification of new unseen patient records is based on the Lagrangian formulation. For many "real-world" practical problems, using the linear boundary to separate the classes may not reach an optimal separation of hyperplanes. However, SVM kernel functions that are capable of performing linear and nonlinear hyperplane separation exist. The outcome of applying the SVM for prediction is that each patient record, represented as a vector Xi, is mapped to exactly one class label yi, where $yi = \pm 1$, such that $(X1, y1)$, $(X2, y2)$, …, (Xm, ym), and hence, yi can take one of two values, either −1 or +1 corresponding to the classes organ-confined disease and extra-prostatic disease.

3.5 ANN IN CLINICAL PHARMACOLOGY

In the pharmaceutical process, NN finds preformulation parameters for predicting the physicochemical properties of drug substances because of its nonlinear relationships. It is also used in applications of pharmaceutical research, medicinal chemistry, quantity structure–activity relationship (QSAR) study, pharmaceutical instrumental engineering. Its multiobjective concurrent optimization is adopted in the drug discovery process and protein structure analysis. This section describes the uses of ANN in clinical pharmacology.

3.5.1 STRUCTURAL SCREENING IN THE DRUG DISCOVERY PROCESS

In the drug discovery process, ANN helps to predicate how active a chemical compound will be against a given target in the development of new medicines. ANN-based QSAR models are used as the forecast strategies in the virtual screening. In patient care, AI helps to find and examine the picture of the drug. This approach is referred to as virtual screening. For example, if any basic structure of the compound is the input for a NN, it displays various structures similar to those compounds screens over 1000 compounds, among them three compounds with high biologic activity can be identified.

3.5.2 TOXICITY PREDICTION

ANN can be used as an integral part of pharmacotoxicology, especially in quantity structure toxicity relationship (QSTR) contemplates. QSTR is a connection between the substance descriptors and its toxicological activity and can be associated to predict the lethality of compounds. Like QSAR, the molecular descriptors of QSTR are predicted from the physicochemical properties of the compounds and related to a toxicological reaction of intrigue through ANN. For example, the topology method used as the input mode of a network; The ANN-QSTR model was approved by 23 substituted benzene derivatives. The connection coefficient amongst anticipated and real toxicological activities of these compounds was observed to be 0.9088.

3.5.3 DESIGNING OF PREFORMULATION PARAMETERS

ANN modeling has been utilized to enhance the preformulation parameter and to estimate the physicochemical properties of amorphous polymers. They predict the absorption, glass temperatures, and viscosities of different hydrophilic polymers and their physical substance. It demonstrated the potential of ANN as a preformulation tool by prediction of the relationship between the composition of polymer substance and the water uptake profiles, viscosity of polymer solutions, moisture content, and their glass transition temperatures. It has been precious in the preformulation outline and would help to decrease the cost and length of preformulation contemplate.

3.5.4 OPTIMIZATION OF PHARMACEUTICAL FORMULATION

It addresses the multiobjective oriented concurrent optimization issues in the pharmaceutical industry to establish the relationship between reaction factors and insignificant factors. The prediction of pharmaceutical responses in the polynomial equation and response surface methodology has been broadly used as a part of formulation optimization. However, this prediction is small scale due to a low success rate of estimation. An optimization method consolidating an ANN has been created to overcome these shortcomings and to anticipate release profile and improve detailing of different drug formulations. The results calculated by the trained ANN model satisfy well with the theoretically observed values including in vitro release pattern that helps to improve the effectiveness of process and formulation variables.

3.5.5 QUANTITY STRUCTURE–ACTIVITY RELATIONSHIP (QSAR)

ANN is a useful tool to establish quantity structure–activity relationship and predict the activities of new compounds. QSAR links the physicochemical parameters of compounds with substance or biological activities. These parameters include molecular weight, log *p*-value, electronic properties, hydrophobicity, steric effects, a hydrogen donor, molar volume, and molar refractivity. Experimental determination of such properties can be the time-consuming process. An initial phase in QSAR contemplates figuring a massive number of structural descriptors that are used as mathematical illustrative of chemical structure. The relation of structure and activity with the physicochemical descriptors and topological parameters can be controlled by computational techniques. For example, ANN associates to foresee the quantitative structure QSPR of the beta-adrenoceptor antagonist in humans. In the examination, ANN with the concentric arrangement of ten beta-blockers having high set up pharmacokinetic parameters is developed and tried for its ability to predict the pharmacokinetic parameters. Testing an enormous number of possible combinations of descriptors might take a lifetime to succeed. The BP algorithm with topological indices, molecular connectivity, and novel physicochemical descriptors helps to predict the structure–activity relationship of a large series of analogs. It generates valuable models of the aqueous solubility inner arrangement of fundamentally related medications with necessary auxiliary parameters. ANN-predicted properties exhibit a better correlation with the experimentally determined values than those predicted by various multiple regression methods. ANN is valuable in QSAR investigation of the antitumor movement of acridinone subordinates. Moreover, a developed mode is allowed to recognize the critical variables adding to the antitumor movement such as lipophilicity. Therefore, ANN is not just valuable in predicting QSARs yet additionally in distinguishing the part of the particular components relevant to the action of interest. ANN can also be helpful in predicting synthetic properties of compounds. Predicting neural system models have been distributed for alkanes, alkenes, and assorted hydrocarbons. These models commonly demonstrate great fitting and expectation insights with necessary descriptors.

3.5.6 PHARMACOKINETICS AND PHARMACODYNAMICS

ANN monitors human pharmacokinetics parameters from the set of data on the physicochemical properties of medications such as partition coefficient, protein binding, the dissociation constant, and animal pharmacokinetic parameter. Medication doses and drug choices are resolved by the information of the drug's pharmacokinetics and pharmacodynamics. They are adapted to estimate the pharmacodynamics profiles precisely for a broad assortment of pharmacokinetic and pharmacodynamics relationship without requiring any data of active metabolite. As they do not require any structural information, it provides an idea over usual dependent conventional methods. ANN is a quick and straightforward method for predicting and identifying covariates. For example, the rate of clearance, protein-bound fraction of drug, and also volume distribution can be determined.

3.6 CONCLUSION

In this chapter, the branches of AI are explored within the field of biomedical and informatics. The information is presented in a very concise way and the performance of some AI systems that are employed in the biomedical and healthcare domain is investigated. By this chapter, we explore the various AI techniques in different domains in an appropriate way and make the field of AI more robust and applicable is the sense of performance in healthcare. Especially, this chapter describes the uses AI and their related techniques in biomedical and healthcare.

KEYWORDS

- **artificial intelligence**
- **diagnosis**
- **medical imaging**
- **waveform analysis**
- **outcome prediction**
- **clinical pharmacology**

REFERENCES

Duda, R; Hart, P; Stork, D; Pattern Classification, second edition. New York, NY, USA: John Wiley & Sons, Inc., **2001**.

Koprowsk, I. R; Zieleźnik, W; Wróbel Z, Małyszek, J; Stepien, B; Wójcik, W; Assessment of significance of features acquired from thyroid ultrasonograms in Hashimoto's disease, BioMedEngOnLine, **2012**, 11, pp. 1–20.

Kunchewa, LI; Combining Pattern Classifiers, Methods and Algorithms. Hoboken, New Jersey, USA: John Wiley & Sons, Inc., **2004**.

Olaniyi, EO; Oyedotun, OK; Heart diseases diagnosis using neural networks arbitration, International Journal of Intelligent Systems and Applications, **2015**, 7(12), pp. 75–82.

Pratyush, Rn. M; Satyaranjan, M; Rajashree, S; The improved potential of neural networks for image processing in medical domains, International Journal of Computer Science and Technology, **2014**, pp. 69–74.

Rajesh, G; Muthukumaravel, A; Role of artificial neural networks (ANN) in image processing, International Journal of Innovative Research in Computer and Communication Engineering, **2013**, 4(8), pp. 14509–14516.

Sayad, A.T; Halkarnikar, PP; Diagnosis of heart disease using neural network approach, International Journal of Advances in Science Engineering and Technology, **2014**, pp. 88–92.

Shi, Z; He, L; Application of neural networks in medical image processing, Proceedings of the Second International Symposium on Networking and Network Security, China, **2010**, 2, pp. 023–026.

Singh, M; Verma, R.K; Kumar, G; Singh, S; Machine perception in biomedical applications: An introduction and review, Journal of Biological Engineering, **2014**, 1, pp. 20–24.

Smialowski, P; Frishman, D; Kramer, S; Pitfalls of supervised feature selection, Bioinformatics, **2010**, 26, pp. 40–44.

Smita, SS; Sushil, S; Ali, MS; Artificial intelligence in medical diagnosis, International Journal of Applied Engineering Research, **2012**, 7(11), pp. 1539–1543.

Tadeusiewicz, R; Ogiela, MR; Automatic understanding of medical images new achievements in syntactic analysis of selected medical images, Biocybernetics and Biomedical Engineering **2002**, 22, pp. 17–29.

Vyas, M; Thakur, S; Riyaz, B; Bansal, B.B, Tomar, B; Mishra, V; Artificial intelligence: the beginning of a new era in pharmacy profession, Asian Journal of Pharmaceutics, **2018**, 12(2), pp. 72–76.

Yasmin, M; Sharif, M; Mohsin, S; Neural networks in medical imaging applications: a survey, World Applied Sciences Journal, **2013**, 22, pp. 85–93.

CHAPTER 4

HYBRID GENETIC ALGORITHMS FOR BIOMEDICAL APPLICATIONS

P. SRIVIDYA* and RAJENDRAN SINDHU

Department of Electronics and Communication, R.V. College of Engineering, Bangalore 560059, India

**Corresponding author. E-mail: srividyap@rvce.edu.in*

ABSTRACT

In the era of growing technology where digital data is playing a vital role, artificial intelligence (AI) has emerged as a key player. The process of simulating human intelligence using machines is called AI. To accomplish a given task, the machines are trained with activities like reasoning, speech recognition, planning, manipulating, and task solving. Feeding the machines with ample information is the core part of AI. Inculcating common sense, thinking capabilities, and task solving in machines is a tedious task. AI is a combination of various thrust areas like computer science, sociology, philosophy, psychology, biology, and mathematics.

Machine learning, natural language processing (NLP), vision and robotics are the different ways to develop AI in biomedical applications. In machine learning technique, the machine itself learns the steps to achieve the set goal by gaining experience. In NLP, the software itself automatically manipulates the natural language like text and speech. Vision technique enables the machine to see, capture, and analyze the captured image with the image captured from human eyesight. Robots are often used to perform the tasks that humans find difficult to achieve.

AI finds application in different sectors due to the recent progress in digitalization. It has been expanding in areas such as marketing, banking, finance, agriculture, and health care sectors. With the development of data acquisition, data computing, and machine learning, AI is causing a gradual change in medical practice. To act like intelligent systems, machines have to be fed with a huge amount of data. The algorithms play an important role in AI, as they provide instructions to the machine to execute the required task by analyzing the data provided.

In this chapter, the limitations of genetic algorithms (GAs) are discussed and different classification techniques along with Hybrid GAs for biomedical applications will be presented by identifying the challenges in bio-medical applications using AI.

4.1 ARTIFICIAL INTELLIGENCE IN HEALTH CARE SECTOR

Artificial intelligence (AI) in healthcare is the usage of complex algorithms and software to assess human cognition in the analysis of complicated medical data. AI is the capability for algorithms to estimate conclusions without human involvement. Healthcare sectors are under pressure to reduce the cost. Hence, an efficient way to use the data has to be devised. At present very few software and hardware equipments are available to analyze the existing huge medical data. Diagnosing disease and its cure can be simplified if the patterns within the clinical data are identified.

The field of medical diagnostics uses AI and algorithms like genetic algorithms (GAs) for discovering the hidden patterns in the data sets to predict the possibility of a disease. Different types of classification techniques along with data mining have proved to provide useful information for better treatment of diseases.

In the medical field, GAs find extensive use in the field of gynecology, cardiology, oncology, radiology, surgery, and pulmonology.

4.1.1 GENETIC ALGORITHM

The GA (presented by Holland) is a method applied to optimization and search-related problems to provide the most enhanced solution. The basis for the GA is the theory of natural evolution by Charles Darwin. According to the theory, offspring for the next generation will be produced by selecting the fittest individuals at random. A set of solutions for a task will be considered and among these solutions, the best ones will be selected. GA is divided into the following five stages:

1. Evolution
2. Fitness function
3. Selection
4. Crossover
5. Mutation.

The evolution stage starts from a population and it is an iterative process. In each iteration, the fitness of the individual is assessed. The fit individual's genome is used to create the next generation. This new generation is then used in the next iteration. This continues until the desired fitness level is achieved. The crossover points are selected arbitrarily within the genes and to produce offspring using genes exchanged among the parents until the crossover points are reached. The mutation is then applied if the bits in the string are to be flipped. Convergence and degree of accuracy in obtaining a solution are governed by the probabilities of crossover and mutation. The algorithm terminates if the required criteria are almost met or if the required number of generations is produced or by manual inspection. In addition to the above five stages, heuristics can be applied to speed up the process.

Although, GAs are more efficient as they provide a number of solutions for a task when compared with traditional methods. It proves to be better when vast parameters are available and show good performance for a global search. They quite often have more latency while converging to the global optimum. In addition, each time we run the algorithm; the output might vary for the same set of inputs. The problem arises because in the GA the population size is assumed to be infinite. However, in practice the population size is finite. This affects the sampling capacity of a GA and hence its performance. Uniting a GA along with the local search method combats most of the problems that arise due to finite population sizes. The blend of the local search method along with the GA accelerates the optimization process.

Adaptive GAs are a favorable variation to GAs. They are GAs with adaptive parameters. Instead of using fixed values, here crossover and mutation vary based on the solution's fitness values.

Clustering-based adaptive GA is also a variant of the GA. Here, the population's optimization states are judged using clustering analysis. The crossover and mutation depend on the optimization states. For effective implementations, GAs can be combined with other optimization methods to create a Hybrid GA.

4.1.2 HYBRID GENETIC ALGORITHM

Even though the performance of GAs for global searching is superior, it takes a long time to converge to an optimum value. Local search methods on the other hand converge to an optimum value very quickly for smaller search space. Interestingly, though their performance is poor as global searchers.

To improve the performance of GAs, a number of variations have been devised. Hybrid GAs is one such variation. Hybrid GAs are a combination of GA with other optimization and search techniques to produce a hybrid to form the best combination of algorithms for problem solving (El-Mihoub et al., 2006). These algorithms help in improving the genetic search performance by capturing the best of both schemes. Hybrid GAs can be applied to various medical fields like radiology, cardiology, gynecology, oncology, and other health care sectors to find solutions for complex problems. The algorithms can be applied in the screening of diseases and the planning of the treatment.

Hybrid GAs can be used either to enhance the search capability or to optimize the parameters of the GA. Search capability can be improved by combining the GA with a proper choice of local method that is specific to the problem. This helps in improving the quality of the solution and efficiency.

Davis (1991) claims that the hybrid GA can prove to produce the best results only when the GA is combined with correct local search methods, or else the result obtained might be worse than using GA alone.

According to Holland (1975), the quality of the solution can be enhanced by performing the initial search using the GA as shown in Figure 4.1 and later using a suitable local search algorithm to enhance the final population. Local search algorithms should have the ability to identify the local optima with greater accuracy. The efficiency can be in terms of memory or based on time consumed to reach a global optimum. Lamarckian and Baldwinian strategies are found to be the most suitable approaches to combine the local search algorithm with the GA.

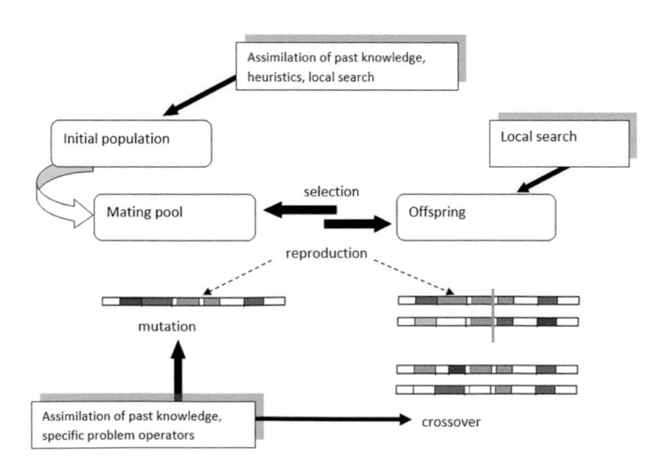

FIGURE 4.1 Genetic algorithm.

4.1.2.1 LAMARCKIAN AND BALDWINIAN APPROACH FOR GENETIC ALGORITHM

The Lamarckian approach is built on learning. The theory states that the characteristics possessed by an individual are acquired from the previous generation. In this approach, the genetic structure reflects the results obtained using the local search. The genetic structure and the fitness of the individual are modified suitably to suit the solution found by the local search technique. Thus, in this approach, the local search technique alters the genetic structure and places the operator back in the genetic population. Thus, the Lamarckian approach helps in accelerating the search procedure of the GA. However, the drawback is that by altering the genetic structure, the GA's discovering abilities are badly affected. This leads to early convergence.

The Baldwinian approach was proposed by Hinton and Nowlan. According to the Baldwinian approach, the characteristics and the genetic structure of the next generation are unaltered. Instead, only the fitness is changed. The local search method is used to create a new fitness value and this is used by the GA to improvise on the individual's ability. Thus, the fitness value is enhanced by applying the local search method. Even though the learning strategies become more effective using this approach, it is slower than the Lamarckian strategy. Baldwinian search can also hamper the evolution process because of the confusion in genetic differences.

The hybridization of Lamarckian and Baldwinian approaches outperforms the individual approaches. This hybridization is either done at an individual level or the gene level. At the individual level, it is done by creating some individuals using Lamarckian and some using the Baldwinian approach. At the gene level, few genes are evolved using Lamarckian and few using the Baldwinian approach.

4.1.2.2 STEPS TO BE FOLLOWED TO DEVELOP A HYBRID GENETIC ALGORITHM

1. Define the fitness function and set different GA parameters like population size, selection method, parent to offspring ratio, required numbers of crossovers, and mutation rate (Wan and Birch, 2013).
2. Generate the current population in random and an objective function for each individual.
3. Using GA operators, create the next generation.
4. For each individual evaluate the objective function.

5. Apply a local search method on each individual of the next generation and evaluate the fitness. If there is some considerable improvement in the solution, then replace the individual.
6. Halt the process if stopping criteria is met.

Many options are available in selecting a local search method. Some of the most popular methods are the classification technique and image processing techniques. Classification technique and image processing techniques are further discussed in Sections 2 and 3, respectively.

4.2 HYBRID GENETIC ALGORITHM DEVELOPED BY COMBINING GENETIC ALGORITHM AND CLASSIFICATION TECHNIQUES

As demand for achieving high accuracy in medical diagnosis grows, the usage of hybrid GAs ensures improved performance over GA. Classification is a technique in data mining to extract the required information from a huge amount of data to group the data into different classes. This helps in identifying the class to which a data belongs. Classification is required when some decisions are to be made based on current situations. Classification is important for preliminary diagnosis of disease in a patient. This helps in deciding the immediate treatment. Machine learning is one of the main approaches for classification.

Machine learning provides automatic learning ability to the machines. It allows the machines to improve from experience without being programmed explicitly and without human intervention. Two different categories of machine learning include supervised and unsupervised learning. Under supervised learning, the training for the machine is provided with some labeled data, meaning the data that has correct answers. Once this is completed, the machine is provided with new examples so that the supervised learning algorithm produces the correct output using the labeled data by analyzing the training data. In unsupervised learning, the machine is trained using the information, which is neither labeled nor classified. The algorithm should act without any guidance on the information. Unsupervised learning is classified into two types: clustering and association.

In the clustering algorithm, the inherent groupings in the data will be identified. Whereas in the association algorithm, rules that describe the greater portions of the data will be discovered.

Different types of machine learning classifiers are given as follows:

- neural network
- decision trees
- support vector machines
- *K*-nearest neighbor
- fuzzy logic

GA can be used for the initial search. To enhance the quality of the search, classification techniques are then applied. The results obtained with the hybrid algorithm show improved classification performance in a reasonable time.

4.2.1 NEURAL NETWORK

The neural network is not an algorithm by itself. It acts as an outline to process complex input data by different machine learning algorithms. As shown in Figure 4.2, neural network comprises of three layers, namely, the input layer, the hidden layer, and the output layer with each layer consisting of nodes. Computations occur at the node. The input from the data combines with the coefficient values in the node. The product from the nodes is then added and passed on to the next layer through the activation function. The hidden layers help in transforming the input into a form that can be used by the output layer. The network is classified as a deep neural network if the network has two or more hidden layers. Each of the hidden layers gets trained depending on the inputs given from the previous layers.

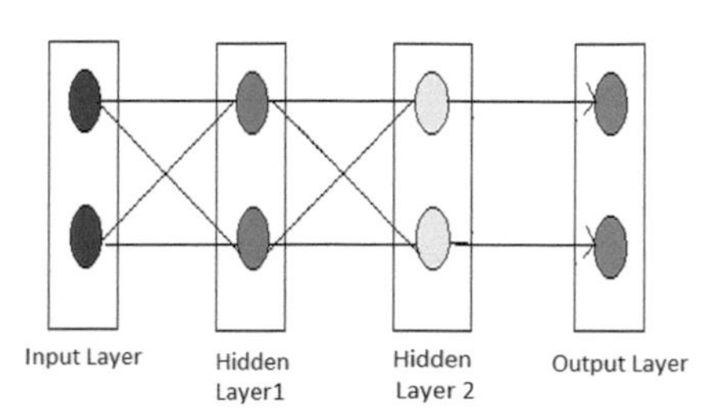

FIGURE 4.2 Interconnections in neural networks.

4.2.2 DECISION TREE ALGORITHM

Decision tree algorithms are controlled learning algorithms that find its use to resolve both regression and classification problems. In decision tree classifiers, depending on the prior data or instances a training model or decision tree is generated (Neelamegam et al., 2013). The model consists of an initial node called the root node. The root node has zero in-coming edges followed by nodes with one inward edge and with outward edges called internal node and nodes with no outgoing edge called terminal nodes as shown in Figure 4.3. Root node and internal nodes have attributes associated with them. However, terminal nodes have classes associated with them. For each attribute, the internal nodes have an outgoing branch. The class for each new instance is determined only at the terminal node after visiting through the internal nodes.

Steps followed in decision tree algorithm are given as follows:

1. The best attribute should be positioned at the root.
2. The training set should be allocated into subsets and every subset should contain the same attribute value.
3. Repeat steps 1 and 2 on all the subsets until the terminal node is reached.

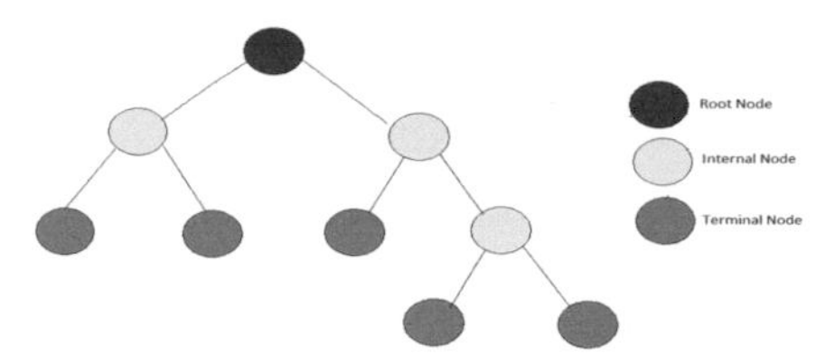

FIGURE 4.3 Structure of a decision tree.

Attributes decide the estimation criterion in decision tree algorithms. The attributes give a measure of how well the input sequence achieves the target classification. Hence, the selection of attributes at each node is a challenge in decision tree algorithms.

Even though decision trees are easy to understand, they have lower prediction accuracy related to other machine learning algorithms. They give a prejudiced response when attributes have a better number of categories.

Chi-square automatic interaction detection (CHAID) algorithm, classification and regression tree (CART) algorithm, iterative dichotomiser-3 (ID3) algorithm, C4.5 algorithm, and C5.0 are the main types of classical decision algorithms. Among them, the C5.0 algorithm provides higher accuracy, consumes less memory, is highly adaptable, has smaller decision trees, and is less complex compared to other algorithms. Owing to their advantages, it is the most preferred algorithm for different applications.

4.2.2.1 CHAID ALGORITHM

It creates all possible cross-tabulations for each predictor until further splitting is unachievable and the best outcome is obtained. The target variable is selected as a root node. The target variable is then split into two or more groups. These groups are called the initial nodes or parent nodes. The groups belonging to this parent node are called child nodes. The last group is called a terminal node. In CHAID, the group that influences the most emanates first and the groups that have lesser influence emanates last

4.2.2.2 CART ALGORITHM

Classification tree algorithms are used when the target variable is fixed. The algorithms then identify the class to which the target variable belongs. The regression tree algorithm is used when the values of the target variable are to be predicted using independent variables.

This is a structured algorithm where a set of questions is asked. The answer to these questions helps in framing the next set of questions. This tree structure continues until no more questions can be formed.

The main stages involved in the CART algorithm are as follows:

1. Based on the value of the independent variable the data at the node is split.
2. The branch is terminated when further splitting is not possible.
3. At each terminal node, the target value is predicted.

4.2.2.3 ID3 ALGORITHM

In Iterative Dichotomiser-3 (ID3) algorithm, the original set acts as a root node. On each reiteration, the algorithm recaps through each unused attribute of the set to calculate the entropy for that attribute. The attribute with smallest entropy value (or maximum information gain) is then selected and the entire set is partitioned using the selected entropy to create the data subsets, that is, the decision tree is made using the entropy. Recur the algorithm by considering the attribute that was never selected earlier.

Example: The root node (exam score) can be fragmented into child nodes depending on the subsets whose marks are less than 40 or greater than 40. These nodes can further be divided based on the marks scored as shown in Figure 4.4.

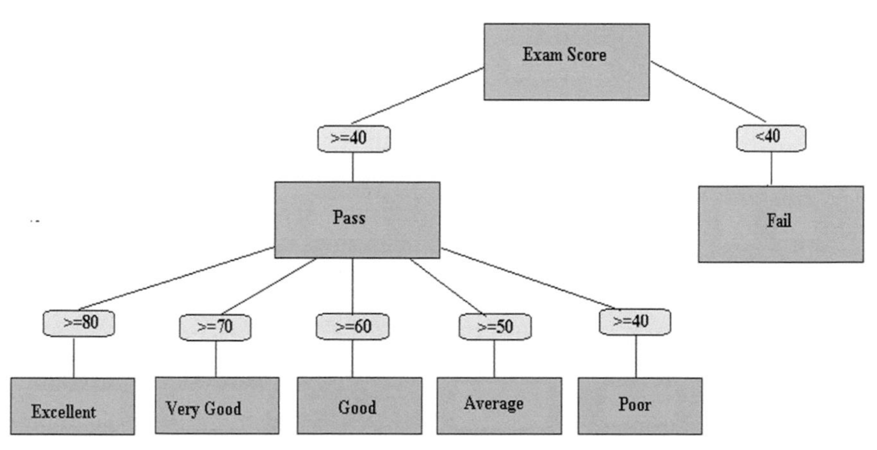

FIGURE 4.4 Example to show the ID3 algorithm.

4.2.2.4 C4.5 ALGORITHM

The extension of the ID3 algorithm is the C4.5 algorithm. ID3 is excessively sensitive to features with large values. To overcome this limitation of the ID3 algorithm, C4.5 can be used. Like in the ID3 algorithm, at every node of the tree the data is sorted to find the best attribute. At each node, one attribute that best separates the data set into subsets is selected. In decision making, the attribute with maximum normalized information gain is selected. To mitigate overfitting inherently it employs the single pass pruning process. It handles both the discrete and continuous attributes. A threshold is created and the attributes are listed that are above or equal or less than to the threshold. It also handles the data with missing attributes.

In short, C4.5 is recursively implemented sequentially as given below:

1. Verify if the termination criteria of the algorithm are satisfied.
2. All attributes are provided with theoretical computer information criteria.
3. Choosing the finest attribute according to the criteria provided by theoretical computer information.
4. A decision node is created on the basis of the finest attribute of step 3.
5. Dataset is induced (i.e., split) on the basis of the new decision node created in step 4.
6. To get a subtree, all the subdatasets in step 5 should call the C4.5 algorithm (recursive call).
7. The decision node created in step 4 is attached to the tree obtained after the execution of step 6.
8. Return tree.

Additional features C4.5 are tree pruning, improvisation in using continuous attributes, handling missing values, and inducing ruleset.

4.2.2.5 C5.0 ALGORITHM

This algorithm is faster than C4.5. It has improved memory usage. The decision trees are smaller when compared with C4.5. The field that provides the highest information is split into subsamples. These subsamples are further split again based on a different field. This continues until a stage when further splitting could not be done. The lowest level subsamples that do not contribute to the model can be removed. Two different models can be built using C5.0 algorithm:

1. Decision tree-based model—here only one prediction is possible for each data value.
2. The rule-set-based model—from the decision trees, rule

sets are derived. The rule sets need not have the same properties of the decision tree. For a particular data value, either one or more rule sets or no rule needs to be applied. If multiple rules are applied, weighted votes are given to the data values and then added. If no rule set is applied, a default value is assigned. The error rates are lower on rule sets thereby helping in improving the accuracy of the result. It also automatically removes attributes that are not helpful.

4.2.3 SUPPORT VECTOR MACHINE

In the support vector machine (SVM) technique, every data point is plotted on an *n*-dimensional space where the number of features is denoted by *n*. The classification is then achieved by finding a hyperplane that clearly splits the data points or features, as shown in Figure 4.5. The selected hyperplane must satisfy the following requirements:

1. Clearly separate the data points.
2. Maximize the distance between the nearest data point and the hyperplane.

SVMs can be used to perform both linear and nonlinear classification. If the data points overlap then either a hyperplane with tolerance or a hyperplane with zero tolerance can be used. The important parameters in SVMs are margin, kernel, regularization, and gamma. By varying these parameters, a hyperplane with nonlinear classification can be achieved at a reasonable time. Finding a perfect class when there are many training data sets consumes a lot of time.

FIGURE 4.5 Linear classification of features using hyperplane.

Although SVMs works well with both structured and unstructured data, gives better result compared to ANNs, solves any complex problem with suitable kernel function, and reduces the risk of overfitting, it has few disadvantages also. Choosing the best kernel function is a difficult task. The time required to train large data sets is more. The memory requirement is more.

4.2.4 BAYSESIAN NETWORK

A Bayesian network is a directed acyclic graph (DAG) belonging to a family of probabilistic distributions that is used to represent not only the variables but also the conditional dependencies of the variables on a directed graph. DAG consists of a probability table, nodes, and edges. Attributes or arbitrary variables are represented as nodes and the conditional dependencies of the variables are represented as edges. Unconnected nodes represent independent attributes. If an edge connects two nodes A and B, then for all the values of A and B the probability P(B|A) should be known to draw an inference. All the probability values will be specified beside the node in a probability table. For example, let two different events A and B cause an event C. Among A and B let B be more dominant, that is, when B occurs, A is not active. The representation of this condition is as shown in Figure 4.6. The variables have two possible values true (T) or false (F).

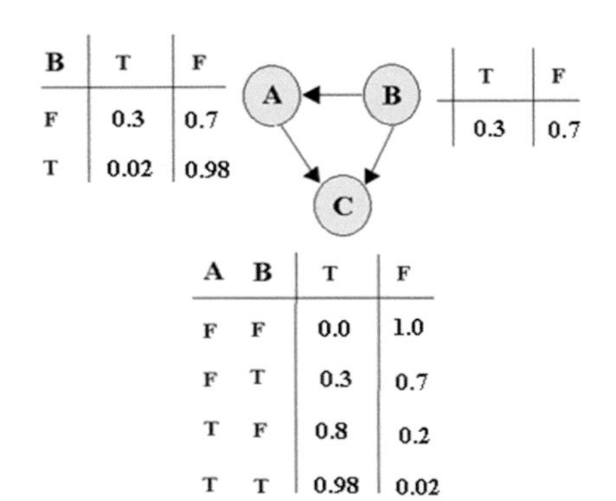

FIGURE 4.6 DAG used in the Bayesian network.

Bayesian networks are probabilistic modeling techniques that are perfect to consider the past event and predict the cause for the happening. For example, if the symptoms are known, the probable diseases are predictable by the network.

4.2.5 *K*-NEAREST NEIGHBOR

K-nearest neighbor is built upon the basis of learning by comparison. The *n*-dimensional space is used for storing training samples. Every sample is denoted by a point in *n*-dimensional space. After a test sample is provided, the *K*-nearest neighbor classifier quests for *K* samples that are close to the test sample and then classify the test sample suitably.

For example, consider that there are two classes 1 and 2, as shown in the Figure 4.7. Let Blue star be the test sample that has to be classified. If $K = 5$, then dotted square 1 is selected. In this square, the number of class 1 samples is more than class 2. Hence, the test sample is assigned to class 1. If $K = 7$, then dotted square 2 is selected. In this square, the number of class 2 samples is more than class 1. Hence, the test sample is assigned to class 2.

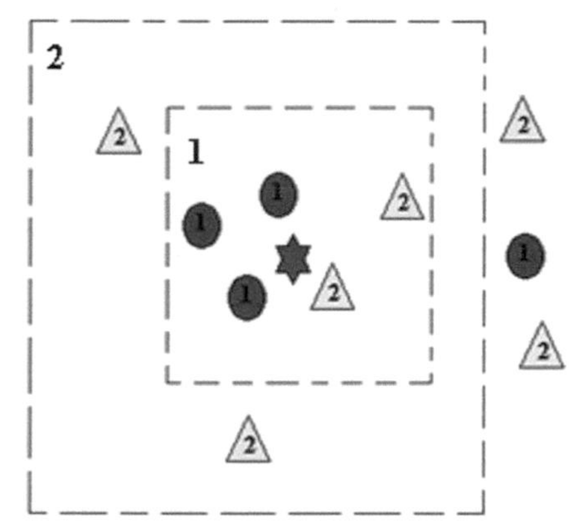

FIGURE 4.7 Class assignment in *K*-nearest neighbor.

4.2.6 FUZZY LOGIC

Fuzzy logic can be used for the problems with uncertain values at the input. It is a multivalued logic in which the true values range from zero to one. Fuzzy logic takes different possibilities of input values to give a definite output. Fuzzy logic deals with uncertainties and provides acceptable reasoning.

Four major modules involved in fuzzy logic system are as follows:

Fuzzifier module: it splits the input into five steps: large positive, medium positive, small, medium negative, or large negative.

Information base module: used to store the guidelines provided by experts.

Inference module: fuzzy inference is made on the input based on the information stored in the information module.

Defuzzification module: output is obtained from the module due to the transformation of the obtained fuzzy set.

The main modules involved in fuzzy logic is shown in Figure 4.8.

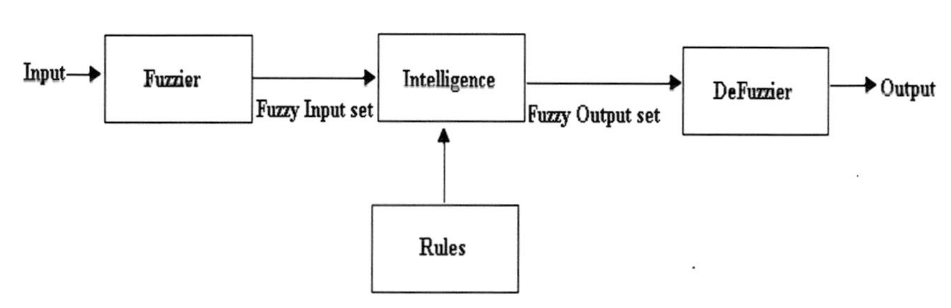

FIGURE 4.8 Fuzzy logic modules.

In fuzzy logic, based on the problem and the fuzzy set, a membership function is devised. This function provides a graphical representation of the fuzzy set (Santhanam et al., 2015). The *y*-axis shows the membership degree in the interval [0,1] and the *x*-axis shows the universe of discourse. Triangular, Gaussian, trapezoidal, and linear are few common shapes for the membership function.

Algorithm:

1. Define the variables for input and output.
2. Build the membership functions for the input and output variables.
3. Construction of base rules.
4. Obtain fuzzy values.
5. Perform defuzzification.

4.3 HYBRID GENETIC ALGORITHM DEVELOPED BY COMBINING GENETIC ALGORITHM AND IMAGE PROCESSING TECHNIQUES

The hybrid medical image retrieval system uses a GA approach for the selection of spatiality reduced sct of features. The development of the system comprises two phases.

1. In phase one, three distinct algorithms are used to extract the important features from the images. The algorithms used for extraction of the features are intrinsic pattern extraction algorithm, Texton-based contour gradient extraction algorithm, and modified shift-invariant feature transformation algorithm.
2. In phase two, it is based on the phase feature selection used to identify the potential feature vector GA. A few of the approaches are "branch and bound algorithm" and "artificial bee colony algorithm" using brain tumor, breast cancer, and thyroid images. In this genetic-based feature selection method, the existing system dimensionality problem is reduced (Shanmugapriya and Palanisamy, 2012; Li et al., 2012).

4.3.1 FEATURE EXTRACTION

To extract the most dominating and important feature that represents the image feature extraction methodologies are used for analyses. The algorithm formulated for the extraction of the features is an intrinsic pattern extraction algorithm, Texton-based contour gradient extraction algorithm, and a modified shift-invariant feature transformation algorithm.

There are different feature extraction techniques available. Some of them are as follows:

1. Intrinsic pattern extraction algorithm

Texture in this context is the intense variation pattern in an image to represent the texture pattern that cannot be analyzed from a single-pixel point. The images require consideration of neighboring pixel point's intensity for which an algorithm called the intrinsic pattern extraction algorithm was introduced originating from the basic of principal component analysis (PCA). Input for intrinsic pattern extraction algorithm is medical images available in the data sets and the output is a positive intrinsic pattern feature vector for each of the input images.

The size of the identified feature vector, which represents the intrinsic pattern of the image provided and computation is reduced in this approach. PCA is a statistical model to classify a few discrete patterns for any given dataset. From the identified pattern design, PCA implements a linear system. This linear system derived by applying linear algebra is used to identify the potential feature vectors from the pattern design. Later, these feature vectors are formalizing a tactic to analyze the continuity in data sets.

2. Texton-based contour gradient extraction algorithm

The gradient of the edge pixel was also analyzed for a more effective feature detecting system. The fundamentals of the gradient are derived from the concept of derivatives, which tells about the variation of the functional variable derived mathematically. The gradient is one such concept that is used to identify the variation in image pixel intensity value in a two-dimensional space at (i, j) location. The gradients are usually vector values whose magnitude determines the change of pixel intensity values and the direction of the gradient specifies the direction where the changes take place. In Texton-based contour gradient extraction algorithm (TCGR), the input is medical images from the datasets and the obtained output is the TCGR feature vector for each input image.

Texton is one of the latest evolving concepts derived from texture analysis to obtain the exact contour gradient of the image provided. It uses spatial filter transformation to extract syntactic features of any user-defined image sets. The complex patterns can be analyzed by Texton and used to develop extensible texture models for an image (Julesz, 1981). These Texton vector features are analyzed and normalized, which can be used for indexing the medical images as per their domain specified.

4.3.2 FEATURE SELECTION

This is a process used to choose a criteria based subset of features. Feature selection is used to remove irrelevant data, reduce the cost of data, improve the efficiency of learning, reduce storage space, and cost of computation. It reduces the evaluation time, increases precision, thereby reducing the complexity of the model, and helps in understanding the data and the model. Hence, the calculation must be powerful in discarding the excess, immaterial, and noisy highlights. A hybrid branch and bound method along with counterfeit bee settlement calculation for the ideal element choice are considered. The calculation builds a binary search tree where the root depicts the set of all specifications and leaves depict the subsets highlights. While crossing the tree down to leaves, the calculation progressively eliminates single features from the present arrangement of "candidates." The algorithm retains the data about the current best subset and the rule value it yields. This value is depicted as a bound value. ABC algorithm begins with the generation of a population of binary strings (or bees). Introduce the population and assess wellness. Select different features from neighborhood features in the underlying population and contrast to assess the wellness. If the chosen features do not fulfill the wellness function then eliminate those features from the population. Thus, every one of the bees is compared with wellness function and structure the ideal list of capabilities. If not even one of the features fulfills the fitness function, find a new fitness function. Then, proceed with hunting for choosing the ideal value. The proposed mixture approach calculation combines the features of both branch and bound algorithm and fake bee colony calculation.

4.3.2.1 BRANCH AND BOUND FEATURE REDUCTION ALGORITHM

This is one of the feature selection algorithm proposed for maximizing the accuracy and reducing the computation time. In this algorithm, the input is taken as a medical images feature vector and the output obtained is nothing but a reduced feature vector. In this algorithm, consecutive tree levels are created and tested with the evaluation function, while traversing down the tree the algorithm removes single features and the bound values are updated. This algorithm allows efficient search for optimal solutions. It is well suited for discrete or binary data.

4.3.2.2 BC FEATURE REDUCTION ALGORITHM

ABC algorithm the input is the same as the branch and bound feature reduction algorithm and the output obtained is also the same as the branch bound algorithm. In this algorithm, the first stage is the generation of a population of binary strings, initialize population, and evaluate fitness. In the initial population, the other features from the neighborhood are compared with evaluated fitness; if the features do not satisfy the function, remove them from the population, and thereby, all the binary strings are compared and discarded; this keeps continuing until the optimal value is obtained.

4.3.3 DIVERSE DENSITY RELEVANCE FEEDBACK

Based on the relevance feedback method in medical image retrieval, the system first receives a query image that the user submits. The system uses Euclidean distance to calculate the similarity between images, to return initial query results. By using the user feedback information the results are marked positive and negative. On the basis of the user's interest and the feedback refine the query image feature selection on a repeated basis. The diverse density algorithm is used to achieve relevance feedback. The input image can be both positive image and negative image and the output obtained would be the features that the user is interested in.

In this algorithm, the image content is considered as a set of features, and the finds the features within the features space with the greatest diversity density. The diversity density is a measure that refers to the more positive examples around at that point, and the less negative examples, The co-occurrence of similar features from different images are measured using a function-defined called DD function. The main objective of DD is to find features that are closest to all the positive images and farthest from all the negative images.

4.4 APPLICATIONS OF HYBRID GENETIC ALGORITHMS IN CLINICAL DIAGNOSIS

As the human population is increasing, the diseases are also increasing in rapid pace. Human death due to

malignancy and heart attack has increased worldwide and accurate clinical diagnosis is the demand of the hour. In most of the cancer patients, the malignant tumors are diagnosed late or they are misdiagnosed leading to the death of the patient.

4.4.1 HYBRID GENETIC ALGORITHM IN RADIOLOGY

Magnetic resonance imaging (MRI), compute tomography (CT) scan, positron emission tomography (PET) scan, mammography, X-rays, and ultrasound are some of the imaging modalities used in the field of medicine. The imaging modalities are used in detecting and diagnosing a disease. The time required for detection and diagnosing has reduced after the invention of computer-aided detection and diagnosis (CAD). CAD systems also help in improving the detection performance of the imaging modalities.

Even though CAD systems assist in diagnosis, the images captured by imaging modalities are affected by noise. This noise affects the diagnosis process by the radiologists. Hence to diagnose the disease, the detection machines have to process and interpret the captured image using algorithms based on classification techniques (Ali Ghaheri, Saeed Shoar et al., 2015; Ghaheri et al., 2015).

1. In treating cancer patients, exact tumor size and its volume determination play a vital role. This can be done using imaging techniques. MRI of the organ can be captured and GA can be applied for image segmentation. Artificial neural network can then be applied to reduce the false-positive results. This technique was adopted by Zhou et al. to predict tongue carcinoma using head and neck MRIs.
2. Genetic algorithm (GA) can be applied for feature extraction from mammograms or PET scans to identify the region of interest (Mohanty et al., 2013). Feature extraction is a process of selecting appropriate features by removing irrelevant features and constructing a model. This helps in reducing the time, complexity, and cost involved in the computation of irrelevant features. By applying feature extraction, the region of interest can be identified to be malignant or not.
3. GAs can also be applied in image fusion to combine two different images captured using different image modalities. An example CT scan image can be merged with an MRI image or a CT scan image can be merged with a PET scan image. This helps in easy diagnosing since each image possesses different information acquired under different conditions.

4.4.2 HYBRID GENETIC ALGORITHM WITH NEURAL NETWORK FOR BREAST CANCER DETECTION

In malignancy detection, GA can be used to set the weights and the neural networks can be used to give an accurate diagnosis as elucidated by Schaffer et al. (1992). The population of *n*-chromosomes along with their fitness values is maintained by GA. Based on the fitness values, the parents are selected to produce the next generation. Crossover and mutation are then applied to obtain the best solution. GA discards all bad proposals and considers only the good once. Thus, the end result is not affected.

Neural networks are used in solving classification problems. They are capable of learning from previous experiences and improvise on the behavior when they are trained. Neural networks mimic the human brain. It consists of neurons joined using connecting links. The weight of each link in the network is multiplied by the transmitted signal.

Every node in the network forms the output node of the network, the lines form the input, and the intermediate layer forms the hidden layer. The output of the hidden layer is the input to the output layer (Alalayah et al., 2018; Ahmad et al., 2010).).

The below steps are involved in classifying malignancy:

1. **GA:**
   (a) Selects the optimal weights and bias values for the neural network.
   (b) Evaluation of fitness value.
   (c) On the basis of fitness value, parents are selected.
   (d) The new population is formed from the parents using a cross over and mutation.
   (e) Stop when a fixed number of generations is reached.
2. **Artificial neural network:**
   (a) Input the training samples and the class of the sample.
   (b) Compare the output with the known class and adjust the weight of the training sample to meet the purpose of classification.
3. **Algorithm for malignant cell detection using GA and neural network:**
   (a) Initial solutions are generated using GA.
   (b) It is then fed as input to neural network.
   (c) The output from the neural network is then evaluated using a fitness function.
   (d) If the stop condition is not reached, a new selection, crossover,, and mutations are performed and fed back to the neural network. Else, the process is stopped.

4.4.3 HYBRID GENETIC ALGORITHM FOR HEART DISEASE DETECTION

Heart diseases usually occur due to improper pumping of the blood, block in arteries, high blood pressure, diabetes, etc. It has become a prominent cause of death these days. Hence, predicting its occurrence has gained prominence in the health field. Shortness of breath, feeling fatigue, and swollen feet are all some of the symptoms for heart disease. Early diagnosis of heart disease helps in reducing the risk of heart attacks and death (Durga, 2015). The common method of diagnosing heart disease is based on examining and analyzing a patient's medical history and the symptoms by cardiologists. This diagnosis is time-consuming and not very precise (Al-Absi et al., 2011).

The difficulties involved in the manual examination can be improvised by using few predictive models based on GA combined with machine learning techniques like SVM, *K*-nearest neighbor, decision tree, naive Bayes, fuzzy logic, artificial neural networks, and others. The data mining repository of the University of California, Irvine also called as the Cleveland heart disease dataset provides the input dataset for the Hybrid Algorithm. The dataset provides samples of 303 patients with 76 features and a few missing values. This can be used for investigations related to heart diseases. This dataset can be taken as input to different algorithms and classification techniques can be applied to refine the investigation.

General steps involved in developing a hybrid algorithm for heart disease diagnosis include the following:

1. The data preprocessing by replacing the missing values in the dataset.
2. Best attributes selection using GA. Attribute reduction helps in enhancing accuracy.
3. Classifier creation using the reduced attributes.
4. Heart disease prediction using the created classifier.

4.4.4 ELECTROCARDIOGRAM (ECG) EXAMINATION USING THE HYBRID GENETIC ALGORITHM AND CLASSIFICATION TECHNIQUE

The electrical activity of the heartbeat is measured using ECG. For every heartbeat, a wave is sent through the heart as shown in Figure 4.9. The wave causes the heart to pump the blood by squeezing the muscle. ECG has the following waves:

P-wave: created by the right and left atria.

QRS complex: created by the right and left ventricles.

T-wave: made when the wave returns to the state of rest.

The time duration taken by the wave to traverse through the heart can be determined by measuring the time intervals on the ECG. This helps in finding out the electrical activity of the heart.

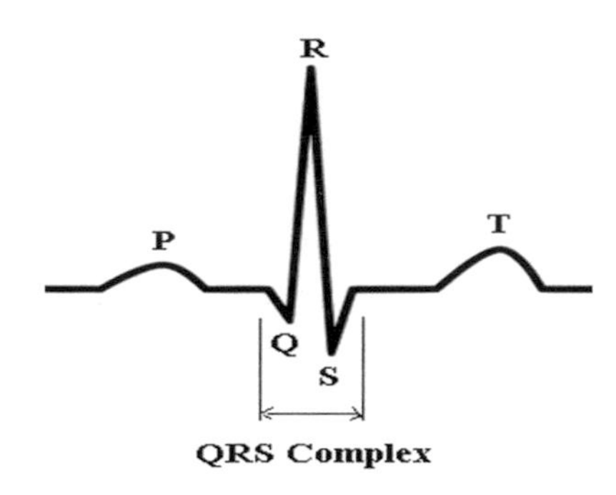

FIGURE 4.9 Normal ECG

GA can be used to detect QRS complex in ECG as explained by Diker et al. (2019). The QRS complex is the main and central spike seen on the ECG line. Interpreting the QRS complex in ECG allows the analysis of heart rate variations. P-waves and T-waves from the ECG can then be examined to diagnose the disorder.

4.4.5 HYBRID GENETIC ALGORITHM AND DEEP BELIEF NETWORK FOR HEART DISEASE DETECTION

Deep belief network is a class of neural networks that helps in performance improvement by maintaining several hidden layers and by passing the information to those layers. It consists of two different types of neural networks—belief networks and restricted Boltzmann machines. Belief networks are composed of stochastic binary units along with weighted connections. Restricted Boltzmann machine (RBM) is organized as layers. Deep belief networks are used in recognizing, clustering, and generating images, motion capture data, and video sequences. Once the RBM receives the input data it is trained and it passes the result to the next layer as input. The initial weights are assigned by DBN and then the error propagation algorithm is performed to obtain optimized performance. However, using only a deep belief network will not result in optimum value for a number of nodes and layers (Lim et al., 2018). Combining DBN with GA will prove to be better in achieving an optimum solution.

Steps involved are as follows:

1. Apply GA to find optimal value by using selection, crossover, and mutation. Feed this training set into deep belief network.
2. Using unsupervised learning, construct the RBM network.
3. Learning the backpropagation algorithm using supervised learning.

4.4.6 HYBRID GENETIC ALGORITHM WITH FUZZY LOGIC FOR PREDICTING HEART DISEASES

GA can be used for feature extraction and fuzzy logic for classification and prediction. GA is applied to extract the relevant information from the entire data set fed into it. It thus helps in reducing the count of the attributes present in the dataset that helps in reducing the search.

Steps involved are as follows:

1. Select the required features using GA.
2. Develop fuzzy rules.
3. Fuzzify the input values.
4. To generate rule strength combine fuzzy input and rules.
5. Generate the output distribution by combining the rule strength with output function.
6. Defuzzify the output.

4.4.7 HYBRID GENETIC ALGORITHM AND ARTIFICIAL NEURAL NETWORKS IN ORTHODONTICS

In growing children, predicting the size of unemerged tooth becomes very important (Moghimi et al., 2012). By predicting the size of the unemerged tooth, the dentists can evaluate whether the vacant space available is sufficient for the growth of the permanent tooth in proper alignment. Three different methods are available for prediction:

1. Using radiographs for measuring the unemerged teeth.
2. Using the prediction equations and prediction tables for calculations.
3. Combination of both of the above methods.

With the considerable development of medical databases, the available traditional methods should be upgraded to more efficient methods for computation.

Artificial neural networks find major application in analyzing medical data and in dental predictions. GA when combined with artificial neural networks helps in predicting the size of unemerged tooth. In the prediction process, the GA finds the best reference and the artificial neural networks predicts the size of the tooth based on the information provided by GA.

The algorithm for prediction is described as follows:

1. GA introduces the reference tooth value during every iteration into artificial neural network.
2. Reference input values are mapped to the target values.
3. Stopping criteria are checked.
4. If the obtained results are satisfactory, the results are displayed. Else, genetic algorithm shifts to the next generation and again

searches for the better match among the reference teeth.

5. The cycle repeats until the stopping criteria are met or until the predefined value is exceeded by the number of generations.

At the end of the training, the GA will not be used during the training process. The reference tooth introduced by the GA will not be used any further. The artificial neural network will further use the data to predict the output by using the best function for mapping.

4.5 CONCLUSION

The need for digitalization has been rising day by day. AI has contributed to many fields such as medical, automobile, education, etc. The research to extract a huge amount of information available in clinical data to improve the diagnosis of a disease is critical. GA helps to find an optimal solution for complex data at a reasonable time. Hence, their usage in the field of medicine helps the physician to solve complex diagnosing problems. The search-ability of GA can be increased by the proper blend of GA with the local search method. This blend is called hybrid GA.

In this chapter, an insight into how hybrid GA can be used to diagnose the presence of malignancy and heart disease in a patient is provided in detail. The explanation about the different GAs used often along with its various approaches and its application. These approaches illustrate that hybridizing is a potential method to build a capable GA that cracks tough problems swiftly, consistently, and precisely. The chapter also talks about the different applications of these algorithms in real-time specifically related to the diagnosis of diseases.

KEYWORDS

- **artificial intelligence**
- **genetic algorithms**
- **hybrid genetic algorithms**
- **classification techniques**
- **image processing**

REFERENCES

Ahmad F., Mat-Isa N. A., Hussain Z., Boudville R., Osman M. K. Genetic algorithm-artificial neural network (GA-ANN) hybrid intelligence for cancer diagnosis. Proceedings of the 2nd International Conference on Computational Intelligence, Communication Systems and Networks, July **2010**, 78–83.

Alalayah K. M. A., Almasani S. A. M., Qaid W. A. A., Ahmed I. A. Breast cancer diagnosis based on genetic algorithms and neural networks. International Journal of Computer Applications (0975-8887), 180, **2018**, 42–44.

Al-Absi H. R. H., Abdullah A., Hassan M. I., Shaban K. B. Hybrid intelligent system for disease diagnosis based on artificial neural networks, fuzzy logic, and genetic algorithms. ICIEIS, 252, **2011**, 128–139.

Cagnoni S., Dobrzeniecki A., Poli R., Yanch J. Genetic algorithm based interactive segmentation of 3D medical images. Image and Vision Computing 17(12),**1999**, 881–895.

Davis L. The Handbook of Genetic Algorithms. Van Nostrand Reinhold, New York, USA, **1991**.

da Silva S. F., et al. Improving the ranking quality of medical image retrieval using a genetic feature selection method. Decision Support Systems, 51(810), **2012**, 810–820.

Diker A., Avci D., Avci E., Gedikpinar M. A new technique for ECG signal classification genetic algorithm wavelet kernel extreme learning machine. Optik, 180, **2019**, 46–55.

Durga Devi A. Enhanced prediction of heart disease by genetic algorithm and RBF network, International Journal of Advanced Information in Engineering Technology (IJAIET), 2(2), **2015**, 29–36.

El-Mihoub T. A., Hopgood A. A., Nolle L., Battersby A. Hybrid genetic algorithms: a review, Engineering Letters, 13(3), **2006**, 124–137.

Ghaheri A., Shoar S., Naderan M., Hosein S. S. The applications of genetic algorithms in medicine. Oman Medical Journal, 30(6), **2015**, 406–416.

Grosan C. and Abraham A. Hybrid evolutionary algorithms: methodologies, architectures and reviews. Studies in Computational Intelligence, 75, **2007**, 1–17.

Haq A. Ul, Li J. P., Memon M. H., Nazir S., Sun R. A hybrid intelligent system framework for the prediction of heart disease using machine learning algorithms. Mobile Information Systems, 2018, **2018**, 1–21.

Holland J. H. Adaptation in Natural and Artificial Systems. The University of Michigan, **1975**.

Idrissi M. A. J., Ramchoun H., Ghanou Y., Ettaouil M. Genetic algorithm for neural network architecture optimization. Proceedings of the 3rd International Conference on Logistics Operations Management (GOL), May **2016**, 1–4.

Julesz B. Textons, the elements of texture perception, and their interactions. Nature, 290(5802), **1981**, 91–97.

Lei L., Peng J., Yang B. Image feature selection based on genetic algorithm. Proceedings of the International Conference on Information Engineering and Applications (IEA), 8(25), **2012**.

Lim K., Lee B. M., Kang U., Lee Y. An optimized DBN-based coronary heart disease risk prediction. International Journal of Computers, Communications & Control (IJCCC), 13(14), **2018**, 492–502.

Lowe D. G. Distinctive image features from scale-invariant keypoints. International Journal of Computer Vision, 60(2), **2004**, 91–110.

Minu R. I., Thyagharajan K. K. Semantic rule based image visual feature ontology. International Journal of Automation and Computing, 11(5), **2014**, 489–499.

Moghimi S. Talebi M., Parisay I. Design and implementation of a hybrid genetic algorithm and artificial neural network system for predicting the sizes of unerupted canines and premolars. European Journal of Orthodontics, 34(4), **2012**, 480–486.

Mohanty A. K., et al. A novel image mining technique for classification of mammograms using hybrid feature selection. Neural Computer and Application, 22(1151), **2013**, 1151–1161.

Nagarajan G., Minu R. I. Fuzzy ontology based multi-modal semantic information retrieval. Procedia Computer Science, 48, **2015**, 101–106.

Neelamegam S., Ramaraj E. Classification algorithm in data mining: an overview.

International Journal of P2P Network Trends and Technology, 3(5), **2013**, 1–5.

Santhanam T., Ephzibah E. P. Heart disease prediction using hybrid genetic fuzzy model. Indian Journal of Science and Technology, 8(9), **2015**, 797–803.

Schaffer J. D., Whitley D., Eshelman L. J. Combinations of genetic algorithms and neural networks: a survey of the state of the art. Proceedings of the International Workshop on Combinations of Genetic Algorithms and Neural Networks (COGANN-92), June **1992**, 1–37.

Shunmugapriya S., Palanisamy A. Artificial bee colony approach for optimizing feature selection. International Journal of Computer Science Issues, 9(3), **2012**, 432–438.

Smith L. I. A Tutorial on Principal Components Analysis, Cornell University, **2002**.

Wan W., Birch J. B. An improved hybrid genetic algorithm with a new local search procedure. Journal of Applied Mathematics. 2013, Article ID 103591, **2013,** 10.

CHAPTER 5

HEALTHCARE APPLICATIONS USING BIOMEDICAL AI SYSTEM

S. SHYNI CARMEL MARY* and S. SASIKALA

Department of Computer Science, IDE, University of Madras, Cheapuk, Chennai 600 005, Tamil Nadu, India

Corresponding author. E-mail: shynipragasam@gmail.com

ABSTRACT

Artificial intelligence (AI) in healthcare is an emerging trend identified as a collection of technologies, programmed to sense, comprehend, act, and learn. Its performance is appreciated for the administrative and clinical healthcare functions and also for research and training purposes. AI includes natural brainpower, instrument learning, deep learning, neural networks, robotics, multiagent organisms, modernization in healthcare, and indistinct reasoning. Health includes healthcare, diagnostics, sanatoriums, and telemedicine, medications, therapeutic utensils and supplies, health assurance, and medical data. Healthcare curriculums are restricted by the unavailability of clinicians and inadequate dimensions, and AI amends these insufficiencies. The importance of AI in India has become a key technology for improving the effectiveness, value, cost, and reach of healthcare. The ethical issues for applying AI in healthcare must be analyzed and standardized. The encounters to the use of AI in healthcare were recognized primarily through an analysis of literature, interviews, and roundtable inputs.

5.1 INTRODUCTION

The new paradigm view of non-natural brainpower is an online-enabled technology that sets guiding principles and recommendations to the community to help those who are involved in decision making in all the domains. Especially in the medical domain, AI is used to evaluate the data and predicts the deceases and prescribes the medication accordingly. It is interlined with the lifestyle of the individuals and their related data. The evaluation of artificial intelligence (AI) technology leads to think and generate knowledge based on the continuous data analysis process and predict as human experts by adopting various algorithmic processes including industry 4.0.

The life span of humans varies in various ways because of this contemporary world. The current upcoming science and technology affects the lifestyle in both fields of advancement and defect. Revolutions happening in this modern era are because of the prominent invention of computer processing. Computing and knowledge processing applications expand the living conditions in many areas and many ways. Medical science is one of the potential domains where the computing process is used for the advancement of human lifestyle. Many research programs are carried out using a combination of computing processes and medical applications for the betterment of human health (Jiang et al., 2017). Most of the medical applications used with computations are aiding practitioners to take decisions on their drug recommendations and the identification of the diseases. In all of the fields, computing technologies are well established and applied. Medical science is also using technology, but it defines proper ways to implement technology, and the processing depends on the requirements of the medical applications, which is a challenging task to the researchers.

The experts use computation processes in the medical science field (He et al., 2019) as a tool, but the tools are advanced by computer specialists with different technical procedures. The technical experts derive the concept and incorporate the same in several medical science applications. Data and images analyses are calculated by the medical field computational process. The research process comprises analysis and design of the solution and implements the identified existing algorithms in medical science.

5.2 APPLICATIONS OF BIOMEDICAL AI SYSTEM

AI is playing a vital role in the Industry 4.0 revolution. AI is the replication of human intelligence processes by machines, especially computer systems. These methods include acquisition of information and procedures for using the information, reaching approximate or definite conclusions and self-correction. The AI technologies are used in Big Data Analytics, Autonomous robots, Simulation, Internet of

Things, etc. The industrial world is adopting revolutionary technologies in the name of "Industry 4.0" for scaling up socioeconomic applications. These technologies are applied to enhance the scientific processes on the existing analysis and predictions in science, business, and healthcare domains.

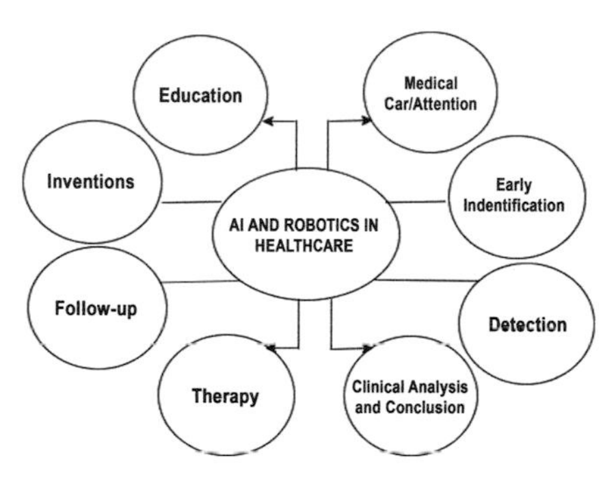

FIGURE 5.1 Applications of AI and robotics in healthcare.

When a model and its simulation states are applied in the biomedical processes, it is called a biomedical system (Salman et al., 2017). Its requirements and process states are different according to the simulation environment integrated with medical applications (Yu et al., 2018). An AI-based system integrated with a biomedical model produces a biomedical AI system. The primary objective of AI is providing a cognitive construct to humans that will mimic human knowledge and also the cognitive process (Pesapane et al., 2018). In healthcare, AI is predominantly used for diagnosis and clinical decisions. It provides diagnosis and therapeutic interventions in different fields of medical science, such as cardiology, genetics, neurology, radiology, etc., for identifying the variation of abnormalities, analyses, and predictions. AI can definitely assist physicians to make effective clinical decisions or even replace human verdict in certain functional areas of biomedical healthcare processes, like radiology (Mazurowski et al., 2019) for image processing.

The AI process aids to keep the humans well based on the analysis and observations of their regular lifestyle. The wellness starts from the new born baby vaccination process and continues throughout the entire life of humans. AI recommends the vaccination process in accordance to the analysis of regional factors and family history. The AI techniques generate a customized individual vaccination record as per the child's immune system. The welfare recommends the food chart, growth chart, and regular physical and mental growth verification for the newborn child. The well-being process is an integrated analysis of the human physical system (Faust et al., 2018) and its ability to perform a sensitive process that is an integrated and trained set of AI algorithmic process as per the regional and common factors. The AI technology recommends a rule-based approach and generates a good process systematic approach for regular assessment and health development process.

Further, AI technology can be applied for early detection based on the AI-rule-based approach and can develop the new rule as per the regional factors. There are two ways the detection process is carried out in the medical phenomena, namely, data and image analysis process. The early detection process collects the set of text and numerical data, matches the similarity of the current existing case, and predicts the possibility of the disease as per the similarity occurrence. The AI logical and predictive approach finds the similarities to determine the possible diseases. This approach adopted and applied predictive algorithms using personal data. Processing personal data involves the use of big data algorithms (Lo'ai et al., 2016) such as clustering and classification for the more accurate process of early detection. It depends on the appropriate data process of the questions of data scope, quality, quantity, and trustworthiness. In this process, the questions must be framed carefully by deciding the objective of the prediction and its process. The algorithmic process determines the appropriateness of the data needed for the research and aims for the detection of diseases. The early detection process adopts supervised learning process algorithms (Ramesh et al., 2016) such as resolution trees, naive Bayes cataloging, ordinary least square regression, logistic deterioration, upkeep vector machine, and ensemble method. It also adopted a few unsupervised algorithms such as clustering algorithms, principal component analysis, singular value decomposition, and independent component analysis.

In the diagnosis process, AI plays a vital role in accuracy and prediction (Paul et al., 2018). The AI techniques diagnose patients without a doctor. A software produces a diagnostic result, detecting the fulfilment or partial fulfilment of the conditions. The AI approach diagnoses via passive and nonpassive approaches. The results are compared with the standard values, and identification of variations in these variants leads to determine the diseases. The AI algorithms such as crunchers, guides, advisor, predictors, tacticians, strategies, lifters, partners, okays, and supervisory approaches are applied for detecting the diseases through diagnoses.

The decision-making process in the medical domain is highly supported by artificial techniques. It is applied based on the training data set and uses the current and captured data to decide the factors, depending on which the decisions are arrived. It is a machine that is able to replicate humans' cognitive functions to solve the problems focused on healthcare issues (Rodellar et al., 2018; Xu et al., 2019; William et al., 2018; Johnson et al., 2018). It helps the well-being industry by two ways, such as making decisions for experts and common people. Common people used these self-recommended decisions based on captured data and with the support of AI for instant remedies. At the same time, it helps the experts to decide a complex case.

Treatment is a process for subscribing medication or surgical processes in the medical industry. The AI techniques are supporting to subscribe medication based on the preceding cases and training. It is highly recommended to rely on the training data set with all possible comparisons of the rules that are applicable according to the cases. The AI system will consider the regional factors, experts' opinions, and previous cases with a similar pattern and will play a major role to subscribe the treatment. AI can be used for life care process with the support of robotics. Robots are used in emergency situations, for rescue purposes, as emergency servants, and in all other possible manners. In the current scenario, they are performing activities as an expert tutor and train the experts too.

According to the availability of the biomedical data and rapid development of device knowledge, profound education, natural language processing (NLP), robotics, and computer visualization techniques have made possible to create successful applications of AI in biomedical and healthcare applications. The authoritative AI techniques can reveal medically relevant knowledge hidden in the massive volume of data, which in turn can contribute to decision making.

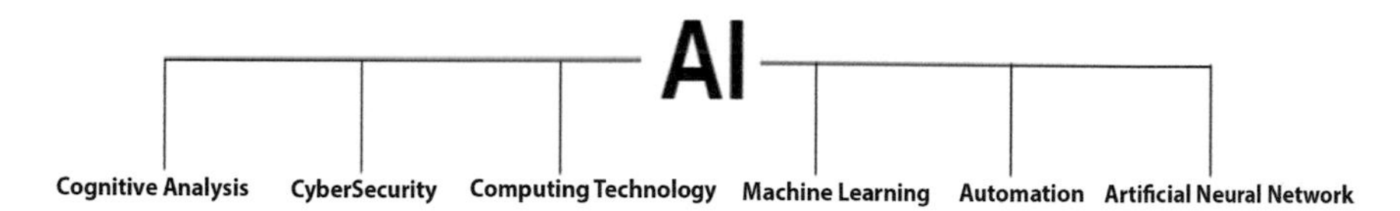

FIGURE 5.2 Industry 2.4 technologies.

Medical data mining used by AI is to search for relationships and patterns within the medical data that will provide useful knowledge for effective medical diagnosis. The probability of disease will become more effective, and early detection of disease will aid in increased exposure to required patient care and improved cure rates using computational applications. The application of data mining methods in the medical domains is to improve medical diagnosis. Some examples include predicting breast cancer survivability by using data mining techniques, application of data mining to discover subtle factors affecting the success, failure of back surgery that led to improvements in care, data-mining classification techniques (Kim) for medical diagnosis decision support in a clinical setting, and data mining techniques used to search for relationships in a large clinical database.

Many research works focused on the development of data mining algorithms to learn the regularities in these rich, mixed medical data. The success of data mining on medical data sets is affected by many factors. Knowledge discovery during training is more difficult if information is irrelevant or redundant or if the data is noisy and unreliable. Feature selection is an important process for detecting and eradicating as much of the inappropriate and redundant information as possible. The necessary preprocessing step for analyzing these data, i.e., feature selection, is often considered, as this method can reduce the dimensionality of the data sets and often leads to better analysis. Research proves that the reasons for feature selection include improvement in performance prediction, reduction in computational requirements, reduction in data storage requirements, reduction in the cost of future measurements, and improvement in data or model understanding.

5.3 HEALTHCARE APPLICATIONS OF THE BIOMEDICAL AI SYSTEM

The AI approach is used for prediction from a massive amount of data for generating a configuration and to draw knowledge from experience independently. Medical practitioners try to analyze related groups of subjects, associations between subject features, and outcomes of interest. In the diagnostic domain, a substantial quantity for the existing AI-analyzed records of indicative imaging, hereditary testing, and electrodiagnosis is used to determine the diseases based on available clinical data for the benefits of human's healthcare systems.

In this chapter, the healthcare applications of the biomedical AI system are reviewed in the following aspects.

1. The motivation behind using AI in healthcare.
2. Different techniques of AI.
3. Role of AI in various healthcare applications.

Knowing the importance of AI in healthcare with the technologies that are supporting different aspects of healthcare will help to get more understanding of how things are developed and utilized, and this will also emerge thing to new innovations.

5.3.1 THE MOTIVATION OF USING AI IN HEALTHCARE

In the recent era, medical issues challenge the society as well as the practitioners in identification and decision support-related treatments. There are many reasons for the emergence of AI in the field of medicine. The factors considered as most important are an increase in available healthcare data, requiring a greater efficient practitioner and healthcare system, due to rising external burden in medicine, for the diagnosis, and people demanding fast and more custom-made care. So, for this, AI is introduced in the field of healthcare that uses a collection of algorithms. This provides sophistications to learn and develop a pattern from the large volume of data more accurately, aids the physician to get more knowledge and understanding about the day-to-day updates in the field, and reduces the errors or the flaws occurring due to human clinical practices in diagnostics and therapeutics.

5.3.2 DIFFERENT TECHNIQUES OF ARTIFICIAL INTELLIGENCE

To understand AI, it is more useful to know the techniques and algorithms and how they are processed for the implementation of solving complicated tasks or the mistakes done by human interventions. Implementation of AI in healthcare includes machine learning (ML) as a division of AI that is used for clustering, classification, and predictive analysis of medical data. Its enhanced application of neural networks is introduced with deep learning algorithms and genetic algorithms.

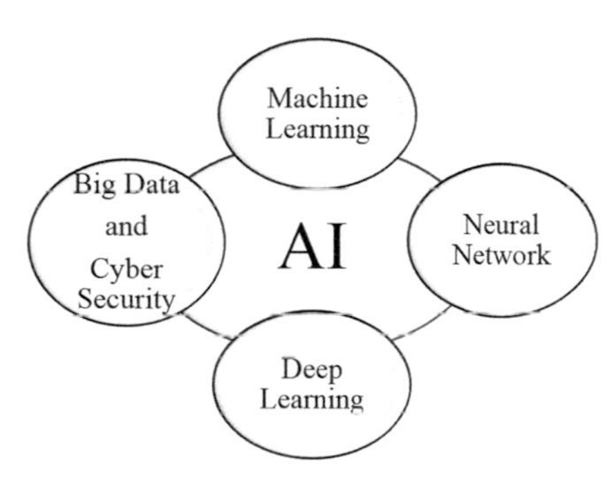

FIGURE 5.3 Applications of AI techniques in healthcare.

Big data analytics is in the recent trend to handle the medical healthcare data that are collected from mass screening, diagnosis, and treatment monitoring and evaluation through methods of demographics, therapeutic notes, electronic recording of medical strategies, and physical examinations. Cyber security is an ethic of AI to be applied in healthcare. The most important techniques of artificial intelligence are described below.

5.3.2.1 MACHINE LEARNING

ML is a technique that must be understood to assist AI. This algorithm represents the data structure for the clustering and classification of abnormalities. It consists of statistical inference to give more accuracy in prediction and classification. The algorithm in ML is categorized based on supervised and unsupervised algorithms. The supervised algorithm is learning with input and output that produce the prediction and classification result by mapping. So, based on the trained data, the new data result will be produced. However, the unsupervised algorithm is ideal for the dispersal of data to learn more about the data. So, there is no output, and thc algorithm by itself has to discover and deliver the interesting pattern in the data.

The popular image analysis and ML algorithms are water immersion, Hough transform, seed-based region growing, genetic algorithm, active contour, *K*-means, fuzzy logic, etc. Each algorithm works efficiently and gives a unique result for different applications. Water immersion is a segmentation technique used to find the closed boundaries of the region of interest in the data, but it is not suitable for overlapping cells. Hough transform is used to find similarities in shapes, and it is suitable only for round-shaped images. Seed-based region growing algorithm identifies the edges, but it depends on user-defined parameters. Genetic algorithm is very slow in execution, but it tries to search the best solution. Active contour is used to detect the boundaries of the image, and its drawback is it takes much time for energy minimization. *K*-means algorithm detects the threshold, but then, its overlapped cells cannot be separated from the background. Fuzzy logic will discover the robustness and uncertainty in the data, but it not flexible because of firm data. Classification algorithm such as decision tree performs well only when data are trained carefully. Bayesian network understands the data based on statistical inference; *k*-nearest neighbor (KNN) and artificial networks are very fast. Although these are fast, they face difficulty when used for large attributes and tolerance of noise data. Finally, support vector machine is suitable for only linear data, unless using multiple parameters and kernel trick multiple classes of classification are not possible.

5.3.2.2 NEURAL NETWORKS

Information technology and the biological system that has distributed communication nodes are stimulated neural networks. There are many algorithms used in a neural network. Very few are important, such as radial basis function network, perceptron, back propagation, logistic regression, gradient descent, and hop field network. Radial basis function network has an activation function called radial basis function in the hidden layer. It has an input layer, hidden layer, and linear output; also, it is used for the time series prediction, classification, and system control. Perceptron is a linear unsupervised binary classifier. There are two layers of perceptron such as a single layer and a multilayer. The multilayer is called the neural network. It has input layer weights and bias, nct sum, and activation function.

Backpropagation is used for the classification that is essential for neural network training. It back propagates the information about the error through a network that is made by wrong guesses of the neural network. So, immediately according to the error information, the parameter passed by the neural network will be changed in one step at a time. Logistic regression is a nonlinear binary classification method that translates the signal into space from 0 to 1. It is used to calculate the probability of the set of input with the label that is matched. Gradient descent is a neural network algorithm used to access the weight based on the error rate and to find the local minimum of the function that is known as an optimum function. Hop field network has the feedback connection that is called recurrent. They are fully interconnected neural networks and applied for image segmentation only if it has an optimization problem.

5.3.2.3 DEEP LEARNING

Deep learning is an extension of ML based on neural network concepts that can be a supervised network, a semi-supervised network, or an unsupervised network. There are many deep learning algorithms available such as deep neural network (DNN), deep belief network (DBN), recurrent neural network (RNN), and convolutional neural network (CNN), restricted Boltzmann machine (RBM), auto encoder network, and long short-term memory (LSTM). This LSTM is a special type of RNN ML technique. The advantage is that the expected output compares with models' output with updates in weights. All of these algorithms are used in compute vision, speech recognition, and NLP, drug design, and medical image analysis. DNN is a feedforward network with a complex nonlinear relationship, where data flows through input to output without looping back. It has a multilayer between the input and output. DBN is an unsupervised probabilistic algorithm that consists of a multilayer of the stochastic latent variables. Latent variables are binary and also called as feature detectors and hidden units.

RNN is a type of neural network that has an in-built memory so that all the information about the sequence is stored and the output result is produced as an input of another network. This process is not there in other traditional neural networks, and it is also used to reduce the complexity of the parameter. CNN is the deep artificial neural network that is used for classification done only after by clustering the image similarity with image recognition. RBM is a simple type of neural network algorithm that is used to minimize the dimensionality, classification, and reduction, and it also works like perceptron. Auto encoder is an unsupervised algorithm that works like a back propagation algorithm. It is trained to feed input to its output with the help of a hidden layer that has the code to represent the input layer. This has been used for dimensionality reduction and feature extraction. LSTM is an artificial neural network that works like a RNN. It is used for the classification, very particularly for the predictions based on the time series.

Deep learning is the most popular technique in the recent era because of its advancements. It has the following advantages that are features used as an input for the supervised learning model. All these algorithms are used for the predictive representation of patients in an unsupervised learning method from the electronic health record. However, there are some limitations, such as it requires a large volume of data with high expensive computation to train the data and interpretability is difficult.

5.3.2.4 BIG DATA

The word big data represents a huge volume of data. These data can be in any of the following form: the data cannot be stored in memory and the data that cannot be managed and retrieve because of its size. Big data give great hope in healthcare for the data collected from the biological experiments for the medical drugs and new ways of treatment discovery, efficient patient care, in and outpatient management, insurance claim, and payment reimbursement. So, there are different types of healthcare data. The different types of biomedical data are integrated electronic health record (EHR), genomics-electronic health record, genomics-connectomes, and insurance claims data. Big data algorithm and its architecture are used to manage the collection of data to promise the tremendous changes for effective results.

Some of the important big data concepts and their applications commonly used are discussed here.

The map reduce method with the stochastic gradient descent algorithm is used for the prediction purpose. Logistic regression is used to train EHR with the stochastic descent algorithm. There are many architectures developed for monitoring personal health such as Meta Cloud Data Storage architecture, which is used to transform the data collected into cloud storage. To store a huge set of data, the Hadoop Distributed File System is used. To secure big data that are all collected and transformed into the cloud, the data are protected by using an integrated architecture model called the Grouping and Choosing (GC) architecture with the Meta Fog Redirection architecture. Logistic regression implanted with the MetaFog architecture that is used for the prediction of diseases from the historical record when the dependent variable has existed.

5.3.2.5 COGNITIVE TECHNOLOGIES

Cognitive innovations are exceptionally and famously presented in well-being care to diminish human decision making and have the potential to correct for human blunder in giving care. Restorative mistakes are the third driving cause of death all over India and additionally in the world, but they are not for the most part due to exceptionally awful clinicians. Instep, they are frequently ascribed to cognitive blunders (such as disappointments in discernment, fizzled heuristics, and predispositions), anon attendance or underuse of security nets and other conventions.

The utilization of AI advances guarantees to diminish the cognitive workload for doctors, in this way moving forward care, symptomatic exactness, clinical and operational productivity, and the in general quiet involvement. Whereas there are reasonable concerns and dialogs around AI taking over human occupations, there is restricted proof to date that AI will supplant people in well-being care. For illustration, many considerations have been proposed that computer-aided readings of radiological pictures are fair as precise as (or more than) the readings performed by human radiologists.

Rather than large-scale work misfortune coming about from the computerization of human work, we propose that AI gives an opportunity for the more human-centric approach to increase. Different from mechanization, it is increasingly presumed that savvy people and savvy machines can coexist and make way for better results than either alone. AI frameworks may perform a few well-being care errands with constrained human intercession, subsequently liberating clinicians to perform higher level assignments.

5.3.2.6 CYBER SECURITY

The AI applications are applied within the restorative choice making with tall unmistakable comes about since of the advancement of progressed calculation combined with adaptably getting to computational assets. Its framework plays a major role within the process information and makes a difference for the authorities to create the right choices. Doctors who handle the restorative and radiological gadgets that give the touchy administrations require appropriate preparation and certification. Obviously, the applications are considered by their standards based on which artificial systems are held so that the procedures and techniques will prove their standards. Profoundly speaking, the AI system provides forward thinking about every potential scenario and every possible consideration, whereas the humans, within the limitations, consider what is apparent to their brains alone.

AI medical devices facilitate the medical field for various reasons. They may be adopted for generating revenue for manufacturers and the physicians will possibly have additional tools at their disposal. Whenever an AI device is proposed, an appropriate policy infrastructure must be developed for introducing the device effects due to lack of stability, unanimous definition, or instantiation of AI. In the event that the AI gadgets ought to work and perform more rapidly and precisely, indeed, on the off chance that they have similarity to the human neural systems and think in an unexpected way unlike humans. AI medical devices have enough and more opportunities in every available situation, but in contrast, humans are incapable of doing proper processing and making the right decisions. Quickly, it starts making a self-governing determination in connection to analyze medicines and moving ahead of its work as just a bolster instrument and an issue comes up as to whether its designer can be held dependable for its choice. The primary question to address is: who will be sued in case an AI-based gadget makes a botch?

In this AI framework, the question remains to be unanswered about the guardian relationship between the patients and restorative frameworks with most recent approach activities, the direction of information assurance and cyber security, and the talk about almost the bizarre responsibilities and duties. So, it has to address the trustworthiness of the data. Of course, AI is a good application because of its power, assistance, and value, but on the contrary, it must be taken seriously so that no bad and unethical uses of the technology are formulated. If there is any unethical use of technology, this will be dangerous, so all the authorities, especially physicians, patients, and other technicians, must work together to prevent any evil in this scenario. If there is a good spirit of cooperation in this healthcare system, we could find a balanced situation in providing security and privacy protection. It is also important to maintain the ethical use of sensitive information to give the surety about both the human and structured management of thc patients. It is mandatory that the ethical views and challenges have to be very clear to the point and be safeguarded against some social evils. There is also a possibility of ethnical biases that could be built in the medical algorithms since the healthcare output already varies by ethnicity.

So, the use of AI leads to two problems with regard to data collection from various devices. One is providing security for the data collected and stored using these devices, and the second one is that the same data is endangered by the cyber hackers when using such devices. So, there is an urge to provide data protection and cyber security for the AI device system applications.

As of now, the European and American Government administrative formulated an enactment for information assurance and cyber security. The European Government shaped common information assurance control over information assurance and cyber security orders, restorative gadget control, and in vitro demonstrative therapeutic gadget directions. In the United States, the well-being protections movability and responsibility act (HIPAA) is a person right to security, over well-being data and preparation for how to handle the touchy information. The government nourishment medicate and restorative act (FDA) for cyber security permitted the producer to depict the hazards and steps to limit the defenselessness. So, the rapid development of AI should provide safety and trustworthiness in healthcare.

5.3.3 ROLE OF AI IN VARIOUS HEALTHCARE APPLICATIONS

AI approaches are used for prediction from a large amount of data to generate a pattern to draw knowledge from experience independently. Its techniques are effectively used to develop biomedical healthcare systems. It is applied for the detection, diagnosis, and characterization of diseases and interpreting medical images.

Wealthy semantics may be a connection of characteristically depicted strategies for data extraction from clinical writings; the extraction of events is to analyze clinical medicines or drugs primarily for the reason of clinical coding, location of sedate intuitive, or contradictions.

ML and NLP are the successful techniques in providing semantics that includes a description of clinical events and relations among clinical events (e.g., causality relations). Additionally, underpins in subjectivity, extremity, feeling or indeed comparison for alter in well-being status and unforeseen circumstances or particular restorative conditions that affect the patient's life, the result or viability of treatment and the certainty of a determination. DNN is used for speech recognition with a Gaussian mixture model. It is utilized to extricate highlights from crude information and for discourse feeling acknowledgment. For each emotion segment, the emotion probability distribution is created.

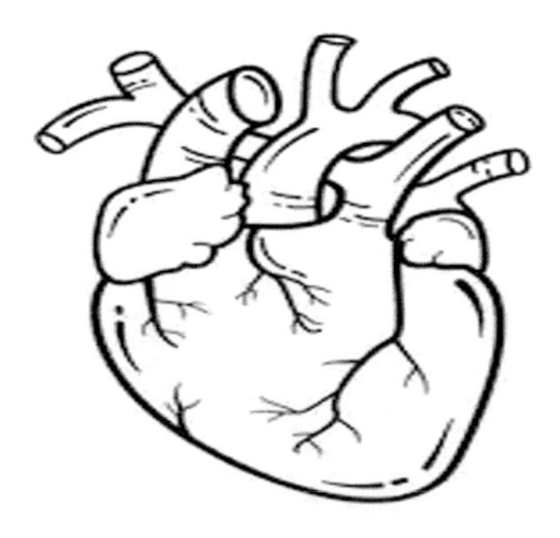

FIGURE 5.4 Heart disease.

The manufactured neural arrange has moreover demonstrated its capacity by working on the classification of heart diseases. In this procedure, for the classification of a stroke, the input of the sensor is given to the framework that employs a bolster forward arrange with the run the show of back engendering within the conceivable strategy. The viable result of classification is given by the re-enactment framework.

Heart beat classification and congestive heart failure detection are done using a convolution neural network deep learning method. It also helps in two-dimensional (2D) image classification for three-dimensional (3D) anatomy localization. Heart diseases of both invasive and noninvasive types are detected using deep learning algorithms. Indicators of heart diseases are end-systolic volume and end-diastolic volume of the cleared out ventricle (LV) and the discharge division (EF). These are all portioned utilizing U-Net profound learning engineering. It may be a convolution neural arrange that is commonly connected for the division of biotherapeutic pictures.

MRI brain tumor examination is conducted by utilizing manufactured neural arrange procedures for the classification of pictures in demonstrative science. A common relapse neural arrange (GRNN) is utilized as a 3D method of classification and pooling-free completely convolutional systems with thick skip associations for the picture of the brain tumor.

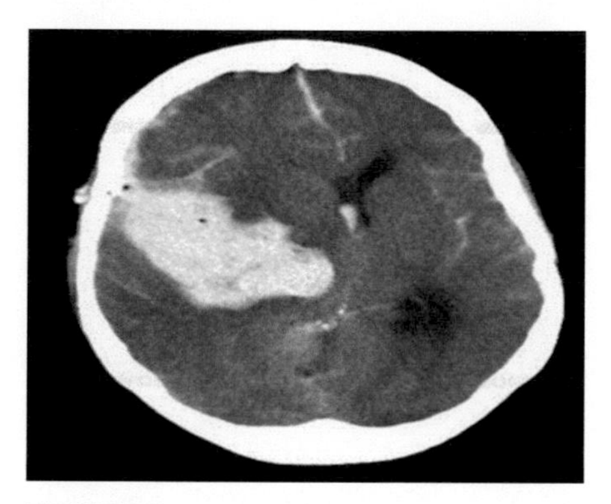

FIGURE 5.5 MRI brain tumor.

Many methods and algorithms are used for the classification of deformations in the acquired image. Exceptionally, the late slightest squares bolster vector machine (LSSVM) could be a well-known strategy utilized for the conclusion of ordinary and irregular ranges of the brain from the information of MRI. By the independent way of classifying the MRI picture, it gives a successful result with more prominent exactness than other classifiers.

Prognosis of Alzheimer's disease is diagnosed by sparse regression and deep learning combination called the deep ensemble sparse regression network. The leftover profound neural arrange (ResNet) design is utilized to assess the capacity in foreseeing 06-methylguanine methyltranferase quality status for the division of the brain tumor. The profound convolutional encoder system is connected for the division of sclerosis. Profound learning convolution neural organizes for the brain structure division with a distinctive fashion of engineering such as patch-wise convolution neural organize design, semantic-wise convolution neural arrange design, and cascaded convolution neural arrange design. These designs contain pooling, enactment, and classification (completely associate). The convolutional layer yields feature maps by employing a kernel over the input picture. The pooling layer within the design is utilized to down test that comes about. The actuation capacities are amended straight unit (ReLU) and defective ReLU.

Brain glioma segmentation can be done using a deep neural network with max pooling and without using pooling. Cervical cancer discovery and classification of pap-smear test pictures are done utilizing ML strategies. Numerous sorts of enquires have experienced utilizing ML calculations for the highlight division, including choice and classification. Very common unsupervised segmentation and classification algorithms are the K-NN method, C4.5 and logical regression, SVM, random forest, DT, Bayesian, SVM, FCM clustering, and Otsu's method. All these algorithms are applied to this type of data and achieved the maximum accuracy. The calculations such as K-NN, metaheuristic calculation with a hereditary calculation, moment arrange neural arrange preparing calculations Levenberg–Marquard with versatile force id known as LMAM and optimized Levenberg–Marquard with versatile energy is known as OLMAM calculations are used. Different methods applied for identifying cervical cancer are two level cascade classification utilizing C4.5 and consistent relapse, KNN, fluffy KNN, SVM, and irregular timberland-based classifier, iterative edge strategy, and pixel level classification such as pixel level examination.

Classification results can be obtained not only by the classification algorithm but also after implementing the segmentation algorithms such as SVM block-wise feature extraction with the Otsu's segmentation method. Cervical cell segmentation is generally performed by the scene segmentation method, decision tree, Bayesian, SVM, and combination of the three individual posterior probabilities. Profound learning calculations play a critical part in extricating data from the physiological signals such as electromyogram (EMG)-, electroencephalogram (EEG)-, electrocardiogram (ECG)-, and electrooculogram (EOG)-based healthcare applications. EMG flag measures the electrical sensors called as anodes put on the skin to detect muscle actuation, constrain, and state of the muscles. EEG is measured by placing terminals on the cranium to discover the brain movement. ECG is measured by setting the anodes on the chest to record the human heart action. EOG measures the corneoretinal potential by placing terminals within the back and front of the human eye developments.

Deep learning algorithm used with EMG signals for the limb movement estimation employs RNN that performs well and estimates a 3D trajectory. CNN is used for the neuroprosthesis control, movement intension decoding, and gesture recognition. Deep learning calculation with EEG signals are for the applications such as EEG interpreting and visualization, translating energized developments, but do the expectation for patients with a post-anoxic coma after cardiac capture, segregate brain movement, brain computer interface, seizure discovery, following of neural elements, forecast of drivers cognitive execution, reaction representation, highlight extraction, epileptic seizure forecast, and engine symbolism classification connected with convolution neural arrange.

Auto encoder is applied to identify the sleep state. RBM is used for the application of motor imagery classification and effective state recognition. Deep learning network is applied for emotion recognition. DBM performs well in the applications such as emotion classification, motor imagery, and detecting target images. DL calculation with EGG signals is applied for applications like arrhythmia discovery by utilizing distinctive interims of tachycardia EGG portions, coronary supply route infection flag distinguishing proof, coronary course illness discovery, screening paroxysmal atrial fibrillation, checking and identifying atrial fibrillation, discovery of myocardial violation, and distinguishing proof of ventricular arrhythmias are comprise of convolution neural organize. LSTM is utilized for coronary supply route malady flag distinguishing proof.

RNN is applied for sleep apnea detection. The RBM algorithm is implemented for signal quality classification and heart beat classification. Auto encoder is used with the signal EOG and EEG for the driving fatigue detection application. CNN is used with EOG, EEG, and ECG for the momentary mental workload classification and with EOG signal for drowsiness detection. DBN is applied with EOG, EEG, and EMG for the classification of sleep stage. Profound learning is connected from one-dimensional physiological flag to 2D therapeutic pictures for picture investigation and utilizing these signals to determine and to derive a conclusion with computerized organizing and discovering of Alzheimer's infections, breast cancer, and lung cancer.

Lesions in various parts of the human body are detected and classified using different neural network architectures such as convolution layers, CNN, Alex net, and VGG with small changes in the weight and layers. Layer performances are tested using area under curve (AUC) to reduce the overtraining. ResNet and the Inception architecture are the best networks used for presurgical MRI of the brain tumor. Inception V3 network seems to be the best one for the feature extraction rather than the GoogLeNet, which is used for the classification.

Breast mass lesion classification is done using deep learning methods such as RNN with LSTM for training the data, and the features are extracted using VGGNet. Also, the performance is tested using the ROC curve (AUC) method to distinguish benign and malignant lesion. Classification of genomic subtypes is performed by utilizing three distinctive profound learning approaches that are used to classify the tumor according to their atomic subtypes: learn the tumor patches as a pre-prepared strategy and extricate standard highlights by neural systems utilized for classification with a back vector machine. The design of neural systems is connected in GoogleNet, VGG, and CIFAR. The 10-fold cross approval strategy is utilized for approval, and the region beneath the collector working characteristic (AUC) is utilized to degree its execution.

Computerized mammograms are among the foremost troublesome therapeutic pictures to be studied due to their low contrast and contrast within the sorts of tissues. Vital visual clues of breast cancer incorporate preliminary signs of masses and calcifications clusters. Tragically, within the early stages of breast cancer, these signs are exceptionally subtle and varied in appearance, making conclusion troublesome and challenging indeed for pros. This is often the most reason for the advancement of the classification system to help masters in restorative teaching. Due to the importance of a robotized picture categorization to assist physicians and radiologists, much inquiry within the field of restorative picture classification has been done as of late. With all this exertion, there is still no broadly utilized strategy to classify restorative pictures. This is often because the therapeutic space requires tall exactness and particularly the rate of untrue negatives to be exceptional.

In expansion, another vital calculates that impacts the victory of classification strategies is working in a group with therapeutic masters, which is alluring but frequently not achievable. The results of mistakes in discovery or classification are exorbitant. Mammography alone cannot demonstrate that a suspicious zone is threatening or kind. To choose that, the tissues should be evacuated for examination by utilizing breast biopsy procedures. An untrue positive discovery may cause superfluous biopsy. Measurements show that 20–30 rates of breast biopsy cases are demonstrated as cancerous. In an untrue negative discovery, a real tumor remains undetected that might lead to higher costs or indeed to fetch a human life. Here is the tradeoff that shows creating a classification framework that might specifically influence human life. In expansion, the tumor presence is distinctive. Tumors are of diverse shapes and a few of them have the characteristics of the ordinary tissues. The density level of tumors decides the level of the stage. The stage determination process is explained in the next session.

Lung cancer is recognized and analyzed by utilizing computed tomography pictures that have been utilized with CNN for a computerized classifier, recognizing prescient highlights. Profound highlight extraction with preprepared CNN is proficient than the choice tree classifier. VGG-f pretrained CNN is used for linear unit feature extraction, and the result will be better when feature ranking algorithm is tailed with a random forests classifier. Fine-tuning a CNN is the strategy that has been pretrained with a huge set of named preparing information, which is the issue confronted in the profound convolutional neural arrange (CNN). In lung knob classification, multi-edit convolution neural organize (MC-CNN) is connected to extricate knob notable data by employing a novel multi-crop pooling approach that crops different locales from CNN including maps with max-pooling. Programmed division of the liver and its injury utilizes a profound learning cascade of completely convolutional neural systems (CFCNs) and thick 3D conditional arbitrary areas. CT gut pictures of fifteen hepatic tumors are recognized and approved by utilizing twofold cross-validation.

The probabilistic boosting tree is used to classify lesions or parenchyma in the liver. Convolutional layer with a fully connected layer detect lesions and liver morphology. Drug-induced liver damage (DILI) is recognized with the DL engineering comprising profound neural organize (DNN), convolutional neural arrange (CNN), and repetitive or recursive neural arrange (RNN). Liver tumor from CT is distinguished by utilizing convolution neural organize that can discover fluffy boundaries, inhomogeneity densities, and shapes and sizes of injuries. Irregular timberlands, K-NN, and back vector machine (SVM) with spiral premises work less and the lattice looks are utilized for central liver injuries with contrast-enhanced ultrasound.

The convolutional neural arrange is used to identify and classify districts of intrigued as threatening or kind. A profound convolution neural arrange (DCNN) is utilized with FP, which incorporates four convolutional layers and three completely associated layers connected to mammogram breast pictures for mass discovery. The problem occurring in layer implementation is overfitting, and this can be solved using Jittering and dropout techniques. For the DCNN-based approach and FP reduction, the prescreening stage is the same in both systems in the FP reduction stage. CAD is a feature-based system consisting of 3D clustering, an active contour method for segmentation, morphological, gray level, and texture features for the linear discriminant classifier to mark the detected masses. Breast tumor is identified by utilizing the vicinal back vector machine (VSVM), which is an improvement learning calculation. The preparing information is clustered into diverse delicate vicinal zones in highlight space through the part-based deterministic tempering (KBDA) strategy. The collector working characteristics (ROC) are utilized to determine the degree of execution of the calculation.

Segmentation, quantitative features, and classification of the *malignant blood cell* and blast cell and irregular lymphoid cells are called abnormal lymphoid cells

(ALC). The lymphocytic such as unremitting (chronic) lymphocytic leukemia (CLL), splenic minimal zone lymphoma, hairy cell leukemia, mantle cell lymphoma, follicular lymphoma, B and T prolymphocytic lymphoma, large granular lymphocyte lymphoma, Sezary syndrome, plasma cell leukemia, blast cell leukemia (it is interrelated with both myeloid and lymphoid acute leukemia), reactive lymphocytes (RL) linked to viral infection, and normal lymphocytes (NL) are using ML algorithms. For the segmentation, two regions of interests (ROIs) are considered from the image such as the nucleus and cytoplasm. Morphological operations with watershed segmentation algorithms are used to find the ROIs. Highlights are extricated with the gray-level co-occurrence lattice (GLMC). At that point, it is prepared by employing an administered calculation called back vector machine (SVM).

Robotics: Biomedical AI systems can be trained in interpreting clinical data such as screening, diagnosis, and treatment assessment. Commonly used robots are improved and exhibit enhanced proficiencies for the biomedical field, which have proved their greater mobility and portability. Robots in medical science are used for surgical, rehabilitation, biological science, telepresence, pharmacy automation, companionship, and as a disinfection agent. Differences between AI and robots are as follows:

AI programs

- These are used to function in a computer-simulated universe.
- An input is given within the frame of images and rules.
- To perform these programs, we require common reason computers.

Robots

- Robots are utilized to function within the genuine physical world.
- Inputs are given within the shape of the simple flag and within the frame of the discourse waveform.
- For their working, uncommon equipment with sensors and effectors are needed.

Mechanical arm direction is done with seven motions by utilizing a CNN profound learning organize. Surgical robots are a type of robots utilized to help in surgical methods either by way of automated surgery, computer-assisted surgery, or robotically assisted surgery.

Robotically assisted surgery is used to improve the performance of doctors during open surgery and introduce advanced minimally invasive surgery like laparoscopy. It permits a broad extend of movement and more accuracy. Ace controls permit the specialists to control the rebellious arms, by deciphering the surgeon's characteristic hand and wrist developments to compare, extract, and scale. The objective of the automated surgery field is to plan a robot to be utilized to perform brain tumor surgery. There are three surgical robots that have been used as of late but were for heart surgery:

- da Vinci Surgical System,
- ZEUS Automated Surgical System, and
- AESOP Automated System.

The automated specialists such as da Vinci surgical framework and ZEUS mechanical play a critical part in the heart surgery framework.

Rehabilitation robotics provides rehabilitation through robotic devices. It assists different sensory motor functions such as arm, hand, leg, and ankle. It also provides therapeutic training and gives therapy aids instead of assistive devices. Biorobots are planned and utilized within different fields, famously in hereditary designing to constrain the cognition of people and creatures. Telepresence robots give a remote organize back to communicate in farther locations.

Pharmacy automation handles pharmacy distribution and inventory management and also involves pharmacy tasks such as counting small objects, measuring the liquids for compounding, etc. *A companion robot* emotionally engages and provides a company to the users and alert the users if there is any problem with their health. *A disinfection robot* acts as an agent that resists the bacterial spores in the place or room rather killing the microorganisms.

In *Automated Fraud Detection in the Healthcare sector*, the system monitors the employer in the medical billing. Based on the prescription of the experts or the practitioners, the bill has to be claimed for the delivery of the medicine. If wrongly keyed-in or misspelt, medicines will lead to a greater consequences. Medical product invoice trustworthiness is important because it has made it as online delivery in recent years. So, the analysis must be done carefully for the data integrated with different medicine issued based on different problems. For online medical bills and medicine delivery systems, implementing AI efficiently will produce the remarks as well as the common people will intelligently learn automatically by their own. This biomedical application is integrated with the health indicator and their working environment.

In *pharmaceuticals*, AI is implemented as an automatic machine that is used in manufacturing or discovering medicines. The deep neural network is implemented with the machines with big data concepts with a huge set of molecule compound ratios for each medicine with a different combination. Medicines are formulated based on the chemical compounds. The number of molecules with the ratio value helps to produce the medicine. It is very difficult to produce a huge volume of medicine within the aspect of man power based on the necessity of medicine availability. So, by making this production as an automated process, time will be saved and human errors will be avoided. Although the automated system is introduced in the pharmaceuticals, monitoring is very important and this can be done only by the humans.

5.4 SUMMARY OF HEALTHCARE APPLICATIONS OF BIOMEDICAL AI SYSTEM

Healthcare applications of the biomedical AI framework are dissected. Based on the writing survey of the numerous investigates, an outline of major application of biomedical AI frameworks is arranged in Table 5.1.

TABLE 5.1 Survey of Healthcare Applications of Biomedical AI System

Application	AI Algorithms	Description
Rich semantics	ML and NLP	Diagnoses, clinical treatments, or medications
Speech recognition	Deep learning network	Speech emotion recognition
Heart diseases	Artificial neural network, CNN, Feed forward network	Classification of stroke, heartbeat classification and congestive heart failure detection, heart diseases of invasive and noninvasive
Brain tumor	Artificial neural network, GRNN, CNN, LSSVM, regression and deep learning, ResNet	Classification of pictures, ordinary and irregular ranges of the brain from information of MRI, diagnosis of prognosis of Alzheimer's disease, segmentation of brain tumor
Cervical cancer	ML techniques K-NN, C4.5 and logical regression, SVM, random forest, DT, Bayesian, decision tree, SVM, FCM, Otsu	Detection and classification of pap-smear test images
Physiological signals	Deep learning algorithm RNN Deep learning algorithm used with EMG signals	Skin to find the muscle activation, force and state of the muscle and heart activity limb movement

TABLE 5.1 *(Continued)*

Application	AI Algorithms	Description
Sleep state	RNN, CNN, RBM, DBM, LSTM, DNN	Emotion classification, arrhythmia detection, coronary supply route, Malady Flag Recognizable proof, coronary course infection location
Breast mass lesions	RNN with LSTM VGGNet	Benign and malignant lesions Classify the tumor
Breast tumor	Neural networks (GoogleNet, VGG, and CIFAR), SVM DCNN ,VSVM	Mammogram breast
Lung cancer	CNN, random forest classifier, MC-CNN	Detection and diagnosis
Liver and its lesions	CFCNs, DNN, CNN, RNN, random forest, K-NN, SVM, DCNN	Liver tumor, liver lesions
Malignant blood cells	ML Algorithms, GLCM, SVM	Abnormal lymphoid cell and chronic lymphocytic leukemia diagnosis
Robotics	CNN	Surgical, rehabilitation, telepresence, pharmacy, companion, disinfection
Fraud detection	DNN	Medical billing
Pharmaceuticals	DNN	Produce the medicine

5.5 CONCLUSION

AI techniques are applied to the biomedical healthcare system for analysis and prediction where the technical support required for human decision making a process or automated system. In the current era, AI has formed a steady improvement in providing greater support and assistance for a better lifestyle and living. The key point in the start of the development of AI has to be necessarily known so that new ideas emerge for the production of new and efficient AI systems. Basically, in AI, the decisions are taken by the use of algorithms. The usage of the algorithm has proved itself in giving the right results. Invariably, the AI plays a major role in healthcare to take the right decision very especially within the short span of time. An AI machine is a secured user with the predefined algorithms system to secure or protect the humans from the unsure situations; unlike the human beings, the machines are faster and long runners. Since machines would not get tired, they can be functioning continuously without having a break.

The data that is collected out of the implementation of the AI machine will be huge, but only a few of them can be used. Since it is a machine and when its usage becomes massive, the human's dependency keeps increasing. Universally, the massive usage machines make humans to fade away in their research, experience, and scientific knowledge. Unlike the other countries, India still manages in giving job to maximum humans without giving the much importance. Once machines are used for every work and for automatic functioning, the humans will have high crises in finding a job, which will create chaotic situations in their livelihood. A machine functions out of commands where you can find no creativity. Practically speaking, creativity alone can create a new and vibrant society. Machines are a creativity of humans. If we are going to miss the crown of creativity by using machines, we are going to miss a vibrant and innovative society. Although it has little disadvantages, by the perfect functions of AI, the human mind gets stimulated in cognitive processes and behaviors. It is the machine but it is the creation of the human effort and expertise. Although humans may be experts for a particular period of time, it is the machines that capture and preserve the human expertise. Also, machines have the ability to comprehend large amounts of data quickly and also can give a fast response for any related queries. Finally, AI has to provide the trustworthiness when data are reused for training or research purposes to discover new things or to analyze the existing ones for the prediction.

KEYWORDS

- **machine learning**
- **artificial intelligence**
- **deep learning**
- **big data**
- **cognitive technologies**
- **neural networks**
- **natural language processing**
- **cyber security**
- **robotics**
- **electronic health record system**
- **clinical decision support**
- **imaging modalities**

REFERENCES

Faust, Oliver, et al. "Deep learning for healthcare applications based on physiological signals: a review." *Computer Methods and Programs in Biomedicine* 161 (2018): 113.

He, Jianxing, et al. "The practical implementation of artificial intelligence technologies in medicine." *Nature Medicine* 25, 1 (2019): 30.

Jiang, Fei, et al. "Artificial intelligence in healthcare: past, present and future." *Stroke and Vascular Neurology* 2, 4 (2017): 230243.

Johnson, Kipp W., et al. "Artificial intelligence in cardiology." *Journal of the American College of Cardiology* 71, 23 (2018): 26682679.

Kim, Jae Won. "Classification with Deep Belief Networks."

Lo'ai, A. Tawalbeh, et al. "Mobile cloud computing model and big data analysis for healthcare applications." *IEEE Access* 4 (2016): 61716180.

Mazurowski, Maciej A., et al. "Deep learning in radiology: an overview of the concepts and a survey of the state of the art with focus on MRI." *Journal of Magnetic Resonance Imaging* 49, 4 (2019): 939954.

Paul, Yesha, et al. "Artificial intelligence in the healthcare industry in India." The Centre for Internet and Society, India (2018).

Pesapane, Filippo, et al. "Artificial intelligence as a medical device in radiology: ethical and regulatory issues in Europe and the United States." *Insights into Imaging* 9, 5 (2018): 745753.

Ramesh, Dharavath, Pranshu Suraj, and Lokendra Saini. "Big data analytics in healthcare: a survey approach." Proceedings of the 2016 International Conference on Microelectronics, Computing and Communications (MicroCom). IEEE, 2016.

Rodellar, J., et al. "Image processing and machine learning in the morphological analysis of blood cells." *International Journal of Laboratory Hematology* 40 (2018): 4653.

Salman, Muhammad, et al. "Artificial intelligence in bio-medical domain: an overview of AI based innovations in medical." *International Journal of Advanced Computer Science and Applications* 8 8 (2017): 319327.

William, Wasswa, et al. "A review of image analysis and machine learning techniques for automated cervical cancer screening from pap-smear images." *Computer Methods and Programs in Biomedicine* 164 (2018): 1522.

Xu, Jia, et al. "Translating cancer genomics into precision medicine with artificial intelligence: applications, challenges and future perspectives." *Human Genetics* 138, 2 (2019): 109124.

Yu, Kun-Hsing, Andrew L. Beam, and Isaac S. Kohane. "Artificial intelligence in healthcare." *Nature Biomedical Engineering* 2, 10 (2018): 719.

CHAPTER 6

APPLICATIONS OF ARTIFICIAL INTELLIGENCE IN BIOMEDICAL ENGINEERING

PUJA SAHAY PRASAD[1*], VINIT KUMAR GUNJAN[2], RASHMI PATHAK[3], and SAURABH MUKHERJEE[4]

[1]*Department of Computer Science & Engineering, GCET, Hyderabad, India*

[2]*Department of Computer Science & Engineering, CMRIT, Hyderabad, India*

[3]*Siddhant College of Engineering, Sudumbre, Pune, Maharashtra, India*

[4]*Banasthali Vidyapith Banasthali, Rajasthan, India*

[*]*Corresponding author. E-mail: puja.s.prasad@gmail.com*

ABSTRACT

Artificial intelligence (AI) can now be very popular in various healthcare sectors. It deals with both structured and unstructured medical data. Common AI techniques in which machine learning procedures are used for structured data are neural network and the classical support vector machine (SVM) as well as natural language processing and modern deep learning for unstructured data. The main disease areas where AI tools have been used are cancer, cardiology, and neurology. The development of pharmaceuticals via clinical trials can take more time even decades and very costly. Therefore, making the process quicker and inexpensive is the main objective of AI start-ups. AI thus has a wide application in the field of biomedical engineering. AI can also help in carrying out repetitive tasks, which are time-consuming processes. Tasks such as computed tomography (CT) scans, X-ray scans, analyzing different tests, data entry, etc. can be done faster and more precisely by robots. Cardiology and radiology and are two such areas where analyzing the amount of data can be time-consuming and overwhelming. In fact, AI will transform healthcare in the near future. There are various health-related apps in a phone that use AI like Google Assistant, but there are also some apps like Ada Health Companion that uses AI to learn by asking smart questions to help people feel better, takes control of their health, and predicts diseases based on symptoms. As in expert systems, AI acts as an expert in a computer system that emulates the decision-making ability of a human expert. Expert systems like MYCIN for bacterial diseases and CaDET for cancer detection are widely used. In image processing, it is very critical when it comes to healthcare because we have to detect disease based on the images from X-ray, MRI, and CT scans so an AI system that detects those minute tumor cells is really handy in early detection of diseases. One of the biggest achievements is a surgical robot as it is the most interesting and definitely a revolutionary invention and can change surgery completely.

However, before AI systems can be arranged in healthcare applications, they need to be "trained" through data that are generated from clinical activities, such as screening, diagnosis, treatment assignment, and so on, so that they can learn similar groups of subjects, associations between subject features, and outcomes of interest. These clinical data often exist in but not limited to the form of demographics, medical notes, electronic recordings from medical devices, physical examinations, and clinical laboratories. AI has been intended to analyze medical reports and prescriptions from a patient's file, medical expertise, as well as external research to assist in selecting the right, separately customized treatment pathway. Nuance Communications provides a virtual assistant solution that enhances interactions between clinicians and patients, overall improving patient experience and reducing physician stress. The platform enables conversational dialogue and prebuilt capabilities that automate clinical workflows. The healthcare virtual assistant employs voice recognition, electronic health record integrations, strategic health IT relationships, voice biometrics, and text-to-speech and prototype smart speakers customized for a secure platform. IBM Medical Sieve is an ambitious long-term exploratory project that plans to build a next-generation "cognitive assistant" that is capable of analytics and reasoning with a vast range of clinical knowledge. Medical Sieve can help in taking clinical decisions regarding cardiology and radiology—a "cognitive health assistant" in other terms. It can analyze the radiology images to detect problems reliably and speedily.

6.1 INTRODUCTION

Artificial intelligence (AI) technology has exposed massive opportunity in the medical healthcare system. The advantages that AI have fueled a lively discussion that whether AI doctors shall in due course replace human doctors in the near future. However, it is very difficult to believe that human doctors will not be replaced by AI machines in likely future, as AI definitely provides assistance to physicians so that they can make better clinical decisions. AI may also or even replace human judgment in certain functional areas of healthcare like radiology. The growing availability of medical healthcare data and fast development of different big data analytic methods have made possible to give recent fruitful applications of AI in medical healthcare. Mining medical records is the most obvious application of AI in medicine. Collecting, storing, normalizing, tracing its lineage are the first steps in revolutionizing existing healthcare systems. Directed by relevant clinical queries, influential AI techniques can solve clinically relevant information concealed in the huge amount of data, which in turn assist medical decision making.

This chapter discusses the current status of AI in healthcare, as well as its importance in future. This chapter has four subsections:

- Motivations of applying AI in healthcare
- AI techniques
- Disease types that the AI communities are currently tackling
- Real-time application of AI
- Future models

6.2 MOTIVATIONS OF APPLYING AI IN HEALTHCARE

A number of advantages of AI have now been widely found in the medical literature. AI can use refined algorithms to "learn" different features from a large volume of healthcare data and then use to build models that help in medical practice. Learning and self-correcting abilities are also equipped to improve its precision based on comments. An AI system can help physicians by giving up-to-date medical evidence from textbooks, journals as well as clinical practices to notify good patient care. One more benefit of an AI system is that it help to decrease therapeutic and diagnostic errors that are predictable in the human medical practice. Furthermore, an AI system extracts valuable information from a large number of patients to help in making real-time implications for health outcome prediction and health risk alert. There are numerous advantages of using AI machines in healthcare. AI service has found its application in many industries due to its advantages. Some of the advantages of AI are as follows:

- AI machines reduce the error in the operations. They are highly accurate and have a high degree of precision.
- Robotics and AI machines have helped in surgical treatments, predicting disease from symptoms, monitoring patients, etc. As AI machines can work in any environment, exploration becomes easier.
- AI machines reduce the risk factor as they can perform tasks that involve risks to the lives of humans.
- AI is a good digital assistant also. It can interact with the user in the form of text, speech, or biometrics. It can perform simple tasks that a human assistant would do.

Medical data before AI systems can be organized in health related applications, these data need to be "trained" using data that are produced from different clinical activities, such as diagnosis, screening, treatment assignment, and so on. The main objective is that they may be able to learn similar groups of subjects, relations between different subject features, and outcomes that they needed. These medical data frequently exist in but not limited to the form of medical notes, electronic recordings, demographics, from medical devices, physical examinations, clinical laboratories, and images. Especially, in the diagnosis stage, a considerable proportion of the AI literature helps in analysing data from genetic testing, diagnosis imaging, and electrodiagnosis.

6.3 AI TECHNIQUES

The AI techniques that are mostly used in health care system are

- machine learning (ML) and
- natural language processing (NLP).

In this section, the main focus is on different AI techniques that have been found useful in all different types medial applications. We categorize them into three subgroups:

- classical ML techniques,
- recent deep learning techniques, and
- NLP methods.

Classical ML builds data analytical algorithms that extract features from different set of data. Inputs to ML algorithms include qualities or traits of patient "traits" as well as sometimes medical results or outcomes of interest. A patient's traits include disease history, age, allergies gender, and so on, as well as disease-specific data like gene expressions, diagnostic imaging, physical examination results, EP tests, medications, clinical symptoms, etc. Not only the traits but also patients' medical results are often collected in medical research. This consists of patient's survival times, disease indicators, and quantitative disease levels like tumor sizes. ML algorithms can be divided into two major classes:

- unsupervised learning and
- supervised learning.

Unsupervised learning is well known for reducing dimension and clustering, while supervised learning is suitable for classification and regression. Semi-supervised learning is hybrid between unsupervised learning and supervised learning, which is suitable for circumstances where the result is not present for certain subjects.

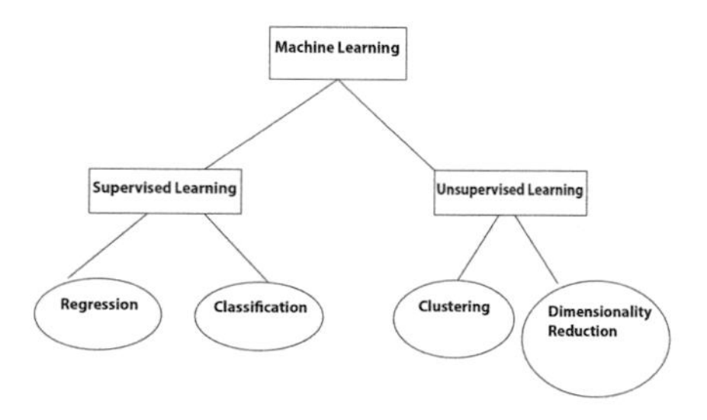

FIGURE 6.1 Different algorithms of machine learning.

The main aim of supervised learning is to classify data and to train the data for a particular outcome. Regression and classification are the two main methods for supervised learning. The best example of supervised learning is used by a cardiologist for the automated analysis of the EKG, in which pattern recognition is performed to select from a limited set of diagnoses, which is a classification task. In the field of radiology, automatic detection of a lung knot by seeing chest X-ray is also supervised learning. In this, the computer is approximating what a trained doctor is already done with high accuracy.

Supervised learning is also used to estimate risk. Framingham risk score for coronary heart disease can predict the 10-year cardiovascular risk, and it is a supervised gender-specific ML algorithm. Principal component analysis (PCA) and clustering are two major methods of unsupervised learning. Regression is a significant statistical method for the analyzing medical data. It helps in identification and description of relationships among different multiple factors.

Clustering is the method of grouping subjects having similar characters together into groups, without using the outcome information. Clustering algorithms output the cluster labels for the patients through minimizing and maximizing the similarity of the patients within and between the clusters. General clustering algorithms include hierarchical clustering, k-means clustering, and Gaussian mixture clustering. PCA is mainly for dimension reduction, especially when the trait is recorded in a large number of dimensions, such as the number of genes in a genome-wide association study. PCA schemes the data onto a few PC directions, without losing too much information about the subjects. Sometimes, one can first use PCA to decrease the dimension of the data and then using clustering algorithm to cluster the subjects. While supervised learning reflects the subjects' consequences together with their behaviours it drives through an assured training process to define the best yields or output that are associated with the corresponding inputs and that are closest to the results on average. Usually, the output designs vary with the results of interest. For example, the outcome can be the expected value of a disease level or probability of receiving a particular clinical event, or the expected survival time.

It is seen that compared to unsupervised learning, supervised learning delivers more clinically applicable results, and this is the reason supervised learning is more popular AI applications in healthcare. Significant techniques comprise logistic regression, naïve Bayes, linear regression, decision tree, random forest, discriminant analysis, nearest neighbour, SVM, and neural network.

6.3.1 DEEP LEARNING

Deep learning is the extension of classical neural network. Deep learning has the ability to discover more complex nonlinear patterns in the data. Another cause for the fresh popularity of deep learning is due to the increase in complexity as well volume of data. Deep learning is different from the typical neural network, it consists of more hidden layers so that the algorithms can lever complex data having various structures. Convolutional neural network (CNN) is the most commonly used in after that recurrent neural network, and deep neural network. The convolutional neural network is established in viewing of the ineffectiveness of the classical ML algorithms when handling complex and high dimensional data, means data having number of qualities. Traditionally, the ML algorithms are planned to analyze data when the number of traits is small. However, the image data are obviously high-dimensional because individually image contains thousands of pixels as traits. Dimension reduction is one solution, but there is a possibility of loss of information in the images by using ML algorithms. However, heuristic feature selection procedures may lose information in the images. Unsupervised learning methods such as PC analysis or clustering methods can be cast-off for data-driven reducing dimension. CNN was first proposed and supported for the analysis of high-dimensional images. The inputs for CNN are the appropriately normalized pixel values on the images. CNN then shift the pixel values in the image from side to side weighting in the convolution layers as well as sampling in the subsampling layers otherwise. The final output is a recursive function of the weighted input.

6.3.2 NATURAL LANGUAGE PROCESSING

The main focus of NLP is to process the narrative text into machine understandable form. A number of clinical data or information like physical examination, laboratory reports, discharge summaries, and operative notes are incomprehensible and unstructured for the computer program. In this type of unstructured data, NLP's main targets is to extract meaningful information from this narrative text so that clinical decision will become easy. ML algorithms are more useful in case of genetic data and EP as these data are easily understandable by the machine for quality control processes or preprocessing. So, the main aim of NLP is to assist in decision making by processing the narrative text. Two main components of NLP are

(a) text processing and
(b) classification.

In text processing, disease-significant keywords are identified based on historical data. After that, a keyword subset is selected through inspecting their effects on the classification of the abnormal and normal cases. The confirmed keywords then move in and enrich the clinical structured data to assist in medical decision making. The NLP pipelines have been established to assist medical decision making on monitoring adverse effects, alerting action arrangements so on. On introducing NLP to analyze the X-ray reports of chests, it would help the antibiotic assistant system to alert doctors for the need for anti-infective treatment. Laboratory-based adverse effects canbe also automatically monitored by using NLP. NLP also helps to diagnose diseases. For example, 14 variables that are associated with the cerebral aneurysm disease are found to be successfully used for classifying persons with cerebral diseases and normal persons. NLP also used to mine the outlying arterial disease-associated keywords from narrative clinical notes. This keyword is then used for classification between the normal person and the patients with peripheral arterial disease having 91% accuracy.

6.4 DISEASE TYPES CURRENTLY TACKLED BY AI COMMUNITIES

Stroke is a frequently occurring disease that affects more than 500 million persons worldwide. China and North America are one of the leading countries having high death rate due to strokes. Medical expenses due to stroke are also high ad put heavy burden on families and countries. So, research on avoidance as well as treatment for stroke has good significance. AI methods have been used in more and more in stroke-related studies as it is one of the main causes of death throughout the country. Three main areas of strokes are predicting the disease by early diagnosis, cure, and outcome prediction as well as prognosis evaluation. Though there is a lack of finding or judgement of early stroke symptoms, only a few patients might receive treatment on time. For predicting early stroke, movement-detecting devices are already present. PCA and genetic fuzzy finite state ML algorithm were implemented into the device for building solutions. The detection procedure included a stroke-onset detection stage and human movement recognition stage. The movement recognition stage helps to recognize abnormal behavior having movement different from the normal pattern. Collecting data about pathological gaits is also helpful for predicting a stroke. Hidden Markov models and SVM are used here, and they could appropriately classify 90.5% of the subjects to the correct group. MRI and CT are good neuroimaging techniques for disease evaluation. In some of the literature works, it was found that apply ML methods to neuroimaging is useful for diagnosis. Some used support vector machine for MRI data, which helps in identifying end phenotypes of motor disability after stroke.

Some researchers also use three-dimensional CNN for finding lesion segmentation in brain MRI (multi-modal). In this, for postprocessing of CNN segmentation maps, a fully conditional random field is used. Gaussian process regression is also used for stroke anatomical MRI images. ML is also used to analyze the CT scan of patients.

After stroke, a free-floating intraluminal thrombus may form as a lesion, which is hard to notice with a carotid plaque in CT imaging. For this, a researcher uses three ML algorithms for classification of these two kinds by the quantitative shape, including SVM, linear discriminant analysis, and artificial neural network. Treatment using ML has been useful for analyzing and predicting the stroke treatment performance. Intravenous thrombolysis outcome has strong relationship with prognosis and the survival rate. Some researchers use SVM for predicting whether patients having tPA cure would mature a symptomatic intracranial hemorrhage by the CT scan. For this, full-brain images are used as the input into the SVM, which achieved more improvement than the conventional radiology-based approaches. Another researcher proposed a stroke treatment model to improve the medical or clinical decision-making procedure of tPA cure. Another stroke treatment model for analyzing practices strategies, clinical trials, and meta-analyses by using the Bayesian belief network. This model consists of 57 different variables and three conclusions to analyze the process of diagnosis, treatment, and outcome prediction. One researcher subgroup analyzes using interaction trees to discover suitable tPA dosage formed on patient features, taking into account both the risks of treatment efficacy and bleeding. Prognosis evaluation and outcome prediction are the factors that can affect disease mortality and stroke prognosis. Compared to the conventional approaches, ML procedures have benefits to improving forecast performance. For better support medical decision-making processes, a model is developed that analyzes the three-month treatment result by analyzing physiological factors during 48 hours after stroke by using a method called logistic regression. Some researchers accumulated a database of clinical infomation of 110 patients with acute posterior and anterior circulation stroke who went for intra-arterial therapy.

6.5 REAL-TIME APPLICATIONS OF AI

The increasingly growing number of applications of ML in healthcare allows us to glimpse at a future where data, analysis, and innovation work hand-in-hand to help countless patients without them ever realizing it. Soon, it will be quite common to find ML-based applications embedded with real-time patient data available from different healthcare systems in multiple countries, thereby increasing the efficacy of new treatment options that were unavailable before.

6.5.1 ELECTRONIC HEALTH RECORDS

Handling medical records and other medical data are very important part of the medical field. Clinical records are one of the significant parts for managing patients. Two reasons are necessary to put records safely. It helps in evaluating the patients and to plan treatment protocols, important for the doctor so that they properly maintain the clinical records of the number of patients. One more advantage is that it also assists the legal system that trusts mostly on documentary proofs in cases of any medical negligence. So, it is important that clinical records should be properly preserved and written to aid the attention of the doctor and his patient.

A digital form of a patient's prescription chart named electronic health records (EHRs) are patient-cantered, real-time records that prepare the available information securely and instantly to the authorized users. An EHR contains a patient's diagnoses, medical history, treatment plans, medications, treatment history of patients, laboratory and test results, radiology images, immunization dates. An EHR system allows access to the evidence-based practice network tools that help in taking decision for patients. It also streamlines and automates provider workflow.

One of the significant characteristics of an EHR is that facts related to patient can be made and managed by certified or authorized providers in a proper digital format that can be shared with additional providers across more than one health care organization. The key aim of EHRs is made to share information with other healthcare organizations and providers—like specialists, medical imaging facilities, laboratories, pharmacies, emergency facilities, schools, and workplace clinics. The main aim behind is that they encompass information from all clinicians that are involved for patient's care.

EHRs, as a very large and networked healthcare delivery system, are regularly observed as monolithic, inflexible, costly to configure, and difficult to use. They are practically obtained from commercial vendors and require significant time, consulting assistance, money, and support for implementation.

The most popular systems are often built around older underlying technologies, and it often shows in their ease of use. Many healthcare systems find these systems complex and difficult to navigate, and it is rare that the EHR system is a good fit with their preferred care delivery processes.

As delivery networks grow and deploy broad enterprise EHR platforms, the challenge of making them help rather obstruct clinicians is increasing. Clinicians' knowledge extends far beyond their clinical domain—care procedure knowledge, patient context knowledge, administrative process knowledge—and it is rare that EHRs can capture all of it efficiently or make it easily available (Simpson and Demner-Fushman, 2012).

Introducing AI to the existing EHRs makes the EHR system more effective, intelligent, and flexible (Bernstam et al., 2010). A number of works are introducing AI capabilities in the medical field, but this requires more work in this direction. The text data of individual patients, their specific illnesses, as well as treatment methods for those diseases are in a developing stage to use the ML algorithm and make more effective

Biotech and pharmaceutical groups already have the pressure to use resources as proficiently as possible. This pressure forces them to search for any opportunity to streamline processes that are more secure, act faster, as well as of low costs. Many life science companies are targeting biologics and precision medicine therapeutics, with focus shifting toward smaller, geographically distributed patient segments (Patel et al., 2009; Wilson, 2011). This shift has resulted in an increasing mandate to capture data from previously unavailable, nontraditional sources. These sources include mobile devices, IoT devices, and in-home and clinical devices. Life science companies merge data from these sources with data from traditional clinical trials to build robust evidence around the safety and efficacy of a drug. Clinical decision an important decision support, which endorses treatment strategies and was rule-based and generic in the past (Zeng et al., 2010). To enable more personalized care, companies such as Change Healthcare, IBM Watson, and Allscripts are using the learned data obtained from different clinical sources.

The main aims to introduce AI in an EHR system are to personalize treatment and in data discovery and extraction as well as to make it more friendly. This is very complicated and hard to use and often seen as contributing to doctor's burnout. Nowadays, customizing EHRs to make it easier for doctors is mainly a manual and time-consuming process; as also, systems' rigidity is an actual obstacle in their improvement. AI and ML specifically could help EHRs to continuously get used to users' choices, improving both outcomes and clinicians' life (Alexander, 2006).

Most current AI choices are "encapsulated" as separate aids and are not available as integrated ones and also require physicians to learn working of new interfaces. However, nowadays, EHR vendors are beginning to add AI capabilities to make their systems easier to use. Firms like Epic, Cerner, All Scripts, and Athena are adding capabilities like NLP and ML for clinical decision support, integrating with telehealth technologies and automated imaging analysis. This will provide integrated interfaces, access to data held within the systems, and multiple other benefits—although it will probably happen slowly (Lymberis and Olsson, 2003).

Future EHRs should also be developed with the integration of telehealth technologies in mind (as is the EHR at One Medical). As healthcare costs rise and new healthcare delivery methods are tested, home devices such as glucometers or blood pressure cuffs that automatically measure and send results from the patient's home to the EHR are gaining momentum. Some companies even have more advanced devices such as the smart t-shirts of Hexoskin that can measure several cardiovascular metrics and are being used in clinical studies and at-home disease monitoring. Electronic patient reported outcomes and personal health records are also being leveraged more and more as providers emphasize on the importance of patient-centered care and self-disease management; all of these data sources are most useful when they can be integrated into the existing EHRs.

Most delivery networks will probably want to use a hybrid strategy—waiting for vendors to produce AI capabilities in some areas and relying on a third party or in-house development for AI offerings that improve patient care and the work lives of providers. Starting from scratch, however, is probably not an option for them. However necessary and desirable, it seems likely that the transition to dramatically better and smarter EHRs will require many years to be fully realized.

6.5.2 ROBOTIC APPLICATIONS IN MEDICINE

A wide range of robots has been developed for providing helps in different roles within the clinical environment. Two main types of robots called rehabilitation robots and surgical robots are nowadays specializing in human treatment. The field of therapeutic robotic and assistive robotic devices is also growing rapidly. These robots help patient to rehabilitate from severe or serious conditions like strokes; empathic robotsalso help in the care of mentally/physically challenged old individuals, and in industries robots help in doing a large range of routine tasks, such as delivering medical supplies and medications and sterilizing rooms as well as equipment. The areas where robots are engaged are given below.

- ***Telepresence***

In telepresence, doctors uses robots to help them to examine and treat patients in remote locations and rural areas. Consultants or specialist can be on call use robots to answer health-related questions and also guide therapy from far or remote locations. These robotic devices have navigation capability within the electronic record and built-in sophisticated cameras for the physical examination.

- ***Surgical Assistants***

Assisting surgical robots are very old and present since 1985 for remote surgery, also called unmanned surgery and minimally invasive surgery. Robotic surgical assistance has too many advantages including decreased blood loss, smaller incisions, quicker healing time, less pain, as well as ability to pinpoint positions very precisely. In this type of surgery, the main trademark is that remote-controlled hands are manipulated by an operating system while seating outside the operation theatre. Further applications for surgical-assistant robots are constantly being developed to give surgeons more enhanced natural stereovisualization with augmented technology as well as spatial references required for very complex surgeries.

- ***Rehabilitation***

People having disabilities, including improved mobility, coordination, strength, and low quality of life are assisted by rehabilation robots that can be automated to adapt to the situation or condition of each patient separately as they recover from traumatic brain strokes, spinal cord injuries, or neurobehavioral as well as neuromuscular diseases like multiple sclerosis. Virtual reality having integrated by rehabilitation robots also improves gait, balance as well as motor functions.

- ***Medical Transportation Robots (MTRs)***

This type of robots supply meals and medications transported to staff and patients. By using MTRs, there is optimized communication between hospital staff members, doctors, and patients. Self-navigation capability is one of the important characteristics of this type of robots (Horvitz et al., 1988). There is still a need for extremely advanced as well as indoor navigation property that is based on sensor fusion location technology so that navigational capabilities of transportation robots become more robust.

- ***Sanitation and Disinfection Robots***

These robots can disinfect a clinic or room containing viruses and bacteria within minutes. With the outbreaks of infections as well as increase in antibiotic-resistant strains like Ebola, there is a requirement of more healthcare facilities having robots to disinfect surfaces and clean surfaces. Presently, the major methods that are used for disinfection are hydrogen peroxide vapors and ultraviolet light, so there is a need to introduce robots in this area.

- ***Robotic Prescription Dispensing Systems***

The major advantages of robots include accuracy and speed, and these are two features that are very important for pharmacies (Reggia and Sutton, 1988; Patel et al., 1988). robots in automated dispensing systems handle liquids, powder, as well as highly viscous materials with much higher accuracy and speed.

6.5.3 VIRTUAL NURSES

AI is a very big driving force that started to make changes the way we do our day-to-day activities. Virtual nurses are one of the examples of application of AI in nursing job as there is always a shortage of nursing staff. Due to a lack of time and long distance from the hospital, people can get assistance and medical advice immediately through video conferencing, phone, e-mail, chat as well as instant gratification.

Healthcare applications combined with AI can save healthcare economy of United States as much $150 billion by five years, and among them, virtual nursing aides could save around $20 billion.

There is a big challenge for healthcare business to upgrade commitment and care quality to the patients by working together with developers to make solutions using AI that could be easily set up into the care delivery process.

Nowadays, people are too busy in their day-to-day activities, so requirement for home services for routine check-ups are in more demand. In this busy schedule, people want that nurses should visit their homes and provide services. Medical assistant apps may be very useful in this type of situations. AI apps learn the behavior of nurses and medical staff and give them proper required care and also connect them with the proper care providers. Virtual Nurse is an AI that can provide you with useful and educational information on hundreds of illnesses, injuries, and medications. Medications are now listed on Virtual Nurse. Virtual Nurse can tell a person about illness and also give information about it. We can find detailed information about medications including side effects and instructions. We can also get first-aid advice, also ask for medical advice, and know exactly what to do in your next emergency with Virtual Nurse on your devices.

6.5.3.1 CHARACTERISTICS OF VIRTUAL NURSE

- Helps in assisting answering medical and health and wellness-related questions.

- This also gives reminders for appointments/follow-ups and medication administration.
- Symptom checkers are also present, which help to diagnose related medical problems by enquiring symptoms and also search symptoms by using medical dictionary and reach at probable diagnosis.
- Virtual Nurse also provides cardiopulmonary resuscitation (CPR) instructions to the person using it.
- This provides information about what will be required for an expecting mother at an every stage of pregnancy in terms of food, medicine, and fluid.
- It also instructs patients about the side effects caused by an active drug and also advices to cope with the drug.
- It helps patients by using visual and audio aids to do basic nursing work like changing bandages.
- It also books an appointment of a general practitioner.
- It improve treatment observance rates, by 24 × 7 availability.
- It helps to make lifestyle change as well as make corrections to encourage healthier living.
- It ensures data collection, helps in analysis as humans just cannot make so many calls, records abundant information, and conducts analysis.

Chronic diseases like heart failure, cancer, diabetes, and chronic obstructive pulmonary disease have a continuous presence in person's life. One of the virtual avatar apps monitors patient's health, also helps patient with hospital readmission revisits, etc. It also helps to get medicinal instructions after interacting with doctors (Tiwana et al., 2012; Poon and Vanderwende, 2010). For helping patients that have chronic diseases, virtual nursing apps could be a blessing for patients who spend their lifespan in and round healthcare facilities.

AI-based virtual Nurse application is compatible to work with various different age groups. This app is based on a rule-based engine having algorithms based on normally acknowledged medical protocols for diagnosing and dealing with specific chronic diseases. The protocols and contents for the app can be delivered by partner and clinic hospitals. The patient's mood and modulations are used to qualify the app to answer. Such apps are being developed to collect signals about a patient's health and help as an unnamed database on symptoms. One of the benefits of this app is that these are able to integrate emotions analytic tools, conversational chat and voice assistants, and emotion recognition tools into the artificial platform.

The virtual avatar app of nurses can also be programmed to perform detailed counselling of a behavioral health problem. This type of app talks to patients through a smartphone about their health condition. The patient has no need to type: they only talk to the virtual avatar about their health condition, then this conversation can be rolled and transcripted to record, and afterward it would be reviewed by the health provider. The virtual avatars will speak to person empathetically and naturally, which might also benefit people who are ailing elderly and having chronic diseases.

By hearing unusual voice or detecting unusual emotional tone of a patient such as depression and anxiety, the app does emotional analysis and provides alert to the health provider who may prescribe medications.

6.6 FUTURE MODELS

Research for developing advanced robots continues for an ever-expanding variety of applications in the healthcare area. For instance, a research team led by Gregory Fischer is developing a high-precision, compact surgical robot that will operate only within the MRI scanner bore, as well as the software and electronic control systems that are present in it, to improve the accuracy of prostate biopsy.

In this, the main aim is to develop a robot for an MRI scanner that can work inside it. However, there are number of physical challenges for placing a robot inside the scanner, as scanner uses a powerful magnet, so it is necessary that the robot should be made up of nonferrous materials. Most of the technical difficulties have already been overcome by this team. Besides this, they need to develop the software interfaces and communication protocols for properly controlling the robot with planning systems and higher level imaging (Holzinger, 2012). For the nontechnical surgical team, the robot must be easily sterilized, easily placed, and easily setup in the scanner, which are also some of the requirements. Because of all this, it is a huge system integration assignment that requires many repetitions of the software and hardware to get to that point.

In other projects, a rehabilitation robot is integrated with virtual reality to expand the variety of therapy exercise, increasing physical treatment effects and motivations. Nowadays, discoveries are being made using nanomaterials and nanoparticles. For example, in "blood–brain barrier," nanoparticles easily traverse. In the coming future, nanodevices can be filled with "treatment payloads" of required medicine, can be injected into the body, and automatically guided to the exact target sites inside the body. Very soon, digital tools that are broadband-enabled will be accessible, which will use wireless technology to monitor inner reactions to medications. AI is playing a prominent role in the healthcare industry, from wearable technologies that can monitor heart rate and the number of calories you have burnt to apps that combat stress and anxiety. The number of advancements that AI pioneers have made in the healthcare industry is nothing short of spectacular. Are you wondering who are the innovators and creators behind the technology that is vastly changing the medical industry? Well, here are the top 10 companies that are using their advanced technology to make a positive difference in healthcare.

6.6.1 GOOGLE DEEP MIND HEALTH

Google is obviously one of the most renowned AI companies in the world. The AI division of the company is focusing more on applying AI models to advance healthcare. These new models are capable of evaluating patient data and forming diagnoses. Google believes that their technology will be able to develop entirely new diagnostic procedures in the near future.

6.6.2 BIO BEATS

Bio Beats is a leading app developer that specializes in improving patient well-being. Their apps focus on dealing with stress, which has a debilitating effect on health. Bio Beats uses sophisticated AI algorithms to better understand the factors that contribute to stress and the steps that people need to take to mitigate it. Their app is programmed to help people develop better resilience to stress, which will help improve their long-term health. The app offers a variety of features, which include helping promote better sleep, encouraging more physical activity, and conducting a regular assessment of your mood.

6.6.3 JVION

Jvion has a similar approach to Bio Beats. It is a company that focuses on improving long-term physical health by addressing stress and other cognitive factors.

According to their website, the Jvion Machine has been used to help over two million patients. While the machine uses AI to identify physical risk factors that were previously undetected, the majority of risk factors are cognitive. The technology used to run this machine has found interesting correlations between various psychological factors and physical ailments. One

54-year-old man was suffering from deep vein thrombosis. He was at the risk of exacerbating the problem to become an embolism. Fortunately, the machine was able to identify the socioeconomic problems that were contributing to his disease. The doctors were able to develop a customized treatment plant to help save him.

6.6.4 LUMIATA

Lumiata is a company that offers a unique cost and return analytics technology that is used to improve the efficiency and cost-effectiveness of healthcare. While most other cutting-edge AI companies focus on developing AI models to improve the effectiveness of healthcare diagnostics, this company focuses on the financial side of the equation.

Lumiata is a relatively new company, but it is already making a splash in AI-based health care. They secured $11 million in funding from several prominent investors this summer. The company is expected to get even more funding in the future as a growing number of investors see opportunities to use deep learning to improve healthcare analytics. This should significantly change the economic models behind the healthcare industry.

6.6.5 DREAMED

Managing diabetes can be incredibly frustrating and complex. The biggest problem is that every diabetic patient's needs are different. They have to account for gender, weight, and metabolic factors. Their diet also plays a key role in their diabetes management plan.

DreaMed is a new company that has found new ways to improve the delivery of diabetes management services. It uses carefully tracked biometrics a data to help develop a custom treatment plan for every patient.

6.6.5 HEALINT

Healint is a company that focuses on helping patients manage chronic diseases. Their technology uses a wide range of data sources, which include information from mobile phones, wearable fitness devices, and patient records. This technology is able to provide some of the timeliest healthcare services available.

6.6.6 ARTERYS

Arterys is one of the most sophisticated AI solutions for medical imaging. This technology leverages cloud technology to perform some other fastest medical computations imaginable. The medical imaging analytics platform is known for its unprecedented speed and quality of imaging.

6.6.7 *ATOMWISE*

Atomwise is helping pharmaceutical companies make breakthroughs in rolling out new drugs. They have already helped companies find a number of new drugs much more quickly.

The company is forming partnerships with other leading healthcare organizations to use AI to improve healthcare models. Last month, Atomwise entered into a relationship with Pfizer. A few months ago, they sponsored a new project that is helping over 100 academic institutes around the world to develop better data.

6.6.8 *HEALTH FIDELITY*

Processing payments is a key aspect of any healthcare system. Unfortunately, some healthcare payments involve a higher level of risk than others. Health Fidelity uses data to help mitigate the risks of taking these types of payments.

6.6.9 *GINGER.IO*

Ginger.io is one of the most cutting-edge data-driven solutions to improving mental health. They keep track of different patient inputs to establish causal links between various mental health issues and create custom patient plans.

6.6.9 *IBM WATSON SYSTEM*

IBM Watson System is good pioneer in this field. The Watson System includes both ML and NLP.

Processing modules have made encouraging progress in the field of oncology. Treatment recommendations for cancer patients from Watson are about 99% coherent with the doctor decision. For AI genetic diagnostics, Watson collaborated with the Quest Diagnostic for offering better solution. In addition, the collaboration shows the impact on real-life clinical practices. By analyzing the genetic code in Japan, Watson effectively identified the rare leukemia (secondary) that is triggered through myelodysplastic syndromes.

CC-Cruiser is a web-based application that supports proposals for dealing with ophthalmologists as it enables high-quality medicinal care as well as individualized treatment for the person of developing areas. Besides, this software can also be used in giving teaching activities ophthalmology junior students. One more research prototype called cloud-based CC-Cruiser connects AI system back end clinical activities with front end input data. All the clinical data like images, blood pressure, genetic results medical notes, and so on and demographic information like age, sex, etc. are collected into the AI system. By using this information, AI app results some suggestion and then these suggestions are sent to the physician to assist in clinical decision making. Feedback related to suggestions like wrong or right will also be collected together and input back into the AI system as a result to keep improving accuracy.

6.7 CONCLUSION AND DISCUSSION

This chapter gives motivation of using AI for healthcare organization. By using various categories of healthcare data. AI has analyzed and charted the key diseases that take advantages of AI. After that, the two most classical techniques of AI called SVM and neural networks and modern deep learning techniques discuss the process of developingdifferent AI apps for healthcare industry. For developing any successful AI app, the ML component is used for handling different categories of data like genetic data as well as the NLP component is used for mining or handling unstructured data. The healthcare data is used for training using more sophisticated algorithms before giving apps or systems to assist physicians for diagnosing diseases and providing medical treatment suggestions.

KEYWORDS

- **principal component analysis**
- **naïve Bayes**
- **linear regression**
- **decision tree**
- **random forest**
- **NLP**

REFERENCES

Alexander, C.Y., 2006. Methods in biomedical ontology. *Journal of Biomedical Informatics*, *39*(3), pp. 252–266.

Bernstam, E.V., Smith, J.W. and Johnson, T.R., 2010. What is biomedical informatics. *Journal of Biomedical Informatics*, *43*(1), pp. 104–110.

Holzinger, A., 2012. On knowledge discovery and interactive intelligent visualization of biomedical data. In *Proceedings of the International Conference Int. Conf. on Data Technologies and Applications, DATA* (pp. 5–16).

Horvitz, E.J., Breese, J.S. and Henrion, M., 1988. Decision theory in expert systems and artificial intelligence. *International Journal of Approximate Reasoning*, *2*(3), pp. 247–302.

Lymberis, A. and Olsson, S., 2003. Intelligent biomedical clothing for personal health and disease management: state of the art and future vision. *Telemedicine Journal and e-Health*, *9*(4), pp. 379–386.

Patel, V.L., Groen, G.J. and Scott, H.M., 1988. Biomedical knowledge in explanations of clinical problems by medical students. *Medical Education*, *22*(5), pp. 398–406.

Patel, V.L., Shortliffe, E.H., Stefanelli, M., Szolovits, P., Berthold, M.R., Bellazzi, R. and Abu-Hanna, A., 2009. The coming of age of artificial intelligence in medicine. *Artificial Intelligence in Medicine*, *46*(1), pp. 5–17.

Poon, H. and Vanderwende, L., 2010, June. Joint inference for knowledge extraction from biomedical literature. In *Human Language Technologies: The 2010 Annual Conference of the North American Chapter of the Association for Computational Linguistics* (pp. 813–821). Association for Computational Linguistics.

Reggia, J.A. and Sutton, G.G., 1988. Self-processing networks and their biomedical implications. *Proceedings of the IEEE*, *76*(6), pp. 680–692.

Simpson, M.S. and Demner-Fushman, D., 2012. Biomedical text mining: a survey of recent progress. In *Mining text data* (pp. 465–517). Springer, Boston, MA, USA.

Tiwana, M.I., Redmond, S.J. and Lovell, N.H., 2012. A review of tactile sensing technologies with applications in biomedical engineering. *Sensors and Actuators A: Physical*, *179*, pp. 17–31.

Wilson, E.A., 2011. *Affect and Artificial Intelligence*. University of Washington Press.

Zeng, D., Chen, H., Lusch, R. and Li, S.H., 2010. Social media analytics and intelligence. *IEEE Intelligent Systems*, *25*(6), pp. 13–16.

CHAPTER 7

BIOMEDICAL IMAGING TECHNIQUES USING AI SYSTEMS

A. AAFREEN NAWRESH[1*] and S. SASIKALA[2]

[1]*Department of Computer Science, Institute of Distance Education, University of Madras, Chennai, India*

[2]*Department of Computer Science, Institute of Distance Education University of Madras, Chennai, India*

[*]*Corresponding author. E-mail: anawresh@gmail.com*

ABSTRACT

Artificial intelligence (AI) can be defined as the one that makes machines to think, work, and also achieve tasks that generally humans do. AI built for medical diagnosing systems has promised to work at the rate of performing 10^{25} operations per second. AI will be helpful in providing much unconditional help that usually takes long hours/day to be done by humans. AI will automate a heavy amount of manual work and speed up the processing. The digital world would experience simple yet powerful systems that make tasks done in a jiffy. AI not only enhanced medical diagnosis but also groomed the relevant areas by developing applications in biotechnology, biophysics, bioinformatics, genomics, and so on. One steady development is seen in medical imaging systems since there is an increase in the requirement that is satisfied such that the AI-based systems become highly valuable. AI will help doctors, surgeons, and physicians to see more and do more at an earlier stage so that patient outcomes can be improved. AI-incorporated systems and applications would surely bring a great change in the field of the medical world and its evolutionary practices.

7.1 INTRODUCTION

Biomedical imaging focuses on capturing the images useful for both diagnostic and healing purposes. Imaging techniques provide unique important details about tissue composition, skeletal damage, and breakage, and also an explanation of many elemental biological procedures. In recent time, biomedical imaging science has transformed into a diverse and logical set of information and model, and it has attained a position of central importance in the field of medical research.

There were times when people used to travel long distances from their village or town to take the blood test, scans, and furthermore wait for the reports to come that used to take a week or even a month. Magnetic resonance imaging (MRI) machines were less available in the nearby hospitals, and many hospitals do not favor running them continuously for 24-h in a day due to huge electricity bills. Computed tomography (CT) scans were used only in emergency situations, in the case of trauma or accidents. A CT scan generally costs a little more than the existing X-ray and also takes time in producing results. On the contrary, to get a clear and deep image of the tissues, nerves, a full-body MRI machine was invented, which was named "indomitable," that means impossible to defeat. It sure by name is the best imaging system since then. In this era of vast development, we can see that there are essential components in a medical device where patients can monitor themselves. Devices are being equipped with advanced techniques that help people perform tasks that were once done only in clinical test centers.

There are several promising applications of AI systems in the medical devices sector, and professionals are looking for an advantage of the impact of this technology. The attempt to build medical devices as more reliable, compact, accurate, and automated is producing a growing interest in finding ways to incorporate AI systems. Medical imaging is an area that is progressively developing and improving devices to support the management and treatment of chronic diseases and will most likely continue to be a major area of focus.

In general, an AI system is a machine that will achieve jobs that a human can perform on their own. The task of achieving or completing objectives depends on the training phase that is given to the systems where it learns to follow certain rules or protocols when input data is given. It also involves targeting self-realization of errors and correcting it to attain proper and accurate results. Artificial intelligence (AI) is an emerging field that invokes researches for human reasoning in AI systems. Researchers have tried to create a major pace in the development of successful AI applications and systems. Apart from creating exquisite machines to make work of capturing the region of interest easier, AI has also entered into creating possibilities of making other areas of research in medicine fascinating. Among the potential and efficient discoveries of AI, the establishment has been in the field of medicine, the various products discovered in the field of biomedical research, transactional research, and clinical practice are given in the figure below (Figure 7.1).

The scope involved in making tasks easier is by establishing an interface between the user and the application/system. This ensures a promising platform for making adaptability to the tech world. Some of the applications that made tasks easier in the domain of medical science and research are being discussed in the below section.

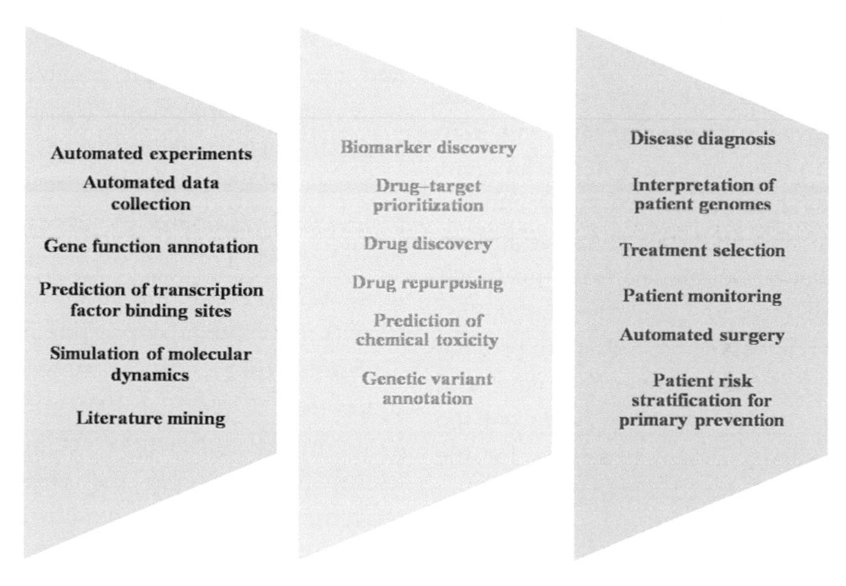

FIGUE 7.1 Applications of AI in medicine.

7.1.1 BASIC BIOMEDICAL RESEARCH

Automated experiments: Accessing scientific data is clearly a very vast task, which needs efficient and rapid systems. If the systems are equipped then there is an efficient and quick recording of data allowing experiments to perform exercises effectively. During the development phase, scientists and researchers augmented by computers were able to retrieve information from the data by creating an interface or bridge between science and data science. And at this point of time, the complete size, dimensionality, and the amount of scientific data have become very enormous that the dependence on the intelligent and automated systems is becoming the need of the hour. Algorithms are predominantly integrated to provide faster accessing and accurate results, the more the requirement, the more efficient the algorithms to provide better discovery in the processing of data.

Automated data collection: AI has started showing a great impact in the field of health care; it is expected to provide machines that will help doctors, nurses, and other technicians to save time on tasks. Techniques such as voice-to-text transcriptions will now help order tests, prescribe medicines, and provide chart notes. IBM'S Watson has started to provide an opportunity to mine data and help the physicians to bring out the best and efficient treatment for the patients. It renders help by analyzing millions of medical papers using natural language processing to provide treatment plans.

Gene function annotation: The major task involved in gene function annotation is to identify enhanced biological subjects, to ascertain the improved function-related gene groups, the group repeated annotation expressions, categorize and display the related many-genes-to-many-terms in a 2D view, search for existing related or similar genes that are not available in the list, define the proteins that interact with one another, provide names of genes in groups, list out the genes connected with diseases, emphasize protein functional domains and their motifs, highlight related literature reviews, and conversion of gene identifiers.

Prediction of transcription factor binding sites: The transcription factors are essential gene regulators that have a characteristic task in the improvement, cell signaling and cycling, and their association with various diseases. There are about thousands of position weight matrices (PWM) that are accessible to choose for the detection of explicit binding sites. This process is mainly used for the prediction based on the PWMs that have false-positive rates.

Simulation of molecular dynamics: Molecular dynamics is a computer simulation technique useful for analyzing the physical movements of atoms and molecules. Molecular dynamics simulation allows the study of complex and dynamic procedures that happen in a biological system. The study includes confrontational changes, protein stability, protein folding, ion transport in biological systems, molecular recognition such as of proteins, DNA, and membranes and also provides an urge to perform the other studies such as drug designing, structure determination of X-ray, and nuclear magnetic resonance.

Literature mining: There is an increase in research studies in the biomedical field every day; there is a need to automate systems that can be helpful to retrieve unknown facts and information from the already published articles and journals. This automated system is known as text mining, which helps in preprocessing of the documents, natural language processing, searching and retrieval of documents, techniques useful for clustering, and classification of the gathered information. Text mining methods will ease the mining of a huge amount of information on a topic from the published research publications/articles and provide a concise conclusion that is impossible to get.

7.1.2 TRANSLATIONAL RESEARCH

Biomarker discovery: Biomarkers are biological descriptions that are efficiently measured and evaluated as indicators of biological procedures, pharmacologic responses to a therapeutic intervention. Biomarkers are effectively used to provide disease occurrences, progress, and effectiveness of the medicinal treatment, patient vulnerability to develop a particular disease, or even predict the usefulness of treatment at the stage of a particular disease. The most popular biomarker is protein molecular biomarker because of the availability of a large range of systematic instrumentation, which are used to identify and calculate proteins in the compound biological sample.

Drug-target prioritization: Target prioritization helps to identify suitable targets like proteins, genes, nonpeptide gene products for classification, since it is the most difficult step in biology especially in annotating the gene function, discovery of drugs, and provide a molecular base on diseases. For a gene, the treatment versus control discrepancy expression level will be determined. To integrate information on the functional associations between proteins, the differential expression values are mapped to the STRING network. Genes are being ranked with respect to their differential expression of their nearest neighbors in the network using the kernel diffusion method. The steps through the network resolve on the size of the neighborhood considered. The correlation diffusion method is used to rank the genes using the differential expression of them using the tough connectivity correlation.

Drug discovery: In the existing research, drugs were discovered either by identifying the active compound from existing remedies or through unexpected or fortunate discovery. At present, new approaches have been made, first, to recognize how the disease and the particular infection can be controlled at the molecular and physiological level and to target the attributes on the information gathered. The process involves recognition of candidates, synthesis characterization, screening, and attempts for therapeutic value. If a compound has revealed its value in the tests, it will start the procedure for the development of drugs before trials.

Drug repurposing: Drug repurposing (also known as drug reprofiling or drug repositioning) is the technique of redeveloping a compound for usage in a different disease. This method is based on the fact that many of the recognized drugs and neglected compounds have already been tested by humans and their relevant information will be made available on their pharmacology, formulation, dosage, and possible toxicity. Drug repurposing has advantages over the established existing drug discovery approaches where it considerably reduces the cost and creation time, also provides a need to undergo clinical trials by humans.

Prediction of chemical toxicity: The prediction of complex toxicities is a significant component of the drug design development procedure. Computational toxicity inferences are not only quicker than the determination of toxic dosage measures in animals, which can also help to decrease the number of animal trials. As these are helpful in providing prediction of compound drug-like features rapidly in the present decision making system of drug discovery, AI technologies are being extensively accepted in creating fast and high throughput in silico analysis. It is accepted that early screening of the chemical attributes can effectively decrease the heavy costs related to late stage failures of drugs due to poor properties. The toxicity prediction procedures and structure–activity correlation depends on the exact estimation and depiction of physicochemical and toxicological properties.

Genetic variant annotation: Genetic variant annotation and its management offer a solution for collecting and annotating genetic data result with the help of a huge impact on public resource domains. The system is efficient in handling thousands of millions of rows, filtering out from the input data of SNPs and genetic variants of interest. The queries that arise are to display genetic variants that are identified to be probable pathogenic from a repository database while limiting the search to only those possibilities present in exon coding regions. They present rare genetic varieties that contain population possibilities that are lesser than 5% in the 1000 genomes when restricting the search variants to display their associations in public sources studies. It is also helpful in annotating the association analysis results, rapidly displaying particular genetic types in a particular interest, also with linkage disequilibrium plots.

7.1.3 CLINICAL PRACTICE

Disease diagnosis: As a well-known factor, diseases like cancer, meningitis, heart failures, brain tumor, lung cancers or strokes, kidney failure, liver sclerosis, and many other infectious diseases are being diagnosed effectively. With the rapid development of diagnostic techniques and systems, people get the information associated with their symptoms on the Internet easily. Provided one must not get confused with the common symptoms and conclude that it could be so and so disease, rather visit a doctor immediately when the symptoms go high. Nowadays, it is a common practice by physicians to give either an antibiotic or a sedative based on the patient's situation and balance their condition for a particular period. One must be aware of the fact that some diseases like meningitis need to be taken care of immediately than wait for severe symptoms to project out, as it has to be checked immediately and correctly using CT scan and lumbar puncture to confirm whether it is caused by a bacteria or virus. The application of machine learning in diagnostics has just begun, where the determined systems access a disease or its growth with the help of multiple data sources such as CT, MRI, genomics and proteomics, patient data, and even handwritten medical chart records. AI systems will be essential in differentiating between normal, cancerous, and malignant lesions/patterns, which will be helpful to the doctors to give an inference from those predictions.

Interpretation of patient genomes: The interpretation of genomic data is an even more complex task than producing and organizing the data. The genome does not signify the identical thing to each person at every stage of time. The implication of particular variations depends on the age, health conditions, and other contextual aspects through many life phases. Using the cancer genomes, the sequence reveals a huge amount of data on the variations that can be used in classification, prediction, and therapeutic management. Therefore, it is practical to know whether in the future the easy, cost-effective, and the efficient assay will be the predictive power of tumor genome sequencing or even sequencing key variants. An important research aspect is whether every cancer is diverse if it is true then it is compulsory to skim through huge amounts of genomic data to know each patient's disease.

Automated surgery: Currently, surgical robots are human–slave machines without data-driven AI. Computer vision systems will soon encompass surgical depth imaging for difficult tasks. The robots will use sensors to accumulate data during a surgical process, and hence, they become semi-autonomous. Particularly, robots must have sensors that will help to analyze the surgical setting and recognize the placement of tools and instruments with respect to the setting in real-time. Surgical automation also requires to understand how and where to navigate the instruments with a good level of accuracy at the time of operation. The three-dimensional (3D) integrated technologies will be used to provide precise depth interpretation of the surgical surface and also estimate the position of the surgical instruments and the camera that is in contact with the surface. Such type of information/facts are required to build a data-driven platform for surgical robots. The information received from computer vision systems will ensure a better digital understanding of the surgical field. The information gathered creates a base for computer-driven robotic surgery, as robots will now be able to see through and comprehend the area where it has been operating. This process surely makes the robot-assisted surgeon to perform the surgery securely without the risk of destructing vital structures caused by accidental crash or unknown anatomy. In the upcoming advancement, surgeons will be able to use robots to perform recurring jobs like providing first aid, suturing, which may provide time for the more essential area of the operations. The patients belonging to rural hospitals who attain fewer facilities will also get benefited from this safe and complication-free surgery.

Patient monitoring: Uninterrupted readings of patient's parameters like blood pressure, respiratory rate, heart rate and rhythm, blood–oxygen diffusion, and other parameters are essential factors to look for in critical patients. Quick and accurate decision making is essential for efficient patient care, where electronic monitoring is done consistently to gather and display physiological information. The data are gathered using harmless sensors from patients at hospitals, medical centers, delivery and labor wards, clinics, or even from patient's homes to sense any unpredicted serious life situations or even to document routine necessary data capably. Patient monitoring is generally defined as frequent or nonstop remarks of observations or measurements of the patient, their physiological functioning, and also the functioning of the life support device, which is used for conducting and managing decisions like therapeutic recommendations and evaluation of those interventions.

Patient risk stratification for primary prevention: Risk stratification helps providers to recognize the accurate level of concern and service for diverse groups of patients. It is generally a process to assign a jeopardy strategy to a patient and then using the accumulated facts to direct care and improvise the health conditions. The main aim of risk stratification is to partition/group patients into separate groups of similar complications and concern needs. The groupings may be classified into highly complex, high-risk, rising-risk, and low-risk patients. Special care modules and strategies are used for every group.

1. *Highly complex*: It is a group containing a small number of patients, which needs intensive care. This group may contain a count of about 5% who has multiple complex illnesses, which may include psychosocial needs or barriers. Care models for this group of patients need exclusive, proactive intensive care management. The objective of this group is to make use of low-cost care management facilities to attain better health results when avoiding high-cost crisis or unwanted acute care service.
2. *High risk*: This group contains 20% population count and includes patients with numerous risk factors that when left unseen will consequence the shifting of patients into a high complex group. This cluster of patients is suitable to hold in a planned care management program that will offer one-to-one support in the supervision of medical, community, and social requirements. A care manager assigned to work for this group ensures that every patient gets appropriate disease supervision and also preventive measures.
3. *Rising risk*: This group consists of patients who suffer from one or more chronic conditions or threats and those who do not have a balanced condition, i.e., sometimes they are fine and sometimes they go down in their conditions. The analysis done on this group showed that rendering care management services to this group of patients will reduce the count of patients who has moved to high-risk groups by 12%, which is a reduction compared to a 10% decrease

in the overall costs. The general risk factors include smoking, obesity, blood pressure, and cholesterol level monitoring. Recognizing these risk issues will provide the team to target the root cause of multiple conditions.

4. *Low risk*: This group belongs to the patients who follow a proper and healthy diet and are stable in condition. The patients in this group have slight conditions that will be managed effortlessly. The main objective of this low-risk model is to maintain the health care system and to keep the patients healthy.

7.2 LITERATURE REVIEW

Biomedical imaging in the field of radiography had initiated the development of X-ray, CT scan, positron emission tomography (PET), MRI, and ultrasound, which produced good visualization of the affected areas.

X-ray: Medical imaging was initiated during the year 1895 when Wilhelm Conrad Roentgen discovered X-ray.[1] X-ray is the basic machine commonly used to analyze bone and chest fractures (Figure 7.2). Many times radiologists found it difficult to correctly analyze the area of interest because of the interference of tissue or muscle mass, so fluoroscopy was used to overcome such situations . In the late 1920s, radiologist gave patients radio-opaque barium to swallow so where the barium traversed as it entered the gastrointestinal tract. This was helpful in analyzing cancers formed in the stomach, esophagus, bowel also ulcers, diverticulitis, and also appendicitis. With the invention of fluoroscopy, many diseases that were diagnosed are now being easily analyzed using a CT scan. X-ray tomography was established in the 1940s where “tomograms” or slices were obtained through tissues without an over or underlying tissue being captured. It was attained by rotating the X-ray tube such that only the required region of interest can be focussed and captured during the rotation of the tube. In the current era, tomography is no longer being used, which is being replaced with CT scans. To get to the point, both CT and MRI are tomography techniques that are useful to display the anatomy in slices than through projections like the working of X-ray. In the late 1950s, a new technique known as nuclear medicine entered the diagnostic imaging procedures. In these procedures, the source of X-rays was not X-ray tubes but radioactive compounds, which emit gamma rays as they start to decay. The test that is generally used today is PET, where the isotopes emit positrons (positively charged electrons) instead of emitting gamma rays. Commonly, PET is based on the positron-emitting isotope of fluorine that is integrated into glucose called fluoro-deoxyglucose.

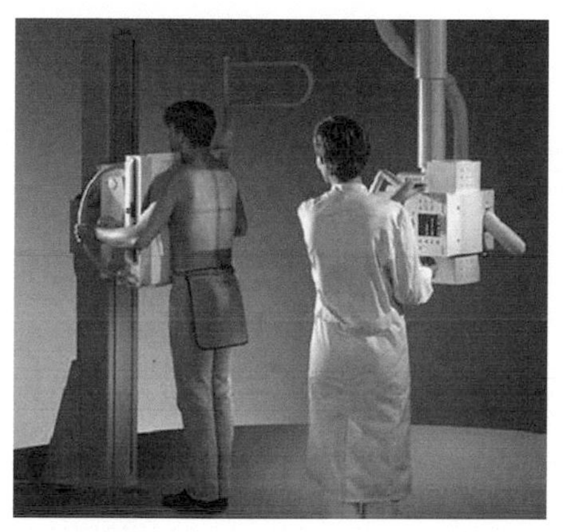

FIGURE 7.2 X-ray procedure of the back.

Computed tomography (CT): Before the initiation of CT in 1973, there were only plane films of the head showing the bones or angiography showing the masses when the vessels of the brain were banished from their original position. Fundamentally, there was no way to directly image the brain. In CT, an X-ray tube rotates around the patient and various detectors pick up the X-rays that are not absorbed, reflected, or refracted as they pass through the body (Figure 7.3). Early CT units produced crude images on a 64 × 64 matrix. Early computers took all night to process these images. Today's multidetector row CTs acquire multiple submillimeter spatial resolution slices with processing speeds measured in milliseconds rather than hours. Iodinated contrast agents are used with CT since they block X-rays based on their density compared with that of normal tissue.

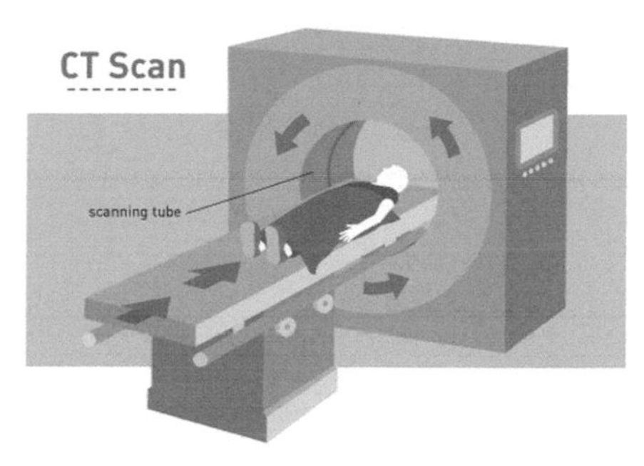

FIGURE 7.3 Computed tomography scanning procedure.

Magnetic resonance imaging (MRI): MRI evolved in the 1970s, initally producing images with low spatial resolution through the resistive magnets that had weak magnetic fields.[2] The soft tissue evaluation of MRI was better than that of CT when an early diagnosis was made. MRI had an advantage where it did not involve ionizing radiation like the X-ray-based CT scanners. Most medical MRI nowadays uses the hydrogen nucleus since it is so plentiful in water and for the reason that its nucleus has a property known as spin. Functional MRI is another imaging method called "magnetoencephalography" or "MEG." MEG is alike to the known electro-encephalography (EEG) also it is better than EEG for localizing signals coming from the brain. The electrical signals of EEG are disfigured by the scalp and other tissues between the brain and the electrodes on the skin. The electrical currents pulled out up by EEG also create weak magnetic fields picked up by MEG but without the interference of the scalp. The magnetic fields from the brain are of various orders of magnitude, which are fewer than the earth's magnetic field; MEG needs to be carried out in a special magnetically protected room. Functional MRI is based on blood flow, the resolution is on the order of seconds, on the other hand, MEG works on the order of milliseconds. The magnetic signals identified by MEG are typically presented on a 3D MRI that has been blown up like a balloon.

Magnetic resonance imaging (MRI) is a diagnostic system used to generate images of the body (Figure 7.4).3 It requires no radiation since it is based on the magnetic fields of the hydrogen atoms in the body. MRI is proficient to present computer-generated images of the human's internal organs and tissues. MRI generally scans the body in an axial plane surface (splitting the body into slices starting from front to back). Generally, the images are in 2D, meaning the MRI images are accessible in slices—top to bottom. Nevertheless, by means of useful computer computation, the 2D slices can be fixed together to construct a 3D model of the area of interest that was scanned, thus it is called 3D MRI. MRI is available for a huge variety of analysis like heart-vessel functioning, liver-bile ducts diseases, chest imaging, nerve conditions in the brain, orthopedic situations like shoulder and hip injury. 3D MRI will give clear information on conditions like cardiovascular pathology and hepatobiliary pathology. The 3D reconstruction of the nervous system is also done. MRI utilizes strong magnets that create a powerful magnetic field that forces protons in the body to line up with that field.[4] When radiofrequency current is pulsed into a patient, the protons get enthused, and turn out of equilibrium, straining against the pull of the magnetic field. When the radiofrequency current is turned off, the MRI sensors will be able to perceive the energy that is released as the protons realign with the magnetic field. The time required for the protons to realign with the magnetic field and the amount of energy released will change according to the environment and the chemical nature of the molecules. The difference between various types of tissues are distinguished by the physicians based on the magnetic properties.

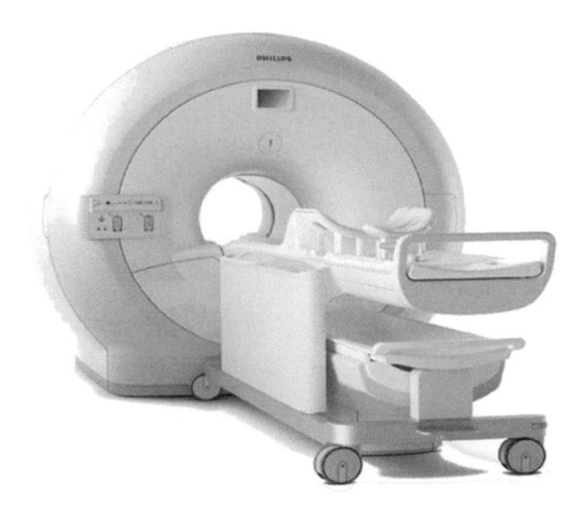

FIGURE 7.4 Magnetic resonance imaging machine.

Ultrasound: Ultrasound was initially used medically during the 1970s. Unlike X-ray and nuclear medicine, ultrasound uses only sound waves and not ionizing (Figure 7.5). When the sound waves pass through the tissue and then are reflected back, tomography images will be produced and tissues will be categorized. For example, a lump or a mass found on a mammogram can be additionally categorized as solid (cancer) or cystic (benign). Ultrasound is, in addition, valuable for the noninvasive imaging of the abdomen and pelvis, as well as imaging the fetus during pregnancy. Early medical ultrasound units were large equipments with articulated arms that formed low-resolution images. These days ultrasound is performed by a portable unit no outsized than a laptop.

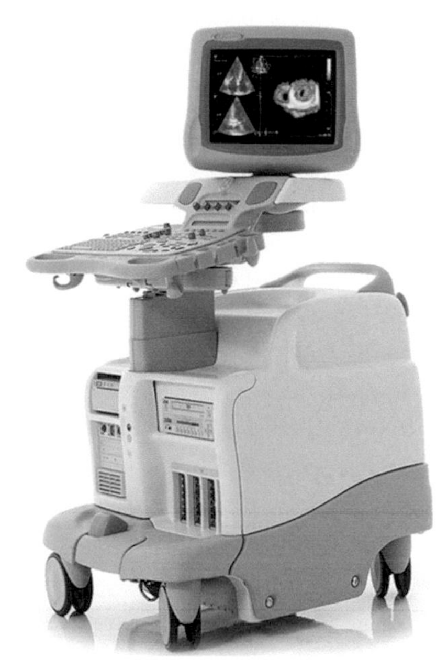

FIGURE 7.5 Ultrasound machine.

Sphygmomanometer: A sphygmomanometer is a device that is used to measure blood pressure.[5] It consists of an inflatable rubber cuff, wrapped around the arm (Figure 7.6). A measuring device specifies the cuff's pressure. A bulb inflates the cuff and a valve liberates pressure. A stethoscope is further being used to listen to arterial blood flow sounds. Blood is forced through the arteries as the heart beats and this causes a rise in pressure known as systolic pressure, which is being followed by a decrease in the pressure as the heart ventricles get ready to perform another beat, this low pressure is known as diastolic pressure. The systolic and diastolic pressures are stated as systolic "over" diastolic, which is 120 over 80. Blood flow sounds are known as Korotkoff sound. There are three types of sphygmomanometers: mercury, aneroid, and digital (Figure 7.6). Digital sphygmomanometers are automated to provide blood pressure reading without making someone to operate the manually used cuff or even listen to the sound of blood flow through a stethoscope. Though physicians use a digital sphygmomanometer for testing, they still prefer manual sphygmomanometers for validating the readings in some situations. On the other hand, the manual sphygmomanometers comprise of aneroid and mercury devices. The operation of the aneroid and mercury devices is the same, except for the aneroid device requiring periodic calibration.

At present, we can see that many android applications are being created that predict the blood pressure and monitor the heart rate with the flash of the mobile. Applications, namely, heart rate monitor, blood pressure diary, instant heart rate, cardiac diagnosis, and many more provide support at an emergency situation but how accurate they are in giving accurate results is still a question to the digital world.

Computers provided aide to the world of medical imaging from the early 1970s with the start of CT scan and then with the MRI scan. CT was the main advancement that primarily authorized numerous tomography images (slices) of the brain to be obtained. As technology advancement started to develop, there have been many changes both dimensionally and algorithmically, only to increase the speed and accuracy

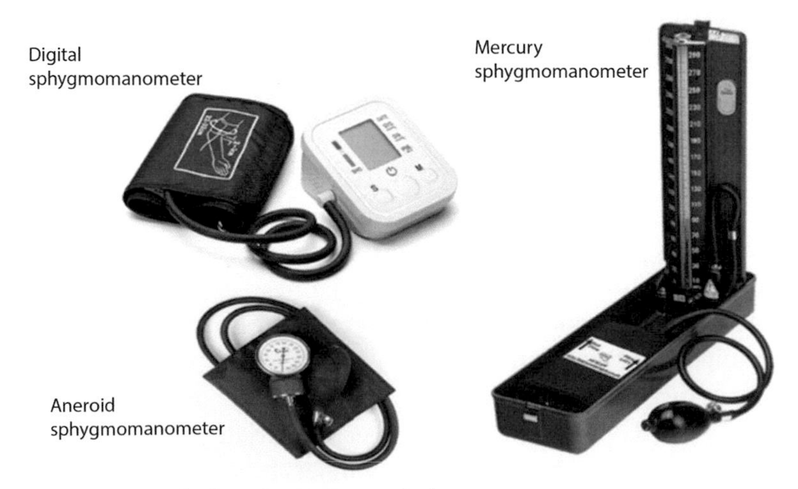

FIGURE 7.6 Types of sphygmomanometer devices.

of the system and application, and also to decrease the workload of the physicians and doctors. The technologies that have provided good services to the medical field are being discussed below.

7.2.1 BIOMEDICAL IMAGING TECHNIQUES USING AI

Companies are incorporating AI-driven proposals in medical scanning devices to improve image clarity and clinical results by reducing the exposure to radiation emission. General Electric (GE) Healthcare in collaboration with the NVIDIA AI platform claims to improve the speed and accuracy of the existing CT scans. The algorithms are powered up to reorganize small patterns of the damaged organ that seems to have been failed to notice when the physician was skimming the scan. The finer details getting captured can help in supporting faster diagnosis and reduced error rates. The GE healthcare system claims that the CT scan system developed will be two times faster than the existing system and it is probably useful to quickly detect liver and kidney lesions, because of the high volume of data accessible through NVIDIA's AI platform.

Machine learning algorithms can be trained to see patterns in the same way doctors see them. The main difference is that algorithms require a lot of real examples—thousands and millionsto learn. Also, these examples are to be precisely digitizedas machines cannot comprehend between the lines in textbooks. Hence, machine learning is mainly useful in areas where the analytical diagnostic data that a doctor examines is already digitized. Like, identifying lung cancer using *CT scans,* considering the risk of unexpected cardiac death or other heart ailment using *electrocardiograms* and *cardiac MRI images,* classifying skin abrasion in *skin images,* finding markers of diabetic retinopathy in *eye images.*

The newly developed devices tend to provide quick and accurate medical decisions to avoid collapsing of life. One such handy device that came into the world is the Accu-Chek,[6] it provided a complete solution for diabetes monitoring, where people can test frequently for a better knowledge and also control it accordingly. It gives the result in just a few seconds at temperatures from 57 to 104°F. The device potentially works on testing diabetes, where the patient inserts a biosensor into the meter and drops in a tiny drop of blood. The biosensor collects the drop of blood, it performs a sequence of tests, which undergoes an enzymatic chemical reaction and then followed by an electrochemical reaction. On the other hand, the electronic meter is used for measurement, storage, and communication. It applies potential differences in a programmed manner to the sensor, collects biamperometric current data, later records, and displays the results.

Medtronic is one of the companies manufacturing medical devices that are creating their step into the AI world, aiming to help diabetic patients handle their situation more efficiently.[7] In September 2016, Medtronic announced its *Sugar. IQ app*, developed in collaboration with IBM Watson, a mobile assistant that helps to track the amount of glucose content in food items, specify diet chart, therapy-related actions through the sensors. It works to monitor continuously the blood glucose levels by examining the data generated from Medtronic glucose sensors and insulin pumps. The devices will be attached to the patient's or user's body. The application characteristics consist of a smart food logging system, motivational insights, a glycemic aide, an information tracker. The analysis is done with the help of machine learning algorithms and Internet of Things to predict the outcomes. The company also developed a *MiniMed 670G* system that will deliver insulin level (self-adjust baseline insulin) needed at a given time every five minutes. The Sugar.IQ app is made accessible to consumers of the Medtronic Guardian Connect system. It is the first smart standalone Continuous Glucose Monitoring system4 intended to empower individuals with diabetes consuming multiple daily injections by means of actionable equipment to help them get forward with the high- and low-glucose levels. It provides support for customizable prognostic alerts up to 60 min before an unpredictable level high or low, and the Sugar.IQ support, guardian connect continuous glucose monitoring offers individuals with additional options to stay in scope than any other continuous glucose monitoring system.

Apple Watch users can now take a look at their heart rhythm just by holding the crown of the device.[8] The software update provided to the Apple watch series 4 provides a new feature, to identify atrial fibrillation, also provides extra passive monitoring. People over the age of 22 and above can utilize the features provided to differentiate between a normal heart rate or with atrial fibrillation and sinus rhythm. An optical heart sensor uses green LED lights paired with light-sensitive photodiodes helpful to detect blood volume pulses in the wrist of a human using the "photoplethysmography" based algorithm, which is an easy and inexpensive optical method that is used to detect blood volume variation in the micro-vascular area of tissue. To verify heart rate variability, Apple Watch captures a tachogram, a plot of the time between heartbeats, every 2–4 h. It also allows the user to send messages to friends showing their level of heart rate in case of emergency and also recommends the user to visit a doctor to seek proper medical care if irregular symptoms are seen. The watch works in models like the iPhone 5 and above, though it can work with fewer features when not connected to the phone, it is necessary that iPhone has to be connected for it to work efficiently. The Apple Watch integrates innovation in hardware, software, and user interface design such that users can interact with it through sight, sound, and touch. Apart from monitoring heartbeat rate, people can answer calls, send messages, keep track of appointments and meetings, listen to music, find routes to destinations, and also even unlock doors. It is customizable with a lot of user desired accessories.

Entrepreneur Jonathan Rothberg of *Butterfly Network* proposed to create a new handheld medical-imaging device that aims to take both MRI and ultrasounds scans easier and in a cheaper rate.[9] Butterfly iQ's device uses semiconductor chips, instead of piezoelectric crystals. It will use ultrasound scanners to create 3D images, then sends the images to a cloud service, which will further work on enhancement, zooming in on identifying features in the images and help to automate the diagnoses. Measurements on images can be performed with up to four linear measurements and one elliptical dimension. Butterfly iQ has a built-in battery to reduce drawing power away from the mobile. A wireless charging base is provided, where one needs to place the iQ on the charger base where the battery indicator should face upwards and the charging will be completed in less than 5 h. The battery lasts up to a 10-h shift or nearly 2 h of continuous scanning. The service will incorporate deep learning techniques into its device. The ultrasound-on-chip technology will replace the existing transducer and the system with a silicon chip. The plan is to even automate many of the medical imaging procedures, the device will be made available for usage in clinics, retail pharmacies, and in poorer regions of the world with affordable pricing of about $1999. One added advantage is that Butterfly Cloud has been given to the users of Butterfly iQ mobile app users, where the cloud is a storage web application that helps users to upload the case study to an Internet-based storage space system. Butterfly Cloud is available for purchase to an individual and also for a team. The preference comprises of limitless archiving, anonymous sharing, and secure access through an iPhone mobile or a laptop. All local and in-country privacy and security regulations are provided that certify that the data is protected and secure.

DeepMind has bought many healthcare projects across the world, now in collaboration with UCL's radiotherapy department, it has initialized to reduce the amount of time taken to plan treatments.[10] Through machine learning, DeepMind has given access to 1 million images of the eye scans, along with their patient data. It sets to train itself to read the scans and predict spot early signs that may indicate the occurrence of the degenerative eye and also reduce the time taken for diagnosis is reduced to one fourth. The convolution neural network that was built for this system was the first deep learning model that was intended to effectively learn the control guidelines straight from a high-dimensional sensory input using the reinforcement learning algorithm. This remarkable accomplishment was soon improved by succeeding forays into gaming. During the year 2015, DeepMind in collaboration with Royal Free NHS Trust had been used to create a patient safety application that was called "Streams," it reviews the clinical test results to check for signs of sickness and sends notifications and alerts to staffs instantly if an emergency examination is required. The application also helps physicians to rapidly check for other critical or serious conditions like the acute kidney injury and also provide results of blood tests, X-rays and scans at the press of a button. Nurses and other assistants said that the application saved their time of up to 2 h in a day.

Google Glasses initially helpful for recognizing text and translating it, recognizing objects, and searching for its relevant match, looking at posters and playing videos, getting directions on the go, all of this happening in front of the eye.[11] Some editions of Google Glasses had no lenses in them; what all editions had is a thick area of the frame over the right eye, it was where Google had inserted the screen for the glasses. To look upon the screen, one has to peek up with the eyes. The region of placement was quite important, since the screen inserted in the direct line of vision may result in serious problems. The display has a resolution of about 640 × 360 pixels, making it as on the low side for mobile devices. The camera has about 5-megapixel quality and it also records videos at about 720 pixels. The only issue is in the battery life, which lasts for about 5 h for average usage, for taking a longer video or using the glass for a longer time might drain the battery quickly. Google Glass has a storage capacity of about 16 GB of storage and it also synchronizes with Google Drive for an added accessibility to the videos and photos taken by the user. It is also equipped with a micro-USB port for transferring files and charging the device. The frame is generally lightweight and it has a replaceable nose pad in case of any accidental breaking. Sounds of phone calls and other notifications are produced through bone-conduction transfer, or even by passing some vibrations directly to the skull, thus transmitting sound to ears. The glass is an optical head-mounted display worn as a pair of spectacles. With its multitasking capability and responsiveness to hands-free voice and motion commands gained acknowledgment in the medical field, where doctors can actually perform surgery as a surgical navigation display. The first-ever surgery using Google Glass was done by Dr. Marlies P. Schijven in the year 2013, at the Academic Medical Centre, Netherlands. In the operation theatre, doctors can see the medical data without even having to turn away from the patients. Researchers have found that the navigating options were helpful in finding tumors by the doctors who perform surgery and can also venture a form of a tunnel vision or blindness on a part that could make them miss unconnected lesions or the inconvenience around them. Google Glass will further be helpful in recording the surgery to maintain it for documentation purposes to keep a track of the patient's medical record and to assess the surgical competency to check the ongoing professional skill development and certification. The start of Google Glass will provide a technical change in the way people get to understand the world.

An ultrasonography exam takes quite a lot of time in identifying the planes in the brain, which needs an ample amount of training and manual work.[12] There could also be a missed or delayed diagnosis. Now, with AI systems, users will just need to find a starting point in the fetal brain and the device will automatically take measurements after identifying the standard planes of the brain. The data or the documentation is maintained as the patient may visit for examination some other day; this will help in a more positive diagnosis.

EchoNous has developed a convolutional neural network for the automatic detection of the urinary bladder with the help of high-quality ultrasound images captured with Uscan, using the advantage of the "high spatial density fanning technique".[13] With the help of the captured image, one can compute the urinary bladder volume with much higher accuracy. Uscan actively recognizes the contours of the bladder, where the measurements are very accurate than the results got from the existing scanners. *EchoNous Vein* is an ultrasound-based device intended particularly for nurses to improve peripheral IV catheter placements. It is being developed for handling a wide variety of patients including both adults and children. EchoNous Vein offers immediate, crisp images at depths from 1 to 5 cm for rapidly visualizing superficial and deeper veins with just two buttons to control.

Cancer can be diagnosed promptly with the help of deep learning and AI concepts.[14] A Chinese start-up named "*Infervision*" uses image recognition technology and deep learning to efficiently diagnose the signs of lung cancer with X-rays the same as Facebook recognizes faces in photographs. Infervision trains its algorithms routinely with data got from varied sources. Usually, it takes doctors almost 20 min to analyze an image, but with Infervision AI helps in processing the visuals available and generate a report within 30 s. Data is passed through networks of nodes, where these networks adapt to the data that has been processed from node to node. In this manner, the neural networks effectively process the data (next bit) that comes along keeping a record of the data that came before it. This type of ability to learn from the data and effectively teach the system by itself is what made deep learning become dominant. As in the initial training phase, an X-ray image was used to teach the system to predict whether it was normal or abnormal.

In May 2015, a Canadian ophthalmologist announced that he has created a bionic lens that could correct a person's vision for life, effectively resulting in vision three times better than 20/20.[15] The product came into life after eight years of research and about $3 million funding; the Ocumetics Bionic Lens is said to require a painless 8-min in-office procedure that requires no anesthesia. The researcher folds up the custom-made lens like a tiny taco (a Mexican dish) to fit it into a saline-filled syringe. Then, uses the syringe to place it in the eye through a super-small incision, then leaves it there to unravel over about 10 s. And, it is finally done. The bionic lens is made to replace our natural eyes; the process may get rid of any risk of cataracts in the future. As cataracts may release chemicals that raise the risk of glaucoma and other issues, it also helps in protecting the overall eye health.

Diagnosing a disease or illness needs a lot of facts about the patient, AI seems to be well suitable in collecting information as it has a lot of memory, energy, and it does not even need to sleep or take rest.[16] One such AI system has been developed in the University of California, San Diego where Professor Kang Zhang and his team trained an AI system on the medical records from over 1.3 million patient data who have visited medical clinics and hospitals in Guangzhou, China. The patient's data taken for training belonged to age groups of all under 18 years old, who visited doctors during the period of January 2016 and January 2017. The medical charts associated with every patient had medical charts that were text written by doctors and also a few laboratory test inferences. To ease the work performed by AI, Zhang and his team made doctors annotate the records to recognize the part of the text linked with the patient's issues, the period of illness, and also tests performed. When the testing phase began using unseen cases, AI was efficient enough to identify roseola, chickenpox, glandular fever, influenza, and hand–foot–mouth disease giving an accuracy rate of about 90%–97%. It may not be a perfect score but still, we should know that even doctors cannot predict correctly at times. The performance measure was compared with some 20 pediatricians who have various years of clinical experience. AI outperformed the junior doctors, and also the senior doctors performed well than the AI. When doctors are so busy while looking upon about 60–80 patients in a day, they can only accumulate little information, as that is where the doctors lack interest and provided might make mistakes in recognizing the seriousness of the disease or illness. That is where an AI can be counted on. AI can be efficiently used to check out the patients in emergency sections, provided AI should be able to predict the level of illness with sufficient data, and also confirm whether the patient needs to visit doctor or it is just a common cold. It is likely that junior doctors who depend on this AI system could possibly miss out on their learning and check patterns in the patient's queries. The team looks forward to training AI systems that can also diagnose adult diseases.

A team from Beth Israel Deaconess Medical Center and Harvard Medical School have developed an AI system to predict disease based on the training given to the systems to investigate the pathological images and perform pathological diagnosis.[17] The AI-powered systems are incorporated with machine learning, deep learning algorithms, where it trains the machines to understand the complex patterns and structures experienced in real-time data by creating multilayer perceptron neural network, is a procedure that is used to show similarities with the learning step that occurred in layers of neurons. In the evaluation where researchers were given slides of the lymph node cells and required to identify if it was cancerous or not, the automated diagnosis method gave an accuracy of about 92%, which nearly matched the success efficiency rate of the pathologist who gave an accuracy of about 96%. Recognizing the presence or absence of metastatic cancer in the patient's lymph nodes is an important work done by pathologists. Looking into the microscope to skim through several normal cells to find out a few of the malignant cells is a tremendous task using conventional methods. This task can be efficiently done by a computer that has high power, and this was the case that was proved. The system was trained to differentiate between normal and cancerous tumor regions through a deep multilayer convolutional network. The system extracted millions of small training samples and used deep learning to build a computational model to classify them. The team later recognized the specific training samples for which the system fails to classify and hence re-train them with a greater number of difficult training samples, therefore improving the performance of the computer. It is a good sign that AI will change the way we see pathological images in the coming years.

Emedgene is the world's first fully automated genetic interpretation platform, incorporated with advanced AI technology to considerably restructure the interpretation and evidence presentation processes.[18] Emedgene is connected with the advanced AI technology to radically scale genetic interpretations and findings. The AI knowledge graph codifies the complex and endlessly updated web of variants, genes, mechanisms, and phenotypes that reside at the heart of the interpretation process. At the center of this determined mission lies Emedgene's capability to ingest and analyze unstructured information from present scientific publications with the addition of a wealth of data to be accessible structured data. A set of convention machine learning algorithms that frequently discover clinically significant associative models within the AI knowledge graph. With the help of these algorithms, Emedgene can effectively identify the potentially contributing mutation for both recognized and unidentified genes. Science is intensely indulged in each and every aspect of the Emedgene platform. A meticulous standard operating procedure guarantees the quality and accuracy of the AI knowledge graph. The genomic research department leads to the improved development of clinically victorious variant discovery algorithms. Emedgene keeps on working to solve the unsolved cases routinely. The samples are reanalyzed when applicable new scientific results enter the knowledge graph. Regularly updating the AI knowledge graph, integrating the structured and unstructured data from the latest scientific literature. It contains an automatic evidence builder that presents all of the available data points that guide to a particular variant identification, such that it can offer the most time-effective decision support system for geneticists. The AI interpretation engine locates the reason behind the genetic disease and provides proof for an obvious path to clinical conclusions. The clinical workbench is a fully featured lab solution that provides the user from analysis to reporting. Using Emedgene, healthcare suppliers are able to offer individual care to more and more patients with the help high-resolution rates and probably improved yield.

7.2.2 CHALLENGES

As there are progressive developments in the medical AI systems, there will be a foreseeable demand in their medical use and operation, which may create new social, economical, and legal problems. Geoffrey Hinton, one of the great researchers in neural networks predicts that AI will bring extreme changes in the field of medical science and practice. AI will improve the value of care by decreasing human mistakes and reducing doctor's tiredness caused by habitual clinical practice. But still, it may not decrease the workload of a doctor, since clinical rules may recommend that diagnosis should necessarily be carried out very quickly for high-risk patients. Even though AI's systems and applications are being well equipped to provide quick solutions, for every support and decision-making situation nothing is possible without the intervention of a doctor. Getting to the point, AI can never replace a doctor at any point of time but can only provide support.

7.3 CONCLUSION

The field of AI has modified its knowledge over the last 50 years. AI is hovering to transform many portions of present medical customs in the predictable prospect. AI systems can improve medical decision-making, ease disease analysis, recognize formerly unrecognized imaging, or genomic prototypes related to patient phenotypes, and aid in surgical interventions for a variety of human diseases. AI applications even include the probability to carry out medical knowledge to an isolated area where experts are inadequate or not accessible.

Even though AI assures to change medical tradition, many technical disputes lie forward. As machine-learning-based processes depend deeply on the accessibility of a huge quantity of premium-quality training dataset, the concern must be engaged to collect data that is representative of the end patient group. Data from various healthcare centers will contain different types of noise and bias, which will cause a trained model on a hospital's data to fail to generalize to a different one. Motivated by knowledge disparities got from the past and future human civilizations, AI will quickly discover the type of potential that cliché would provide to human development for a decade, century, or millennium starting from now. Many high-end machine-learning models produce outcomes that are complicated to understand by unaided humans. An AI would absorb an enormous quantity of extra hardware subsequent to the attainment of some edge of proficiency. Faster-emerging artificial intelligent system represents a better scientific and technological challenge.

In this era of advancement, we can see that machines are learning while people are being hooked up with their mobile; knowing the fact that it is simply a tool we fiddle with. As the advancement gradually moves up just to ease the humungous task with the help of machines, people will now become jobless. AI-based applications and machines are now being used to clean the dishes, serve as a waiter for 24 h with a single charge, check for cash balance at the bank, talk when bored, and answer your mysterious questions, help you in cooking, recite a poem or even sing a lullaby song when you are insomniac, suggest you to drink water, check for the recent missed calls and dial a number, give you the recent weather updates, drive you home safe in the driverless car, check for the shortest route to reach your destination quickly, and to do what not.

AI has improved medical analysis and decision-making performance in numerous medical undertaking fields. Physicians may necessarily adapt to their fresh practice as data accumulators, presenters, and patient supporters, and the medical education system may have to give them the equipment and technique to do so. How AI-enhanced applications and systems performance will provide a great impact on the existing medical practice including disease analysis, detection, and treatments will probably be determined on how AI applications will co-combine with the healthcare systems that are under revolutionary development financially with the adaptation of molecular and genomic science. The people who get benefited, or get controlled from the AI applications and systems are yet to be determined, but the balance of rigid regulations of safeguards and market services to certify that people/patients get advantage the most should be if great priority. AI is a one road challenge, and it is the road we will end up taking.

In this article, we came across a plenty number of applications and systems that have started to create wonders in the field of medical science, such as the drug discovery, automated data collection, literature mining, simulation of molecular dynamics, prediction of chemical toxicity, disease diagnosis, automated surgery, patient monitoring, and many more. Every potential discovery approached has helped a lot of researchers and medical practitioners in making their work easier and also providing competent efficiency.

This is just the beginning of a storm. The more the medical data is digitized and unified, the more the AI will be used to help us discover important patterns and featuresthese can be used to make accurate, cost-effective decisions in composite analytical processes.

KEYWORDS

- **artificial intelligence**
- **medical diagnosis**
- **digital world**

ENDNOTES

1. Bradley, William G. History of medical imaging. *Proceedings of the American Philosophical Society* 152(3) (2008): 349–361. Retrieved from http://www.umich.edu/~ners580/ners-bioe_481/lectures/pdfs/2008-09-procAmerPhilSoc_Bradley-MedicalImagingHistory.pdf
2. Retrieved from https://www.lanl.gov/museum/news/newsletter/2016-12/x-ray.php
3. Retrieved from https://www.myvmc.com/investigations/3d-magnetic-resonance-imaging-3d-mri/
4. Retrieved from https://www.khanacademy.org/test-prep/mcat/physical-sciences-practice/physical-sciences-practice-tut/e/the-effects-of-ultrasound-on-different-tissue-types
5. Retrieved from https://www.practicalclinicalskills.com/sphygmomanometer
6. Hill, Brian. Accu-Chek Advantage: Electrochemistry for Diabetes Management. CurrentSeparations.com and Drug Development 21(2) (2005). Retrieved from http://currentseparations.com/issues/21-2/cs21-2c.pdf
7. Retrieved from https://www.mobihealthnews.com/content/medtronic-ibm-watson-launch-sugariq-diabetes-assistant
8. Retrieved from https://electronics.howstuffworks.com/gadgets/high-tech-gadgets/apple-watch.htm
9. Retrieved from https://www.butterflynetwork.com
10. Retrieved from https://deepmind.com
11. Retrieved from https://electronics.howstuffworks.com/gadgets/other-gadgets/project-glass.htm
12. Retrieved from https://www.diagnosticimaging.com/article/how-ai-changing-ultrasounds
13. Retrieved from https://www.businesswire.com/news/home/20180711005836/en/EchoNous-Vein-Receives-FDA-Approval-New-Innovation
14. Retrieved from https://www.bernardmarr.com/default.asp?contentID=1269
15. Retrieved from http://ocumetics.com/
16. Retrieved from https://www.newscientist.com/article/2193361-ai-can-diagnose-childhood-illnesses-better-than-some-doctors/

17. Retrieved from https://healthcare-in-europe.com/en/news/artificial-intelligence-diagnoses-with-high-accuracy.html
18. Retrieved from https://emedgene.com/

CHAPTER 8

ANALYSIS OF HEART DISEASE PREDICTION USING MACHINE LEARNING TECHNIQUES

N. HEMA PRIYA,* N. GOPIKARANI, and S. SHYMALA GOWRI

Department of Computer Science PSG college of Technology, Coimbatore -14 Tamil Nadu

Corresponding author. E-mail:nhemapriya@gmail.com

ABSTRACT

Analytical models are automated with the method of machine learning, a predominant evolution of artificial intelligence. In health care, the availability of data is high, so is the need to extract the knowledge from it, for effective diagnosis, treatment, etc. This chapter deals with heart disease, which is considered to be an obvious reason for the increase in mortality rate. A method to detect the presence of heart disease in a cost-effective way becomes essential. The algorithms considered are *K*-nearest neighbors, decision tree, support vector machine, random forest, and multilayer perceptron. The performances of these algorithms are analyzed and the best algorithm for heart disease prediction is identified. The results show that the machine learning algorithms work well for heart disease prediction and thc model is trained by applying different algorithms. Each algorithm works on the model to fit and trains the model using a training dataset and the model is tested for a different dataset, test data for each of the algorithms. The accuracy increases with the random forest algorithm, being 89% that is considered to be better than other techniques.

8.1 INTRODUCTION

Health care is one of the most important areas of huge knowledge. Extracting medical data progressively becomes more and more necessary for predicting and treatment of high death rate diseases like a heart attack. Hospitals can make use of appropriate decision support systems, thus minimizing the cost of clinical tests. Nowadays, hospitals employ hospital information systems to manage patient data. Terabytes of data are produced every day. To avoid the impact of the poor clinical decision, quality services are needed. Hospitals can make use of appropriate decision support systems, thus minimizing the cost of clinical tests. Huge data generated by the healthcare sector must be filtered for which some effective methods to extract the efficient data is needed. The mortal rate in India increases due to the noncommunicable diseases. Data from various health organizations like World Health Organization and Global Burden of Disease states that most of the death is due to the cardiovascular diseases.

Heart disease is a predominant reason for the increase in the mortality rate. A method to detect the presence of heart disease in a cost-effective way becomes essential. The objective of the article is to compare the performance of various machine learning algorithms to construct a better model that would give better accuracy in terms of prediction.

8.2 OVERVIEW OF HEART DISEASE

Heart diseases or cardiovascular diseases are a class of diseases that involve the heart and blood vessels. Cardiovascular disease includes coronary artery diseases (CADs) like angina and myocardial infarction (commonly known as a heart attack). There is another heart disease called coronary heart disease, in which a waxy substance called plaque develops inside the coronary arteries that is primarily responsible for supplying blood to the heart muscle that is rich in oxygen. When plaque accumulates up in these arteries, the condition is termed as atherosclerosis. The development of plaque happens over many years. Over time, this plaque deposits harden or rupture (break open) that eventually narrows the coronary arteries, which in turn reduces the flow of oxygen-rich blood to the heart. Because of these ruptures, blood clots form on its surface. The size of the blood clot also makes the situation severe. The larger blood clot leads to flow blockage through the coronary artery. When time passes by, the ruptured plaque gets hardened and would eventually result in the narrowing of the coronary arteries. If the blood flow has stopped and is not restored very quickly, that portion of the heart muscles begins to die.

When this condition is not treated as an emergency, a heart attack occurs leading to serious health problems and even death. A heart attack is a common cause of death worldwide. Some of the symptoms of the heart attack (Sayad & Halkarnikar, 2014) are listed below.

- *Chest pain*: The very common symptom to easily diagnose heart attack is chest pain. If someone has a blocked artery or is having a heart attack, he may feel pain, tightness, or pressure in the chest.
- *Nausea, indigestion, heart-burn, and stomach pain*: these are some of the often overlooked symptoms of a heart attack. Women tend to show these symptoms more than men.
- *Pain in the arms*: The pain often starts in the chest and then moves towards the arms, especially on the left side.
- *Dizziness and light headed*: Things that lead to the loss of balance.
- *Fatigue*: Simple chores that begin to set a feeling of tiredness should not be ignored.
- *Sweating*: Some other cardiovascular diseases that are quite common are stroke, heart failure, hypertensive heart disease, rheumatic heart disease, cardiomyopathy, cardiacarrhythmia, congenital heart disease, valvular heart disease, aortic aneurysms, peripheral artery disease, and venous thrombosis.

Heart diseases develop due to certain abnormalities in the functioning of the circulatory system or may be aggravated by certain lifestyle choices like smoking, certain eating habits, sedentary life, and others. If the heart diseases are detected earlier then it can be treated properly and kept under control. Here, early detection is the main objective. Being well informed about the whys and whats of symptoms that present will help in prevention summarily (Chen et al., 2011).

The health-care researchers analyze that the risk of heart disease is high and it changes the life of the patients all of a sudden. There are various causes for heart disease some of them arc due to change in lifestyle, gene, and smoking. Numerous variations of genetic will increase heart disease risk. When the heart disease is accurately predicted, the treatment of the same that includes the intake of cholesterol-lowering drugs, insulin, and blood pressure medications is started.

Predicting is not easier; the accurate prediction of heart attack needs a constant follow up of cholesterol and blood pressure for a lifetime. A foundation of plaques that cause heart attack needs to be identified that is much more sensitive to the patient. The health-care industry had made its contribution to early and accurate diagnosis of heart disease through various research activities. The function of the heart is affected due to various conditions that are termed as heart disease. Some of the common heart diseases are CAD, cardiac arrest, congestive heart failure, arrhythmia, stroke, and congenital heart disease.

The symptoms to predict heart disease depend upon the type of heart disease. Each type will have its own symptoms. For example, chest pain is one of the symptoms for coronary artery; all the people would not have the same symptoms as others, some may have chest pain as a symptom of indigestion. The doctor confirms the heart disease with the diagnosed report of the patient and various other parameters. Some of the most common heart diseases are listed in Table 8.1 with their description.

TABLE 8.1 Most Common Cardiac Diseases[a]

Sr. No	Cardiac Disease	Explanation
	Coronary artery disease (CAD)	The condition where circulatory vessels that supply oxygenated blood to the heart get narrowed. This occurs due to a deposition of plaque
	Cerebrovascular disease (CVD)	A type of CVD associated with circulatory vessels that supply blood to the brain, causing the patient to have a stroke
	Congenital heart disease (CHD)	Most commonly identified as birth defects, in the new born children
	Peripheral arterial disease (PAD)	A condition caused by reduced blood supply to limbs due to atherosclerosis

[a]Src: Sachith Paramie Karunathilake and Gamage Upeksha Ganegoda, 2018, "Secondary Prevention of Cardiovascular Diseases and Application of Technology for Early Diagnosis," Hindawi Bio Med Research International, Article ID 5767864

8.3 MOTIVATION

It is evident from the recent statistics that the major cause for the death of both males and females is heart disease. In 2017, nearly 616,000 deaths have been caused due to heart disease. Hence, the need for an efficient and accurate prediction of heart disease is increasingly high.

The motivation to do this problem comes from the World Health Organization estimation. According to the World Health Organization estimation until 2030, very nearly 23.6 million individuals will die because of heart diseases. So, to minimize the danger, the expectation of coronary illness ought to be finished. Analysis of coronary illness is typically in view of signs, manifestations, and physical examination of a patient. The most troublesome and complex assignment in the medicinal services area is finding the right ailment.

Providing quality service at a cost that is affordable lies as a major challenge in front of the health-care organization and medical centers. The quality service includes the proper diagnosis of the patients leading to the administration of effective treatments. The main objective of the chapter is to analyze the performance of various machine learning algorithms for heart disease prediction. The future is expecting the usage of the above techniques for eliminating the existing drawbacks and improving the prediction rate thus providing a way for improving the survival rate for the well being of mankind.

8.4 LITERATURE REVIEW

Each individual could not be equally skilled and hence they need specialists. Each specialist would not have similar talents and hence we do not have authorized specialists' access without effort.

8.4.1 MACHINE LEARNING

Machine learning (Hurwitz, 2018) is a powerful set of technologies that can help organizations transform their understanding of data. This technology is totally varied from the ways in which companies normally present data. Rather than beginning with business logic and then applying data, machine learning techniques enable the data to create the logic. One of the greatest benefits of this approach is to remove business assumptions and biases that can cause leaders to adopt a strategy that might not be the best. Machine learning requires a focus on managing the right data that is well prepared. Proper algorithms must be selected to create well-designed models. Machine learning requires a perfect combination of data, modeling, training, and testing.

In this chapter, we focus on the technology underpinning that supports machine learning solutions. Analytical models are automated with the method of machine learning. A machine learns and adapts from learning. Machine learning helps computers to learn and act accordingly without the explicit program. It helps the computer to learn the complex model and make predictions on the data. Machine learning has the ability to calculate complex mathematics on big data. The independent adaption to new data is achieved through an iterative aspect of machine learning (Ajam, 2015). Machine learning helps to analyze huge complex data accurately. The heart disease predicting systems build using machine learning will be precise and it reduces the unknown risk. The machine learning technique will take completely different approaches and build different models relying upon the sort of information concerned. The value of machine learning technology is recognized in the health-care industry with a large amount of data. It helps the medical experts to predict the disease and lead to improvised treatment.

Predictive analytics (Hurwitz, 2018) helps anticipate changes based on understanding the patterns and anomalies within that data. Using such models, the research must be done to compare and analyze a number of related data sources to predict outcomes. Predictive analytics leverages sophisticated machine learning algorithms to gain ongoing insights. A predictive analytics tool requires that the model is constantly provided with new data that reflects the business change. This approach improves the ability of the business to anticipate subtle changes in customer preferences, price erosion, market changes, and other factors that will impact the future of business outcomes.

The machine learning cycle creates a machine learning application or operationalizing a machine learning algorithm is an iterative process. The learning phase has to be started as clean as a whiteboard. The same level of training while developing a model is needed after a model is built. The machine learning cycle is continuous, and choosing the correct machine learning algorithm is just one of the steps.

The steps that must be followed in the machine learning cycle are as follows:

Identify the data: Identifying the relevant data sources is the first step in the cycle. In addition, in the process of developing a machine learning algorithm, one should plan for expanding the target data to improve the system.

Prepare data: The data must be cleaned, secured, and well-governed. If a machine learning application is built on inaccurate data the chance for it to fail is very high.

Select the machine learning algorithm: Several machine learning algorithms are available out of which best suitable for applications to the data and business challenges must be chosen.

Train: To create the model, depending on the type of data and algorithm, the training process may be supervised, unsupervised, or reinforcement learning.

Evaluate: Evaluate the models to find the best algorithm.

Deploy: Machine learning algorithms create models that can be deployed to both cloud and on-premises applications.

Predict: Once deployed, predictions are done based on new input data.

Assess predictions: Assess the validity of your predictions. The information you gather from analyzing the validity of predictions is then fed back into the machine learning cycle to help improve accuracy.

Figure 8.1 demonstrates the relationship between consumers and service providers—hospitals.

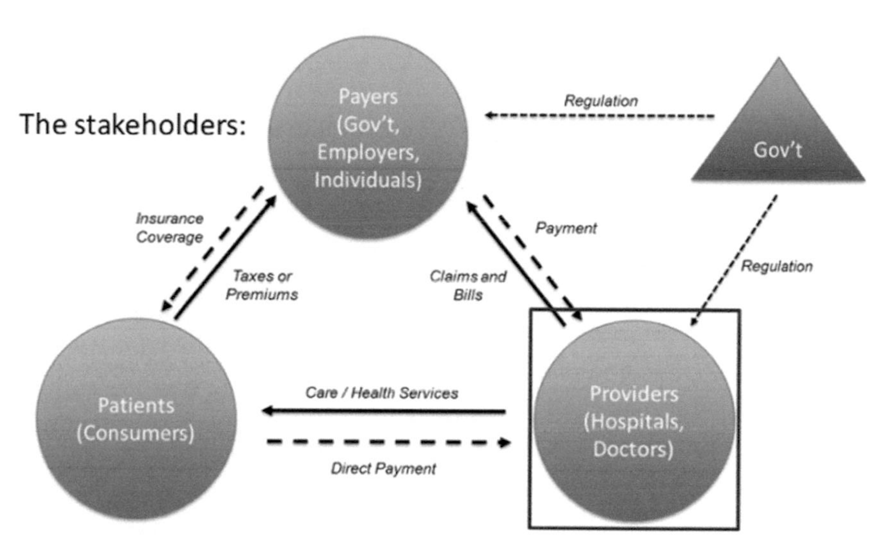

FIGURE 8.1 Transformation in health care.
(*Source:* http://www.mahesh-vc.com/blog/understanding-whos-paying-for-what-in-the-healthcare-industry)

8.4.2 DEEP LEARNING

A system that makes its own decision based on intelligence is highly needed in today's life. The breakthrough is one of the biggest defeats in artificial neural networks (ANNs) that becomes a major reason for the rise of the new approach called deep learning. A new approach, deep learning, arises from machine learning to meet the inner goals of artificial intelligence. The world needs a deep learning system that thinks like a neo-cortex. Various types of data is been learned by the system and the patterns are recognized as the same as that of the real sense. To train a large neural network, deep learning is the best-situated approach. The performance measure considered with large data set is high in deep learning rather than other approaches. The advantage of deep learning is that it allows automatic feature extraction and makes the feature learning easier. This approach helps to discover the structure in the data. The problem in unsupervised and supervised learning is tackled with the help of a deep learning algorithm.

Machine learning algorithms are far away from other algorithms. With most algorithms, a programmer starts by inputting the algorithm. However, this is not the case with machine learning. In machine learning, the above-mentioned process is reversed. Here, the data that is to be processed creates a model. When the size of the data becomes large, the algorithm becomes more refined and polished. As the machine learning algorithm is exposed to more and more data, it is able to create an increasingly accurate algorithm.

8.4.3 HEART DISEASE PREDICTION

In (Khemphila, 2011), heart disease is classified by hybridizing the backpropagation and multilayer perceptron algorithm. To select the appropriate features from the patient dataset, information gain along with biomedical test values is used. A total of 13 attributes were chosen to classify the heart disease. The experimental results show that 13 attributes were reduced to 8 attributes using information gain. The accuracy obtained as a result of this work is 1.1 % for the training data set and 0.82% for the validation data set.

In “Heart Attack Prediction System Using Data Mining and Artificial Neural Network” (2011), the weighted associative classifier (WAC) is used. WAC (Soni et al., 2011) is introduced to diagnose if the patient is affected with any cardiovascular disease or not. GUI-based interface was designed to enter the patient record. They have taken 13 attributes and 303 records of data for training and testing. The author has incorporated a little modification in the dataset. Instead of considering 5 class labels (4 – different types of heart disease and 1—no heart disease) they have considered only 2 class labels “1” for the presence of heart disease and “0” for the absence of heart disease. In this work, a new technique of WAC has been proposed to get the exact significant rule instead of flooding with insignificant relation. From the results, it is observed that WAC outperforms than other associative classifiers with high efficiency and accuracy. WAC classifier obtained an accuracy of 81.51% with a support and confidence value of 25% and 80%, respectively.

Kim and Lee (2017), have used the dataset taken from the sixth Korea National Health and Nutrition Examination Survey to diagnose the heart-related diseases. The feature extraction is done by statistical analysis. For the classification, the deep belief network is used that obtained an accuracy of 83.9%.

Olaniyi and Oyedotun (2015) have taken 270 samples. They were divided into two parts that are the training dataset and the testing dataset. This division is based on 60:40, that is, 162 training dataset and 108 testing dataset for the network input. The target of the network is coded as (0 1), if there is presence of heart disease and (1 0) if heart disease is absent. The dataset that is used is taken from the UCI machine learning repository. The feedforward multilayer perceptron and support vector machine (SVM) is used for the classification purpose of the heart disease. The results obtained from the work are 85% in the case of feedforward multilayer perceptron, and 87.5% in case of SVM, respectively.

In the work, "Prediction of Heart Disease Using Deep Belief Network" (2017), the deep belief network is utilized for the prediction of heart disease that is likely to occur for the human beings. It was developed in MATLAB 8.1 development environment. This proposed solution is then later on compared with the results of the CNN for the same. Results yield 90% accuracy in the prediction of heart diseases, whereas CNN achieves only 82% accuracy, thus enhancing the heart disease prediction.

Ajam (2015) chooses a feed-forward back propagation neural network for classifying the absence and presence of heart disease. This proposed solution used 13 neurons in the input layer, 20 neurons in the hidden layer, and 1 output layer neuron. The data here is also taken from the UCI machine learning repository. The dataset is separated into two categories such as input and target. The input and target samples are divided randomly into 60% training dataset, 20% validation dataset, and 20% testing dataset. The training set is presented to the network and the network weights and biases are adjusted according to its error during training. The presence and the absence of the disease are known with the target outputs 1 and 0, respectively. The proposed solution had proved to give 88% accuracy experimentally.

In "Diagnosis of heart disease based BCOA," UCI dataset is used to evaluate the heart attack. This dataset includes test results of 303 people. The dataset used in this work contains two classes, one class for healthy people and the other class for people with heart disease. In this work, a binary cuckoo optimization algorithm (BCOA) is used for feature selection and SVM is used for constructing the model. The final model of this work has accuracy equal to 84.44%, sensitivity equal to 86.49%, and specificity equal to 81.49%.

Chitra (2013) has used the cascaded neural network for the classification purpose. For the accurate prediction of the heart disease the cascaded correlation neural network was considered. The proposed work has taken a total of 270 data samples among which 150 was taken for training and remaining for testing to simulate the network architecture. The number of input neurons for the proposed work is 13 and the number of output neurons is 1. The training set accuracy of 72.6% and the testing accuracy of 79.45% is obtained by using ANN with a back-propagation algorithm. It is 78.55% and 85% for testing and training, respectively, in the case of CNN. Experimental results prove that the accuracy of CNN increased 3% than ANN. When the performance of the above is analyzed, the Cascaded correlation neural network provided accurate results with minimum time complexity in comparison with ANN.

The work by Lu et al. (2018) proposes a cardiovascular disease prediction model based on an improved deep belief network (DBN). The independent determination of the network depth is done by using the reconstruction error. The unsupervised training and supervised optimization are combined together. A total of 30 independent experiments were done on the Statlog (heart) and heart disease database data sets in the UCI database. The result obtained includes the mean and prediction accuracy of 91.26% and 89.78%, respectively.

Ajam (2015) has studied that ANNs show significant results in heart disease diagnosis. The activation function used is tangent sigmoid for hidden layers and linear transfer function for the output layer.

8.5 METHODOLOGY

There is a lot of data put away in stores that can be utilized viably to guide medical practitioners in decision making in human services. Some of the information obtained from health care is hidden as it collects a huge amount of medical-related data of patients which in turn is used for making effective decisions. Hence, the advanced data mining techniques are used for the above for obtaining the appropriate results.

Here, an effective heart disease prediction system is developed. It uses a neural network for accurate prediction of the risk level of heart disease. The 14 attributes including age, sex, blood pressure, cholesterol, etc., are used by the system. The quality service includes the proper diagnosis of the patients leading to the administration of effective treatments.

The process flow is depicted in Figure 8.2.

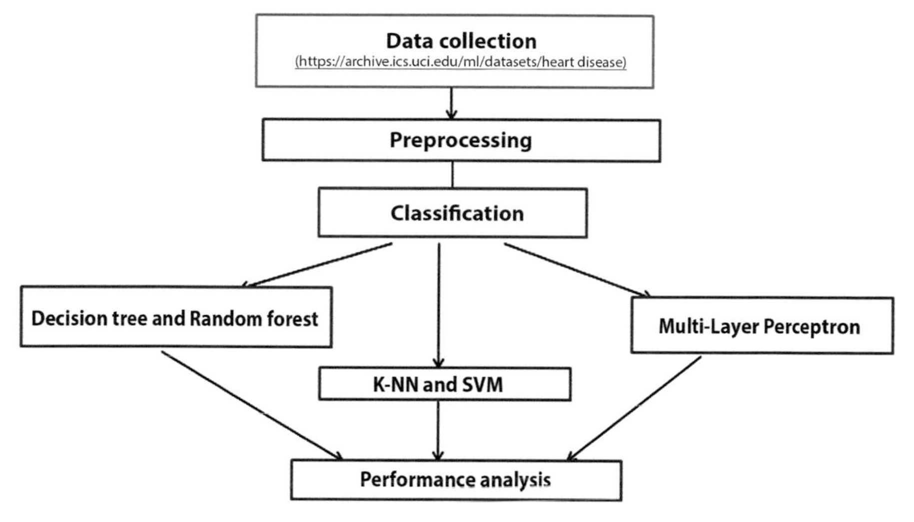

FIGURE 8.2 Flow diagram.

The dataset has been collected from UCI machine learning repository (Cleaveland clinic dataset). The dataset includes 303 records including the 14 attributes. The types of the attributes present in the dataset chosen are as follows (Table 8.2):

- Input attributes
- Key attribute
- Predictable attribute

TABLE 8.2 Dataset Description

Input Attributes	Key Attributes	Predictable Attributes
1. Age in years		
2. Sex represented by values		
0—males		
1—Females		
3. Chest pain type		
4. Resting blood		

TABLE 8.2 *(Continued)*

Input Attributes	Key Attributes	Predictable Attributes
5. Serum cholesterol in mg/dL		
6. Fasting blood sugar		
value 1: >120 mg/dL		
value 0: <120 mg/dL		
7. Resting Electrocardiographic results (values 0: normal, 1: having ST-wave abnormality, 2: showing probable or definite left ventricle hypertrophy)	Patient's ID: Patient's identification number	Diagnosis: Value 1 = <50% , no heart disease Value 0 = >50% , heart disease
8. Maximum heart rate achieved		
9. Exercise induced angina		
(value 1: yes, value: 0: no)		
10. Old peak is the ST depression induced by exercise relative to rest		
11. The slope of the peak exercise ST segment (value 1: unsloping, value 2: flat, value 3: down sloping)		
12. No. of major vessels colored by flouroscopy (value 0–3)		
13. Thal (value 3 = normal; value 6 = fixed defect; vlaue7 = reversible defect)		
14. Obesity		
15. Smoking		

8.5.1 PREPROCESSING

Data can be very intimidating for a data scientist. When working with data, there are so many ways in which data can be used for analysis, out of which preprocessing is the mandatory initial step. Preprocessing is done to transform raw data into an understandable format. Raw data (real-world data) is always incomplete and that data cannot be sent through a model as it would lead to errors.

Normally, some of the following steps must be followed in data preprocessing:

1. Import libraries
2. Read data
3. Checking for missing values
4. Checking for categorical data
5. Standardize the data
6. Principal component analysis (PCA) transformation
7. Data splitting.

8.5.2 IMPORT DATA

As main libraries, Pandas, Numpy, and time can be used.

Pandas: Use for data manipulation and data analysis. *Numpy:* a basic package for scientific computing with Python.

As for the visualization *Matplotlib* and Seaborn are generally used. For the data preprocessing techniques and algorithms, *Scikit-learn* libraries can be made use of.

```
# main libraries
import pandas as pd
import numpy as np
import time
# visual libraries
from matplotlib import pyplot as plt
import seaborn as sns
from mpl_toolkits.mplot3d import Axes3D
plt.style.use('ggplot')
# sklearn libraries
from sklearn.neighbors
import KNeighborsClassifier
from sklearn.model_selection
import train_test_split
from sklearn.preprocessing import
normalize
from sklearn.metrics import
confusion_matrix,accuracy_score,
precision_score,recall_score,f1_
score,matthews_corrcoef,
classification_report, roc_curve
from sklearn.externals import joblib
from sklearn.preprocessing import
StandardScaler
from sklearn.decomposition import
PCA
```

Read Data

Read the data in the CSV file using pandas

```
df = pd.read_csv('../input/creditcard.
csv')
df.head()
```

Checking for missing values

```
df.isnull().any().sum()
> 0
```

Checking for categorical data

The features not in numerical format are converted to categorical data, nominal and ordinal values. Seaborn distplot() can be used to visualize the distribution of features in the dataset.

Standardize the data

The dataset contains only numerical input variables which are the result of a PCA transformation. Features V1, V2, … V28 are the principal components obtained with PCA, the only features that have not been transformed with PCA are "time" and "amount.". So, PCA is effected by scale so we need to scale the features in the data before applying PCA. For the scaling I am using Scikit-learn's StandardScaler(). To fit to the scaler the data should be reshaped within −1 and 1.

```
# Standardizing the features
df['Vamount'] = StandardScaler().fit_transform(df['Amount'].values.reshape(−1,1))
df['Vtime'] = StandardScaler().fit_transform(df['Time'].values.reshape(−1,1))
df = df.drop(['Time','Amount'], axis = 1)
df.head()
```

Now all the features are standardize into unit scale (mean = 0 and variance = 1)

PCA transformation

PCA is mainly used to reduce the size of the feature space while retaining as much of the information as possible. Here, all the options are remodeled exploiting PCA.

```
X = df.drop(['Class'], axis = 1)
y = df['Class']
pca = PCA(n_components=2)
principalComponents = pca.fit_transform(X.values)
principalDf = pd.DataFrame(data = principalComponents
, columns = ['principal component 1', 'principal component 2'])
finalDf = pd.concat([principalDf, y], axis = 1)
finalDf.head()
```

Data splitting

```
# splitting the feature array and label array keeping 80% for the training sets
X_train,X_test,y_train,y_test = train_test_split(feature_array,label_array,test_size=0.20)
# normalize: Scale input vectors individually to unit norm (vector length).
X_train = normalize(X_train)
X_test=normalize(X_test)
```

For the model building, *K*-nearest neighbors (KNN), the algorithms that are considered are KNN, decision tree, support vector machine, random forest, and DBN. The performances of these algorithms are analyzed and the best algorithm for heart disease prediction is identified (Table 8.3).

After we build the model we can use predict() to predict the labels for the testing set.

TABLE 8.3 Prediction Model

```
The model for heart disease prediction is as follows:
deftrain_model(X_train, y_train, X_test, y_test, classifier, **kwargs):
 # instantiate model
 model = classifier(**kwargs)
 # train model
model.fit(X_train,y_train)
 # check accuracy and print out the results
fit_accuracy = model.score(X_train, y_train)
test_accuracy = model.score(X_test, y_test)
 print(f"Train accuracy: {fit_accuracy:0.2%}")
 print(f"Test accuracy: {test_accuracy:0.2%}")
 return model
```

8.5.3 DECISION TREE

Decision trees is a classification algorithm where the attributes in the dataset are recursively partitioned. Decision tree contains many branches and leaf nodes. All the branches tell the conjunction of the attributes that leads to the target class or class labels. The leaf nodes contain class labels or the target class that tells to which class the tuple belongs.

This serves as a very powerful algorithm for the prediction of breast cancer. There are several decision tree algorithms available to classify the data. Some algorithms include C4.5, C5, CHAID, ID3, J48, and CART. Decision tree algorithm is applied to the dataset and accuracy obtained is 94.6% that proves that the decision tree algorithm is a powerful technique for breast cancer prediction.

The tree can be built as:

- The attribute splits decides the attribute to be selected.
- The decisions about the node to represent as the terminal node or to continue for splitting the node.
- The assignment of the terminal node to a class.

The impurity measures such as information gain, gain ratio, Gini index, etc. decides the attribute splits done on the tree. After pruning, the tree is checked against overfitting and noise. As a result, the tree becomes an optimized tree. The main advantage of having a tree structure is that it is very easy to understand and interpret. The algorithm is also very robust to the outliers also. The structure of the decision tree is shown in Figure 8.3.

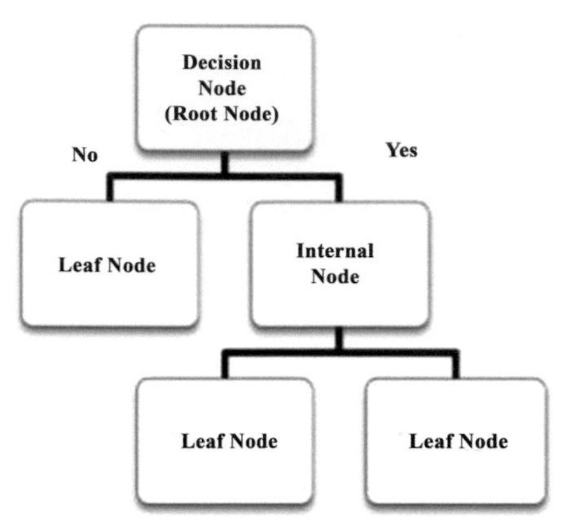

FIGURE 8.3 Binary DS.

The equation for the decision tree is:

model = train_model(X_train, y_train, X_test, y_test, DecisionTreeClassifier, random_state=2606)

8.5.4 *K*-NEAREST NEIGHBOR

The KNN algorithm which is known as KNN is a nonparametric method used for various purposes like regression and classification. The output of the classification from KNN is a class label that describes to which class or group it belongs to. In KNN, the membership is assigned based on the majority vote by the neighbors which is decided by the *K* value.

The input consists of the *k* closest training examples in the feature space. The output depends on whether KNN is used for classification or regression.

The KNN algorithm assumes that similar things exist in close proximity. In other words, similar things are near to each other.

In other words, the object is assigned to the class which is most common among the neighbors. For example, if $k = 2$ then the query point is assigned to the class to which the nearest two neighbors belong to. Being the most simplest machine learning algorithm the explicit training step is not required. The neighbors are taken from the set of the objects for which the class is known that can be considered as the training step. The algorithm is very delicate or sensitive when it comes to the local data. Euclidean distance for continuous variables and Hamming distance for discrete variables are most widely used.

However, the usage of the specialized algorithms like large margin nearest neighbor or neighborhood components analysis helps in the improvement of the accuracy. The *k* value is chosen based on the data. The appropriate *K* value must be chosen so that most of the tuples are classified correctly. Heuristic techniques like hyperparameter optimization are used because the larger *k* value reduces the effect of noise but makes the boundary between the classes less distinct.

The performance gets degraded when the noise in the data gets increased. The accuracy achieved from KNN is about 95% with an appropriate *K* value. KNN is shown in Figure 8.4.

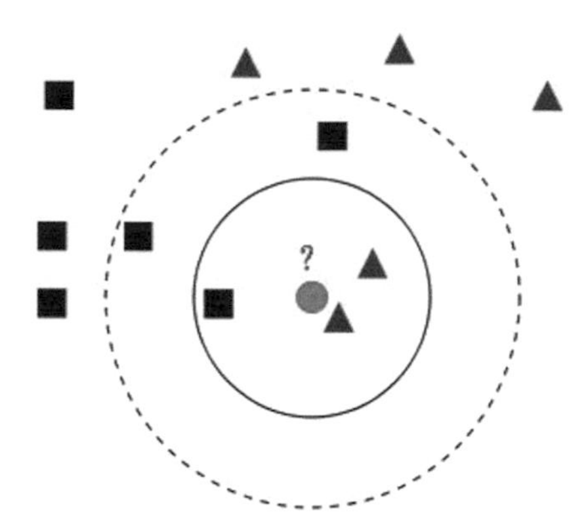

FIGURE 8.4 *K*-nearest neighbor.

Source: https://www.analyticsvidhya.com/blog/2018/03/introduction-k-neighbours-algorithm-clustering/

The equation is as given below:

model = train_model(X_train, y_train, X_test, y_test, KNeighborsClassifier)

8.5.5 SUPPORT VECTOR MACHINE

The support vector machine is also known as SVM in machine learning is a supervised learning model that analyzes the data used for classification and regression analysis using associated learning algorithms. The SVM, an emerging approach is a powerful machine learning technique for classifying cases. It has been employed in a range of problems and they have a successful application in pattern recognition in bioinformatics, cancer diagnosis, and more.

In a SVM, an *N*-dimensional space is considered for plotting all the data points. SVM not only performs linear classification but also the nonlinear classification where the inputs are implicitly mapped to the high dimensional feature space. SVM is considered to be advantageous because of a unique technique called Kernel Trick, where a low dimensional space is converted to high-dimensional space and classified. Thus, the SVM constructs a hyperplane or set of hyperplanes. The construction is done in a high or infinite-dimensional space for usage in classification, regression, or outlier detection.

The hyperplane that has the largest functional margin is said to achieve a good functional margin. Thus, the generalization error of the classifier is reduced. The SVM gives an accuracy of 97% when employed on the dataset. The hyperplane that is used for classification purposes is shown in Figure 8.5.

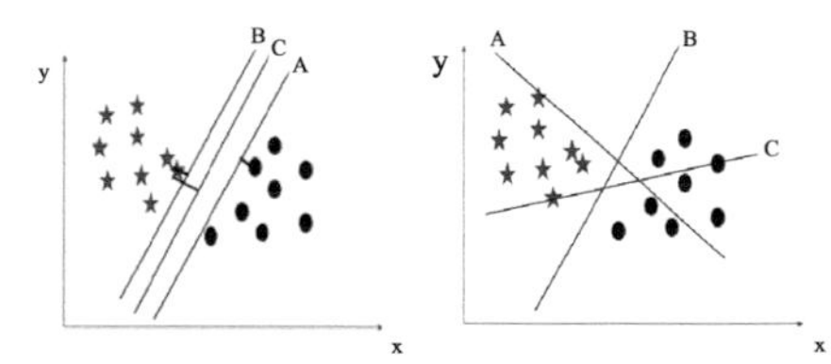

FIGURE 8.5 Support vector machine (SVM).

The equation is given as

model = train_model(X_train, y_train, X_test, y_test, SVC, C=0.05, kernel='linear')

8.5.6 RANDOM FOREST

For the classification of whether it is malignant or benign the random forest algorithm could be employed. Based on some type of randomization, random forest has to be built that is an ensemble of decision trees. The random forest is most widely used for its flexibility. The random forest is a supervised learning algorithm. Here, the algorithm creates a forest with many trees.

The accuracy increases with a large number of trees. The advantage of using random forest is that it could be used for both classification and regression. It could also handle the missing values and it will not overfit in case of more number of trees. Random forest works by taking the test features and predicting the outcome of the randomly created trees based on rules and then stores the result. The votes for each predicted target is calculated since each tree results in different prediction.

Finally, the target receiving high vote will be considered as the final prediction. The random forest is capable of processing a huge amount of data at a very high speed. The structure of each tree in random forest is binary that is created in a top-down manner. The random forest achieves a faster convergence rate. The depth and number of trees are considered to be the most important parameters. Increasing the depth is evident to increase the performance. Thus, random forest is considered as the best classification algorithm in terms of processing time and accuracy. The random forest algorithm is applied on the dataset and the accuracy obtained is 97.34%. The overall view of the random forest algorithm is shown in Figure 8.6.

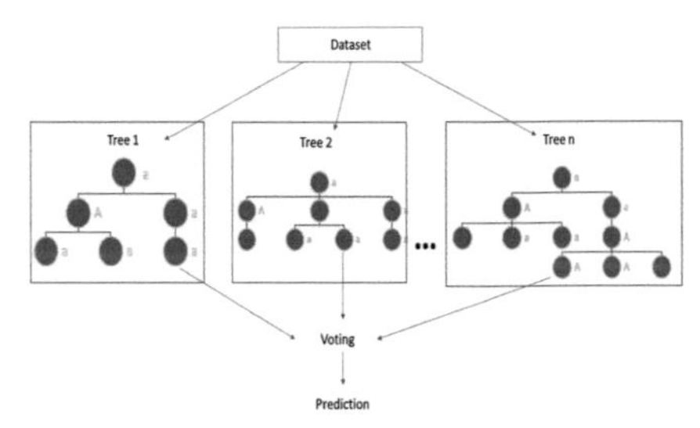

FIGURE 8.6 Random forest.

The equation is as:

model = train_model (X_train, y_train, X_test, y_test, RandomForestClassifier, n_estimators=110, random_state=2606)

8.5.7 MULTILAYER PERCEPTRON

A multilayer perceptron is a feed-forward ANN that is used for classification. The layers present in the multilayer perceptron are an input layer, a hidden layer, and an output layer. Multilayer perceptron adapts a supervised learning technique for the training where the data are associated with the class label. The MLP differs from the other traditional models with its multiple layers and nonlinear activation. The MLP consists of a minimum of three layers. In some cases, it may be more than three with an input and an output layer with one or more hidden layers. Since MLPs are fully connected, each node in one layer connects with a certain weight to every other node in the succeeding layer.

The amount of error is calculated at each iteration and the connection weights are updated. Therefore, it is an example of a supervised learning algorithm. The backpropagation that is carried out is based on the least mean square algorithm in the linear perceptron. A multilayer network contains input, hidden, and output neurons called ANN. ANN is a computing system that performs the task similar to the working of the brain. It is used for solving problems. The collection of nodes connected in a network transmits a signal and process the artificial neurons. The activation node contains the weight values. A neural network is trained to teach the problem-solving technique to the network using training data. The accuracy obtained from the constructed model using a multilayer perceptron is 96.5% (Figure 8.7).

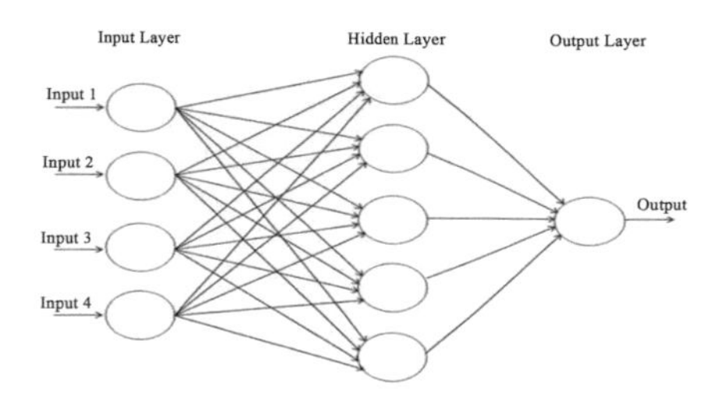

FIGURE 8.7 Multilayer perceptron.

```
for i in range(1,8):
print ("max_depth = "+str(i))
train_model(X_train, y_train, X_test, y_test, MultilayerPerceptronClassifier, max_depth=i, random_state=2606)
```

8.6 RESULT ANALYSIS

Decision Tree:

```
Train accuracy: 85.38%
Test accuracy: 85.71%
```

FIGURE 8.8 Accuracy of decision tree.

The prediction model is trained using a decision tree algorithm and the obtained results are shown in Figure 8.8. The attributes of the dataset are recursively partitioned and we obtained the accuracy as 85% for the selected training dataset.

K-nearest neighbor (KNN):

The k value is chosen based on the data. The appropriate K value must be chosen so that most of the tuples are classified correctly. The accuracy obtained is 83% as shown in Figure 8.9, which can also be improved by using other algorithms.

```
Train accuracy: 90.57%
Test accuracv: 83.52%
```

FIGURE 8.9 Accuracy of KNN.

Support vector machine:

All the data points are plotted in an *N*-dimensional space. SVM uses linear classification, and in addition to that, it also uses nonlinear classification. The obtained accuracy is 87% as shown in Figure 8.10.

```
Train accuracy: 84.91%
Test accuracy: 87.91%
```

FIGURE 8.10 Accuracy of SVM.

Random forest:

```
Train accuracy: 100.00%
Test accuracy: 89.01%
```

FIGURE 8.11 Accuracy of random forest.

Based on some type of randomization, random forest has to be built that is an ensemble of decision trees. The random forest is used most widely due to its flexibility. The accuracy increases with a large number of trees and obtained 89% as shown in Figure 8.11.

Multilayer perceptron:

A neural network is trained to teach the problem-solving technique to the network using training data. The accuracy obtained from the constructed model using a multilayer perceptron is 60% as shown in Figure 8.12.

```
Train accuracy: 78.77%
Test accuracy: 60.44%
```

FIGURE 8.12 Accuracy of MLP.

8.7 DISCUSSION

The setup was built on a hardware arrangement of Intel i7 GPU with the capacity of 16GB RAM using Python. The results show that the machine learning algorithms work well for heart disease prediction and the model is trained by applying different algorithms. Each algorithm works on the model to fit and trains the model using a training dataset and the model is tested for the different datasets, test data for each of the algorithms. The accuracy that is obtained is listed. By examining the results, we found that random forest produced the highest accuracy when compared to all other algorithms (Figure 8.13). Since it constructs a multitude of decision trees and chooses the subset of features for classification, it has the highest accuracy of all others (Table 8.4).

TABLE 8.4 Accuracy Evaluation

Techniques	Train Accuracy	Test Accuracy
Random forest	100	89.01
SVM	84.91	87.91
MLP	78.77	60.44
Decision tree	85.38	85.71
KNN	90.57	83.52

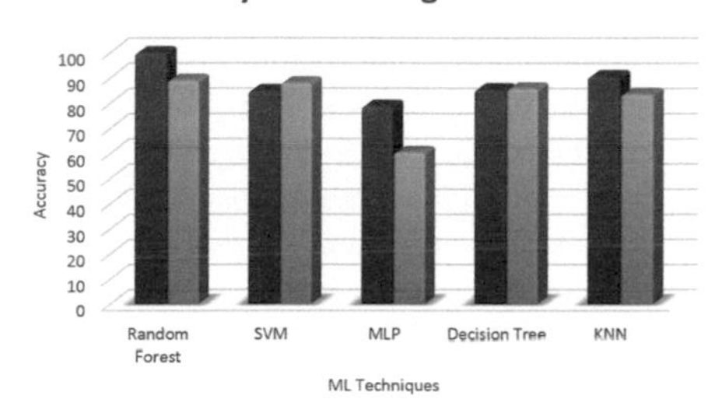

FIGURE 8.13 Accuracy for various ML techniques.

Time complexity is measured between all the algorithms and the same is as given in Table 8.5.

TABLE 8.5 Time Complexity Evaluation

Techniques	Time Complexity (ms)
Random forest	550
SVM	608
MLP	700
Decision tree	730
KNN	1000

The graphical representation is shown in the form of a graph in Figure 8.14

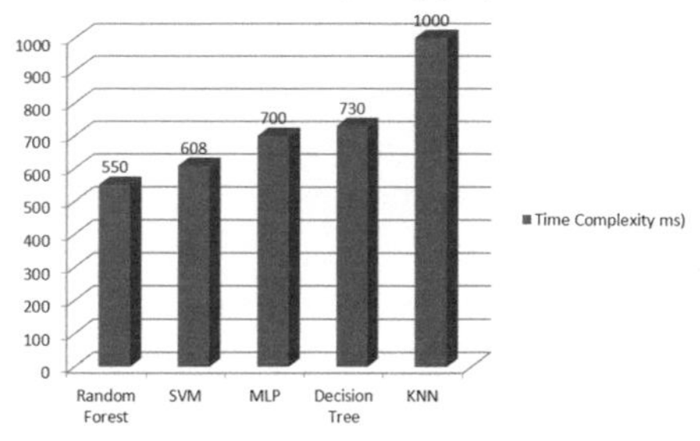

FIGURE 8.14 Analysis of time complexity.

The graph shows the response time evaluation of various machine learning techniques and random forest proves to suit well than other methods. The hybridization of SVM with the random forest is expected to serve the purpose better.

8.8 CONCLUSION

Data from various health organizations like the World Health Organization and Global Burden of Disease states that most of the death is due to the cardiovascular diseases. When the emergency treatment is given at the correct duration than the chances of survival increase. A lot of work has been done already in making models that can predict whether a patient is likely to develop heart disease or not. The various analysis and studies were made on the research papers of heart disease prediction. The random forest has the most wider usage. The accuracy increases with a large number of trees and shows 89% that is considered to be better than other techniques. Prediction accuracy of the existing system can be further improved, so in the future, new hybrid algorithms are to be developed that overcomes the drawbacks of the existing system. In the future, an intelligent cognitive system may be developed that can lead to the selection of proper treatment methods for a patient diagnosed with heart disease, well ahead. Also, the same methodologies can be applied for prenatal, oncological, and rare disease predictions to enhance the overall health care.

KEYWORDS

- **health care**
- **heart disease**
- **prediction**
- **machine learning**
- **accuracy**

REFERENCES

Ajam, N. "Heart diseases diagnoses using artificial neural network," Network and Complex Systems, vol. 5, no. 4, 2015, pp. 7–11, ISSN: 2224-610X (Paper), ISSN: 2225-0603(Online).

Chen, A. H., Huang, S. Y., Hong, P. S., Cheng, C. H., and Lin, E. J. "HDPS: Heart Disease Prediction System," Computing in Cardiology, 2011, pp. 557–560, ISSN: 0276-6574.

Chitra, R. and Seenivasagam, V. "Heart attack prediction system using cascaded neural network," Proceedings of the International Conference on Applied Mathematics and Theoretical Computer Science, 2013.

Chitra, R. and Seenivasagam, V. "Heart disease prediction system using supervised learning classifier," Bonfring International Journal of Software Engineering and Soft Computing, vol. 3, no. 1, 2013.

http://www.mahesh-vc.com/blog/understanding-whos-paying-for-what-in-the-healthcare-industry. Reference cited in https://medium.com/@fkigs/the-healthcare-system-1c2733d88aa9

https://www.saylor.org/site/wpcontent/uploads/2012/06/Wikipedia-Decision-Tree.pdf.

Hurwitz, J., and Kirsch, D. "Machine Learning for Dummies, IBM Limited Edition," John Wiley & Sons, Inc. 2018.

Khemphila, A. and Boonjing, V. "Heart disease classification using neural network and feature selection," Proceedings of 21st International Conference on Systems Engineering, 2011.

Kim, J., Kang, U., and Lee, Y. "Statistics and deep belief network-based cardiovascular risk prediction," vol. 23, no. 3, 2017, pp. 169–175.

Lu, P., Guo, S., Zhang, H., Li, Q., Wang, Y., Wang, Y., and Qi, L. "Research on improved depth belief network-based prediction of cardiovascular diseases," Journal of Healthcare Engineering, Hindawi, vol. 2018, 2018.

Olaniyi, E. O. and Oyedotun, O. K. "Heart diseases diagnosis using neural networks arbitration," International Journal of Intelligent Systems and Applications, vol. 12, 2015, pp. 75–82.

Sayad, A. T. and Halkarnikar, P. "Diagnosis of heart disease using neural network approach," International Journal of Advances in Science Engineering and Technology, vol. 2, no. 3, Jul. 2014.

Soni, J., Ansari, U., and Sharma, D. "Intelligent and effective heart disease prediction system using weighted associative classifiers," International Journal on Computer Science and Engineering (IJCSE), vol. 3, no. 6, Jun. 2011.

CHAPTER 9

A REVIEW ON PATIENT MONITORING AND DIAGNOSIS ASSISTANCE BY ARTIFICIAL INTELLIGENCE TOOLS

SINDHU RAJENDRAN[1*], MEGHAMADHURI VAKIL[1], RHUTU KALLUR[1], VIDHYA SHREE[2], PRAVEEN KUMAR GUPTA[3], and LINGAIYA HIREMAT[3]

[1]*Department of Electronics and Communication, R. V. College of Engineering, Bangalore, India*

[2]*Department of Electronics and Instrumentation, R. V. College of Engineering, Bangalore, India*

[3]*Department of Biotechnology, R. V. College of Engineering, Bangalore, India*

[*]*Corresponding author. E-mail: sindhur@rvce.edu.in*

ABSTRACT

In today's world, digitalization is becoming more popular in all aspects. One of the keen developments that have taken place in the 21st century in the field of computer science is artificial intelligence (AI). AI has a wide range of applications in the medical field. Imaging, on the other hand, has become an indispensable component of several fields in medicine, biomedical applications, biotechnology, and laboratory research by which images are processed and analyzed. Segmentation of images can be analyzed using both fuzzy logic and AI. As a result, AI and imaging, the tools and techniques of AI are useful for solving many biomedical problems and using a computer-based equipped hardware–software application for understanding images, researchers, and clinicians, thereby enhancing their ability to study, diagnose, monitor, understand, and treat medical disorders.

Patient monitoring by automated data collection has created new challenges in the health sector for extracting information from raw data. For interpretation of data quickly and accurately the use of AI tools and statistical methods are required, for interpretation of high-frequency physiologic data. Technology supporting human motion analysis has advanced dramatically and yet its clinical application has not grown at the same pace. The issue of its clinical value is related to the length of time it takes to perform interpretation, cost, and the quality of the interpretation. Techniques from AI such as neural networks and knowledge-based systems can help overcome these limitations. In this chapter, we will discuss the key approaches using AI for biomedical applications and their wide range of applications in the medical field using AI. The different segmentation approaches their advantages and disadvantages. A short review of prognostic models and the usage of prognostic scores quantify the severity or intensity of diseases.

9.1 INTRODUCTION

Artificial intelligence (AI), termed as smart, today's technology, has entered broadly across all possible sectors—from manufacturing and financial services to population estimates. AI has a major hold in the healthcare industry as they have a wide level of applicability because of their accuracy in a variety of tasks. It has helped to uncover some of the hidden insights into clinical decision-making, communicate with patients from the other part of the world as well as extract meaning from the medical history of patients, which are often inaccessible unstructured data sets. The productivity and potentiality of pathologists and radiologists have accelerated because of AI. Medical imaging is one of the most complex data, and a potential source of information about the patient is acquired using AI.

Health management includes fault prognostics, fault diagnostics, and fault detection. Prognostics has been the latest addition to this game-changing technology that goes beyond the existing limits of systems health management.

The main reason for the exponential boom of AI is that its scope of innumerable applications in care-based applications has widened within healthcare. It is seen that the healthcare market using AI has a growth rate of almost 40% in a decade. Right from delivering advanced vital information to physicians promptly to initiating informed choices, customized real-time treatment facilities are the basis for the revolutionizing care of the patients that are appraised as the applications of AI.

An extensive quantity of data is obtained to clinical specialists, starting from details of clinical symptoms to numerous styles of organic chemistry knowledge and outputs of imaging devices. Every form of knowledge provides info that has got to be evaluated and assigned to a selected pathology throughout the diagnostic process. To contour the diagnostic method in daily routine and avoid misdiagnosis, the computing method may be utilized. These adaptive learning neural network algorithms will handle various styles of medical knowledge and integrate the classified outputs.

The variety of applications in today's world using AI is as follows:

Diagnosis: Diagnosis of disease in healthcare is one of the challenging parts. With the assistance of AI, machines are powered up with the flexibility to search large data from existing medical images, indicating early detection of the many disorders, through diagnostic imaging victimization neural networks. It has numerous applications in proactive diagnosing of the probability of tumor growth, stroke, etc.

Biomarkers: For choosing ideal medications and to assess treatment sensitivity, bookmarkers are used that automatically provide accurate data of the patients as both audio and video of the important health parameters. The precision and fastness of biomarkers make them the most ideal diagnosis tools, which is used to highlight the possibilities of any disorder. By using these, there is promptness in diagnosing the diseases.

AI and drug discovery: The study of multiple drug molecules and their structure accurately and promptly is done using AI to predict their pharmacological activity, their adverse effects, and potency, which are the most cost-effective routes of drug discovery. This is mainly used across pharmaceutical companies, thereby reducing the cost of medications drastically. AI-based drug discovery has led to the auxiliary treatment of neurodegenerative disorders and cancer.

AI-enabled hospital care: Smart monitoring of intensive care unit and IV solutions in hospitals using AI has simplified care delivery. Robot-assisted surgeries have been booming with the intervention of AI in routine phlebotomy procedures. Other applications of AI among the hospital are patient medication chase, nursing worker performance assessment systems, patient alert systems, and patient movement chase. The main advantage is a decrease in dosage errors and an increase in the productivity of nursing staff.

In this chapter, we will emphasize on the applications of AI in different sectors of healthcare.

9.1.1 BENEFITS OF AI IN THE MEDICINE FIELD

The importance of AI is booming, especially in the medical field; considering that the email address details are extremely precise and fast, these are a few advantages of this stream, some of them are because there is technical advancement in all sectors, as follows:

- *Quick and precise diagnosis*: For diagnosis of certain diseases, immediate action measures should be taken before they become serious. AI has the power to find out from past situations, and also, it has tried to prove that these networks can quickly diagnose diseases such as malignant melanoma and eye-related problems.
- *Diminish human mistakes*: It is normal for humans to make small mistakes, as the profession of doctors is extremely perceptive since they focus on a great deal of patients that can be very exhaustive, thus leading to a lack of activeness that compromises patient safety. In this case, AI plays a major part in assisting physicians by reducing human errors.
- *Low prices:* With the increase in technological advancements, the patients can get assistance by doctors without visiting clinics/hospitals. AI provides assistance to patients via online and with their past wellness documents, thereby reducing the cost.
- *Virtual presence*: Technology is rolling out so much that this assistance can be given to clients living in remote locations by utilizing a technology called Telemedicine.

9.1.2 ARTIFICIAL NEURAL NETWORKS

Synthetic networks such as artificial neural networks (ANNs) have various applications in branches of chemistry, biology, as well as physics. They are also used extensively in neuroscientific technology and science. ANNs have actually an extensive application index that is utilized in chemical dynamics, modeling kinetics of drug release, and producing classification that is agricultural. ANNs are also widely used in the prediction of behavior of industrial reactors and plants that are agricultural determination. Generally speaking, an understanding of biological item classification and knowledge of chemical kinetics, if not clinical parameters, are handled in an identical way. Machine strategies such as ANN apply different inputs in the initial stage (Abu-Hanna and Lucas, 1998). These input files are forms of processes that are well within the context of the formerly known history of the defined database to generate an appropriate output that is expected (Figure 9.1).

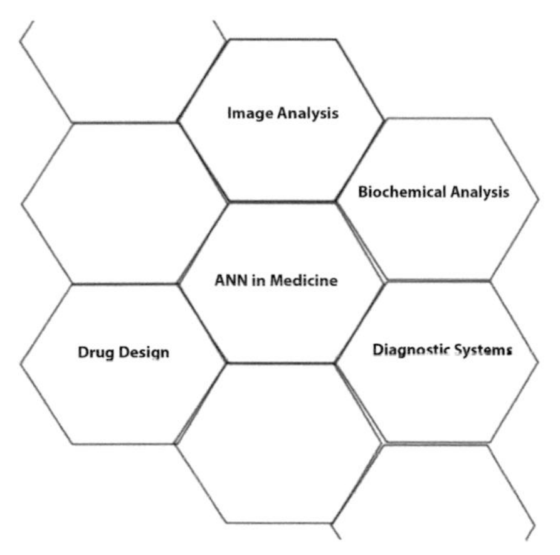

FIGURE 9.1 Application of ANN in the medical sector.

(*Source:* Reprinted from Abu-Hanna and Lucas, 1998.)

9.1.2.1 FUNDAMENTAL STEPS IN ANN-BASED MEDICAL DIAGNOSIS

The flow diagram describing the analysis of ANN starts with clinical situations and is shown in Figure 9.2. This flow diagram gives us an overview of the involved processes or steps that utilize ANN in medical diagnosis (Ahmed, 2005).

The first step is initiated with the network receiving a patient's data to make a prediction of the diagnosis. The next step would be feature selection. Once the diagnosis is completed, feature selection takes place. Feature selection provides the necessary information to differentiate between the health conditions of the patient who is being evaluated. The next step is building of the database itself. All the data available are validated and finally preprocessed. With the help of ANN, the training and verification of database using training algorithms can be used to predict diagnosis. The end diagnosis as predicted by the network itself is further evaluated by a qualified physician.

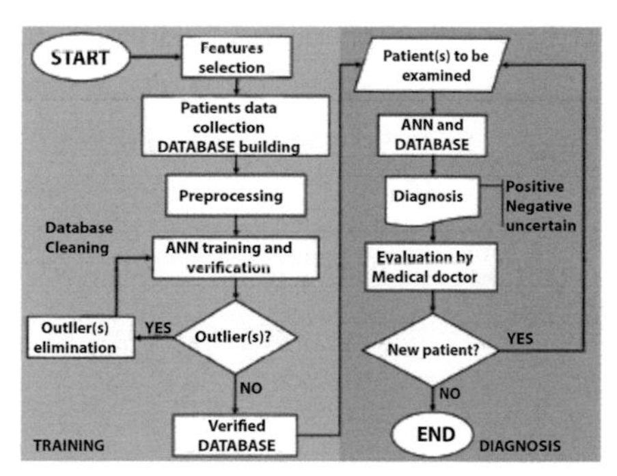

FIGURE 9.2 Flow diagram of steps in ANN-based diagnosis.

(*Source:* Reprinted from Abu-Hanna and Lucas, 1998.)

1. *Feature Selection*: The exact recognition designation of a disease is normally predicted using varied and mostly incoherent or confusing

data. The factors that significantly affect any kind of diagnosis are instrumental data and laboratory data and largely determined by the convenience of the practitioner itself. Clinicians are provided with ample training to enable them to extract the required relevant information from every sort of data and point out any kind of diagnosis that can be done. In ANN, this particular information is referred to as features. Features may range from being biochemical symptoms to any other information that gives insight into what the ailment could possibly be. The last diagnosis is linked to the level of expertise associated with the skilled clinician. ANNs have actually higher flexibility, and their ability to compare the information with formerly stored samples is what has enabled fast medical diagnosis. Varieties of neural networks are feasible to solve sensory activity problems, whereas some are adapted for purposeful information modeling and approximation. Irrespective of the features chosen, the people selected to train the neural system must be robust and clear indicators of a given clinical scenario or pathology. The choice of features depends upon medical expertise choices done formerly. Thus, any short, nonspecific information that is redundant to the investigation itself is avoided. Selection/extraction of appropriate features among other ones that are obtainable is sometimes allotted a victimization varied approach. The primary tools that can be utilized for variable selection are as follows:

a. Powerful mathematical means of information mining.
b. Principal component analysis.
c. A Genetic algorithm program.

Utilizing the help of suitable trainee examples, we train the network with the "example" data of one patient that is fed, examined, and collected as a feature. The major component that affects the prediction of diagnosis, quality of training, and the overall result is the training sample used. Enough number of samples whose diagnosis is well known must be within the database used for training to enable the community to extract the provided information hidden within the database. The network employs this knowledge that is extremely assessing the new cases. Despite this, laboratory data received from clinics should be easily transferable with other programs for computer-aided diagnosis (Aleksander and Morton, 1995).

2. *Building the database*: Multilayer feed-forward neural networks, such as Bayesian, stochastic, recurrent, and fuzzy are used. The optimal neural community architecture for the maximum values for both training

and verification should be selected in the first stage. These are obtained by testing sites that have multiple layers that have concealed nodes in them.

3. *Training algorithm*: There are multiple techniques for obtaining necessary training. The most common algorithm is backpropagation. The backpropagation rule requires two training parameters, namely, learning and momentum rates. In many cases, below par generalization ability of the network is the total result of high values of parameters. High-value parameters cause learning instability. Therefore, the performance is generally flawed. The training parameter values rely on the studied system's complexity. The worth of momentum is not up to that of the learning rate; in addition, the sum total of their values should be equal to 1.

a. *Verification*: ANN-based medical diagnosis is confirmed by means of a dataset that is not utilized for training.

Robustness of ANN-based approaches: It has been well established that ANNs have the potential to tolerate a noise that is particularly within the data and usually offer sufficient accuracy associated with the end result or in cases like prediction of diagnosis. This sound can possibly cause deceptive results on the bright side. This situation or anomaly is discovered to be occurring while modeling complicated systems, for example, human health. This noise would not only impact the uncertainty that is conventional of measured data but also influence secondary facets, including the presence of more than one disease. Crossed effects cannot be foreseen unless they need to be considered throughout the development of this training database (Alkim et al., 2012). Any issue that alters or affects the outward indications of the condition under study should be considered by including instances that are in the database. By this means, the network can precisely classify the in-patient. Of course, a technique to avoid this can be a combination of the expertise associated with the clinical specialist and the power of ANN-aided approaches.

4. *Testing in medical practice*: The last step in ANN-aided diagnosis is testing in medical practice. The outcome of this system is carefully examined by a practitioner for every patient. Medical diagnosis data of patients without any error may be finally enclosed inside the training database. Nonetheless, an in-depth analysis of ANN-assisted diagnosing applications in the clinical environment is important even throughout completely different establishments. Verified ANN-aided diagnosis that have medical applications in clinical settings is a necessary condition for additional growth in drugs.

9.2 APPROACHES OF AI IN BIOMEDICAL APPLICATIONS

AI allowed the development of various algorithms and implemented them within the field of medical imaging system. Based on physical, geometrical, statistical, and functional strategies, a wide range of algorithms and methods are developed using AI, and these strategies are used to solve the problems of feature extraction, image, texture, segmentation, motion measurements, image-guided surgery, method anatomy, method physiology, telemedicine with medical images, and many by using pattern image datasets. In recent studies, the medical information analysis is increasing because various and large data sets are collected and these large data sets are analyzed on a daily basis (Amato et al., 2013). The manual analysis of such vast information can mechanically result in human errors. So, to avoid such mistakes, automated analysis of the medical information needs to be implemented. Thus, the automated medical information analysis is currently in demand. As a result, usage of AI techniques specifically in medical proves useful because it will store huge information, retrieve data, and provide most fascinating use of knowledge analysis for higher cognitive processes in finding issues (Shukla et al., 2016).

9.2.1 MEDICAL IMAGE ANALYSIS

Many image processing techniques are employed in automatic analysis of medical images. Before segmenting the image, operations related to preprocessing have to be undertaken, for example, noise removal, image improvement, edge detection, and so on. After the preprocessing operations, the image is now ready for analysis. Mining of the region of interest (ROI) from the image is completed within the segmentation section by a combination of intelligent ways. Later, feature extraction or even selection of features is performed to spot and acknowledge the ROI, which can be a neoplasm, lesion, abnormality, and so on. Figure 9.3 shows the automated medical image segmentation mistreatment intelligent methods.

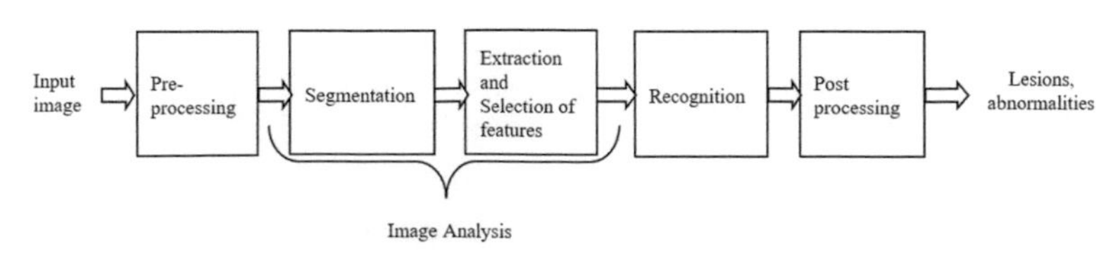

FIGURE 9.3 Automatic medical image segmentation by intelligent methods.
(*Source:* Adapted from Rastgarpour and Shanbehzadeh, 2011.)

9.2.2 *IMAGE SEGMENTATION*

In the healthcare system, for medical imaging it is necessary to diagnose the disease accurately and provide correct treatment for the disease. To solve such a complex issue, algorithms for automatic medical image analysis are used, which provide a larger and correct perceptive of medical images with their high reliability (Costin and Rotariu, 2011). As a result, by applying these intelligent methods, correct analysis and specific identification of biological options will be done. AI of such type is suggested to use digital image processing techniques and medical image analysis along with machine learning, pattern recognition, and fuzzy logic to improve the efficiency.

Figure 9.4 shows the final theme of the medical image analysis system.

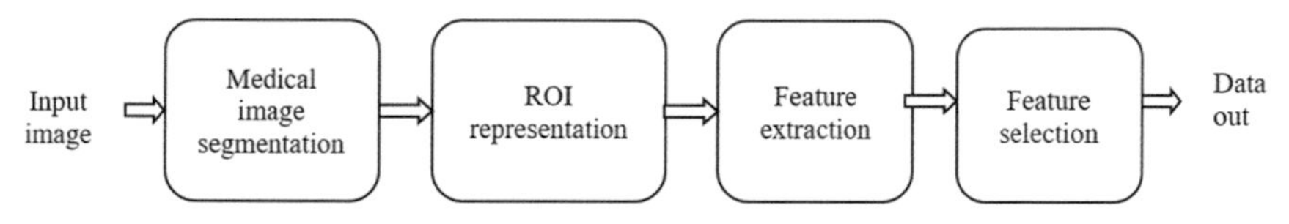

FIGURE 9.4 General representation of medical image analysis system.
(*Source:* Adapted from Rastgarpour and Shanbehzadeh, 2013.)

It is tough to investigate the disease in automatic medical image analysis. As a result of an incorrect image, segmentation can usually result in incorrect analysis of the image, and also, improper segmentation affects the later steps, particularly ROI illustration and extraction in medical image analysis. So, to get correct diagnosis of diseases or sequent lesions in medical image analysis, segmentation of the ROIs must be done accurately. Thus, a correct segmentation methodology is important. In medical applications, segmentation identifies the boundaries of ROIs as well as tumors, lesions, abnormalities, bony structures, blood vessels, brain elements, breast calcification, prostate, iris, abdomen, respiratory organ fissure, gristle knee, and so on.

9.2.3 *METHODS OF IMAGE SEGMENTATION*

1. Image Segmentation Using Machine Learning Techniques

(a) Artificial neural network (ANN): ANN is a statistical model that works on the principle of neural networks. There are three layers in the neural network model; the first one is the primary layer and called as input layer, the next one is the middle layer that includes a number of hidden layers, and the last one is the output layer. As the number of hidden layers grows, the capability of solving the problem accurately also increases (Deepa and Devi, 2011). ANN uses an iterative training method to solve the problems by representation of weights. Hence, this method is a very popular classification method. It has a simple physical implementation arrangement, and its class distribution is complex but can be mapped easily. There are two types of techniques used by a neural network: (1) supervised classification techniques and (2) unsupervised classification techniques (Jiji et al., 2009). Medical image segmentation based on clusters and classification is achieved by using "self-organizing maps" of neural networks, which are very helpful in the application for decision making of "computer-assisted diagnosis" as well as categorization. In ANN, the supervised classification technology uses various methods such as Bayesian decision theory, support vector machine (SVM), linear discriminant analysis (LDA), etc. Each method uses different approaches to provide a unique classification result. Usually, the information of the data is divided into two phases, training and test phases to classify and validate the image (Figure 9.5).

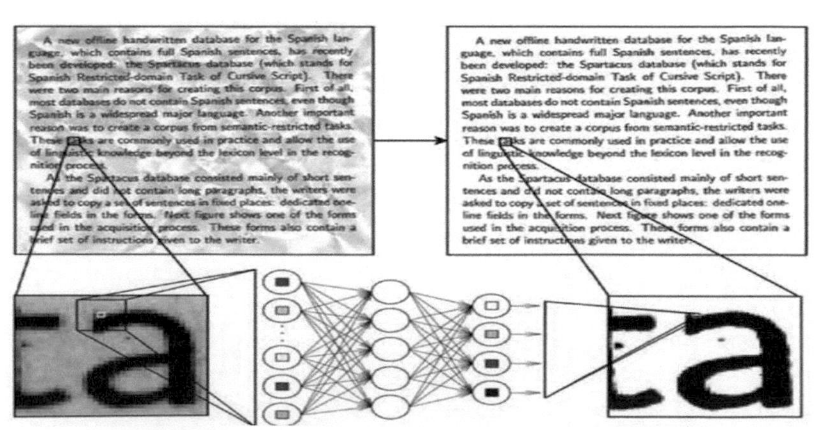
A new offline handwritten database for the Spanish language, which contains full Spanish sentences, has recently been developed: the Spartacus database (which stands for Spanish Restricted-domain Task of Cursive Script). There were two main reasons for creating this corpus. First of all, most databases do not contain Spanish sentences, even though Spanish is a widespread major language. Another important reason was to create a corpus from semantic-restricted tasks. These tasks are commonly used in practice and allow the use of linguistic knowledge beyond the lexicon level in the recognition process.

As the Spartacus database consisted mainly of short sentences and did not contain long paragraphs, the writers were asked to copy a set of sentences in fixed places: dedicated one-line fields in the forms. Next figure shows one of the forms used in the acquisition process. These forms also contain a brief set of instructions given to the writer.

FIGURE 9.5 ANN applied on an image document. The left figure shows the noise scanning of the image, and the right figure is a clear image after using ANN.

(*Source:* Adapted from Shuka, et al., 2016.)

(b) Linear discriminant analysis (LDA): The LDA method is used as a dimensionality reduction technique that reduces the feature space that is very large. LDA computes very best transformation by reducing the within-class distance and increasing the between-class distance at the same time, consequently achieving most of the category discrimination. By applying the Manfred Eigen decomposition on the scatter matrices, the simplest attainable transformation in LDA may be pronto-computed (Amin et al. (2005). It has been widely used in numerous applications relating to high-dimensional information. On the opposite hand, classical LDA needs the disputed total scatter matrix to be necessarily nonsingular. In some applications that involve high-dimensional and low-sample-size information, the overall scatter matrix may be singular. As a result, the info points are from a really high-dimensional area, and therefore, the sample size might not surpass this dimension. This is often called singularity downside in LDA (Mandelkow et al., 2016).

(c) Naive Bayes (NB): The NB method could be a straightforward technique for constructing classifiers models; the class labels are assigned to the instances of the problem delineated as vectors of feature values; from some finite sets, the class labels are drawn. This type of classifier does not use a single algorithm, but there are a bunch of algorithmic techniques that have their own principles. All NB classifiers assumes that, given the class variable, the features are independent of each other, that is, they have different feature values at different feature sets. In simple words, it can be explained with an example, such as by using NB method, a fruit can be recognized as an apple based on its shape if it is round, its color if it is red, and its diameter if it is about 10 cm, which means every feature value such as color, diameter, and roundness is independent of each other (Zhang, 2016).

(d) Support vector machine (SVM): Supervised learning models are referred to as SVM of the machine learning technique, which understands the data with the help of different learning algorithms for classification and multivariate analysis. SVM behaves as a nonprobabilistic binary linear classifier, such as when there exists a given set of partially or fully trained examples, the SVM method models a new example for one or the other category of the training sets. The SVM method develops a model like points in a space, that is, the examples of two different categories are kept apart by some distance in a space so that when a new example is fed it should fall under one of the two categories that are s by some distance.

2. Image Segmentation Using Fuzzy Logic

Image processing fuzzy logic is the cluster of all methodologies that represent, recognize, and process medical pictures, as a fuzzy sets, their options, and segments. The process and illustration depend upon the chosen fuzzy sets and the difficulty to be settled. The three elementary levels are as follows:

- image fuzzification,
- image defuzzification, and
- modification of membership values (Amin et al., 2005).

Fuzzification refers to the writing of medical image data and deciphering of the results, referred to as defuzzification, which are the two steps that produce a conceivable method to convert a medical image in fuzzy sets. The central step is that the elementary force of fuzzy image processing (adjustment of the values of membership). Once the dynamics of image data changes from a gray-level plane to a membership plane (called as fuzzification), correct fuzzy approaches modify the membership values.

A fuzzy logic has two completely different approaches, one is region-based segmentation and another is contour-based segmentation. In region-based segmentation, one looks for the attributes, division, and growth of the region by using a threshold type of classification. In contour-based segmentation, native discontinuities like mathematical morphology, derivatives operators, etc. are examined (Khan et al., 2015).

3. Image Segmentation Using Pattern Recognition

Supervised ways embrace many pattern recognition techniques. Several pattern recognition ways are forward explicit distributions of the options and are known as constant ways. For instance, the maximum likelihood (ML) technique normally assumes variable mathematical distributions. This means that the covariant E-matrices for every tissues that are calculable from a user provided coaching set, typically found by drawing regions of interest

(ROI) on the pictures. The remaining pixels are then categorized by conniving the chance of every tissue class and selecting the tissue sorted with the very best chance. Constant ways are solely helpful once the feature distributions for the various categories are well known, so that the case for main pictures is not essential. Statistic ways, like *k*-nearest neighbors (kNNs) do not depend on predefined distributions, rather on the particular distribution of the coaching samples themselves. kNN has given superior results in terms of accuracy and duplicability compared to constant ways. Conjoint reporting is the use of ANNs and a call tree approach (Secco et al., 2016).

4. Image Segmentation Using Textural Classification

Based on the texture of the image, textural classification of image segmentation is divided into four categories as geometrical, statistical, model-based, and signal processes. In this classification method, a number of sample images are trained based on their texture, and the testing image is assigned to one of the trained texture images for accurate classification and segmentation. This type of classification method is very much efficient for almost all types of digital image classifications. Another important part of the textural classification is wavelet transform, in which spatial frequencies are evaluated at multiple scales by designing wavelet function. "Markov random fields" are the most accepted methods to model the images to extract special information of an image in which the method says that every element of pixel intensity depends only on the neighborhood pixel intensity. These models assume that the intensity at every picture element within the image depends on the intensities of solely the neighbor pixels. The next important step is the feature selection or feature extraction. This is very important because the segmentation of image and classification is purely based on which type of feature is selected; the machine learning algorithm uses the statistical feature extraction method, for example, features such as entropy, energy, standard deviation, mean, homogeneity, and so on are extracted to achieve the best classification result.

5. Image Segmentation Using Data Mining Technique

The data mining technique is used for extracting the features of the image, segmenting the image, and also for the classification of the segmented image. As this technique is also part of AI, it performs the classification along with other methods such as pattern recognition, machine learning, databases, and statistics to extract the image information from the large set of data. The data mining technique is used for many applications that have a large set of images.

Such a large set of images is analyzed by one of the popular methods called cluster method in which the features that have the same characteristics are combined as one large set unit for analysis. The C-means algorithmic rule is the preferred method based on cluster analysis.

The decision tree algorithm method is another method used for classification of images by the data mining technique. This type of method uses association rules. The algorithm follows the tree pattern that helps the user to segment and classify the image efficiently and makes the process easy.

9.3 MEDICAL DIAGNOSIS USING WORLD WIDE WEB

Figure 9.6 indicates the web-based medical diagnosis and corresponding prediction. It involves four components, namely, a diagnosis module, a prediction module, a user interface, and a database. The model can both diagnose and predict. It consists of two sets of databases, namely, a patient database and a disease database. The patient database comprises personal information such as name, address, and medical history of the patient if any. The disease database consists of all the information regarding patient's illness, which includes the type of disease, treatments taken and suggested, and tests encountered. This is mainly bifurcated to improve the storage of patients' records, which helps the other departments to utilize the records when the patients are referred to them, thereby providing centralized information access to secure the information from unauthorized users. The prediction module makes use of the neural network technique for prediction of illness and condition of the patient based on similar previous cases. The data obtained via databases will be used for training and testing. On the other hand, the diagnosis module comprises specialist systems and fuzzy logic techniques for performing the tasks. The specialist system follows a set of rules defined based on the two databases and information about the disease. This system uses these rules to identify patient's ailments or diseases based on their current conditions or symptoms. The fuzzy logic techniques are used along with the system to improve the performance and enhance the reasoning. In Figure 9.6, World Wide Web (WWW) is an essential part for interaction, and the figure also describes how WWW is acting as an interface between the system and patients, taking data from the specified databases.

9.4 AI FOR PROGNOSTICS

Out of the various methods being currently utilized and developed, prognostic models are also constructed.

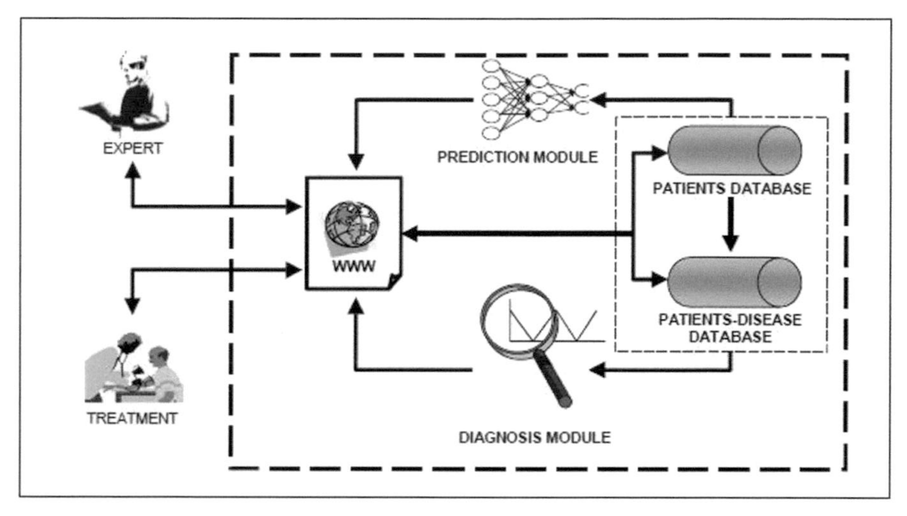

FIGURE 9.6 Web-based diagnosis and prediction.
(*Source:* Reprinted from Ishak and Siraj, 2008.)

An integrated system health management (IHSM) takes certain input values from thc signals and is able to provide fault detection, fault identification, and ultimatcly fault prognostics. An effective diagnostic should provide both fault identification and isolation. This is rather effective than just notifying an unusual pattern in the sensor values. Prognostics may be defined as a method of detecting a failure and also the time period before which an impending failure occurs. To reduce damage, we must be able to carry out a diagnosis before performing prognostics. ISHM also gives a response command to the detected failure, and these responses may vary from changing the configuration of the hardware components to even recalibration of sensor values (Figure 9.7 and Table 9.1).

A very crude model of prognostics has been used. This makes use of statistical information regarding the time required for a failure to occur in a certain component of a system. This statistical information can also make the life prediction of all the other components present in that system. Howevcr, thcsc arc only to help predict the life of a component. The reason for failure is not predicted using this basic model (Asch et al., 1990).

Modeling the system via ANN is the widely used approach to prognostics. ANN models create a mapping or a relationship between the input and the output. These relationships or parameters may be modified to get the optimum result. This modification process is usually done by exposing the network to a

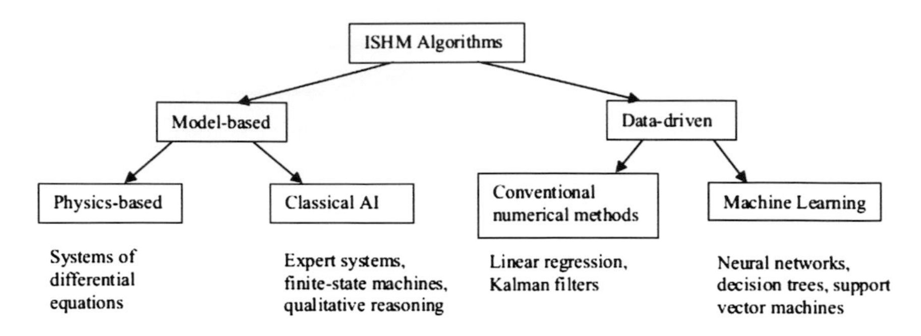

FIGURE 9.7 Taxonomy of ISHM algorithms.

(*Source:* Reprinted from Schwabacher and Goebel, nd.)

TABLE 9.1 Sample Method for Each Set of Algorithm Type (Rows) ISHM Problem (Columns) in Data-Driven Prognostics

	Faculty Detection	Diagnostics	Prognostics
Physics based	System Theory		Damage propagation models
AI-model based	Expert systems	Finite state machines	
Conventional numerical	Linear regression	Logistics regression	Kalman filters
Machine learning	Clustering	Decision trees	Neural networks

(*Source:* Reprinted from Schwabacher and Goebel, nd.)

number of examples and recording its response to those situations. This helps minimize the errors. A popular method to reduce these errors is gradient descent. Another approach is novelty detection. This learns optimal behavior of the system, and when the sensor output fails to tally with the given or expected data, an anomaly is indicated. Other data mining techniques also try and discover a pattern in the dataset.

Fuzzy logic is also a popular method used in AI. Using this, we can translate or derive a meaning regarding the qualitative knowledge and solve the problem in hand. Using linguistic variables, dynamic systems are modeled. This variable is interpreted as an elastic constraint that is propagated by fuzzy interference operations. The mechanism of reasoning gives fuzzy logic incredible robustness regarding the variations and disturbances in the data and parameters. This can be applied to prognostics in addition to machine learning algorithms. This would help reduce the uncertainty that most prognostics estimations fail to address. The learning algorithm to nullify any uncertainty is implemented, such as lazy or Q-learning process.

9.4.1 *APPLICATION OF PROGNOSTICS*

Several systems have seen an upward growth when it comes to utilizing prognostic models. These include different engineered systems, actuators, bearings, aircrafts, electronic devices and turbines, helicopter gear boxes, hydraulic pumps, etc. A joint strike fighter is being developed, which will be used by the US Air Force. The idea is to implement prognostics and fault detection mechanisms into every system available in an aircraft. Using rule-based architecture and model-based and data-driven methods, we can improve safety and also significantly reduce the damage and cost.

The prognostic models are used in various applications, such as

(a) deriving healthcare policy by creating global predictive scenarios,
(b) supporting relative review among hospitals by (case-mix modified) mortality predictions,
(c) defining research eligibility of clients for ground breaking remedies,
(d) outlining addition requirements for medical studies to manage for variation in prognosis, (age) price reimbursement programs, and
(e) Selecting appropriate tests and methods of therapy in certain client management systems with supportive decisions on withdrawing or withholding treatment.

As supportive decision making tools, it is beneficial to differentiate between two kinds of models based on prognostics: models throughout the patient populace in addition to the level that is person specific. Both of these amounts imply different needs regarding the model concerning which prognostic techniques can be employed for building and assessing the model.

The population versus certain degree aspects gives a fair idea of two quantities of Shortliffe's typology of decision-support functions in several cases. Shortliffe differentiates between three types of decision help functions, which are as follows:

- information management,
- concentrating attention, and
- patient-specific assessment.

A prognostic model to the population degree found an excellent assurance system that can bring about the recognition of the discrepancy between your anticipated and real prices of problems originated after surgery of a group of patients, which therefore help spot the reasons why this discrepancy is used, for example. A prognostic model can be used as a foundation for providing treatment advice to the patient, which is for certain connected with patient-specific assessment functions. The thought of "group" is, needless to state, coarse as you team with circumstances where two Method Inform Med clients are stratified in a large number of various subgroups, for example, danger teams, since it covers circumstances where the entire populace is customer-taken.

9.4.2 MODEL BUILDING

Many ways can be suggested to represent prognostic models, for example, quantitative and probabilistic approaches. Models need to be integrated in the medical management of the subject or patient that needs assessment.

There already exists a massive body of experience required to develop prognostic models for the field of medical statistics. Some of the techniques that can be used are decision-based rules that depend on a prognostic score and the Bayes rule.

The simplest prognostic instrument used in medicine is assessment using the prognostic score that helps in the classification of a patient into a future risk category. This score system gives us a method to quantify the seriousness of the diseases in hand. The higher the score, the greater the severity. The factors that contribute or influence this scoring system are as follows:

1. physiological variables,
2. demographic variables, and
3. laboratory findings.

Each factor accounts to "penalty points" for the prognostic scoring system. This prognostic score may be used independently in medical diagnosis.

Survival analysis consists of a collection of statistical methods for time-dependent data. A few methods have been employed for this type of survival analysis. We are basically only interested in the timeline, that is, when the event will occur. Although predicting the survival of patients' sounds condescending, this technique can be used to predict the time of complete recovery after treatment or usual expected time of initiation of a particular symptom after being subjected to certain environments to study the patients and diseases well.

9.4.3 ASSESSMENT OF PROGNOSTIC MODELS

Over the past few years, the art of evaluating prognostic models has become crucial. The one question that arises for this model is, will this work on a population that is different from that used to develop the model itself? This is the main challenge faced by this model. Let us understand how to solve this problem.

There are two types of evaluations, namely, clinical and laboratory evaluations. The criteria for a model to pass a laboratory test is the comparison with the statistical tests previously collected. This kind of analysis has so many disadvantages, listed as follows:

a. inaccuracy,
b. imprecision,
c. inseparability, and
d. resemblance.

In clinical evaluation, the model is checked to satisfy its clinical purpose and whether its performance is efficient enough to deal with clinical prediction. The other factor that governs the choice of models is whether the model is needed to operate on the individual or population level. For example, prediction of infection causing death over a large risk group or various risk groups and the classification of the vital status of an individual patient cannot be evaluated under the same bracket.

These evaluation methods require a great deal of care. In the machine learning community, many models are being developed, each better than the one before. The next challenge that arises is the one using the correct statistical test? Thus, tests have to be modified and adapted according to the requirement.

9.5 APPLICATIONS OF PATIENT DIAGNOSIS AND MONITORING

All aspects of the universe have some very favorable advantages. However, they also have their own characteristic disadvantages. Medical diagnosis requires complex systems to run errands for applications such as fitness maintenance, and these systems are guided by ANNs. The flow of procedure is given below.

ANNs have the following advantages:

- They deliver work room to accomplish inspection, association, illustration, and categorization of medicinal information.
- They generate innovative tools for supporting medicinal decision making and investigation.
- They combine actions in health, workstation, intellect, and also sciences.
- They provide amusing content for the upcoming technical therapeutic field.

There were a lot of challenges in the field of education, reasoning, and language processing. The wide range of applications of AI has solved all of these challenges by introducing a modern and digital way of getting things done. Some of the applications of AI in the medical field are discussed below.

1. **Virtual inquiry system:** This modern technology is used to teach hospitals, medical aspirants in colleges, as well as inhabitants. Clinicians make use of this online system to interact with the patients. Basically, in this system, the software

accumulates data of thousands of patients; it is then interpreted by experts and AI as subject cases, which gives an overview of clinical issues. This system provides opportunity for the medical students for diagnosis and provides treatment plans and develop problem solving skills in clinical aspects, and the teachers understand the student's perception and adjust the course accordingly.

Through interactions with the multiple cases, the students are able to understand the skills involving disease diagnosis. Simultaneously, the system has the ability to detect any error that the students make while studying the case. The system can solve all of these difficulties with the help of deep learning and analysis. The psychological steps can be tracked by using a tool named intelligent tutor system; this tool diagnoses the concepts wrongly interpreted and approximates the student's understanding extent in the field and guides accordingly based on the feedback provided.

2. Medical distance learning: Web-based teaching methods are used for communication, sharing, and learning. In the field of medicine, mobile nursing and clinical practice teaching play a vital role. Some of the teaching methods such as microblogging and virtual simulation training have seen their applications in remote transmission technology of pathological films and imaging. Their application is also seen in instant transfer technology active monitoring, online storage technology, integrated platform technology, self-healing technology, and computer-aided diagnosis. All of these have had a significant impact on the methods of teaching. At present (as of 2019), 50 state-level medical education projects have been approved, and more than 4000 projects have been announced every year. Most of the medical colleges have already been equipped with this facility. Each year, more than 1500 experts contribute to distance continuing medical education covering more than 20 secondary disciplines and 74 tertiary disciplines.

3. Influence of AI on distance medical education: Using information technology, resource libraries and data centers can be constructed for recruitment of students. This is can be also used for the training process management as well as evaluation. This can help improve service level of medical education and efficiency.

9.6 CONCLUSION

With the growing population and the need for technology, AI has played a vital role in the healthcare sector by providing different technological developments covering both urban and rural sectors in terms of assistance given by doctors. ANNs have tried appropriately for reasonable identification of a variety of diseases; additionally, their use makes the diagnosis a lot consistent and thus will increase patient satisfaction. However, despite their extensive application in modern identification, they need to be thought of solely as a tool to facilitate the ultimate call of a practitioner, who is eventually liable for crucial analysis of the output obtained. In this chapter, we discuss about the need for AI, benefits in the medical industry of using AI, one of the budding fields, prognostics, and the prototype systems for making predictions, and the different approaches and the neural network algorithms used mainly for diagnosis of a wide variety of diseases. The chapter is concluded by the applications of AI technology in the field of healthcare.

9.7 FUTURE VISION

With the recent advancement in the techniques of AI, discussed in the chapter, fuzzy logic, image processing, ML, and neural networks have focused on superior augmentation of diagnosis data by computer. Key features such as reliability, less dependency on operator, robustness, and accuracy are achieved using image segmentation and classification algorithms of AI. Blue Brain, Human Brain, Google Brain are some of the several on-going projects in the field of AI. Integration of science and mathematics has invariably result in ground-breaking advancements within the medical trade. AI technology has seen notable advancements, particularly in medical specialty analysis research and medicines, thereby raising the hopes of society to a higher level. It has far more to supply in the coming years, given that support and comfortable funding are formed on the market.

KEYWORDS

- **artificial intelligence**
- **fuzzy logic**
- **neural networks**
- **prognostics**
- **biomedical**
- **applications**

REFERENCES

Abu-Hanna A, Lucas PJF. Intelligent Prognostic Methods in Medical Diagnosis and Treatment Planning. In: Borne P, Iksouri, el Kamel A, eds. *Proceedings of Computational Engineering in Systems Applications (IMACS-IEEE)*, UCIS, Lille, 1998, pp. 312–317.

Ahmed F. Artificial neural networks for diagnosis and survival prediction in colon cancer. Mol Cancer. 4: 29, 2005.

Aleksander I, Morton H. An Introduction to Neural Computing. International Thomson Computer Press, London, 1995.

Alkim E, Gürbüz E, Kiliç E. A fast and adaptive automated disease diagnosis method with an innovative neural network model. Neural Netw. 33: 88–96, 2012.

Amato F, López A, Peña-Méndez E M, Vaňhara P, Hampl A, Havel J. Artificial neural networks in medical diagnosis. J ApplBiomed. 11:47–58, 2013. doi 10.2478/v10136-012-0031-xISSN 1214-0287

Amin S, Byington C, Watson M. Fuzzy Inference and Fusion for Health, 2005.

Asch D, Patton J, Hershey J. Knowing for the sake of knowing: the value of prognostic information. Med Decis Mak. 10: 47–57, 1990.

Costin H, Rotariu H. Medical Image Processing by using Soft Computing Methods and Information Fusion. In: *Recent Researches in Computational Techniques, Non-Linear Systems and Control*. ISBN: 978-1-61804-011-4

Deepa SN, Aruna Devi B. A survey on artificial intelligence approaches for medical image classification. Indian J Sci Technol. 4(11): 1583–1595, 2011, ISSN: 0974-6846.

Goebel K, Eklund N, Bonanni P. Fusing Competing Prediction Algorithms for Prognostics. In: *Proceedings of 2006 IEEE Aerospace Conference*. New York: IEEE, 2006.

Ishak, WHW, Siraj, F. Artificial Intelligence in Medical Application: An Exploration. 2008. https://www.researchgate.net/publication/240943548_artificial_intelligence_in_medical_application_an_exploration

Jiji GW, Ganesan L, Ganesh SS. Unsupervised texture classification. J Theor Appl Inf Technol. 5(4): 371–381, 2009.

Khan A, Li J-P, Shaikh R A, Medical Image Processing Using Fuzzy Logic. IEEE, 2015.

Mandelkow H, de Zwart JA, Duyn JH. Linear discriminant analysis achieves high classification accuracy for the bold FMRI response to naturalistic movie stimuli. Front Hum Neurosci. 10: 128, 2016.

Rastgarpour, M, Shanbehzadeh, J. Application of AI Techniques in Medical Image Segmentation and Novel Categorization of Available Methods and Tools. In: Proceedings of the International Multiconference of Engineers and Computer Scientists, Vol 1, IMECS, 2011. http://www.iaeng.org/publication/IMECS2011/IMECS2011_pp519-523.pdf

Rastgarpour, M, Shanbehzadeh, J. The Status Quo of Artificial Intelligence Methods in Automatic Medical Image Segmentation. International Journal of Computer Theory and Engineering, Vol.5, No. 1, February 2013. https://pdfs.semanticscholar.org/e5f7/e56f19cf3efc460ba5d4ce0188cee664a735.pdf

Schwabacher M, Goebel K. A Survey of Artificial Intelligence for Prognostics. NASAAmes Research Centre. https://ti.arc.nasa.gov/m/pub-archive/1382h/1382%20(Schwabacher).pdf

Schwabacher M, Goebel K. A survey of artificial intelligence for prognostics. In: *Proceedings of the AAAI Fall Symposium—Technical Report*, 2007.

Secco J, Farina M, Demarchi D, Corinto F, Gilli M. Memristor cellular automata for image pattern recognition and clinical applications. In: *Proceedings of the 2016 IEEE International Symposium on Circuits and Systems (ISCAS)*, 2016, pp. 1378–1381. doi:10.1109/ISCAS.2016.7527506

Shiyou L. Introduction to Artificial Intelligence (2nd edition). Xi'an University of Electronic Science and Technology Press, 2002.

Shukla S, Lakhmani A, Agarwal AK. Approaches of Artificial Intelligence in Biomedical Image Processing. IEEE, 2016.

Smitha P, Shaji. L, Mini MG. A Review of Medical Image Classification Technique. In: *Proceedings of the International Conference on VLSI, Communication & Instrumentation (ICVCI)*, 2011.

Taoshen L. Artificial Intelligence. Chongqing University Press, 2002

Zhang Z. Naive Bayes classification in R. Ann Transl Medic. 4(12): 241, 2016.

Zixing C, Guangyou X. Artificial Intelligence and Its Applications (2nd edition). Tsinghua University Press, 1996.

CHAPTER 10

SEMANTIC ANNOTATION OF HEALTHCARE DATA

M. MANONMANI* and SAROJINI BALAKRISHANAN

Department of Computer Science, Avinashilingam Institute for Home Science and Higher Education for Women, Coimbatore 641043, India

Corresponding author. E-mail: manonmaniatcbe@gmail.com

ABSTRACT

In recent times, there has been more research aiming at providing personalized healthcare by combining heterogeneous data sources in the medical domain. The need for integrating multiple, distributed heterogeneous data in the medical domain is a challenging task to the data analysts. The integrated data is of utmost use to the physicians for providing remote monitoring and assistance to patients. Apart from integrating multiple data, there is a large amount of information in the form of structured and unstructured documents in the medical domain that pose challenges in diagnosing the medical data. Medical diagnosis from a heterogeneous medical database can be handled by a semantic annotation process. Semantic annotation techniques help to assign meaningful information and relationships between different sources of data. Incorporating semantic models in artificial intelligence (AI) expert system for prediction of chronic diseases will enhance the accuracy of prediction of the disease since the meaning of the data and the relationships among the data can be clearly described. This chapter aims to address the challenges of processing heterogeneous medical data by proposing a lightweight semantic annotation model. Semantic models have the capability to tag the incoming healthcare data and attach meaningful relationships between the user interface and the healthcare cloud server. From the relationships, prediction and classification of chronic diseases can be easy and at the earliest possible time with enhanced accuracy and low computation time. The main objective envisaged in this chapter consists of proposing a semantic annotation model for identifying patients suffering from chronic kidney disease (CKD). The purpose of the semantic annotation is to enable the medical sector to process disease diagnosis with the help of an Ontograf that shows the relationship between the attributes that represent the presence of CKD represented as (ckd) or absence of CKD represented as (not_ckd) and to attach meaningful relationships among the attributes in the dataset. The semantic annotation model will help in increasing the classification accuracy of the machine learning algorithm in disease classification. A collaborative approach of semantic annotation model and feature selection can be applied in biomedical AI systems to handle the voluminous and heterogeneous healthcare data.

10.1 INTRODUCTION

Semantic annotation of healthcare data aids in processing the keywords attached to the data attributes and deriving relationships between the attributes for effective disease classification (Du et al., 2018). Semantic annotation implemented on the basis of artificial intelligence (AI) expert systems will bring accurate and timely management of healthcare data due to the ever-increasing quantity of medical data and documents. Effective knowledge discovery can be envisaged with the foundation of AI expert systems in the field of medical diagnosis. The complexity of the medical data hinders communication between the patient and the physician that can be simplified with semantic annotation by providing a meaningful abstraction of the features in the diagnosis of chronic diseases.

Proper implementation of semantic annotation in medical diagnosis will ensure that every tiny detail regarding the health of the patient is taken care of and important decisions regarding their health are delivered to the patients through remote access. Heterogeneous medical information such as pharmaceutical's information, prescription information, doctor's notes or clinical records, and healthcare data is generated continuously in a cycle. The collected healthcare data need to be harnessed in the right direction by providing integrity of heterogeneous data for diagnosing chronic illness and for further analysis. Otherwise, the integrity of the vast medical data poses a major problem in the healthcare sector. Semantic annotation of the incoming healthcare data forms the basis of research motivation so that the problem of late diagnosis can be overcome.

10.2 MEDICAL DATA MINING

Medical data mining deals with the creation and manipulation of medical knowledgebase for clinical decision support. The knowledgebase is generated by discovering the hidden patterns in the stored clinical data. The generated knowledge and interpretations of decisions will help the physician to know the new interrelations and regularities in the features that could not be seen in an explicit form. However, there are many challenges in processing medical data like high dimensionality, heterogeneity, voluminous data, imprecise, and inaccurate data (Househ and Aldosari, 2017).

10.2.1 SEMANTIC ANNOTATION OF MEDICAL DATA

10.2.1.1 MEANING

Clinical care and research are increasingly relying on digitized patient information. A typical patient record consists of a lot of composite and heterogeneous data that differ from lab results, patients' personal data, scan images including CT, MRI, physician notes, gene information, appointment details, treatment procedures, and follow-ups. The main challenge encountered in using this huge amount of valuable medical records lies in integrating the heterogeneous data and making use of the results which in turn helps the data analysts to exploit the advantages of medical research. Harnessing this wide range of information for prediction and diagnosis can provide more personalized patient care and follow-up. In many organizations, the data about a patient is maintained in different departments, and at the receiving end, there is a lack of interconnection between the departments that leads to incomplete, unreliable information about the patient. At the physician's end, the data about a particular patient is updated only with regard to details concerning the treatment procedures and follow up only for their department. This results in a fragmented view of a particular patient, and there is a lack of coordination among the different departments in the hospital to get an overall picture of the health of the patient. On the other hand, the clinical decision support system aims to provide increased healthcare assistance with enhanced quality and efficiency (Du et al., 2018). Efficient access to patient data becomes a challenging work with the use of different formats, storage structures, and semantics. Hence, it becomes reasonably advantageous to the medical sector to capture the semantics of the patient data to enable seamless integration and enhanced use of diverse biomedical information.

Semantic annotation is based on the concept of tagging the data with conceptual knowledge so that it can be represented in a formal way using ontology so that the information is understandable and can be used by computers. The basis of the semantic annotation is the creation of tags that are computer-understandable and gives a clear picture of the role of the features in a given dataset. In the medical domain, semantic annotation models can help to overcome the issue of integration of different hardware components by using high-end ontology structure so that the interaction between the user interfaces can be handled easily (Liu et al., 2017).

Semantic annotation can be used to enrich data as well as it can also be a viable solution for the application in semiautomatic and automatic systems interoperability (Liao et al., 2011). Moreover, semantic feature selection processes when applied to medical datasets, provide inference about feature dependencies on the dataset at the semantic level rather than data level so that a reduced feature set arrives that in turn aids in early detection of diseases for patients suffering from chronic illness.

10.2.1.2 NEED FOR SEMANTIC ANNOTATION

When a patient is admitted to the hospital, he is entangled by many medical devices including glucose level sensing devices, heart monitors, blood pressure monitors, and IVs. But, the management of these devices and recording of information from these devices consumes time and are sometimes even prone to errors.

In the present scenario, with the advent of biomedical devices and AI expert systems, handling of patient's data can be done automatically with the help of electronic health record (EHR) systems. These systems provide accurate information about the patient and save time for the nurses so that they can spend valuable time with the patient in providing more care. Added to the advantages provided by biomedical devices and AI expert systems, semantic annotation of the medical data can enhance the wide range of solutions provided in the medical field. Physicians are loaded with a lot of patient information and proper diagnostic solutions can arrive if the data received from heterogeneous devices are semantically annotated. Semantic analysis of medical data provides meaningful relationships between the symptoms of any disease and these relationships aid in the proper diagnosis of the disease (Zhang et al., 2011).

10.2.1.3 INTEGRATING MACHINE LEARNING TECHNIQUES AND BIOMEDICAL EXPERT SYSTEMS TO SEMANTIC ANNOTATION PROCESS

Semantic Annotation of medical data plays an important role in healthcare organizations enabling ubiquitous forms of representing the information that is collected from various heterogeneous sources. By integrating machine learning techniques and biomedical expert system information with the semantic annotation process, complex queries regarding emergency medical situations can be handled with ease and reliable information sharing can be achieved (Pech et al., 2017).

Effective healthcare delivery systems require automation of data sharing among computers that are interconnected so that information sharing is done quickly and at high-speed response time. In the healthcare sector, data interconnection systems and the underlying schema of the devices enable data sharing between and across the stakeholders of the medical domain surpassing the vendor and application details. However, lack of interoperability between the organizational boundaries hinders information exchange between systems via a complex network that is developed by divergent manufacturers.

In medicine, data mining techniques applied in ontology-based semantic annotation processes enable accurate data transfer, secure data sharing, and consistent data exchange surpassing the underlying hardware and physical devices involved in data management. The clinical or operational data are preserved with meaningful relationships so that these data can be accessed and processed at any time when required.

The semantic annotation process provides a clear classification scheme of the features that are relevant in the interpretation of medical data. Semantic annotation models describe the resource provided by the user by the process of annotation of the data as represented in Figure 10.1.

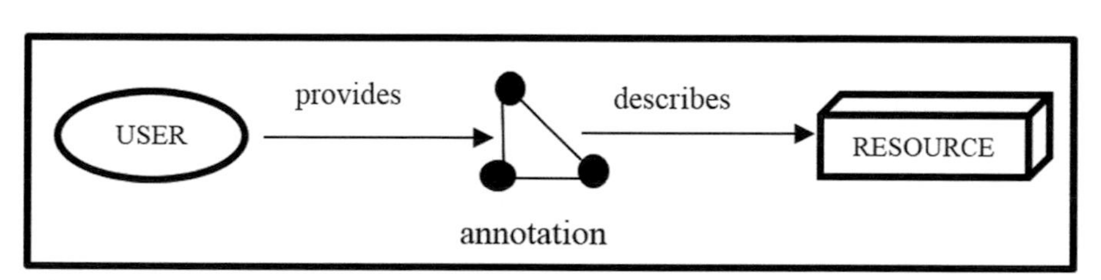

FIGURE 10.1 Generic annotation model.

In a semantic annotation model, the meanings and relations of the data terms are delivered to the computer using ontology objects that, in fact, enriches resource information. Web Ontology Language (OWL) is considered as a standard language for ontology representation for semantic web (Jabbar et al., 2017). The proposed research work is implemented on Protégé 5.0 ontology editor. The Protégé 5.0 ontology editor provides various options for querying and structuring. The query process inbuilt in the Portege tool enables the users to perform a detailed analysis of the input and output data. The results of the semantic annotation process are an OWL instance file, which is available to the end-users of the system and the output file can be processed further for more pattern analysis and knowledge discovery.

10.3 LITERATURE SURVEY

Ringsquandl et al. (2015) have conferred two semantic-guided feature selection approaches for different data scenarios in industrial automation systems. This model was executed to handle the high dimensional data produced by automation systems. To eliminate the highly correlated features and to increase the processing time, the authors have presented the semantic-guided feature selection model. The authors have developed semantic guidance for feature selection that is in fact an intensive task in semantic annotation models. The authors have analyzed the cases of high dimensional data and have operated the features in two different approaches. In the first approach, a large number of features and instances are found. In this case, the authors have used the Web Ontology Language Version 2 Query Language (OWL 2 QL) for Ontology-Based Data Access that uses the T Box level feature selection instead of instance-level approach. In the second approach, there are a fewer number of instances than features. In this case, the authors have introduced an embedded feature selection approach that provided a semantic annotation model for handling engineering knowledge. The results of the semantic feature selection have been represented in a Resource Description Framework (RDF) graph that specifies the reduced feature set that can be fed to the machine learning models for further analysis.

Jabbar et al. (2017) have proposed an IoT-based semantic interoperability model to provide semantic annotations to the features in the medical dataset that helps to resolve the problem of interoperability among the various heterogeneous sensor devices in the healthcare sector. The health status of the patients is monitored and informed to the medical assistants with the help of the sensor devices that operate remotely. The data that is collected through the sensors is semantically annotated to provide meaningful relationships between the features and the class that determines whether the patient is affected with heart disease or not. The dataset that is taken for evaluation is the heart disease dataset that is available in the UCI repository is taken for analysis and the RDF graph is generated to show the relationships between the features and the final class.

Gia ct al. (2015) have used fog computing at smart gateways to provide a high-end health monitoring system. The authors have exploited the advantages of embedded data processing, distributed storage, and notification service in their work. In many of the cardiac disease, electrocardiogram (ECG) feature extraction plays an important role, and hence ECG feature extraction was selected as the case study in this research work. The vital signs of the patients were recorded including ECG signals, P wave, and T wave. Lightweight wavelet transform mechanism was adopted to record the ECG signals, P wave, and T wave. Experiments conducted reveal 90% bandwidth efficiency using fog computing. At the edge of the network, the latency of real-time response was also good for the experiments that were undertaken using fog computing. In this paper, the author has provided an augmented health monitoring system based on fog computing. Apart from proving fog computing at a gateway, the authors have offered better interoperability, graphical user interface with access management, distributed database, a real-time notification mechanism, and location awareness. In addition, they have introduced a versatile and lightweight template extracting ECG features. The results reveal that fog computing at a gateway has in fact produced better results compared to previous achievements.

Chui et al. (2017) have presented the method of diagnosing diseases in smart healthcare. This paper gives a summary of the recent algorithms based on optimization and machine learning algorithms. The optimization techniques discussed in this paper include stochastic optimization, cvolutionary optimization, and combinatorial optimization. The authors have focused their research on diagnosing diseases like cardiovascular disease, diabetes mellitus, Alzheimer's disease and other forms of dementia, and tuberculosis. In addition, the issues and challenges that are encountered in the classification of disease in the healthcare sector have also been dealt with, in this paper.

Antunes and Gomes (2018) have presented the demerits of

the existing storage structure and analytical solutions in the field of data mining and IoT. An automated system that can process word categories easily was devised with an extension to the unsupervised model. They have aimed to provide a solution for semantic annotation, and the Miller–Charles dataset and IoT semantic dataset were used to evaluate the undertaken research work. Among human classification, the correlation achieved was 0.63. The reference dataset, that is, the Miller–Charles dataset was used to find the semantic similarity. A total of 38 human subjects have been analyzed to construct the dataset tath comprises 30 word-pairs. A scaling system of 0 for no similarity and 4 for perfect synonym was adopted to rate the word pairs. Then, 20 frequently used terms were collected and they were ordered into 30-word pairs. Each pair was rated on a scale from 0 to 4 by five fellow researchers. The correlation result was 0.8 for human classification. Unsupervised training methods were used by the author to mark the groups and also to improve accuracy. The model was evaluated based on the mean squared error. For a given target word u and different neighborhood dimensions, the performance of distributional profile of a word represented as DPW (u) and distributional profile of multiple word categories represented as DPWC (u) was calculated for both with affinity and without affinity. Co-occurrence cluster metrics and cosine similarity cluster metrics were evaluated. To take into consideration multiple word categories, the model was extended further to accomplish the multiple words and a novel unsupervised learning method was developed. The issues like noisy dimension from distributional profiles and sense-conflation were taken into consideration and to curb this, dimensional reduction filters and clustering were employed in this model. By this method, the accuracy can be increased, and also this model can be made used more potentially. A correlation of 0.63 was achieved after evaluating the results against the Miller–Charles dataset and an IoT semantic dataset.

Sharma et al. (2015) have presented an evaluation of stemming and stop word techniques on a text classification problem. They have summarized the impact of stop word and stemming onto feature selection. The experiment was conducted with 64 documents, having 9998 unique terms. The experiments have been conducted using nine documents with frequency threshold values (sparsity value in %) of 10, 20, 30, 40, 50, 60, 70, 80, and 90. The threshold is the proportion value instead of the sparsity value. Experimental results show that the removal of stop-words decrease the size of the feature set. They have found the maximum decrement in feature set at sparsity value 0.9 as 90%. Results also indicate that the stemming process affects significantly the size of the feature set with different sparsity values. As they increased the sparsity value the size of the feature set also increased. Only for sparsity value 0.9, the feature set decreased from 9793 to 936. The results further reveal an important fact that stemming, even though it is very important, is not making a negligible difference in terms of the number of terms selected. From the experimental results, it could be seen that preprocessing has a huge impact on performance of classification. The goal of preprocessing is to cut back the number of features that were successfully met by the chosen techniques.

Piro et al. (2016) in their paper have tried to overcome the difficulties encountered in computing the challenging part of the Healthcare Effectiveness Data and Information Set (HEDIS) measures by applying semantic technologies in healthcare. RDFox was used in the rule language of the RDF-triple store and a clean, structured, and legible encoding of HEDIS was derived. The reasoning ability of RDFox's and SPARQL queries was employed in computing and extracting the results. An ontology schema was developed using the RIM modeling standard that depicts the raw data in an RDF-graph format. The authors have implemented a data model in RDFox-Datalog. The RDFox-Datalog was implemented to express HEDIS CDC in rule ontology that is in fact close to the specification language. The data format of RDF proved to be flexible during the development of the rule ontology. Experimental results in fact exceeded the HEDIS data analyst's expectations due to the ability of the semantic technology employed in the work. The results seemed to be exceeding good when the rules were evaluated on the patients' records. By using the high efficiency of RDFox triple store technology, it was possible to fit the details of about 466,000 patients easily into memory and the time taken was 30 min that was low compared to previous works. The discrepancies in the raw data were easily traced using the explanation facilities in RDFox. This enabled reduction in the number of development cycles, and the problems in the vendor solution were resolved. The HEDIS analyst has approved the results and solutions derived by the author based on the application of RDFox.

Lakshmi and Vadivu (2017) have used multicriterion decision analysis for extracting association rules from EHRs. In this paper, the authors have arrived at a novel approach by extracting association rules from patients' health records based on the results of the best association rule mining algorithm implemented for different criteria. The main aim of the research work is to identify the correlation between diseases, diseases and symptoms, and diseases and medicines. Apriori algorithm, frequent pattern (FP) growth, equivalence class clustering and bottom-up lattice traversal, apriori algorithm for transaction database (apriori TID), recursive elimination (RElim) and close algorithms are employed to extract the association rules. Once the association rules are generated the best algorithm is chosen using multicriteria decision analysis. The method that was used for multicriteria decision analysis is the ELECTRE 1 method. The three parameters that are used to evaluate the algorithms are: P1: performance, P2: memory space, and P3: response time. Based on the average number of rules that are generated for different values of support and different values of confidence, the performance metric is calculated. Results indicate that the rules that are generated by the association rule mining algorithms are more or less the same. Initially, the maximum and minimum support values for generating rules were 100 and 10, respectively. The support values were increased and the confidence values were decreased in the next stage. Compared to other algorithms, RElim, apriori TID, and FP-growth algorithms proved to be more effective using multiple criteria decision analysis techniques. To get better classification results, the lift that is a correlation measure is employed to the generated rules of these algorithms.

10.4 SEMANTIC ANNOTATION MODEL FOR HEALTHCARE DATA

The main objective of this research is to provide a semantic annotation model in medical diagnosis that identifies the healthcare data, analyzes keywords, attaches important relationships between the features with the help of knowledge graph, and disambiguates similar entities. The proposed research work is depicted in Figure 10.2.

In the proposed research work as depicted in Figure 10.2, the healthcare data from the UCI is loaded to the ontology editor. Then, fuzzy rule mining with if-then condition is applied to the dataset to identify the keywords in the dataset. The keywords that are extracted from the dataset are in fact tokens that deliver information regarding the medical data in the healthcare sector. The tokens are sent to the tagging section that associates tags for the attribute values. These tags are semantically annotated using the fuzzy rules in the ontology. The rule mining algorithm checks each feature in the dataset and generates a keyword for the satisfied condition. There exists a relationship between the selected features and the binary classification type "yes" if the feature selected satisfies the given if–then condition specified in the rule mining algorithm. If the condition is not satisfied, then a relationship is expressed between the selected feature and the binary classification of type "no." The result is a context-aware data represented in the form of Ontograf that aids in the analysis of the medical diagnosis.

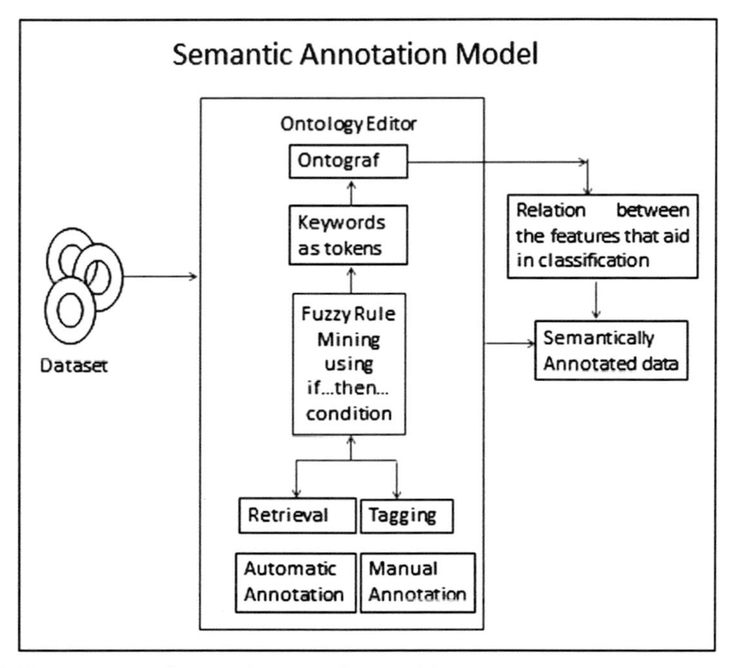

FIGURE 10.2 Proposed semantic annotation model.

10.5 DATASET DESCRIPTION

Experiments were undertaken on CKD dataset taken from the UCI repository that consists of clinical data of patients suffering from kidney disease. The dataset consists of a record of 400 patients suffering from CKD. There a total of 25 features, of which 11 are numeric and 14 are categorical. The features that are present in the CKD dataset are age, blood pressure, specific gravity, albumin, sugar, red blood cells, pus cells, puss cell clumps, bacteria, blood glucose random, blood urea, serum creatinine, sodium, potassium, hemoglobin, packed cell volume, white blood cell count, red blood cell count, hypertension, diabetes mellitus, coronary artery disease, appetite, pedal edema, anemia, and class. The last feature is a binary output variable represented as "ckd" or "notckd."

10.6 EXPERIMENTS AND RESULTS

The dataset is analyzed in the OWL language by creating semantic relationships between the features and the class. The output of the semantic annotation process is an Ontograf that shows the relationships in an ontology graph. Every attribute in the Ontograf has a specific color for identification. The results of an asserted hierarchy of the entire dataset are shown in Figure 10.3.

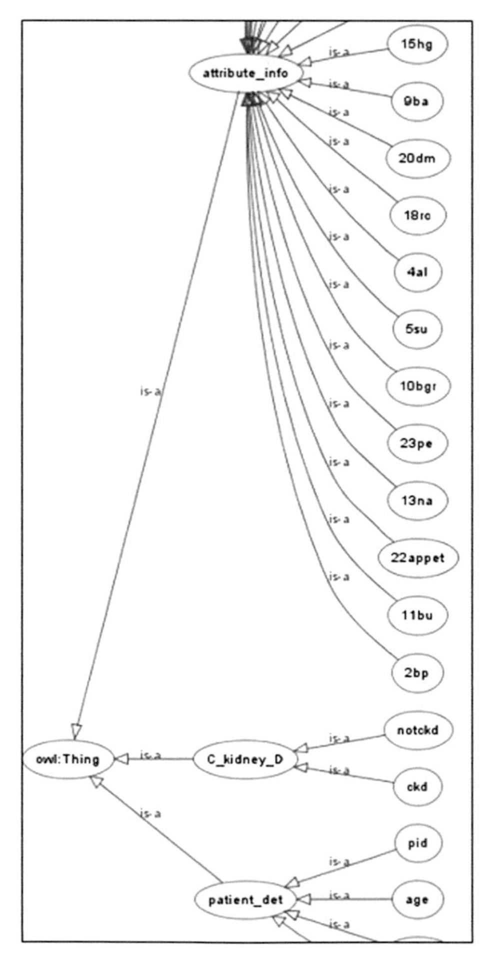

FIGURE 10.3 Ontograf showing the relationship between the attributes and class.

In Figure 10.4 the overall relationship between the main OWL instance, the binary class of CKD, the patient details, and the features related to these classes is depicted.

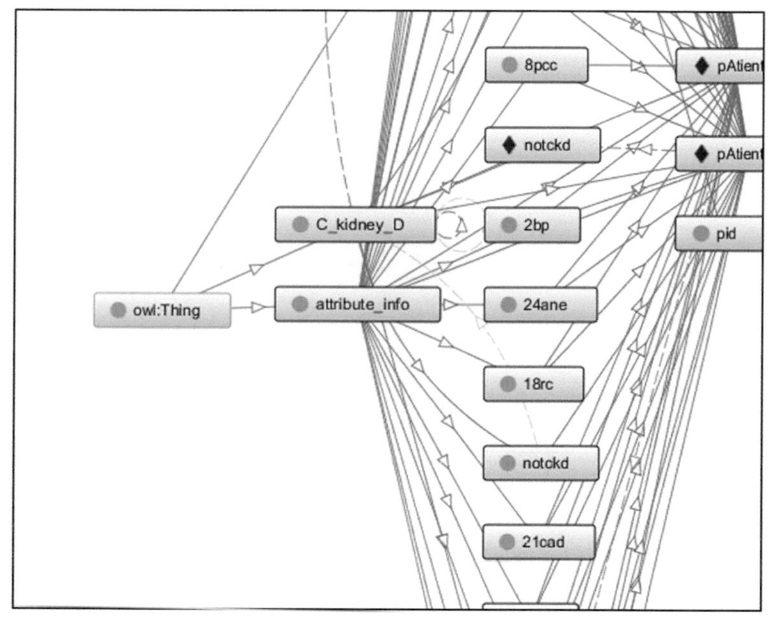

FIGURE 10.4 Ontograf showing overall relationship between the main class, subclass, and features.

In Figure 10.5, the Ontograf is simulated based on the conditions satisfied in the rule mining algorithm. All the features in the dataset are semantically annotated using a fuzzy rule mining algorithm, which shows the interrelationship between the features and the classes.

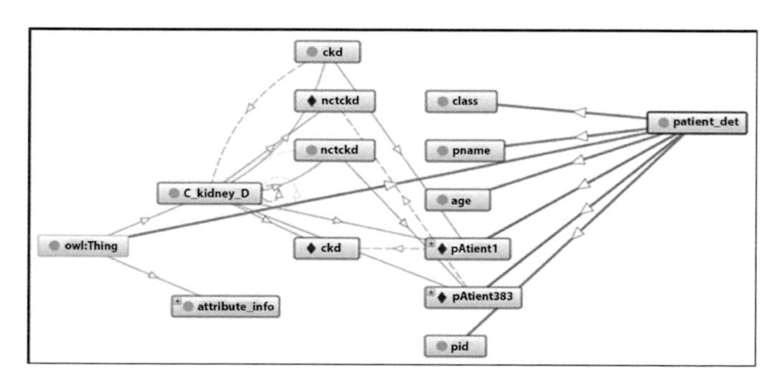

FIGURE 10.5 Ontograf that depicts the semantic relationship between the class and the binary outputs "ckd" and "not ckd."

In Figure 10.6, the ontograf shows the semantic relationship between the patient with id 383 and patient with id 1, who are having ckd and who are not having ckd, respectively.

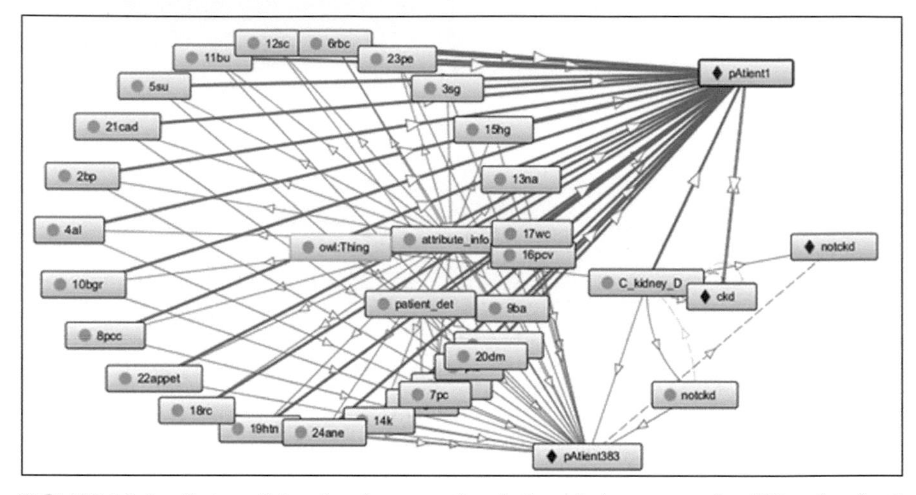

FIGURE 10.6 Ontograf showing the semantic relationship between patient383 and patient1 and binary outputs "ckd" and "notckd."

The clinical data of all patients are annotated semantically based fuzzy rule mining and the relationships indicate the semantic relationship between the features and the binary class "ckd" or "notckd." Thus, semantically annotated data proves to be informative in providing the relationship between the features of the medial dataset that aid in knowledge discovery and early diagnosis of chronic diseases.

10.7 CONCLUSION

This chapter aims to address the challenges of extracting relevant feature subset through the process of semantic annotation in healthcare data by proposing a lightweight semantic annotation model. The CKD dataset was used to analyze the experiments of the semantic annotation process. The features present in the dataset were semantically annotated using fuzzy rule mining with if-then condition. The result is an Ontograf, a graphical representation in OWL that shows the relationships between the features to predict whether the patient is suffering from CKD or not. This semantic annotation model can be used to predict the diagnosis of chronic illness for any incoming patient. Added to this semantic annotation model, data from IoT and sensor devices can be fed to this proposed model to annotate the data and predict if a particular patient is affected or not affected by any chronic disease. The issues of information sharing between the various devices and allied components in the healthcare sector are handled with the help of ontology-based semantic analysis of the healthcare data (Tchechmedjiev et al. 2018). This semantic annotation model will prove to be effective in identifying the relevant features that aid in disease diagnosis mainly for patients who are suffering from chronic diseases. This chapter stresses the need for achieving semantic annotation by surpassing the implementation challenges with the use of ontology. The semantic analysis provides meaningful relationships between the features that help in the early diagnosis of chronic illness.

KEYWORDS

- **semantic annotation**
- **heterogeneous models**
- **medical data mining**
- **artificial intelligence (AI)**
- **expert system**
- **machine learning algorithm**

REFERENCES

Antunes, M., Gomes, D., and Aguiar, R. (2018). Towards IoT data classification through semantic features. Future Generation Computer Systems, 86, 792–798.

Ashrafi, N., et al. (2017). Semantic interoperability in healthcare: challenges and roadblocks. In Proceedings of STPIS'18, vol. 2107, pp 119–122.

Chui et al.(2017). Disease diagnosis in smart healthcare: innovation, technologies and applications. Sustainability. 9(12), 2309.

Du, Y., et al. (2018). Making semantic annotation on patient data of depression. Proceedings of 2018 the 2nd International Conference on Medical and Health Informatics (ICMHI 2018), Association of Computing Machinery, pp. 134–137.

Gefen, D., et al. (2018). Identifying patterns in medical records through latent semantic analysis. Communications of the ACM, 61(6), 72–77.

Gia T. N. et al. (2015). Fog computing in healthcare internet of things: a case study on ECG feature extraction, 2015 IEEE International Conference on Computer and Information Technology; Ubiquitous Computing and Communications; Dependable, Autonomic and Secure Computing; Pervasive Intelligence and Computing, Liverpool, 2015, pp. 356–363.

Guerrero-Contreras, G., et al. (2017). A collaborative semantic annotation system in health: towards a SOA design for knowledge sharing in ambient intelligence. Mobile Informations Systems, 2017, Article ID 4759572, 10 pages.

He, Z., Tao, C., Bian, J., Dumontier, M., and Hogan, W. R. (2017). Semantics-powered healthcare engineering and data analytics. Journal of Healthcare Engineering, 2017, 7983473. doi:10.1155/2017/7983473.

Househ, M. and Aldosari, B. (2017). The hazards of data mining in healthcare. Studies in Health Technology and Information. 238, 80–83.

Jabbar, S., Ullah, F., et al. (2017). Semantic interoperability in heterogeneous IoT infrastructure for healthcare. Wireless

Communications and Mobile Computing, 2017, Article ID 9731806, 10 pages.

Lakshmi, K. S., and Vadivub, G. (2017). Extracting association rules from medical health records using multi-criteria decision analysis. In Proceedings of 7th International Conference on Advances in Computing & Communications, August 2017, Cochin, India, vol. 115, pp. 290–295.

Liao, Y., Lezoche, M., et al. (2011). Why, where and how to use semantic annotation for systems interoperability. 1st UNITE Doctoral Symposium, Bucarest, Romania, pp 71–78. hal-00597903

Liu, F., Li, P., and Deng, D. (2017). Device-oriented automatic semantic annotation in IoT. Journal of Sensors, 2017, Article ID 9589064, 14 pages.

Pech, F., et al. (2017). Semantic annotation of unstructured documents using concepts similarity. Scientific Programming, 2017, 10 pages.

Piro, R., et al. (2016). Semantic technologies for data analysis in health care. In: Groth P. et al. (eds) The Semantic Web—ISWC 2016. ISWC 2016. Lecture Notes in Computer Science, vol. 9982. Springer, Cham

Ringsqunadi, M., et al. (2015). Semantic-guided feature selection for industrial automation systems. International Semantic Web Conference, Springer (2015), LNCS 9367, pp. 225–240.

Sharma, D., Jain, S. (2015). Evaluation of stemming and stop word techniques on text classification problem, International Journal of Scientific Research in Computer Science and Engineering, 3(2), 1–4.

Tchechmedjiev, A., et al. (2018). SIFR annotator: ontology-based semantic annotation of French biomedical text and clinical notes. BMC Bioinformatics, 19, 405, 26 pages.

Zhang, L., Wang, T., Liu, Y., and Duan, Q. (2011). A semi-structured information semantic annotation method for Web pages. Neural Computing and Applications, 2017, Article ID 7831897, 10 pages.

CHAPTER 11

DRUG SIDE EFFECT FREQUENCY MINING OVER A LARGE TWITTER DATASET USING APACHE SPARK

DENNIS HSU[1*], MELODY MOH[1*], TENG-SHENG MOH[1], and DIANE MOH[2]

[1]*Department of Computer Science, San Jose State University, San Jose, CA, USA*

[2]*College of Pharmacy, Touro University, Vallejo, CA, USA*

[*]*Corresponding author. E-mail: dennis.hsu@gmail.com, melody.moh@sjsu.edu*

ABSTRACT

Despite clinical trials by pharmaceutical companies as well as current Food and Drug Administration reporting systems, there are still drug side effects that have not been caught. To find a larger sample of reports, a possible way is to mine online social media. With its current widespread use, social media such as Twitter has given rise to massive amounts of data, which can be used as reports for drug side effects. To process these large datasets, Apache Spark has become popular for fast, distributed batch processing. To extract the frequency of drug side effects from tweets, a pipeline can be used with sentimental analysis-based mining and text processing. Machine learning in the form of a new ensemble classifier using a combination of sentiment analysis features to increase the accuracy of identifying drug-caused side effects. In addition, the frequency count for the side effects is also provided. Furthermore, we have also implemented the same pipeline in Apache Spark to improve the speed of processing of tweets by 2.5 times, as well as to support the process of large tweet datasets. As the frequency count of drug side effects opens a wide door for further analysis, we present a preliminary study on this issue, including the side effects of simultaneously using two drugs, and the potential danger of using less common combination of drugs. We believe the pipeline design and the results present in this work would have helpful implication on studying drug side effects and on big data analysis in general. With the help of domain experts, it may be used to further analyze drug side effects, medication errors, and drug interactions.

11.1 INTRODUCTION

Monitoring drug side effects is an important task for both the Food and Drug Administration (FDA) as well as the pharmaceutical companies developing the drugs. Missing these side effects can lead to potential health hazards that are costly, forcing a drug withdrawal from the market. Most of the important side effects are caught during the drug clinical trials, but even those trials do not have a large enough sample size to catch all the side effects. As for drugs that are already on the market, current reporting systems for those drugs use voluntary participation, such as the FDA Adverse Event Reporting System (FAERS), which monitors reports of drug side effects from healthcare providers 0. Thus, the system only catches side effects that are considered severe while missing side effects that are not reported by the average consumer.

To solve this problem, one solution is to use a much larger database where many more reports of side effects can be found: social media. With the current widespread use of social media, the amount of data provided by platforms such as LinkedIn, Facebook, Google, and Twitter is enormous. Social media has been used in many different fields of study due to both its large sample size as well as its ease of access. For mining drug side effects, social media has many different users who report their daily use of the drugs they are taking as well as any side effect they get, and most of these reports are in the form of communications with other users.

To achieve this goal, machine learning can be used to design and implement a pipeline that will aid in mining Twitter for the frequency of reported drug side effects. The pipeline also has to be fast enough and have the ability to support large datasets through a distributed framework such as Apache Spark. The data used will come from Twitter, which has its own set of unique features. In the pipeline, Twitter was chosen because of its ease of access to the data in the form of tweets through the Twitter Application Program Interface (API). Also, the tweets are only 140 characters long, making them easy to process and store.

Extracting drug side effects from Twitter comes with numerous challenges that are hard to overcome. There are previous works in this regard 00; 0 0, all of which have excellent explorations into different ways of classification and extraction. The work in this chapter expands on extraction, focusing mostly on sentiment analysis (opinion mining) tools. Sentiment analysis is the process of identifying and categorizing opinions expressed in text (0) and in our case using these opinions to classify the tweets as positive or negative. To get the frequency of tweets, identifying tweets with drug side effects is required, and sentiment analysis tools use features such as reactions to taking a drug to provide such identification. Some other challenges of extraction with tweets include reducing the amount of noise in tweets. Tweets usually contain incomplete sentences as well as acronyms and general slang. Tweets also must be filtered properly to remove spam such as advertisements by drug companies or announcements by news organizations. Finally, the dataset mined in this work is larger than earlier works 0; 0; 0. To process this dataset, Apache Spark is used to speed up the pipeline. Apache Spark is an open-source distributed cluster framework that can provide parallel processing to speed up extraction from the dataset 0.

In this chapter, we introduce the importance of monitoring adverse drug events (ADEs), the current lack of methods to report these events, and the potential of social media as new source of reports. Next, we will go into techniques to extract these reports from the social media website Twitter using sentiment analysis and machine learning. Then, we will show how to process and handle large datasets that come from Twitter using distributed computing through Apache Spark. This is followed by an experiment results analysis of best and most efficient ways to correctly extract the frequency of ADEs using the techniques previously described. Afterward, a detailed pharmaceutical analysis is provided for the results, with insight from a domain expert, as well as referring to actual medical and pharmacy databases (WebMD LLC, 2019; Medline Plus, 2017; Shapiro and Brown, 2016). Finally, we will end with a discussion of possible applications of using the frequency of ADEs as well as future research direction and ways to extend our work.

11.2 BACKGROUND AND PREVIOUS WORK

This section will provide background information for understanding the pipeline, including ADEs, the concept of sentiment analysis for text processing, and distributed computing through Apache Spark.

11.2.1 ADVERSE DRUG EVENTS

ADEs refers to any type of injury or harm caused by taking a drug for its intended medical purpose. Catching and monitoring ADEs are extremely important to the FDA to make sure drugs on the market are safe. For further clarification, ADEs can be divided into three types (Shapiro and Brown, 2016). An adverse drug reaction encompasses all unintended pharmacologic effects of a drug when it is administered correctly and used at recommended doses. A side effect is a predicted response to a drug. A medication error occurs when the wrong drug or wrong dose is administered to the patient. However, most of the research and studies into ADEs rely on voluntary self-reports either by the patient or nurses and hospitals. One study focused on finding the incidence rate and preventability of ADEs in hospitals, but relied on doctor and nurse reports 0. The study found most ADEs were common and preventable, and most occurred due to the wrong ordering of the drug, such as incorrect dosage. There has been research in automating identification of ADEs reported in hospital settings 0, but the ADEs still come from voluntary reports while missing out on users who do not visit hospitals or clinics. The pipeline described in this chapter is a new way to get more reports of ADEs.

11.2.2 SENTIMENT ANALYSIS USING N-GRAMS

Sentiment analysis, also known as opinion mining, from text has been a popular tool in extracting text features used in machine learning. Sentiment analysis can be used on the *n*-gram feature, which are sequences of letters or words in the text. *n*-Grams have been around for two decades. Cavnar and Trenkle first introduced the concept of *n*-grams for text categorization of documents 0. There are two types of *n*-grams: word grams and character grams. Word grams convert documents into token count sequences based upon different words in the document while character grams break the document into sets of *n*-character sequences. The reasoning behind using *n*-characters is to be tolerant of errors in the text, especially with spelling. They were able to achieve a high accuracy of 80% in categorizing texts from news articles into groups. Using character *n*-grams is especially useful for Twitter, as tweets from users often have incorrect spelling as well as acronyms and short-hand words. *n*-Grams, from unigrams, which is one word or letter, all the way to four grams, a sequence of four words or letters, are used in our work.

11.2.3 SENTIMENT ANALYSIS ON TWITTER FOR DRUG SIDE EFFECTS

Several previous works have explored mining Twitter for drug side effects 00; 00.

Jiang and Zheng (2013) extracted drug side effects with the use of MetaMap 0. Using five different drugs as their dataset, they developed a machine learning classifier to automate classification of tweets with drug-caused side effects, followed by extraction of drug side effects using MetaMap. They used user experience as the main feature for correct classification of the tweets.

Wu et al. (2015) focused on using opinion lexicons and subjective corpuses as features for classifying tweets. They first constructed a pipeline for extracting drug side effects from tweets, but focused only on a small sample size of four drugs. The features that were used in this approach were syntactic features such as question marks and negations as well as the sentiment scores from the different corpuses. For the four drugs, they were able to achieve an *f*-measure score of 79.5% using support vector machine (SVM) as the machine learning classifier 0.

Yu et al.'s (2016) work took a different approach and focused instead on the cause-effect relations between the drug and the side effect. Tweets containing drugs that directly caused the side effect were the ones identified as positive. To extract this relation, *n*-grams were used as features. Lemmatization of the tweets was also used to reduce the noise of the text to allow for better *n*-gram features. Using unigram and bigram words, a 76.9% accuracy was achieved for a large sample of drugs.

0 work further improved the techniques of 0 approach. They continued to focus more on specifically capturing only tweets related to five different drugs. Their experiment results gave a better detection rate, five times more than the original, as well as simplifying classification techniques.

The current approach in this chapter focuses on combining the techniques from the two earlier approaches 00 for further improvement. The sentiment features from lexicons of the first approach 0 as well as the *n*-gram features of the second approach 0 are both used as features. Also, more machine learning classifiers are explored and combined to test the best combination of these features. In addition, our approach also uses MetaMap to extract drug side effects 0 to calculate the frequency for further analysis and applications.

11.2.4 APACHE SPARK

Apache Spark is the distributed framework used in this chapter to process large datasets. It is a cluster computing system that has become widely used in the recent decade. Spark is an improvement over Hadoop's MapReduce paradigm in terms of speed of batch processing 0. Spark distributes the workload over a cluster for distributed, parallel processing.

Apache Spark's core feature is the resilient distributed dataset (RDD), a read-only dataset over the whole cluster. RDDs can be stored in memory for faster repeat batch processing instead of being stored on the system's hard disk. RDDs are also fault tolerant and can be used in the same tasks that Hadoop can do such as mapping and reducing. Spark has an extensive set of tools supported, and their machine learning library is widely used and integrated well with their RDD paradigm.

Apache Spark is extremely useful when processing large datasets. In the work by 0, Spark is used to improve the speed of identification of potential drug targets to be studied in clinical trials. The original pipeline was changed to process the drug compounds in parallel by running the potential targets through multiple machine learning predictors and calculating a combined score as the identifying feature for the compound. The predictors gave a score for the compound based on how well the compound could target a protein, and this interaction was based on how well the compound's shape complemented the protein shape. They partitioned their data into multiple chunks to process their dataset of compounds in parallel. The results of their work showed that the time for processing their large dataset decreased linearly with the number of nodes used in Spark. Similarly, in this chapter, Spark is used to process the large dataset by splitting the tweet dataset into chunks for parallel processing to improve pipeline speed.

11.3 DESIGN AND APPROACH

This section will go into detail of how the current improved pipeline works based different machine learning classifiers and distributed computing. The pipeline should start by identifying whether a tweet contains a drug-caused side effects and at the end outputting an updated count of the different side effects reported for each drug. There are five parts to the pipeline, as shown in Figure 11.1. First, the tweets are mined and filtered. Then, the tweets are preprocessed before features are extracted. Finally, the classifier uses the features to identify the drug side-effect-related tweets and then the frequency of the side effects is extracted and updated. These steps are explained in the following subsections.

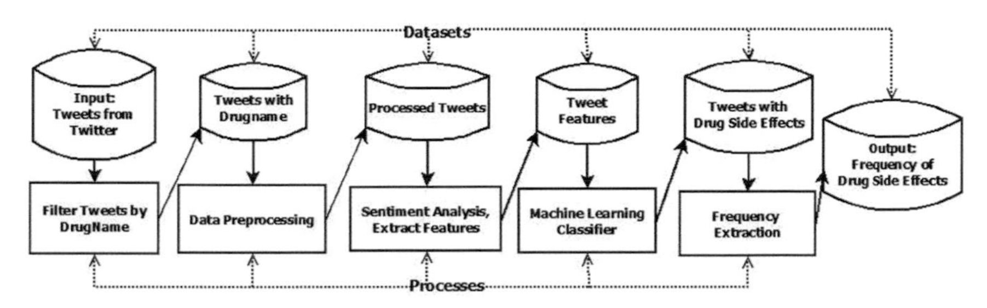

FIGURE 11.1 Pipeline for extracting frequency of drug side effects from Twitter.

11.3.1 MINING AND FILTERING TWITTER THROUGH LIVESTREAM

In the first step, tweets are mined from Twitter through a livestream. Tweepy, a Python library, was used to access the Twitter streaming API 0. The stream was mine for 9 days in December 2016. The tweets were then stored in a csv file for bulk processing. The stream was filtered for keywords containing drug names taken from the top most popular drugs from the drugs website 0, totaling 462 different drug names. The drug names used were their most commonly used names instead of always using their scientific name. The dataset was further cleaned using other filters to further remove spam and to narrow down the tweets we want, using the following:

- No retweets: only tweets from users who are self-reporting drug side effects were mined. Most of the retweets contained advertisements from pharmaceutical companies.
- Tweets with <10,000 followers: users with more were usually organizations or celebrities. Our target was the average consumer.
- Only English tweets were considered for ease of text processing as well as natural language processing.

Filtering by drug names and other filters over the 9 days returned a total of 486,689 tweets as the initial dataset that we need to train our machine learning classifier.

11.3.2 DATA PREPROCESSING

In the second step of the pipeline, the tweets in the csv file were further preprocessed to reduce noise. The preprocess steps that were used on the data included the following:

- Any tweet that started with "RT" was removed. The Twitter API does not completely filter out all retweets, so the tweets had to be checked a second time.
- All hashtag pound symbols and usernames were removed (hashtags remained).
- All nonalphanumeric characters and punctuation were removed to allow for easier text processing. The characters were all converted to lowercase as well.

All drug names in the tweet were replaced with the keyword "drug." Due to the different distribution of drug tweets in the dataset, normalization of the drug name was required to balance the dataset 0.

The words in the tweet were lemmatized. The Natural Language Toolkit 0 was used to lemmatize the words down to their base form to further reduce noise. The words in the tweet were tokenized and labeled with a Part of Speech (POS) tagging before lemmatization.

Stop words were not removed due to the small length of each tweet.

11.3.3 FEATURE EXTRACTION

After the data was preprocessed, the features were then extracted for classification using sentiment analysis, specifically two separate methods. Previous works only used *n*-gram cause-and-effect relations 0 or opinion lexicons 0, but not both. This step of the pipeline here uses both *n*-gram and lexicons as features to train the classifier.

For the *n*-gram classification, the experiment tested using a combination of unigram, bigram, trigrams, and four grams. Both word and character *n*-grams were tested. The *n*-grams used do not have punctuation and the words are all in their base form due to lemmatization. For the machine learning classifier to understand the features, the *n*-grams were converted into vectors based on term-frequency and inverse document frequency (tfidf). The term frequency is how frequent (and thus important) an *n*-gram appears in a document, in this case the document being the tweet. The inverse document frequency is how frequent the *n*-gram appears in different tweets in the whole dataset. Combining these two frequencies together gives us the tfidf statistic feature for how important the *n*-gram is relative to the others. This is one of the features that was passed to the machine learning classifier.

For the opinion mining lexicons features, the number of words covered is small. Thus, multiple lexicons were used to get better coverage of words. The experiment tested out using a combination of four different lexicons:

SentiWordNet 0: This lexicon assigns each word a positive, neutral, or negative score number. SentiWordNet also uses POS tagging to distinguish between different forms of words 0.

AFINN 0: This lexicon rates each word a sentiment score in the range [−5, +5]

MPQA 0: Multipurpose question analysis has its own subjectivity lexicon that rates each word as strong/weak positive or negative. In this experiment, we had a "strong" label be a magnitude of five while a "weak" label be a magnitude of one in ratings.

Bing-Liu 0: This lexicon contains more slang words and jargons than the other lexicons. The lexicon splits the words into positive and negative lists, which in our experiment, we gave a score of positive one and negative one, respectively.

Each of the four lexicons assigns a number based on its sentiment which is used.

11.3.4 MACHINE LEARNING CLASSIFICATION

The features from sentiment analysis were used to train the machine learning classifiers through supervised learning. This requires a labeled training dataset to be run through the classifier, followed by validation and test sets to evaluate how well our classifier has learned. A total of 1000 tweets were manually labeled for the training dataset, with half of the tweets being positively identified as having drug-caused side effects, while the other half were negatively identified as not having drug-caused side effects. This is to provide a balanced training dataset. A total of 1000 tweets were chosen as the dataset for comparison with results from previous works 0. Different combinations of *n*-gram and lexicon features were used to train the following different classifiers:

- Gaussian naïve Bayes (GNB): a simple classifier using Bayes' theorem that is usually used as the baseline
- Logistic regression (LGR): uses a logistic function as a base and is popular for binary classification
- SVM: attempts to separate data using a line (called hyperplane) that maximizes the margin between the data group points

- Stochastic gradient descent (SGD): a linear classifier that attempts to find a minimum one sample at a time
- *k*-Nearest neighbor (kNN): attempts to separate the data into groups that form the nearest neighbors with each other
- Decision tree classifier (DTC): uses a data tree structure where each of data points goes through different decision paths that are used to classify
- Random forest classifier (RFC): uses an ensemble/multiple decision tree classifiers to form the final classification
- Ensemble classifier (NB, LGR, SVM, SGD, kNN, DTC): combination of multiple classifiers making the decision

The ensemble classifier is actually a combination of the first six classifiers taken together in either a majority voting (hard vote) or a prediction of probabilities (soft vote), as shown in Figure 11.2. The ensemble classifier provides a better overall predictive accuracy than any of the classifiers it uses by itself and use of ensemble classifiers has not been previously tested 0; 0. By tweaking the weights of the classifiers, the ensemble's best accuracy can be found.

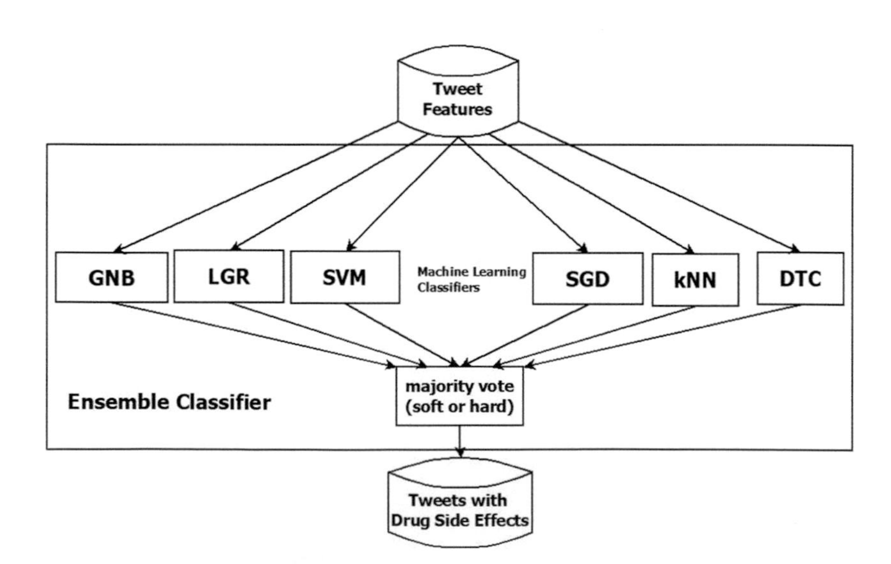

FIGURE 11.2 Ensemble classifier with six classifiers and a soft or hard majority vote of the classifier predictions.

11.3.5 FREQUENCY EXTRACTION

After training the classifier and identifying the drug side-effect-related tweets, a frequency count is taken. To get the frequency, the text extraction of drug side effects in each tweet is done by MetaMap. MetaMap is a tool for recognizing medical concepts from the Unified Medical Language System (UMLS) 0; 0. Currently, there are 15 different semantic group types in MetaMap, each with multiple subcategories, with no change from what was used in the previous work 0, as shown in Table 11.1. MetaMap extracts medical text from the tweet and maps it to a UMLS medical term with a certain confidence. In experimenting on the pipeline, a confidence to 850 out of 1000 was set as the lower bound for accepting a mapping by MetaMap. Once extracted, the side effects were then grouped by each drug for analysis. The most common side effects as well as rare side effects could then be observed.

TABLE 11.1 MetaMap Semantic Groups and Abbreviations

Abbreviation	MetaMap Group
ACTI	Activities and Behaviors
ANAT	Anatomy
CHEM	Chemical and Drugs
CONC	Concept and Ideas
DEVI	Devices
DISO	Disorders
GENE	Genes and Molecular Sequences
GEOG	Geographic Areas
LIVB	Living Beings
OBJC	Objects
OCCU	Occupations
ORGA	Organizations
PHEN	Phenomena
PHYS	Physiology
PROC	Procedures

11.3.6 LARGE DATASET PROCESSING WITH SPARK

After creating the pipeline initially in a Python environment through Scikit-learn, a Spark pipeline was then created to process large datasets, as shown in Figure 11.3. Spark's RDDs as a well-distributed framework allows for parallel processing of all the tweets 0.

For Spark, we first trained Spark's classifier using our previous feature sets used in the original pipeline. Next, the Spark pipeline was implemented in the following steps:

- The large input dataset RDD was partitioned and mapped out to the nodes. The classifier on each node identified if the tweet contained a drug-caused side effect.

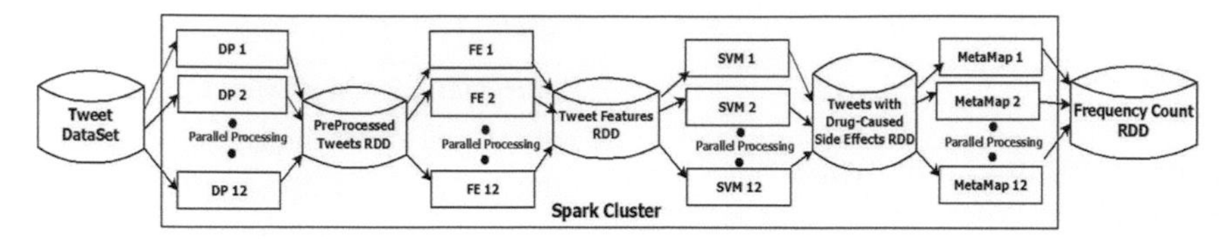

FIGURE 11.3 Apache Spark pipeline with 12 cores for distributed processing. RDDs are stored in memory between each step. Spark is assigns tasks by automatically partitioning at each step: data preprocessing (DP), feature extraction (FE), SVM classifier, and frequency extraction with MetaMap.

- The positively identified tweets were then reduced back into a RDD containing all the tweets with drug-caused side effects.
- The tweets were then labeled with a key that is the drug name associated with the tweet.
- Frequency extraction with MetaMap was then run on the RDD and the frequency counts for each side effect were returned as (side effect, count) pairs.
- The (side effect, count) pairs were then reduced back into one RDD and outputted back to a data text file.

11.4 EXPERIMENT SETUP

To test the pipeline to determine which features and classifiers work the best, an experiment (Hsu, 2017) was setup where two separate pipelines were constructed: one for testing the different machine learning classifiers and does not use parallelism (as shown in Figure 11.1) and the other for testing Apache Spark for large datasets that does use parallelism (as shown in Figure 11.3). The pipelines were then compared for speed from the starting point of the twitter dataset to the final output of the side effect frequencies.

11.4.1 PIPELINE SETUP

For the experiment pipeline in Figure 11.1, the goal was to test which set of features as well as which machine classifier performed the best. This section describes how the Scikit-learn pipeline was setup.

In the initial stream through Tweepy, 486,689 tweets in total were mined over 9 days using the filters mentioned in Section 11.3.1. After removing the retweets not caught by the filter, duplicate tweets were removed. Using the sequence matcher from Python's difflib, all tweet that had 0.6 similarity or above were removed, leaving 226,834 tweets as our dataset. Using regular expressions (Regex), the tweets were preprocessed using the steps shown in Section 11.3.2. NLTK was then used to lemmatize each tweet further to remove noise.

Next, sentiment scores were extracted from each of the four lexicons. For each lexicon, the sum of the sentiment scores for each word in the tweet was calculated as the feature. The sentiment scores were then categorized using a one-hot encoder to provide better feature weight against the *n*-gram features. For the *n*-gram features, Scikit-learn's tfidf vectorizer was used to create unigrams through four grams 0. After the *n*-grams were created, they were then converted to feature vectors based on the tfidf frequencies.

The tweet's extracted features were then run through the machine classifier and were classified as having drug related side effects or not. If it was positively identified, then the tweet was then passed to MetaMap, and the side effects extracted by MetaMap were then stored in a dictionary for the drug along with its count. At the end, the frequency of the side effects for each drug was then outputted.

11.4.2 SPARK SETUP

To test the benefits of distributed computing, a Spark pipeline, as shown in Figure 11.3, was created using Spark's machine learning library MLlib 0. Spark supports Scala, Java, and Python, and for the experiment, PySpark (Python) was used for preprocessing, FE, and machine learning classification, and the tools used were the same as the Scikit-learn pipeline 0. NLTK was used for DP and sentiment score features, while MLlib's vectorizers were used to extract *n*-gram features as well as One Hot Encoding of the sentiment scores. For testing of the Spark pipeline speed, SVM was used as the comparison between the Scikit-learn pipeline and the Spark pipeline 0. The output tweets identified by Spark's classifier were then stored as a permanent RDD in memory with the persist function. The RDD was then passed through the Java API of MetaMap for side effect mapping, and the output was then collected and reduced to get the frequency output of the side effects reported for each drug.

For splitting up the dataset, a Spark configuration of two nodes running on two virtual machines was implemented to allow for parallel processing. The dataset was partitioned automatically over the two nodes, which had a combined total of 12 cores, giving 12 partitions of approximately 18,902 tweets per partition. A map to the two nodes was called to allow Spark to run the predictions in parallel, but an

TABLE 11.2 Classifier Accuracies (f-Measure Score Weighted) for Different Combinations of Features with the Best for Each in Bold

Features	SVC	GNB	LGR	SGD	kNN	DTC	RFC	Ensemble (Soft)	Ensemble (Hard)
f1+f2 (char_wb)	0.6202	0.5080	0.6094	0.5955	0.5273	0.5080	0.5009	0.5389	0.6136
f1+f2+f3 (word)	0.6280	0.6036	0.6096	0.5599	0.5670	0.5036	0.5663	0.5692	0.6066
f1+f2+f3+f4 (char_wb)	0.6532	0.6262	0.6352	0.5767	0.6311	0.6449	0.5338	0.6710	0.6686
f1+f2+f3+f4+f5 (char_wb)	0.6792	0.6768	0.6899	0.5734	0.6576	0.6131	0.6446	0.6468	0.7128
f2+f3+f4+f5+f6+f7 (word)	0.7229	0.6827	0.7099	**0.6913**	0.7036	0.6746	0.6291	0.7097	0.7449
f1+f2+f5+f6+f7 (char_wb)	0.7186	0.6460	0.7260	0.6311	**0.7332**	**0.7467**	0.6873	0.6671	0.7467
f1+f2+f3+f4+f5+f6+f7 (char_wb)	**0.7392**	**0.7219**	**0.7347**	0.6797	0.7032	0.6825	0.6174	**0.7568**	**0.7760**
f1+f2+f3+f4+f5+f6+f7+f8 (char_wb)	0.7028	0.6973	0.7147	0.6907	0.6643	0.5939	**0.6881**	0.7219	0.6992

f1, unigram; f2, bigram; f3, trigram; f4, four grams; f5, SentiWordNet; f6, AFINN; f7, MPQA; f8, Bing-Liu.

inner map call was used to allow the predictions to occur on each node sequentially. The predicted tweets were then run through MetaMap on another Spark job due to MetaMap being supported only with a Java API. The side effect counts were extracted before being merged together into an output text file.

11.5 RESULTS

In the following subsections, results from experiments on the pipeline are presented for the accuracy, processing speed-up, and frequency of drug side effects.

11.5.1 ACCURACY

For testing the Scikit-learn pipeline, a fivefold cross validation was used on different combinations of features. The weighted f1 score was then calculated for each of the machine learning classifiers for comparison, as shown in Table 11.2. The experiment with unigram and bigram was used as a baseline for comparison with the other features.

The best classifier was the ensemble classifier with hard voting with an f1 measure score of 0.7760. Different weights were tested for the ensemble classifier, and the optimal weights were double-weighted for both SVM and LGR compared with the other four classifiers. RFC was excluded from the ensemble classifier as RFC itself is an ensemble classifier. Using Yu et al.'s (2016) work as a baseline of their best f1 score of 0.7690 with SVM 0, our ensemble classifier had a small improvement. The best nonensemble classifier was the DTC with a f1 measure score of 0.7467, which still was a small improvement from the previous work's decision tree classifier f1 score of 0.7447 0.

The best features to use were all four *n*-grams from unigram to four-gram plus three of the lexicons: SentiWordNet, AFINN, and MPQA. The trend of the data shows more features gives better accuracy up to a certain point. Adding the feature of the final lexicon Bing Liu gave a lower accuracy, which most likely is caused by overfitting.

11.5.2 PIPELINE SPEED COMPARISON

Next, the speed between the Scikit-learn pipeline (shown in Figure 11.1) and the Apache Spark pipeline (shown in Figure 11.3) using the SVM classifier were compared. From the dataset, 200,000 tweets ran through the both pipelines and the time was recorded upon completion, as shown in Table 11.3.

TABLE 11.3 Total Time to Extract Frequency of Drug Side Effects for Both Pipelines

Pipeline	Total Time (min)
Scikit-learn	257.88
Apache Spark	105.63

Spark was faster than the Scikit-learn pipeline by around 2.5 times due to Spark's parallel processing capabilities.

11.5.3 FREQUENCY OF DRUG SIDE EFFECTS

The pipeline outputs the frequency of drug side effects. For the experiment, ut of the 200,000 tweets, 78,242 tweets were predicted as tweets containing drug side effects.

Table 11.4 shows the top ten drugs with the most side effects reported.

TABLE 11.4 Top 10 Most Reported Drugs by Twitter Users

Drug	No. of Tweets Predicted	No. of Side Effects Reported
Xanax	12081	27,289
Adderall	7958	16,906
Ibuprofen	7822	16,050
Melatonin	5873	14,274
Benadryl	5259	13,708
Tylenol	5263	13,469
Insulin	5070	12,248
Nicotine	4819	11,763
Aspirin	3185	7638
Morphine	3028	7223

To further investigate these drugs and their side effects, and to compare with the previous work 0, the top five drugs (plus two extras for comparison) are analyzed in Table 11.5, each with their five most reported negative side effects, respectively. These were manually examined and extracted from the list of side effects to remove side effects that were alleviated by the drug and those not caused by the drug. Each of the top five side effects was manually checked to make sure the drug did cause the side effect in their respective tweets.

Most of the side effects were from the MetaMap semantic groups "Disorder" and "Physiology." Note that the side effects reported do not consider if the side effects were directly caused by the drug. The predicted tweets based on the training dataset were geared more toward false positive, as missing side effects were considered more detrimental than over reporting. MetaMap also had problems in extracting side effects due to catching all medical terms, thus requiring the filter of the semantic groups.

The predicted tweets based on the training dataset were geared more toward false positive, as missing side effects were considered more detrimental than over reporting. MetaMap also had problems in extracting side effects due to catching all medical terms, thus requiring the filter of the semantic groups.

TABLE 11.5 Frequency of Side Effects Reported, Showing the Top Five Reported Side Effects per Drug with Number Reported

Drug Name	Drug Use	Side Effect 1	Side Effect 2	Side Effect 3	Side Effect 4	Side Effect 5
Xanax	Anxiety	Drowsiness/ sleep (291)	Abnormally high (76)	Addictive behavior (66)	Blackout (13)	Withdrawal (9)
Adderall	ADHD	Emotions (122)	Addictive behavior (29)	Insomnia (26)	Tired (17)	Binge eating disorder (16)
Ibuprofen	Fever, headache/ pain	Emotions (169)	Drowsiness/ sleep (126)	Binge eating disorder (17)	Abnormally high (16)	Allergic reaction (14)
Melatonin	Insomnia	Emotions (89)	Nightmares (25)	Binge eating disorder (21)	Weight loss (11)	Anxiety (7)
Benadryl	Allergy	Drowsiness (107)	Tiredness (23)	Dry throat (13)	Nausea (3)	Dizziness (2)
Vyvanse	ADHD	Emotions (25)	Abnormally high (12)	Weight loss (6)	Short breath (6)	Chest pain (2)
Gabapentin	Seizure/ pain	Emotions (6)	Insomnia (5)	Hot flushes (2)	Confusion (2)	Dryness (1)

Another problem with MetaMap was side effects not caused by the drug within the tweet were also extracted along with the actual drug-caused side effect. The side effects extracted by MetaMap had to be manually examined to remove noncaused side effects, especially when analyzing side effects shown in Table 11.5 as well as in analysis of multiple drug interactions in Section 11.6. This is further explained in Section 11.5.2.

TABLE 11.6 Side Effects and Number Reported for Xanax in Subcategory "Sign and Symptom" with Keys for Reference

Side Effect	Number Reported	Side Effect	Number Reported
Chills (Relax)	98	Unwanted Hair	3
Spells	39	Agitation	3
Malaise	21	Sleeplessness	3

TABLE 11.6 *(Continued)*

Side Effect	Number Reported	Side Effect	Number Reported
Blackout	13	Muscle Twitch	3
Catch	13	Pruritus	3
Halitosis	11	Sighing Respiration	3
Tired	8	Clumsiness	3
Hunger	6	Headache	2
Blurred Vision	6	Nausea	2
Muscle Cramp	5	Memory Loss	2
Earache	4	Drooling	2
Withdrawal	4	Seizures	2
Vomiting	3	Other	24

Table 11.6 shows further examination of Xanax, the drug with the most reports out of the predicted. Within the semantic group "Disorder" (referring to Table 11.1), Table 11.6 shows all side effects in the subcategory "Sign or Symptom," including side effects that Xanax is suppose alleviate. There were other side effects in other subcategories of "Disorder" such as "Finding" or "Mental or Behavioral Dysfunction" that are not shown here, such as the side effect "Abnormally High." In relation to Table 11.5, the only side effect with the matching number of reports was "Blackout." Other side effects were not caught under the category "Disorder," and the side effect "withdrawal" had reports in multiple categories, with "Disorder" only catching four of them.

The reported frequencies of side effects included both those caused by Xanax as well as side effects caught in the tweet that were not caused by the drug. For example, the side effect for relaxation "chills" was caused by Xanax despite not being a negative side effect, but the side effect was mentioned with an actual negative side effect in the same tweet. "Chills" in the tweets was considered a positive side effect, as people who take Xanax are using it to relax without anxiety, but MetaMap in this case caught this side effect as well. Thus, both side effects were extracted. These are also discussed in Section 11.5.2.

The pipeline was able to output the frequency of drug-caused side effects for all 462 drugs, such as with Xanax, showing both commonly and uncommonly reported side effects, which can be compared with Xanax's known side effects from medical sources 0; 0.

11.5.4 PHARMACEUTICAL ANALYSIS

Based on the results from the pipeline, a domain expert provided some analysis and insight on the different drug side effects and their frequency. The following drugs are discussed based on information from Medscape (WebMD LLC, 2019) and Medline Plus (Medline Plus, 2017). It is not surprising that the Xanax, a benzodiazepine utilized for anxiety, causes drowsiness as a side effect. Its rapid action on the GABA receptors in the limbic system and reticular formation also explains how the dopamine surge in response contributes to addictive behavior, abnormal highs, and withdrawal symptoms. Overdoses of benzodiazepines, especially in the elderly, or use in combination with alcohol or other central nervous system depressants, can cause anterograde amnesia and loss of consciousness.

Adderall is a stimulant used in the treatment of ADHD. Adderall is composed of two amphetamine salts, dextroamphetamine and amphetamine, which are controlled substances due to being highly addictive; thus, it is not surprising that they cause addictive behavior. Similarly, Vyvanse is also an amphetamine salt, lisdexamfetamine, and is also highly addictive. Both Adderall and Vyvanse cause irritability, anxiety, and emotional lability, a.k.a. mood swings. This may occasionally cause those taking Vyvanse to feel high.

Adderall causes insomnia due to its very nature as a stimulant which keeps people awake, similar to caffeine. Tiredness is a side effect of Adderall in the sense that once the stimulant effect wears off, the patient will crash. The binge eating is a paradoxical side effect of Adderall as a stimulant usually suppresses the appetite and causes weight loss, as seen in Vyvanse. Chest pain and shortness of breath are serious but rare side effects seen with Vyvanse, often necessitating emergency medical care.

Ibuprofen is a nonsteroidal anti-inflammatory drug. There is no past precedence of it causing emotional complications, abnormal highs, drowsiness, or binge eating disorders, although allergic rashes up to and including Stevens–Johnson syndrome and toxic epidermal necrolysis have been reported in the literature.

Melatonin is a natural supplement used over the counter for the treatment of insomnia. In the body, it is a hormone produced by the pineal gland. It has been shown to cause irritability, anxiety and nightmares but not to cause binge eating disorder and weight loss.

Benadryl is a first-generation antihistamine utilized not only for allergies but very commonly as a sleep aid. Thus, it causes dizziness, drowsiness and fatigue, and as it also has strong anticholinergic effects, it causes dry mouth and throat as well. Nausea is not defined as a side effect; in fact, Benadryl is used in combination to treat motion sickness.

Gabapentin is utilized for seizure and neuropathic pain. Hot flushes, dryness, confusion, insomnia, and irritability as well as depression have all been reported as well as other emotions. Insomnia has especially been reported upon discontinuation of the drug, along with anxiety.

11.5.5 CHALLENGES AND LIMITATIONS

There have been challenges and limitations to the pipeline concerning the extraction of drug-caused side effects.

First, all of the subcategories for the MetaMap group "Disorder" had to be used, as leaving out any subcategories might cause side effects to be missed. Secondly, MetaMap extracts all side effects from the tweets, both those caused and those not caused by the drug, thus also requiring manual examination to identify the drug-caused side effect. However, an external dataset containing all possible side effects that are alleviated by the drug can be used to remove some of these extra non-caused side effects.

For example, "Chills," the most reported side effect of Xanax, in context means "to relax" but to MetaMap, the concept means "shivers." Thus, extracting negative side effects of the drugs required both reducing by MetaMap category as well as by manual examination, to correctly identify which side effects were negative. Furthermore, each tweet usually contained more than one side effect besides the negative side effect caused by the drug, requiring further manual examination to determine which side effect within the tweet is the one caused by the drug.

Other complications include tweets with multiple drugs, as associating the side effect with the correct drug(s) requires manual examination as well. It is not known whether the side effects in these cases are caused by one of the drugs, both drugs, or some form of interaction between the drugs. These lead to preliminary work in Section 11.6.

11.6 NEXT STEPS: APPLICATIONS OF DRUG SIDE EFFECT FREQUENCY ANALYSIS

Using the frequency extracted from the proposed pipeline, one can make some observations on the most common side effects as well as rare side effects reported. One can also observe the side effects that may be caused by two or more drugs taken together; some may be side effects caused by rare drug pairs that might be potentially dangerous. The following results required manual examination to remove side effects that were not caused by the drug or were alleviated by the drug as well as any other side effect that was incorrectly reported. This required going through the tweets manually to make sure the side effect was caused by the drug(s). Some preliminary studies and observations are reported in the following subsections.

11.6.1 MOST FREQUENTLY REPORTED SIDE EFFECTS

The top three side effects were drowsiness/tiredness, emotions, and being abnormally high. "Drowsiness" is considered a mild side effect that affects most people, thus being commonly reported. People who have reported being emotional can be inferred as being more likely to share their emotions on Twitter, which is probably the cause of large number of reports. Finally, "abnormally high" was largely reported because of the large number of tweets related to drugs that cause this side effect, most notably Xanax and other drugs used for anxiety.

An example of a rare side effect that was less reported but was seen in all the top 10 drugs was nausea. For example, Adderall and Vyvanse are known to cause nausea as they are stimulants that suppress the appetite. Ibuprofen is also known to cause nausea, especially when taken without food, as it is a nonsteroidal anti-inflammatory drug and its effects on inhibiting COX also inhibit stomach protective enzymes. Xanax and Gabapentin cause nausea. Melatonin causes abdominal cramps and would be associated with nausea. Tylenol is not listed as causing nausea but can cause gastrointestinal hemorrhage, so this is new information that would be forwarded to FAERS. Also, Benadryl had two reports of nausea, which were reports that would also be forwarded to 0.

11.6.2 SIDE EFFECTS CAUSED BY MORE THAN ONE DRUG

Next, predicted tweets where more than one drug was used were examined. Having multiple drugs makes it hard to correctly identify which side effect is caused by which drug. Out of the predicted tweets, 2678 contained more than one drug. Table 11.7 lists the top six drugs that were mentioned most out of these tweets containing two or more drugs.

TABLE 11.7 List of Top Five Drugs mentioned with Other Drugs in a Tweet and Top Two Side Effects

Drug	Tweets	Side Effect 1	Side Effect 2
Tylenol	274	Emotions (6)	Drowsiness (3)
Xanax	203	Addiction (4)	Drowsiness (4)
Ibuprofen	196	Drowsiness (4)	Allergic (2)
Adderall	50	Addiction (3)	High (2)
Benadryl	99	Drowsiness (12)	Insomnia (5)

Most of the tweets with multiple drugs did not specify which of the drugs-caused the side effect. Also, some of the tweets focused on one of the drugs not working or causing a side effect that required the second drug (or even third) to solve their problem.

11.6.3 SIDE EFFECTS CAUSED BY THE MOST POPULAR DRUG PAIRS

As shown in Table 11.8, most of the tweets with multiple drugs focused on competing drugs. Ibuprofen, also known as Advil or Motrin, competes with acetaminophen (Tylenol) for relieving pain and headaches. The "emotions" side effect related mostly to anger caused by the ineffectiveness of Ibuprofen or Tylenol at alleviating the pain.

Another pair of drugs, Adderall and Vyvanse, used for attention deficit hyperactivity disorder (ADHD), unfortunately caused insomnia, as the drugs providing focus also stopped the users from sleeping. Same conclusion can be made for Adderall (for ADHD) and Xanax (for anxiety), which caused insomnia on those patients who really need to sleep as well. Although Xanax usually causes drowsiness, paradoxically, it can also cause insomnia, especially at higher doses (WebMD LLC, 2019).

TABLE 11.8 List of Top Five Most Mentioned Drug Pairs with Top Side Effect

Drug Pair	Tweets	Side Effect
Ibuprofen, Tylenol	131	Emotions (2)
Adderall, Vyvanse	54	Insomnia (2)
Adderall, Xanax	36	Insomnia (2)
Mucinex, Tylenol	25	Drowsiness (2)
Benadryl, Melatonin	23	Drowsiness (4)

Finally, another example, Benadryl and Melatonin, had the same common side effect of drowsiness, as people who took Benadryl, used to relieve allergies, usually became drowsy, and they wanted an extra Melatonin, used as a sleeping pill, for extra effect to fall asleep at night. Without manual examination, it would have been hard to figure out if the side effect was caused by multiple or just one of the drugs in the tweet, and this is something that future works might improve on.

TABLE 11.9 Side Effects of Drug Pairs Not Commonly Associated with Either Drug

Drug Pair	Side Effect	Side Effect Count
Klonopin, Zoloft	Emotions	1
Tylenol, Ativan	Abnormally high	1
Adderall, Benadryl	Drowsiness	2

11.6.4 POTENTIAL DANGER: SIDE EFFECTS ASSOCIATED WITH UNCOMMON DRUG PAIRS

Finally, there was an examination of the side effects that were rare and not usually associated with a certain drug, due to taking a combination of drugs. As seen in Table 11.9, three different pairs are shown that had side effects that were considered rare and abnormal for both drugs when taken together.

As the first example, a user took Klonopin to treat his anxiety but at the same time caused him to feel depression. He then took Zoloft for the depression, and instead began to feel emotional, as shown in the tweet:

> *yeah klonopin make it so my depression be way more*
>
> *evident but when I try take zoloft w/it [messed with] me and make me*
>
> *manic so idk*

Trying to treat both depression and anxiety with this drug pair made him feel "manic" and crazy. From this tweet, the fact that the patient mentioned he was going "manic" indicates that he is treating his bipolar depression with Zoloft, sertraline, which he should not be doing because bipolar depression should be treated with a mood stabilizer and atypical antipsychotic or a mood stabilizer in combination with an antidepressant and not solely an antidepressant as treating bipolar depression with an antidepressant will cause a switch from the depressive phase into the manic phase, thus possibly precipitating a manic event and the need for hospitalization per Medscape (WebMD LLC, 2019).

In another pair, Tylenol (used for headaches) and Ativan (used to treat seizures) caused the user to feel "high," which is not a particularly uncommon side effect for Ativan, as it is often abused for this very purpose do to its rapid onset of action. Tylenol has no drug–drug interaction with Ativan in this case, it is merely mentioned in the tweet due to the user's unfamiliarity with the medications:

so apparently mixing tylenol and ativan makes you

extremely high

In the last example, Adderall is used to treat ADHD and is used for focus, but Benadryl made the user fall asleep instead of remaining focused. The tweet, shown below, shows the user saying that the user became drowsy, favoring the side effect of Benadryl (sleepy) instead of Adderall (insomnia, focused). Adderall typically causes insomnia only when used near bedtime, and Benadryl causes drowsiness irregardless of the time of day it is used:

felt a stuffy so I took a benadryl with my coffee and

adderall. I'll be fallin asleep and an inch from death today

From the above examples, we see that finding uncommon side effects from a combination of drugs is important and can be expanded on further in the future.

11.7 CONCLUSION AND FUTURE WORK

Mining the frequency of adverse drug side effects is important for finding side effects that are more common as well as those that are rare but potentially dangerous. In this chapter, an improvement on previous pipelines for extracting drug side effects from Twitter was discussed. A pipeline was created to first identify tweets that contained drug-caused side effects followed by extracting the frequency of those side effects. An increase in the accuracy of the classifier compared to previous works had been achieved. Finally, the pipeline was also implemented in Apache Spark to improve the speed of extraction as well as for processing large datasets.

Future work for this pipeline includes the preliminary study of application of frequency analysis of drug side effects described in this chapter. More studies would be beneficial for finding side effects of concurrently taking two or more drugs, and the proposed Apache Spark-based pipeline may further contribute in this direction.

There are also challenges and limitations of the experiments and analysis, and work may be extended to address and overcome them by involving domain experts and improving the machine learning methods. For example, a domain expert has advised that it is important to split the analysis of adverse drug effects into side effects, medication errors, and adverse drug reactions and categorize them correctly. There may also be drug interactions. In addition, it is wise to properly identify which side effects come from which drug.

In addition, the following may be applied on technically improving the proposed pipelines and experiments. First, have the pipeline be fed live-streams, allowing for constant updates on drug side effects over a certain time period. Next, implementation of our Scikit-learn ensemble classifier in Apache Spark (currently unsupported) can be done to take advantage of distributed processing with the majority vote classifier. More nodes can be added to Apache Spark to speed up the pipeline even further. Also, more tweets with different drug names can be added to the training dataset because those tweets would contain even more different side effects to further improve on our classifier accuracy. Furthermore, for the side effects, a dictionary can be made to remove side effects that are alleviated by the drug instead of being marked as caused by the drug. Tweets with multiple drugs can also be tested specifically to see that the side effect corresponds to the correct drug. Finally, the frequency output of the pipeline can be used to compare with FAERS to see if there are any common side effects that have not been reported to the FDA.

KEYWORDS

- **classification**
- **machine learning**
- **sentiment analysis**
- **opinion mining**
- **Apache Spark**
- **Twitter**
- **natural language processing**
- **supervised learning,**
- **adverse drug event**

REFERENCES

Agarwal, A.; Xie, B.; Vovsha, I.; Rambow, O.; Passonneau, R. Sentiment analysis of Twitter data. In *Proceedings of the Workshop on Languages in Social Media (LSM '11)*. Association for Computational Linguistics, Stroudsburg, PA, USA. 2011. 30–38.

Aronson, A. R. "Effective mapping of biomedical text to the UMLS metathesaurus: The MetaMap program." In *Proceedings of the AMIA Symposium*. 2001. 7–21.

Baccianella, S.; Esuli, A.; Sebastiani, F. "SENTIWORDNET 3.0: An enhanced lexical resource for sentiment analysis and opinion mining." In *Proceedings of the LREC Conference*. 2015.

Banerjee, R.; I. Ramakrishnan, V.; Henry, M.; Perciavalle, M. "Patient centered identification, attribution, and ranking of adverse drug events." In *Proceedings of the International Conference on Healthcare Informatics*. Dallas, TX, USA. 2015. 18–27.

Bates, D.; Cullen, D.; Laird, N.; Petersen, L.; Small, S.; Servi, D. et al. "Incidence of adverse drug events and potential adverse drug events implications for prevention." *JAMA.* 1995; 274(*1*): 29–34.

Bodenreider, O.; Hole, W. T.; Humphreys, B. L.; Roth, L. A.; Srinivasan, S. "Customizing the UMLS metathesaurus for your applications." In *Proceedings of the AMIA Symposium*. November 2002.

Burges, C. "A tutorial on support vector machines for pattern recognition." *Data Mining and Knowledge Discovery*. 1998; 2: 121–167.

Cavnar, W. B.; Trenkle, J. M. "N-gram-based text categorization." In *Proceedings of the 3rd Annual Symposium on Document Analysis and Information Retrieval, SDAIR-94*. Las Vegas, NV, USA. 1994. 161–175.

Deng, L.; Wiebe, J. "MPQA 3.0: An entity/event-level sentiment corpus." In *Proceedings of the NAACL-HLT*, 2015.

Drugs.com. "Popular Drugs" from Drug Index A to Z. from https://www.drugs.com/drug_information.html (accessed December 14, 2016)

FDA Adverse Event Reporting System (FAERS). https://www.fda.gov/drugs/surveillance/fda-adverse-event-reporting-system-faers. (accessed August 12, 2018)

Harnie, D.; Vapirev, A. E.; Wegner, J. K.; Gedich, A.; Steijaert, M.; Wuyts, R.; Meuter, W. D. "Scaling machine learning for target prediction in drug discovery using Apache Spark." In *Proceedings of the 15th IEEE/ACM International Symposium on Cluster, Cloud and Grid Computing*. Shenzhen. 2015. 871–879.

Hsu, D.; Moh, M.; Moh, T.-S. "Mining frequency of drug side effects over a large twitter dataset using Apache Spark." In *Proceedings of the 9th IEEE/ACM International Conference on Advances in Social Networks Analysis and Mining, ASONAM.* Sydney, Australia, July 2017. 915–924.

Jiang, K.; Zheng, Y. "Mining Twitter data for potential drug effects." In *Advanced Data Mining and Applications*. Springer: Berlin, Heidelberg. 2013. 434–443

Liu, B. "*Sentiment Analysis: Mining Opinions, Sentiments, and Emotions*." Cambridge University Press. 2015. https://www.cs.uic.edu/~liub/FBS/sentiment-analysis.html (accessed December 21, 2016).

Medline Plus Database. https://medlineplus.gov/ (accessed December 13, 2017)

Meng, X.; Bradley, J.; Yavuz, B.; Sparks, E.; Venkataraman, S.; Liu, D. et al. "MLlib: Machine learning in Apache Spark." *J. Mach. Learn. Res.*. 2016; *17(1)*: 1235–1241.

Nielsen, F. Å. "A new ANEW: Evaluation of a word list for sentiment analysis in microblogs." In *Proceedings of the ESWC2011 Workshop on 'Making Sense of Microposts': Big Things Come in Small Packages 718 in CEUR Workshop Proceedings*. May 2011. 93–98.

NLTK (Nature Language Tool Kit). www.nltk.org (accessed December 15, 2016)

Pedregosa, F.; Varoquaux, G.; Gramfort, A.; Michel, V.; Thirion, B.; Grisel, O. et al. "Scikit-learn: Machine learning in Python." *J. Mach. Learn. Res.* 2011; *12*: 2825–2830.

Peng, Y.; Moh, M.; Moh, T.-S. "Efficient adverse drug event extraction using Twitter sentiment analysis." In *Proceedings of the 8th IEEE/ACM International Conference on Advances in Social Networks Analysis and Mining, ASONAM.* San Francisco, California. Aug. 2016. 1101–1018.

Pyspark. Spark Python API. http://spark.apache.org/docs/latest/api/python/index.html (accessed December 21, 2016).

Roesslein, J. Tweepy (An easy-to-use Python library for accessing the Twitter API). http://www.tweepy.org (accessed on August 12, 2019).

Shapiro, K.; Brown, S. *Rx Prep Course Book*. 2016. 100–123.

Tabassum, N.; Ahmed, T. "A theoretical study on classifier ensemble methods and its applications". In *Proceedings of the 3rd International Conference on Computing for Sustainable Global Development (INDIACom)*. New Delhi. 2016. 374–378.

Toutanova, K.; Klein, D.; Manning, C. D.; Singer, Y. "Feature-rich part-of-speech tagging with a cyclic dependency network." In *Proceedings of the Conference of the North American Chapter of the Association for Computational Linguistics on Human Language Technology, NAACL'03, Vol. 1. Association for Computational Linguistics*. Stroudsburg, PA, USA. 2003. 173–180.

WebMD LLC, Medscape database. 2019. https://reference.medscape.com/ (accessed August 13, 2019)

Wu, L.; Moh, T.-S.; Khuri, N. "Twitter opinion mining for adverse drug reactions." In *Proceedings of the IEEE International Conference on Big Data (BigData)*. Santa Clara, California. October 2015. 1570–1574.

Yu, F.; Moh, M.; Moh, T.-S. "Towards extracting drug-effect relation from twitter: A supervised learning approach." In *Proceedings of the IEEE 2nd International Conference on Big Data Security on Cloud (BigDataSecurity), IEEE International Conference on High Performance and Smart Computing (HPSC), and IEEE International Conference on Intelligent Data and Security (IDS)*. New York, NY. 2016. 339–344.

Zaharia, M.; Chowdhury, M.; Das, T.; Dave, A.; Ma, J.; McCauley, M.; Franklin, M. J.; Shenker, S.; Stoica, I. "Resilient distributed datasets: A fault-tolerant abstraction for in-memory cluster computing." In *Proceedings of the NSDI'12*. April 2012.

CHAPTER 12

DEEP LEARNING IN BRAIN SEGMENTATION

HAO-YU YANG

CuraCloud Corporation, Seattle, WA, USA

Corresponding author. E-mail: deronmonta@gmail.com

ABSTRACT

Segmentation of normal tissue and lesion is a crucial first step to automated brain image analysis. Segmentation distills key information about the patient such as lesion volume, the affected area, and structural visualization. Traditionally, brain segmentation is done by human annotations which may be time-consuming and requires highly-knowledgeable annotators. While some machine learning methods such as region growing may provide a certain level of automation for segmentation, oftentimes human editing is still required. In recent years, the developments of the deep neural networks have lead to promising results in fully-automated brain image segmentation. Deep neural networks have higher model capacity compared to classical machine learning model, therefore are more suitable to handle highly diversified data such as medical images.

In this chapter, we first introduce some common brain imaging modalities and fundamental concepts in image segmentation. Then, we'll discuss modern deep neural network approaches to brain segmentation. Finally, we'll take a look at state-of-the-art methods in this field.

12.1 INTRODUCTION

Analysis of brain images, specifically segmentation, is a standard procedure for quantitative analysis in clinical diagnosis. Noninvasive imaging techniques such as computed tomography (CT), magnetic resonance imaging (MRI), and positron emission tomography (PET) are some routine imaging modalities for obtaining images of the brain. Brain segmentation can be further divided into two main tasks: segmentation of normal brain tissues and segmentation of brain lesions. Anatomical segmentation of normal brain tissues involves classifying voxels of 3D images or pixels of two-dimensional (2D) images into heterogeneous structures like gray matter, white matter, or cerebrospinal fluid. Segmentation of brain lesions aims to detect abnormal regions of the brain such as a tumor, multiple sclerosis, and stroke lesion.

While the current gold standard for brain segmentation is manual labeling, this approach requires intensive labor from highly experienced experts. Furthermore, it is related to other issues like human error and varying labeling standards from a different institution. Hence, tremendous efforts have been devoted to developing automated algorithms for segmenting brain tissues and lesions with little to no human intervention. Automated image segmentation is a computer vision task that aims to partition an image into meaningful segments for quantitative analysis. It is often the foundation of medical image analysis as useful information, for example, volume and position of anatomical structures, can be extracted from the results of segmentation.

Intensity thresholding, pixel clustering, and histogram-based methods are known as conventional image processing segmentation methods. These methods rely solely on the intensity information and distribution of pixels in the image to segment objects of interests. The convoluted structures of the brain and subtle discrepancy between normal tissues and lesions prove to be too complex for these simple methods to capture. Machine learning, a branch of artificial intelligence (AI) that uses pattern recognition techniques and learnable parameters, has shown the capabilities to replace simple image processing as the method of choice for image segmentation. Methods without stacked layers or deep representation are known as "traditional" machine learning, which includes algorithms like support vector machine (SVM), random forest (RF), and Markov random field (MRF). While demonstrating better performance than image processing methods, traditional machine learning still does not scale well to high-dimensional data such in three-dimensional (3D) or four-dimensional (4D) MRI sequences. Furthermore, it requires meticulous manual feature extraction and may show poor generalization facing different scanners.

Deep neural network (DNN) or deep learning (DL) is a subclass of machine learning algorithms that consists of multiple layers of nonlinear operation. These layers are used to represent different levels of abstraction. In contrast to traditional machine learning models, DNNs eliminate the need for hand-crafted features by incorporating data representation as a part of the training. With the recent rapid development of DL, there has been a growing interest in applying DNNs in medical image analysis. With a diverse choice of network specializations, convolutional neural network (CNN) is by far the most regularly utilized structure in the field of computer vision and image recognition. In its basic form, CNN consists of convolution layers pooling layers and fully connected layers.

This chapter focuses on DL applications in brain image segmentation.

We start off with a brief introduction of brain imaging modalities, image segmentation, followed by the essential image processing procedures. We will then introduce the basic building blocks for CNNs and lay the foundation for modern CNN architectural designs.

Next, we will direct our attention to the publicly available datasets in neuroimaging. Training techniques such as data augmentation and transfer learning will also be discussed in this section. We then review state-of-the-art DL models for brain segmentation and draw comparisons with traditional machine learning methods. Finally, we conclude the chapter with discussions regarding the current state, challenges of clinical integration and future trends of DL in brain segmentation.

12.2 BRAIN IMAGING MODALITIES

Brain imaging (also known as neuroimaging), involves using noninvasive techniques to obtain the structural or functional image of the brain. Structural imaging of the brain refers to obtaining the structure of the central nervous system. Structural imaging is useful for detecting large-scale intracranial diseases such as a tumor or brain trauma. Two of the most common structural brain imaging techniques used in clinical settings are MRI and CT.

Functional imaging is the measurement of a brain's functionality. It is effective in identifying metabolic-related diseases such as Alzheimer's. The functional imaging technique is also often employed in cognitive studies as these researches are concerned with functional connectivity of the brain. Functional MRI (fMRI), positron emission tomography (PET), and functional near-infrared spectroscopy are some examples of functional imaging methods.

Understanding the theoretical background of each imaging modality is a fundamental prerequisite for developing a medical image segmentation algorithm. For example, the pixel values of a CT scan represent physical meanings that correspond to different tissues such as bones, normal tissues, or micro-bleedings. Differentiating the nuances of each modality can help developers build a more efficient and error-proof system. In this section, we provide succinct descriptions and applications of routine imaging modalities in brain imaging. As we introduce each imaging modality, the advantages and disadvantages of each will also be covered.

12.2.1 MAGNETIC RESONANCE IMAGING

MRI is one of the most common modalities for brain imaging. It provides a distinctive contrast between different tissues of the central nervous system. A brain MRI study uses the magnetic gradients to differentiate heterogeneous tissues of the central nervous system. A subject undergoing MRI study is placed inside a confined scanner. The scanner forms a strong magnetic field that excites the hydrogen atoms, which is the most abundant atoms in human and organisms in general. Hydrogen atoms in organisms typically exist in the form of water or fat. Therefore, by alternating hydrogen atoms between excitation and resting state, the scanner can collect gradient signals and map the location of water and fat in human tissue.

The difference magnetization direction during the process of returning from the excitation state to the equilibrium state creates two types of contrasting images: the T1-weighted and T2-weighted MRI. The T1 image is created when the magnetization of the tissue aligns with the static magnetic field, while the T2 image is created by transverse magnetization of the tissue and the field. The two contrasting MRIs demonstrates distinct signal characteristics. A T1-weighted study is useful for establishing a baseline for normal brain structures, while a T2-weighted study can be used to detect inflammation and tumor.

Compared to radioactive imaging modalities that expose patients to small doses of ionizing radiation like X-rays and CTs, MRI studies do not involve radiation and present no known health risk. Moreover, MRI offers better contrast of soft tissues which make up the majority of tissue in the brain. Hence, MRI has gained widespread popularity in neuroimaging since its introduction in the late 1970s.

12.2.2 COMPUTED TOMOGRAPHY

CT integrates multiple X-ray images from various directions to form a tomographic volume image. The “slices” taken from each position are stitched together with digital geometry processing. Although other forms of imaging modalities like PET also utilize computer-generated tomography, the term “CT” typically refers to X-ray CT. Since its introduction in the 1970s, CTs have gained increasing popularity in diagnosing head-related trauma. A CT scan can be used to detect hemorrhage, tumors, and calcifications of the brain. The advantage of CT over MRI is the shorter time of the study. As such, in acute critical conditions like trauma and stroke, CT is the preferred imaging modality. One of the main disadvantages of CT is the high levels of radiation. While providing high-resolution volumetric images, the standard-dose CT scans may present 1000 times of radiation than 2D X-rays.

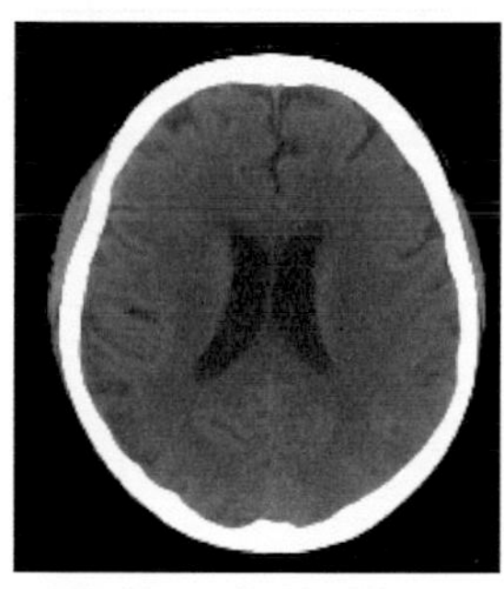

FIGURE 12.1 Example CT image of a stroke patient.

12.2.3 POSITRON EMISSION TOMOGRAPHY

PET is an imaging modality that generates a functional map of the brain by measuring the distribution of radiolabeled chemical markers. PET is one of the mainstream methods besides fMRI for obtaining the functional structure of the brain. PET can also be used for early diagnosis of diseases involving significant changes in the brain's metabolism such as Alzheimer's disease, brain tumors, and strokes. For example, there is little to no noticeable structural shift of the brain in the early stages of Alzheimer's disease. However, PET can measure regional glucose usage and detect subtle changes between patients with Alzheimer's disease and normal patients.

12.3 NEUROIMAGE SEGMENTATION

Image segmentation refers to the task of partitioning the pixels of a 2D image or voxels of a 3D image into semantic meaningful regions. Ideally, the pixels or voxels in the same semantic group should share similar characteristics or present physical significance. The purpose of performing segmentation is to transform the image into a concise representation that can be easily manipulated in further quantitative analysis such as area measurement, object detection, and boundary finding. We refer to the result of image segmentation as "segmentation mask" or simply "mask." Applications of image segmentation span multiple disciplines, ranging from automated video surveillance, face detection, pathology localization, etc.

There are many ways to categorize image segmentation methods.

In the context of this chapter, we classify the segmentation techniques into simple image-processing-based method and machine-learning-based methods with learnable parameters. Pure image processing segmentation techniques involve little to no learning process. This method includes region growing, active snake, and intensity thresholding. In contrast to image processing segmentation techniques, model-based approaches allow learnable parameters to fit the data. Extensive evaluation of a newly developed segmentation algorithm is crucial to understanding its performance. Two metrics commonly applied to evaluate image segmentation are the Sørensen–Dice coefficient and the Hausdorff distance.

Quantitative analysis of a neuroimage requires segmentation of the target of interest. Based on the nature of the target, brain segmentation can be categorized into two task groups. Anatomical brain segmentation is the identification of distinct brain structures such as cerebrospinal fluid, hippocampus, and white matter. Brain lesion segmentation, on the other hand, attempts to segment brain lesions from normal tissue. For instance, stroke, brain tumor, and microbleeding are targets for lesion segmentation in the brain. This section is an introduction to image segmentation as well as its application in neuroimaging.

12.3.1 SEGMENTATION EVALUATION METRICS

12.3.1.1 SØRENSEN–DICE

Sørensen–Dice coefficient, or commonly known as the Dice coefficient, is one of the most popular metrics used for evaluating the segmentation quality of medical images. The coefficient is essentially a similarity measure of two sets X and Y:

$$2 \cdot \frac{|X \cap Y|}{|X| + |Y|} \tag{12.1}$$

where X denotes the predicted mask and Y denotes the ground truth labels. The dice coefficient calculates the intersection of the two sets while taking the sizes of each set into account.

12.3.1.2 HAUSDORFF DISTANCE

In addition to the dice score, the Hausdorff distance is another metric for quantitative evaluation of segmentation performance. The Hausdorff distance of two sets X and Y measures the maximal distance between one point in a set to the nearest point in the other set. It can be calculated with the following equation:

$$d_H(X,Y) = max_{x \in X} min_{y \in Y} d(x,y) \tag{12.2}$$

12.3.2 NEUROIMAGING APPLICATIONS

12.3.2.1 ANATOMICAL BRAIN SEGMENTATION

Segmentation of anatomical brain structures like corpus callosum, lateral ventricle and hippocampus are useful for establishing quantitative assessment for brain developments. A popular method for segmentation of anatomical structures of the brain was the multiatlas approach. In an atlas-based method, several altars that are most similar to the query image are selected and matched. With the recent developments of DL, neural networks have been replacing the atlas-based methods, which also discarded the need for creating atlases.

12.3.2.2 BRAIN LESION SEGMENTATION

Segmentation of brain lesion is the foundation to lesion volume measurement, disease progression, and treatment assessment. Since the lesion area typically takes up a smaller area in proportion to the whole brain, lesion segmentation is often done with detection-based methods. One of the most routine applications of brain lesion segmentation is a brain tumor. Brain tumors are uncontrolled growth of cells in the central nervous system. There are over 100 subtypes of brain tumors with Gliomas being one of the most common subtypes. Depending on the aggressiveness of the cancer cells, Gliomas can be further classified into different grades. Early diagnosis of a brain tumor is crucial to treatment management and recovery. It is important to identify the area affected by the lesion as preoperation segmentation of normal and lesion tissues can help surgeons safely remove the tumor area while avoiding healthy tissue during an operation. Stroke is the cutoff of blood flow to the brain and is another frequent target of brain lesion segmentation.

12.4 IMAGE PROCESSING FOR BRAIN SEGMENTATION

The performance of automated brain imaging analysis is heavily affected by factors like scanner noises, subject-wise variabilities, and manufacture inhomogeneities. The purpose of preprocessing is, therefore, to standardize the images before conducting the automated algorithms in hopes of eliminating as many inconsistencies and artifacts as possible. The preprocessing steps ensure that images acquired from different institution scanners are quantitatively comparable. We introduce common image processing steps used in neuroimage analysis such as intensity normalization, registration, and skull stripping.

12.4.1 IMAGE REGISTRATION

Image registration is the task of aligning images in different coordinates into a unified system. Image registration is an essential step for comparing sequential images obtained from the same source, i.e., sequential CT scans from the same patient. Image registration is a ubiquitous operation in brain imaging since tracking a lesion in the brain will likely require multiple studies in an indefinite time span.

When registering two images, one image is set as the target or fixed image and the other is set as the moving image. The objective is to perform some transformation that matches coordinates of the moving image to that of the fixed image. There are two types of image registration: intensity-based methods and feature-based methods. Intensity-based registration matches the intensity information in the two images with correlation metrics and transforms the moving image accordingly. In a feature-based registration algorithm, the first step is to identify distinctive feature points of both the fixed image and the moving image. The transformation is done by aligning the feature points of the moving image to that of the fixed image.

12.4.2 INTENSITY NORMALIZATION

Intensity normalization is the projection of intensities across different images onto a single comparable scale. As the pixel intensities produced by different modalities present physical meanings, intensity normalization should be handled with the modality information in mind. As an example, the intensity of a CT image may range from negative to 4000. Not selecting the appropriate intensity window is detrimental to the automated algorithm.

12.4.3 SKULL STRIPPING

Skull stripping refers to removing the skull pixels of the head. In some modalities such as CT, the bones present a strong signal in the image and may affect the focus of the segmentation model.

12.5 TRADITIONAL MACHINE LEARNING

Machine learning is a branch of AI that relies on statistical models and pattern recognition to make predictions on unseen data. By fitting the parameters according to the training data, these methods require minimal explicit instructions on how to perform a certain task.

Automated approaches are necessary for large-scale brain segmentation because manual delineation of anatomical structures and lesions are time-consuming and are subject to large observer bias. It is also infeasible to conduct populational-level research involving thousands of image that requires a large number of expert annotators. As mentioned in earlier texts, image processing segmentation methods are not sufficient to capture the complexity of the brain structure. With a learnable parameter that adjusts according to the data, machine learning has been proposed for automated brain segmentation.

The subject of this section is machine learning methods without multilayer representation. We refer to these methods as "traditional" machine learning methods in contrast to the DL models which will be the topic of the next section. While we mainly focus on DL approaches in this chapter, it is imperative to understand the basics of traditional machine learning methods and understand how DL methods compare against the conventional approach.

12.5.1 FEATURE SELECTION

In conventional machine learning, algorithms operate on features extracted from the image rather than the raw image itself. A feature can be thought of as a summary statistics regarding the subject image. Average pixel intensity and standard deviation are some examples of scalar features. Features can also be matrices like the results from a neighboring pixel filter like Gaussian filters, Haar filters, etc. Features can also be study in Section 12.11. In summary, there are no rigorous rules for which feature to extract or to look for and often boils down to expert domain knowledge and previous research experiences.

12.5.2 LINEAR DISCRIMINANT ANALYSIS

Linear discriminant analysis (LDA) is a machine learning method that uses a linear combination of features to determine the class of the inputs. The main principle of LDA can be summarized as projecting data from the extracted feature space onto a single dimension space. The single dimension space can then be classified into two classes by thresholding. The major disadvantage of using LDA is the linear nature of the classifier. In most practical cases, the data cannot be separated in the feature space by a linear function.

12.5.3 SUPPORT VECTOR MACHINE

SVM is a class of supervised learning models that can be used for both classification and regression tasks. With a set of labeled data points, an SVM learns from this dataset and assign an unseen example to a category without assigning probabilities. Hence, SVM is a nonprobability classifier. With SVM, each data is represented as an *n*-dimensional vector. SVM classifies data point by what is known as a "hyperplane" that creates the largest margin between the different classes.

12.5.4 RANDOM FOREST

A decision tree is a simple classifier aimed at performing classification tasks by forming a series of decision nodes and tree-like structure. A decision tree consists of multiple branches whose quantity is based on the number of input features. Typically, each decision node in a decision tree is represented by the values of the input variables. The dependent (target) variable is positioned at the leaf of each tree. Though a simple yet powerful technique, decision trees are prone to poor generalization. *RF* improves upon a single decision tree by assembling the results from multiple independently trained decision tree. An RF is trained by bootstrap aggregating, meaning that each tree node has only a sampled subset of data to train on.

12.6 DEEP NEURAL NETWORKS

DNNs or DL is a specialized branch of machine learning that uses stacked layers of nonlinear operations for processing information. In contrast to the aforementioned traditional machine learning methods, DNNs do not require task-specific nor handcrafted features. Instead, DNN incorporates feature extraction as part of the learning process. The theoretical proofs of neural networks as a universal function approximator have first been introduced in the 1980s. Due to the limitation of computation powers, neural networks have not been given much attention until the early 2010s. With the recent significant advancement of specialized hardware designed to perform matrix multiplication like graphical processing unit, the computational limitation that was hindering the development of neural networks is now gone. A tremendous amount of research has been devoted to DNNs in the past decade, which has been backed-up by state-of-the-art performance in a wide variety of disciplines including natural language processing, computer vision, and medical images.

There exists a wide variety of neural network architectures such as fully connected neural networks, recurrent neural networks (RNNs), convolutional neural networks (CNNs), and restricted Boltzmann machines. These varying structures are designed to optimize the information flow of different data types. RNNs are ideal for processing data with sequential dependencies such as time series and texts. CNNs are the first choice when it comes to data with a spatial relationship such as images. As such, CNNs are by far the most widely adopted in the computer vision domain and will be the focus of our discussions regarding neural

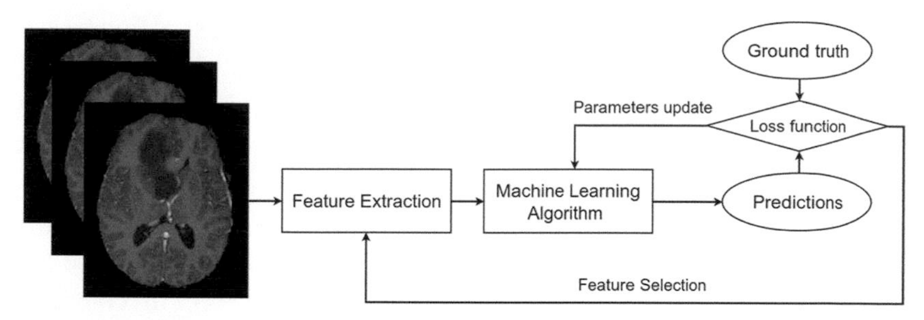

FIGURE 12.2 Training pipeline for conventional machine learning.

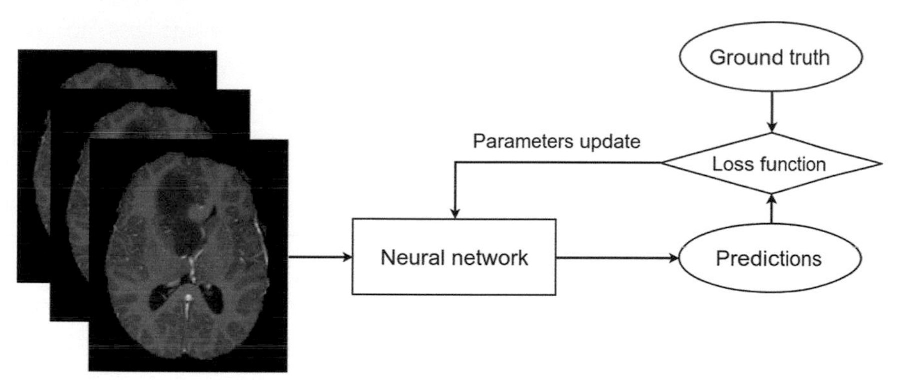

FIGURE 12.3 Training pipeline for deep learning.

networks. In this section, we describe the basic operations that serve as the building blocks for modern CNNs. These layers can be regularly seen across state-of-the-art DL models with different combinations and variants.

12.6.1 CONVOLUTIONAL NEURAL NETWORK

CNNs are a type of neural networks specialized in handling high-dimensional data with spatial resolution such as images. In a traditional multilayer feedforward neural network, the inputs are long vectors with no spatial or temporal correlation between adjacent elements. Using a vanilla multilayer perceptron neural network for analyzing images will destroy the spatial information and requires a large number of redundant parameters. In contrast, CNNs preserve the spatial resolution by using an efficient representation of the encoded information. CNNs are formed by a series of convolution layers, activation functions, and pooling layers. Convolution layers produce feature maps by sliding convolution kernels across the image. Pooling layers are used for downsampling the feature space. The end layer of a CNN is usually a fully connected layer.

12.6.1.1 CONVOLUTION LAYER

The convolution layer is the most essential part of a CNN and typically the layer that accepts the input image. The outputs of a convolution layer are often referred to as the feature map in the DL terminology. A convolution layer consists of multiple convolution filters (also known as kernels). Convolution filters have a fixed spatial arrangement in the same layer, that is, 5 × 5 × 3 pixels. During the forward pass, each filter will be slid across the image and a discrete value convolution commonly denoted with an asterisk (*) will be performed at each sliding position. A general definition of convolution is an operation between two functions that generates a third function. In the case of CNN, the input image (and the input feature maps in the forthcoming layers) can be thought of as the first function and the convolution kernels are therefore the second function. The kernels will learn various visual features and create high activations if the area contains the feature that it is looking for.

The size of the feature maps is determined by three factors: depth, kernel stride, and padding. The depth factor is simply the number of kernels at a specific layer. The kernel stride size is how much the kernel is moved from one convolution position to the next. Higher stride size means more area of the image is skipped and therefore creating a smaller feature map. Padding refers to inserting zeros around the edge of the image. This is useful for retaining the desired feature map size after a convolution layer. In contrast to fully-connected neural networks where connections are made across all neurons, CNN utilizes an idea called parameter sharing. By sharing the same convolution kernel across an image, a good amount of redundant weights can be avoided therefore allowing deeper network structures.

12.6.1.2 ACTIVATION FUNCTION

After a convolution layer, the feature maps are passed through an activation function

The activation function introduces element-wise nonlinearities to the model. Without the activation function, the weights and bias from the previous layer would be passed onto the next merely in the form of some linear combinations. Linear combinations are limited in terms of its ability to fit the data. The choice of activation function remains an open research topic. While the sigmoid function was the standard preference for activation function when DL started to gain attention, the rectified linear unit (ReLU) and leaky ReLU have mostly replaced the sigmoid function as the definitive activation function. The size of the feature map remains unchanged after the activation function.

12.6.1.3 POOLING LAYER

Pooling layers are typically inserted in between convolution layers. The purpose of pooling layers is to downsample the effective spatial resolution and therefore reduces the total number of parameters. In other words, pooling layers provide summary statistics of the previous layers using a different mechanism. For example, the *max-pooling* layer returns the maximum activation within a rectangular area. *Average pooling*, on the other hand, returns the average of pixels within the designated rectangular area. Another benefit of using pooling layers is to introduce translation and spatial shift invariance.

12.6.1.4 FULLY CONNECTED LAYER

The last layer of a CNN is usually a fully connected layer. The feature maps from the previous convolutional layers are reshaped to fit the fully connected layer where the output is an N-dimensional vector. N corresponds to the number of assigned classes in the ground truth labels. For instance, if the task is to classify voxels in an MRI into the background, normal brain tissue, and tumor tissue, then $N = 3$.

12.6.1.5 SKIPPED CONNECTION

Before we introduce the skipped connection (also known as a residual connection), we need to understand the vanishing gradients problem and build the intuition as to why such structural designs are necessary. As mentioned in the earlier section, modern neural networks are becoming deeper with the number of layers well-exceeding hundreds. However, with more layers training neural networks (which we will go into details in the latter section) becomes difficult. This is because the gradients of parameters with respect to the loss become too small after repeated matrix multiplications making earlier layer close the inputs nearly impossible to adjust.

The residual neural networks (ResNet) have been proposed as a solution to the vanishing gradient problem. ResNets utilizes a skip-connection in between neighboring layers. The skip-connection takes the inputs from the previous layer and adds it to the results of the current layer, avoiding computation in the current layer. The skipped connection solves the vanishing gradients problem by letting the gradients propagate through the identity function. The skipped connection can be implemented in almost any CNN structure and proves to be effective in improving CNN training.

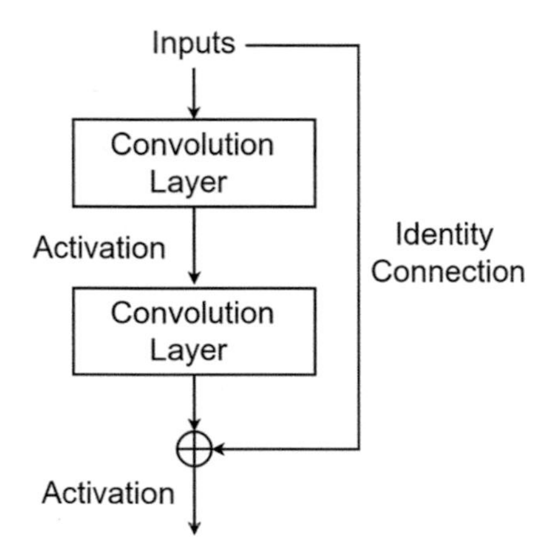

FIGURE 12.4 Example convolution block with skipped connections.

12.6.1.6 TRANSPOSED CONVOLUTION AND UPSAMPLING

Transposed convolutions (sometimes called deconvolution or fractionally strived convolution) and upsampling layers are regularly used in CNNs with a bottleneck structure such as the U-net. Networks with a bottleneck structure share a similar strait as the outputs of these models have the same dimension as the inputs. After repeated convolutions and pooling layers, the feature maps have much smaller heights and widths compared to the input image. Therefore, it is necessary to upsample the feature maps back to the original dimension. Upsampling can be done by parameterless operations like the nearest neighbor interpolation or a learnable function such as transposed convolutions. Transposed convolution can be thought of like a reverse operation to the convolution. With convolution, there is a many-to-one relationship between the inputs and outputs as the pixels under the convolution area are "squashed" into one value. Transposed convolution, on the other hand, shows a one-to-many relationship of the inputs and outputs. A single pixel is mapped to pixels within the kernel size of the transposed convolution. Take a 3×3 transposed convolution, for example, the center pixel is upsampled to nine pixels.

12.7 NEURAL NETWORKS FOR SEGMENTATION

In the previous section, we have discussed the common operations in a convolutional neural network. The topic of this section is the architectural design of CNN specifically for image segmentation purposes. In contrast to image classification task where the outputs are discrete-valued binary labels, image segmentation produces segmented masks that share the same dimension as the input images thus making this task computationally demanding. Furthermore, imaging of the brain often involves three-dimensional images such as CTs and MRIs, which aggravates the issue of the computational burden. Designing a segmentation network for volumetric images, therefore, requires additional attention to the hardware constraints. Common CNN structures used for brain image analysis can be roughly categorized into three subclasses: patch-based, semantic-based, and triplanar. Semantic-based neural networks output dense,

voxel-wise predictions that assign probabilities to each individual voxel. Since the introduction of fully convolutional neural networks (FCNNs) such as U-Net have been the working horse of major segmentation algorithms and have demonstrated state-of-the-art performance.

12.7.1 PATCH-BASED CNN

Patch-based CNN as its name suggests is a CNN training paradigm where the original images are divided into smaller patches. Each patch contains a center pixel and segmentation is done by classifying the center pixel with the neighboring pixels as the context information. Since dealing with volumetric images is the norm in neuroimaging, the patch-based method is an efficient way to handle large memory consumption.

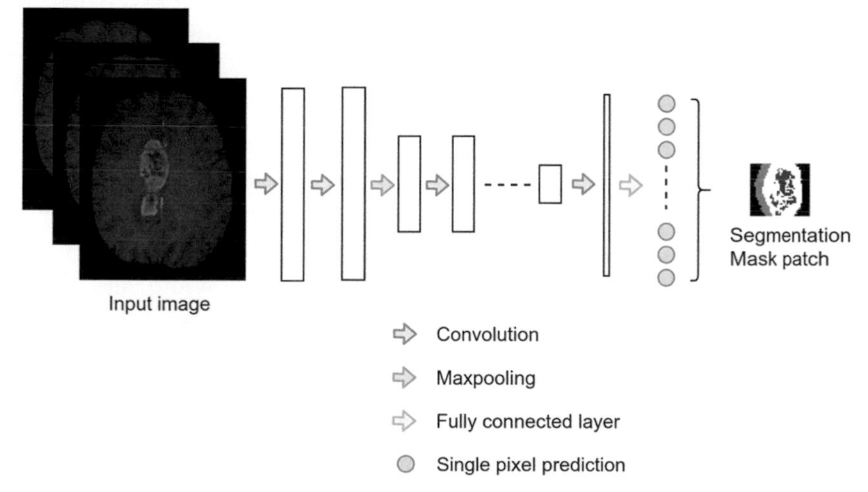

FIGURE 12.5 Schematic overview of a patch-based CNN.

12.7.2 FULLY CONVOLUTIONAL CNN

A fully convolutional CNN model contains only convolution, max-pooling and transpose convolution. This kind of network predicts a whole probability map of each pixel belonging to a certain class. A CNN with a fully-connected layer attached at the end only accepts a fixed dimension of inputs since the size of the last layer determines the size of the final output. The fully convolutional nature of the fully convolutional network allows images with varying dimensions and ideal for performing segmentation.

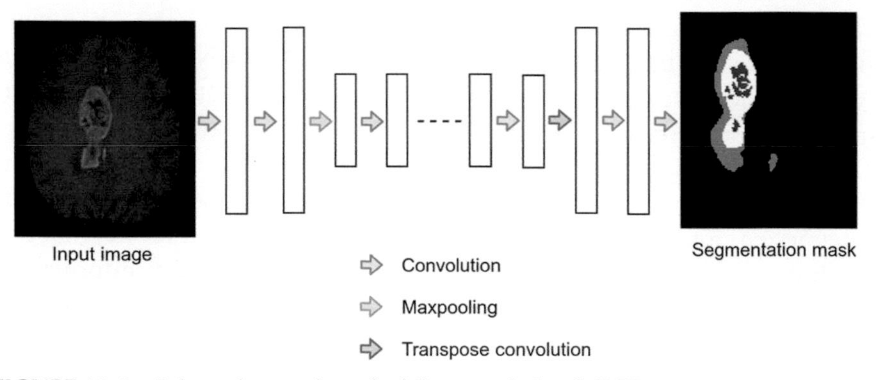

FIGURE 12.6 Schematic overview of a fully-convolutional CNN.

12.7.2.1 U-NET

U-net is a CNN-based neural network introduced by Olaf Ronneberger, Philipp Fischer, and Thomas Brox and was given the name due to its "U"-shaped architecture (Ronneberger et al., 2015). It was originally designed to perform semantic segmentation on biomedical images and is now widely adapted for general segmentation task U-net is an example of the fully convolutional network (FCN) where no fully connected layers are involved. A typical U-net consists of a downsampling path and an upsampling path with a concatenation of the feature maps in between the two paths. The downsampling path contains stacked convolutional layers, activation functions, and pooling layers until it reaches the bottleneck section of the network. With the image gradually passed down the downsampling path, the spatial resolution of the feature map decreases while the feature resolution increases. After another convolution block, the feature maps are passed onto the upsampling path. The upsampling path consists of transpose convolution blocks that up-samples the feature map back to its original dimensions.

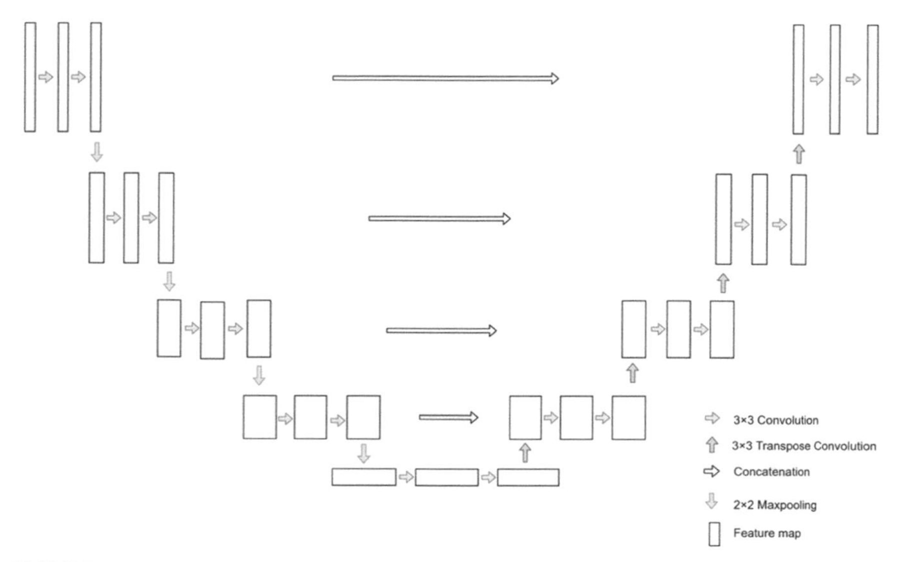

FIGURE 12.7 Schematic illustration of the U-net.

12.7.3 TRIPLANAR CNN

An alternative approach to the patch-based CNN in solving the memory consumption issue is the triplanar CNN. A volumetric medical image (or any 3D image) can be viewed from three planes: sagittal, coronal, and transaxial. The triplanar CNN takes the 2D image from each view and combines the results to form a single 3D segmentation mask. The triplanar CNN is sometimes called a 2.5D method as the third dimension is used in its entirety. Compared to a 3D patch-based method, the tri-planar CNN offers a larger field of view while still incorporating context information along the third dimension.

12.8 TRAINING AND EVALUATION

The two fundamental prospects of DL are the structural design of the network and parameters learning with latter often referred to as "training." Having discussed the structural formation of DNNs in the previous sections, we now shift our attention to the dataset, training and validation aspects of DL.

In the realm of machine learning, a typical paradigm for tuning parameters and validating models is to partition the available data into three nonoverlapping splits: training, validation, and testing. The training dataset is used for fitting the model parameters. The validation set is used for adjusting the hyperparameter such as learning rate, number of kernels, etc. The model's performance on unseen data is evaluated using the test dataset. For the training and validation set, both images and ground truth labels are included. Ground truth labels whether it be binary class labels, bounding boxes or segmentation masks, are obtained from expert annotation as the "gold standard" for the model.

A typical workflow of designing an automated algorithm for brain image analysis can be summarized as follows: First, the developers formulate the problem by examining the dataset and understanding the objective. Second, developers will conduct thorough literature research of current methodologies and design a network architecture tailored for the specific task. Then, model training, debugging and hyperparameter tuning will take place. Finally, the model is evaluated using a test dataset that is not seen by the model.

The aforementioned convolution layers and fully-connected layers all contain learnable parameters in the forms of weight metrics. Training a neural network is effectively adjusting these weight metrics according to the training data. The performance of a neural network undoubtedly depends on the training process. We will be covering a couple of common training techniques for CNNs in this section. In-depth discussions regarding the theoretical details of backpropagation is beyond the scope of this chapter. Readers can refer to Bengio et al. (2017) for detailed theoretical proofs.

12.8.1 TRAINING NEURAL NETWORKS

Training a neural network is formulated as an error-minimization problem where the goal is to minimize the distance between the ground truth label and the predicted results from the model. Training a neural network is essentially a sequential four-step procedure that is repeated iteratively until the network reaches convergence. These steps are the forward pass, loss calculation, backward pass, and parameters update. In the *forward pass* step, an image will be fed as the input to the network and passed through the layers of the CNN. The results of the forward pass will then be compared to the actual label to produce a continuous-valued error. This is called the *loss calculation* step. In the next phase, *backward pass* (also known as backpropagation), each parameter in the networks has attributed a gradient that represents how much the parameter is responsible for the error calculated in the previous step. Finally, *parameters update* adjusts the weights in the convolutional layers and fully connected layers with respect to the gradients from the backward pass step.

12.8.2 HYPERPARAMETERS

The *hyperparameters* in a neural network refer to those parameters that are not attributed in the backward pass and parameters update steps. Since hyperparameters are not learned in the training process, they need to be manually adjusted by the developers. Take the *learning rate*, for example. Learning rate is how much the parameters are adjusted each time during the parameter update step. Balancing the learning rate is important because a low learning rate leads to slow convergence and too high of learning will lead to unstable training. Hence, experimenting with some sensible choice of hyperparameters is crucial to the performance of the model and requires a solid understanding of CNNs. Other hyperparameters include a number of layers, layer arrangements, stride size, kernel size, padding size, etc.

12.8.3 TRAINING TECHNIQUES

12.8.3.1 DATA AUGMENTATION

While having access to more training data improves the performance of a machine learning algorithm, collecting new data may be expensive and sometimes infeasible for the case of medical images. *Data augmentation* is a powerful training technique that extends the existing dataset at hand. It is widely adopted in DNNs and machine learning algorithms in general due to its effectiveness in preventing overfitting and improving model performance. The technique creates small perturbation by performing geometric and color transformation to the input image. For instance, randomly cropping small portions of the image 10 times effectively creates 10 different images to the dataset. Other forms of data augmentation include image translation, random flipping, color jittering, and random affine transformation.

12.8.3.2 TRANSFER LEARNING

Training a neural network from scratch can be time-consuming and random weight initialization will likely lead to models with fluctuating local minimum. Images generally share similar low-level features like edges, lines, and geometric shapes, and these features will likely be learned across any image dataset. *Transfer learning* is an effective training technique that reuses the weights of a CNN previously trained on a large dataset for a secondary task. By reusing the weights from a large dataset, we can assure that the model learns the low-level features on from a robust source. Generally speaking, the fully connected layer at the end of the CNN is the only structural modification that needs to be done for transfer learning.

12.8.3.3 DROPOUT

Dropout is a training technique that is proven to be effective against overfitting.

In simple terms, executing a dropout in training is to ignore neurons by random. The neurons being dropped out will not participate in the forward pass or loss calculation in the back-propagation pass. By dropping some neurons during training, the model is forced to learn from the strongest signal and therefore not likely to overfit.

12.8.3.4 BATCH NORMALIZATION

Batch normalization is a technique for improving the training of neural networks in terms of speed and guarantee convergence. Essentially, batch normalization adjusts each mini-batch by subtracting the batch mean from each sample then dividing by the batch standard deviation. There are two major advantages of using batch normalization. First, by normalizing the activations at each layer, the interlayer interaction can be reduced, meaning that an over-activation in one layer will less likely to affect the up-coming layer. Second, without batch normalization, choosing a high learning rate will likely to cause fluxing training. In short, batch normalization helps stabilize the training process.

12.9 DATASETS AND STATE OF THE ART

Having a large dataset with high-quality labels is just as crucial as the design of the model for data-driven methods like DNNs. There are several challenges regarding the availability of data that are unique in the medical imaging domain. First, compared to tasks involving natural images where abundant data exists, publically accessible medical images are scarce. The lack of data is especially problematic for DL models which are notorious for being "data hungry." Second, developers need to be aware of the privacy issues regarding sensitive patient information when collecting medical data for training machine learning models. The Health Insurance Portability and Accountability Act (HIPPA) is a United States legislation that aims at safeguarding sensitive medical information. Any clinical applications involving patient data needs to be HIPPA compliant.

Recently, online challenges with publicly available dataset are gaining considerable attention. An online challenge is a data science related competition that offers public datasets and invites participants to develop algorithms regarding the dataset. Typically, an online challenge undergoes a sequence of phases. The first phase is usually the "training phase," where the training dataset alongside expert annotated ground truths is released to the participants. Participants will use this data to design and evaluate their automated algorithms. After a designated period of time, the organizers will release a set of test data with no accompanying labels. The participants are asked to either submit their program or the results on the test set to the challenge website. Before the widespread use of online challenges, algorithms are trained and evaluated on different datasets, making it extremely difficult to quantitatively compare the effectiveness of each algorithm. Specifically, it is hard to pinpoint whether the improvements came from a well-designed model or simply having a larger dataset with better quality. Online challenges resolve these issues by offering benchmark datasets and a robust evaluation method for the participants. By using the same dataset for training, different algorithms share the same platform for evaluation

In this section, we introduce several state-of-the-art publically available datasets related to brain segmentation. The modalities of these datasets range from MRI to CT with different segmentation targets such as tumor and stroke. Furthermore, these datasets are often used as the benchmark for task-specific performance. For example, the Multimodal Brain Tumor Segmentation Challenge (BraTS) has been a gold standard for developing automated brain tumor segmentation algorithms.

12.9.1 BRATS

BraTS is an annual online challenge held in conjunction with one of the largest conferences on computer-assisted medical imaging, the International Conference on Medical Image Computing and Computer-Assisted Intervention (MICCAI). The goal of the BraTS challenge is to identify state-of-the-art methods for automated brain tumor segmentation. Each year, the organizers release an increasingly larger number of MR scans with different contrasting methods. The last BraTS challenge (BraTS 2018), featured 135 glioblastoma studies and 108 low-grade glioma studies collect from 19 institutions. The expert-annotated labels included different tumor subtypes like gadolinium enhancing, peritumoral edema, and necrotic and nonenhancing tumor core. The imaging study from a single patient consists of four contrast modalities: native T1, postcontrast T1-weighted (T1Gd), native T2, and T2 fluid attenuated inversion recovery (T2-FLAIR). In the 2018 challenge, the dataset also included overall survival data where the participants are asked to predict the overall survival time using pre-operation scans.

The state-of-the-art method for brain tumor segmentation is the first place method proposed by NVIDIA (Myronenko, 2018) in the BraTS 2018 challenge. The method consists of a regular U-net branch and a variational autoencoder (VAE) branch that shares partial parameters. The VAE branch takes the multimodality MRIs as inputs and seek reconstructs identical images as outputs. The VAE acts as a regularizer to ensure segmentation quality. The dice coefficients for this method are 0.76, 0.88, and 0.81 for enhancing tumor core, whole tumor, and tumor core, respectively.

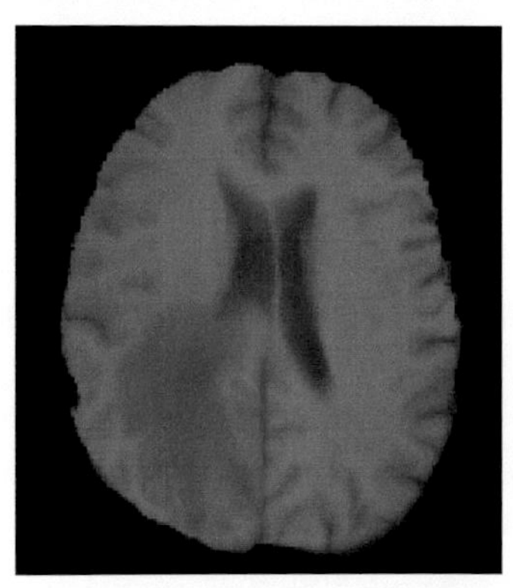

FIGURE 12.8 Example data from the BraTS challenge.

12.9.2 ISLES

Ischemic stroke lesion segmentation (ISLES) is a medical image segmentation challenge aimed at identifying stroke lesions locations. The ISLES challenge is also held in conjunction with MICCAI. While the target of segmentation remained ischemic stroke lesion, the imaging modality for ISLES has changed over the years. For ISLES 2016 and 2017, the dataset provided by the organizers were 51 MRIs from stroke patients. In the 2018 challenge, the organizers provided acute stroke CT and multiple perfusion maps using contrast agents. For instance, cerebral blood volume, cerebral blood flow, and time to peak of the residue function (T_{max}) are some perfusion maps included in the dataset.

The best-performing method for ISLES (Tong, 2018) used a multi-scale 3D U-net with dilated convolution. The multiscale approach combines low-level features and high-level features to ensure precise pixel-wise localization.

12.9.3 MRBRAINS

The grand challenge on *MR brain segmentation* (MRBrainS) is another challenge held with MICCAI. In contrary to BraTS and ISLES, which are both brain lesion segmentation tasks, the MRBrainS challenge is an illustration of anatomical brain segmentation.

In the MRBrainS18 challenge, the organizers provided 30 fully annotated multimodality MRI sequences. The modalities include T1, T1 inversion recovery, and T2-FLAIR. The ground truth covered 11 labels including basal ganglia, white matter, ventricles, brain stem, etc.

The state-of-the-art method for MRBrainS is a 3D patch-based U-net proposed by Luna et al. The modification for this approach is that instead of directly concatenating the feature maps between the downsampling path and the upsampling path of the U-net, the authors added transition layer in the bridge section.

12.10 DISCUSSION

Despite only recently applied to brain image analysis, DNNs have demonstrated immense potential for accelerating clinical workflow, improving segmentation accuracy, and conserving medical resources. DL approaches have surpassed conventional machine learning methods in terms of generalizability and accuracy on numerous occasions. Compared to traditional machine learning methods, DL models do not need hand-engineered features, which lead to faster and more robust development. With the widespread of online challenges and datasets, benchmarking new segmentation algorithms performance can be done with little effort. Methods employing the U-net architecture with a variety of modifications can be found across almost all semantic segmentation challenges. Developers usually start with a U-net as the baseline method for the segmentation task.

Although DL approaches, especially CNN-based models, have demonstrated state-of-the-art performances for segmentation of brain images, it is still challenging to develop a general-purpose model that is robust to scanner, task and modality variations. Currently, these variations are handled by meticulously sought out preprocessing steps that differ from task to task. In the following section, we are going to discuss some challenges DL models face in terms of clinical integration and adapting to the medical domain.

12.10.1 CHALLENGES

One of the main criticisms of the DNN is the lack of interoperability as it has often been coined as a “black box” approach. The black box refers to a model that is difficult to trace the predictions back to the inputs. In other words, it is challenging to know why the black box model made a certain decision by examining the inputs. Interoperability of a computer-aided diagnosis (CAD) system is especially important since applications in the medical domain involve the well-being of human lives. Physicians employing a CAD system should be able to trace the decision process from the input image to the predicted probabilities of disease. While there are other black box machine algorithms such as SVMs, a large number of DNN applications in medical imaging are much more likely to take place in the foreseeable future. Addressing the interoperability issue is therefore imperative to the development of DNNs in the medical domain.

The first convolutional layer perhaps holds the highest interoperability amongst all layers in a CNN. By checking the activations of feature maps from the initial layer, human readers can tell whether a certain kernel detects edges, curved lines or a specific geometric shape. However, the feature maps beyond the first layer are too abstract for human comprehension. Currently, the mainstream approach for interpreting the predictions of CNNs is by the use of attention mechanism. A CNN with attention module will place higher weights on different parts of the image. By examining the feature maps through the form of a weighted heat map, human readers can distinguish which part of the image contributes the most to a certain prediction of the network. However, some drawbacks such as time consumption and limited spatial resolution of the attention maps still exist.

Another issue with a DNN is the potential risk of overfitting. Overfitting occurs when the model parameters fit a limited amount of data points too closely. Real-world data often contains random noises and an overfitting model will likely to consider these noises as also a part of the signal. For instance, machine manufacturer, institutional imaging protocols and meta-information will all affect the neural network as it will mistakenly associate these signal with the true label if not enough data is presented. Furthermore, a neural network that overfitted the training data may produce contradicting results given visually similar inputs. There are two aspects to addressing the overfitting issue. The first solution is to present the model with more data to ensure the model learns from the true data distribution. Though the online challenges have mitigated the lack of medical data, the number of images in the medical dataset is still relatively limited compared to natural image datasets like ImageNet and CIFAR-10. The second solution is to use regularization techniques. The regularization techniques can be training related such as weight decay or learning rate decay. Regularization can also be an explicit design in the neural network like dropout and batch normalization.

12.10.2 FUTURE TRENDS

DL is one of the fastest growing scientific disciplines with substantial breakthrough each year. Since the topic of this chapter is segmentation, we have been focusing on segmentation network specifically CNNs. However, the field of DL is broad and innovations in some other seemingly unrelated domain often have a far-reaching effect. Here, we take a look at some cutting-edge DL trends for DL that were not proposed to solving segmentation task but ended up having a significant impact.

12.10.2.1 REGIONAL CNN

Regional CNN (R-CNN) is a family of models based on region proposal networks where regions of interest are detected through bounding box regression. These methods are widely adopted in the general computer vision community where object detection is much more common than object segmentation. The first generation of R-CNN produces only bounding boxes. However, with the introduction of Mask-RCNN, its application has extended to segmentation as well. There have been recent studies employing Mask-RCNNs in brain segmentation and shown promising results.

12.10.2.2 GENERATIVE ADVERSARIAL NETWORKS

Generative adversarial network (GAN) is a relatively new concept in DL where a generative network and a discriminator network are trained in a min–max game fashion. The purpose of the generator is to produce realistic samples, and it is the discriminator's duty to identify fabricated images. Previously, GANs have been mostly utilized in generating realistic natural images. Recent researches have shown that GANs can also be employed in image segmentation. There are two ways that GANs can be used for image segmentation. The first approach is to use the discriminator as a critique network for refining the segmentation results of a base segmented network. For example, a classification network takes the ground truth segmentation mask and a U-net predicted segmentation mask as inputs and tries to distinguish whether the mask is real or generated. The gradients are then fed back to the U-net to create masks that resemble the ground truth more closely.

The second approach is to use a conditional GAN to transform the original image into a "mask" domain. In contrast to a vanilla GAN where the network produces images based on random Gaussian noises, a conditional GAN generates a new image conditioned on a reference image. In a segmentation scenario, the reference image is the original image that needs to be segmented and the output is a mask with discrete values. The conditional GAN transforms the input image from its original domain to the masked domain. The discriminator network holds a similar objective as the first approach where it tries to differentiate if the transformed mask comes from a true mask distribution.

12.11 SUMMARY

In this chapter, we have provided a comprehensive discussion of DNN applications in brain image segmentation. The first section of this chapter is an introduction to brain imaging and image segmentation. We have then introduced DNNs and frequent operations such as convolution, max-pooling, batch-norm, etc. We then proceed to show how these operations are arranged to form a functioning network for segmentation purposes. Model training and various training techniques were covered in the subsequent section. Finally, an overview of notable public datasets and state-of-the-art models for brain segmentation were discussed.

KEYWORDS

- **brain segmentation**
- **lesion segmentation**
- **deep learning**
- **neural networks**

REFERENCES

Bakas, Spyridon, et al. "Advancing the Cancer Genome Atlas Glioma MRI Collections with Expert Segmentation Labels and Radiomic Features." *Scientific Data*, 5 Sept. 2017, www.ncbi.nlm.nih.gov/pubmed/28872634.

Bengio, Yoshua, et al. *Deep Learning*. MIT Press, 2017.

Brainlesion: Glioma, Multiple Sclerosis, Stroke and Traumatic Brain Injuries: 4th International Workshop, BrainLes 2018, Held in Conjunction with MICCAI 2018, Granada, Spain, September 16, 2018—Part I. Springer, 2019.

Brainlesion: Glioma, Multiple Sclerosis, Stroke and Traumatic Brain Injuries: 4th International Workshop, BrainLes 2018, Held in Conjunction with MICCAI 2018, Granada, Spain, September 16, 2018—Part II. Springer, 2019.

Brant, William E., and Clyde A. Helms. *Fundamentals of Diagnostic Radiology*. Wolters Kluwer/Lippincott Williams & Wilkins, 2012.

Brebisson, Alexandre De, and Giovanni Montana. "Deep Neural Networks for Anatomical Brain Segmentation." *Proceedings of the IEEE Conference on Computer Vision and Pattern Recognition Workshops (CVPRW'15)*, 2015, doi:10.1109/cvprw.2015.7301312.

Brosch, Tom, et al. "Deep 3D Convolutional Encoder Networks With Shortcuts for Multiscale Feature Integration Applied to Multiple Sclerosis Lesion Segmentation." *IEEE Transactions on Medical Imaging*, May 2016, 35, 1229–1239, www.ncbi.nlm.nih.gov/pubmed/26886978.

Caceres, J. Alfredo, and Joshua N. Goldstein. "Intracranial Hemorrhage." *Emergency Medicine Clinics of North America*, Aug. 2012, www.ncbi.nlm.nih.gov/pmc/articles/PMC3443867.

Cancer Imaging Archive Wiki. *Segmentation Labels and Radiomic Features for the*

Pre-Operative Scans of the TCGA-LGG Collection—TCIA DOIs—Cancer Imaging Archive Wiki, wiki.cancerimagingarchive.net/display/DOI/Segmentation Labels and Radiomic Features for the Pre-operative Scans of the TCGA-LGG collection.

Cancer Imaging Archive Wiki. *Segmentation Labels and Radiomic Features for the Pre-Operative Scans of the TCGA-GBM Collection—TCIA DOIs—Cancer Imaging Archive Wiki*, doi.org/10.7937/K9/TCIA.2017.KLXWJJ1Q.

Caselles, Vicent, et al. *Geodesic Active Contours*. Kluwer Academic Publishers, doi.org/10.1023/A:1007979827043.

CIFAR-10 and CIFAR-100 Datasets, www.cs.toronto.edu/~kriz/cifar.html.

Current Methods in Medical Image Segmentation. Annual Reviews, www.annualreviews.org/doi/10.1146/annurev.bioeng.2.1.315.

Deng, Jia, et al. "ImageNet: A Large-Scale Hierarchical Image Database." *Proceedings of the IEEE Conference on Computer Vision and Pattern Recognition*, 2009, doi:10.1109/cvprw.2009.5206848.

Feng, Xue, et al. "Brain Tumor Segmentation Using an Ensemble of 3D U-Nets and Overall Survival Prediction Using Radiomic Features." *Brainlesion: Glioma, Multiple Sclerosis, Stroke and Traumatic Brain Injuries, Lecture Notes on Computer Science*, 2019, pp. 279–288., doi:10.1007/978-3-030-11726-9_25.

Gaillard, Frank. "Ischemic Stroke" *Radiopaedia Blog RSS*, radiopaedia.org/articles/ischaemic-stroke.

Havaei, Mohammad, et al. "Brain Tumor Segmentation with Deep Neural Networks." *Medical Image Analysis*, Jan. 2017, www.ncbi.nlm.nih.gov/pubmed/27310171.

He, Kaiming, et al. "Deep Residual Learning for Image Recognition." *Proceedings of the IEEE Conference on Computer Vision and Pattern Recognition (CVPR'16)*, 2016, doi:10.1109/cvpr.2016.90.

Ho, Tin Kam. "Random Decision Forests." *Proceedings of the 3rd International Conference on Document Analysis and Recognition*, doi:10.1109/icdar.1995.598994.

Kamnitsas, Konstantinos, et al. "Efficient Multi-Scale 3D CNN with Fully Connected CRF for Accurate Brain Lesion Segmentation." *Medical Image Analysis*, 2017, 36, 61–78. www.ncbi.nlm.nih.gov/pubmed/27865153.

Long, Jonathan, et al., "Fully Convolutional Networks for Semantic Segmentation." *Proceedings of the IEEE Conference on Computer Vision and Pattern Recognition (CVPR'15)*, 2015, doi:10.1109/cvpr.2015.7298965.

Mcrobbie, Donald W., et al. *MRI from Picture to Proton*. Cambridge University Press, 2016.

Menze, Bjoern H, et al. "The Multimodal Brain Tumor Image Segmentation Benchmark (BRATS)." *IEEE Transactions on Medical Imaging*, 2015, 34, 1993–2024. www.ncbi.nlm.nih.gov/pubmed/25494501.

Moeskops, Pim, et al., "Automatic Segmentation of MR Brain Images with a Convolutional Neural Network." *IEEE Transactions on Medical Imaging*, 2016, 35, 1252 1261. www.ncbi.nlm.nih.gov/pubmed/27046893.

Morra, Jonathan, et al., "Machine Learning for Brain Image Segmentation." *Machine Learning*, 2012, pp. 851–874, doi:10.4018/978-1-60960-818-7.ch408.

Myronenko, Andriy. "3D MRI Brain Tumor Segmentation Using Autoencoder Regularization." *Brainlesion: Glioma, Multiple Sclerosis, Stroke and Traumatic Brain Injuries, Lecture Notes on Computer Science*, 2019, pp. 311–320, doi:10.1007/978-3-030-11726-9_28.

Pham, D.L., Xu, C., Princ, J.L. *Computer Vision*. Springer, 2014.

Poldrack, Russell A., and Rebecca Sandak. "Introduction to This Special Issue: The

Cognitive Neuroscience of Reading." *Scientific Studies of Reading*, 2018, 8, pp. 199–202, doi:10.4324/9780203764442-1.

Prasoon, Adhish, et al. "Deep Feature Learning for Knee Cartilage Segmentation Using a Triplanar Convolutional Neural Network." *Advanced Information Systems Engineering, Lecture Notes on Computer Science*, 2013, pp. 246–253, doi:10.1007/978-3-642-40763-5_31.

Razzak, Muhammad Imran, et al. "Deep Learning for Medical Image Processing: Overview, Challenges and the Future." *Lecture Notes on Computational Vision and Biomechanics Classification in BioApps*, 2017, pp. 323–350, doi:10.1007/978-3-319-65981-7_12.

Ronneberger, Olaf, et al. "U-Net: Convolutional Networks for Biomedical Image Segmentation." *Medical Image Computing and Computer-Assisted Intervention—MICCAI'15, Lecture Notes on Computer Science*, 2015, pp. 234–241, doi:10.1007/978-3-319-24574-4_28.

Roth, Holger R., et al. "A New 2.5D Representation for Lymph Node Detection Using Random Sets of Deep Convolutional Neural Network Observations." *Medical Image Computing and Computer-Assisted Intervention—MICCAI2014, Lecture Notes on Computer Science*, 2014, pp. 520–527, doi:10.1007/978-3-319-10404-1_65.

Schroff, F., et al. "Object Class Segmentation Using Random Forests." *Proceedings of the British Machine Vision Conference*, 2008, doi:10.5244/c.22.54.

Sergey, et al. "Batch Normalization: Accelerating Deep Network Training by Reducing Internal Covariate Shift." *arXiv.org*, 2 Mar. 2015, arxiv.org/abs/1502.03167.

"Special Section on Deep Learning in Medical Applications." *IEEE Transactions on Medical Imaging*, vol. 34, no. 8, 2015, pp. 1769–1769, doi:10.1109/tmi.2015.2460431.

"Support Vector Machine." *SpringerReference*, doi:10.1007/springerreference_63661.

Tanoue, L.T. "Computed Tomography—An Increasing Source of Radiation Exposure." *Yearbook of Pulmonary Disease*, vol. 2009, 2009, pp. 154–155, doi:10.1016/s8756-3452(08)79173-4.

Wang, Mei, and Weihong Deng. "Deep Visual Domain Adaptation: A Survey." *Neurocomputing*, vol. 312, 2018, pp. 135–153, doi:10.1016/j.neucom.2018.05.083.

Yang, Hao-Yu, and Junlin Yang. "Automatic Brain Tumor Segmentation with Contour Aware Residual Network and Adversarial Training." *Brainlesion: Glioma, Multiple Sclerosis, Stroke and Traumatic Brain Injuries, Lecture Notes on Computer Science*, 2019, pp. 267–278, doi:10.1007/978-3-030-11726-9_24.

Yang, Hao-Yu. "Volumetric Adversarial Training for Ischemic Stroke Lesion Segmentation." *Brainlesion: Glioma, Multiple Sclerosis, Stroke and Traumatic Brain Injuries, Lecture Notes on Computer Science*, 2019, pp. 343–351, doi:10.1007/978-3-030-11723-8_35.

Zhou, S. Kevin, et al. *Deep Learning for Medical Image Analysis*. Academic Press, 2017.

CHAPTER 13

SECURITY AND PRIVACY ISSUES IN BIOMEDICAL AI SYSTEMS AND POTENTIAL SOLUTIONS

G. NIRANJANA* and DEYA CHATTERJEE

Department of Computer Science and Engineering, SRM Institute of Science and Technology, Kattankulathur, Chennai 603203, India

Corresponding author. E-mail: niranjag@srmist.edu.in

ABSTRACT

Machine learning and artificial intelligence (AI) have been rapidly progressing in several fields, and one of these is healthcare and biomedical systems where AI and deep learning algorithms have shown massive success for various applications and use cases, such as virtual healthcare assistants, smart medical homes, automated diagnosis, processing of pathology reports, drug discovery, implantable medical devices, and many more. At this stage of progress, AI can reshape the healthcare workforce and bring about a drastic positive change in the understanding and handling of biomedical data in automated systems. However, as these artificially intelligent systems continue to gain more importance and surpass the state-of-the-art systems in terms of performance in biomedical and healthcare problems, the aspect of preserving the privacy and ensuring the security of our data and the users becomes extremely pertinent. Deep learning techniques may be at the risk of learning and memorization of confidential information instead of generalizing well to it, which, in fact, is the main aim (often termed as overfitting). Biomedical data is inherently complex and involves sensitive and private, often confidential, information of patients and users, which is easily vulnerable in the face of attacks by malicious actors. Often, the handling of clinical data requires more context than the standard for usual applications of deep learning algorithms, such as patient history, patient preferences, social perspectives, etc. Moreover, the "black box" nature of deep learning algorithms results in a lack of model interpretability and gives rise to confusion regarding exactly how an AI model can achieve the kind of performance it does. It is hence not easy to identify the weaknesses of the model or the reasons for the weakness, or even to extract additional biological explanations from the results. This leads to potential misuse of algorithms by attackers and poses potential threats to user/patient security in biomedical AI systems. Moreover, biomedical systems usually leverage third-party cloud platforms due to scalability, storage and performance benefits, and privacy compromisation is also likely to happen in such situations unless secure sharing schemes and suitable encryption techniques are devised. The problem of security also arises in the case of data integration and adoption that is needed to develop large scale biomedical expert systems. Some ways in which user data privacy can be jeopardized include indirect data leakage, data poisoning (i.e., including fake data samples in training set to drastically change the accuracy), linkage attacks (i.e., recognizing the actual identities of anonymized users), dataset reconstruction from published results, adversarial examples (i.e., adding noise to data to mislead the algorithm), transferability attacks, model theft, etc. These are pertinent questions from the perspective of general machine learning security, and is even more important in the biomedical domain, considering the stake of human life and health. As far as biomedical AI systems are concerned, traditional security and privacy mechanisms are not suitable to be adopted due to the ever-changing nature of research and the complexity of medical data. To counter such security threats, many studies have suggested the adoption of a set of best practices when working with biomedical data and also to ensure the optimal use of predictive models in research, especially to discourage inadequate studies with inaccurate results that may compromise the credibility of important and valid research in the field. There have been recent studies pointing to the practice of keeping training data private while simultaneously building accurate AI models. Two important techniques, in this case, are differential privacy and federated learning, which may serve as potential solutions for the problem. To counter linkage attacks and security threats of the similar sort, which is especially important in the wake of healthcare services in mobile devices that can compromise user identity and location data, recent studies have suggested private record linkage and entity resolution techniques, such as deriving unique fingerprints from genomes to preserve patient identity. Finally, it is extremely important to test AI models in real-time clinical situations (which are often complex and noisy) to further understand the fragility of such models and where their vulnerabilities can be exploited, so that better security schemes can be devised to counter the problem. After all, it is always essential to understand the problem thoroughly to come up with actual and effective solutions. In conclusion, addressing the problem of security and privacy in biomedical AI systems is complex, multidisciplinary, and also involves ethical and legal perspectives. As newer and better machine learning and deep learning algorithms are devised to tackle the problems in the healthcare and medical domain, newer security threats will also emerge. We are confident that research in this field will result in quality solutions to achieve the true balance between performance and privacy that is conducive to users and patients in the healthcare and biomedical domains.

13.1 INTRODUCTION

Machine learning and artificial intelligence (AI) have been rapidly progressing in several fields and one of these is healthcare and biomedical systems, where AI and deep learning algorithms have shown massive success for various applications and use cases, such as virtual healthcare assistants, smart medical homes, automated diagnosis, processing of pathology reports, drug discovery, implantable medical devices, and many more. Thus, as biomedical systems become increasingly dependent on technology, and especially AI and data analytics, artificially intelligent systems continue to surpass the state-of-the-art systems in terms of performance in complex clinical problems and healthcare surveillance has become easier and better. AI can reshape the healthcare workforce and bring about a drastic positive change in the understanding and handling of biomedical data in automated systems.

However, patient concerns regarding their sensitive medical information such as medical history, genetic markers, etc. that are fed into AI systems to make predictions about treatment or medical diagnosis, or are given to third-party services or less-trustworthy smaller organizations (Shokri et al., 2017) is likely to rise. As it is, deep learning techniques may be at risk of learning and memorization of confidential information instead of generalizing well to it, which, in fact, is the main aim, which is often termed as overfitting. On top of that, confidential medical data are already available on public platforms and Internet-based research sites, encouraged by open data policies in present times. Although the sharing of healthcare data is extremely important for research purposes, privacy and security also become key concerns in the wake of medical fraud and AI-based attacks.

Hence, the need for improved privacy-preserving techniques and systems to overcome the variety of security and privacy risks for users and patients in biomedical AI systems becomes extremely pertinent.

13.2 BIOMEDICAL AI SYSTEMS

13.2.1 AI IN BIOMEDICAL DATA

Biomedical data is inherently complex and involves sensitive and private, often confidential, information of patients. Moreover, biomedical and healthcare data may be of several types and more often than not, contain a lot of unstructured data, upon which it is even more difficult to carry out conventional privacy-preserving analytics (Ronneberger et al., 2015).

In present times, AI systems are being increasingly used for medical diagnosis and overall health care surveillance and, thus, for the benefit of patients and medical practitioners (for the purpose of furthering research) alike. However, in this regard, we would like to choose the types of AI techniques that do not memorize confidential patient data from the dataset like private biographical details or specific medical histories.

For some reference, the difference between "security" and "privacy"—two key terms that are of interest in this chapter—must be understood first to be able to fully grasp the threats and possible solutions of these. Abouelmehdi et al. suggest that while privacy refers to appropriate utilization of patient information and the authority to decide and restrict, if needed, the flow of that information, security refers to the fact that such decisions should be followed with due respect to all parties involved and it strives to protect the data from external attacks by malicious actors.

Apart from the usual advances in AI for different tasks like image classification, etc., there have also been great advances in research pertaining to AI systems capable of handling complex biomedical data. One of the most important and widely used examples is the U-net, which is an architectural variant of convolutional neural networks developed by Ronneberger et al. (1993, 2014, 2016) and used widely in biomedical image segmentation and which has received several best scores and surpassed the state-of-the-art examples in challenges related to biomedical data.

13.2.2 VULNERABILITIES OF BIOMEDICAL AI SYSTEMS

It is important to understand and pinpoint the vulnerabilities of biomedical AI systems to realize the security threats at large and also how to overcome or resist them. The unique features and complexities of biomedical data, which we have discussed in the previous section not only make the use cases more interesting and complex but also vulnerable to potential security and privacy issues in AI systems.

Often, the handling of clinical data requires more context than the standard for usual applications of deep learning algorithms, such as patient history, patient preferences, social perspectives, etc, which, in turn, makes it more vulnerable to security threats by malicious attackers. Moreover, the "black box" nature of deep learning algorithms results in a lack of model interpretability and gives rise to confusion regarding exactly how an AI model can achieve the kind of performance it does. It is, therefore, not easy to exactly identify the weaknesses of the model or the reasons for the weakness, or even to extract additional biological explanations from the results. This leads to potential misuse of algorithms by attackers and poses potential threats to user/patient security in biomedical AI systems. Moreover, biomedical systems usually leverage third-party cloud platforms due to scalability, storage, and performance benefits, and privacy compromisation is also likely to happen in such situations due to the confidential information being stored in the cloud unless secure sharing schemes and suitable encryption techniques are devised. The problem of security also arises in the case of data integration and adoption that is needed to develop large scale biomedical expert systems.

Thus, it is important for researchers and AI engineers to understand how the sharing of sensitive patient data works. For example, to prevent or reduce the effect of inference attacks by adversaries, the use of polyinstantiation techniques is helpful by separating the dataset into smaller sets and developing "data silos" to avoid disclosing the whole data.

13.3 SECURITY AND PRIVACY ISSUES IN BIOMEDICAL AI SYSTEMS

There are several security and privacy risks associated with biomedical AI systems, and as research advances in this area, the probability of privacy breaches and attacks on sensitive patient data will increase, with more advanced techniques of attacks. Commonly observed types of attacks and security issues have been discussed in this section. Identity theft (Elmisery et al., 2010) and blackmailing schemes, coupled with commonly occurring cases of healthcare reimbursement fraud are some of the problems that plague biomedical systems with respect to patient privacy.

More often than not, hospitals that are stores of highly confidential and immutable biomedical data are targets of the data breach and data stealing, which may be made to cause medical insurance fraud, etc., by malicious actors in intended acts of cybercrime.

In recent times, with the rapid proliferation of deep learning and AI and their increased application in the biomedical field and health-care use cases, AI is being used to combat such cybersecurity threats by proposing novel mechanisms for biomedical systems to defend themselves. For example, deep Q-networks (a deep learning technique) were studied for their use to minimize malware attacks in medical Internet of Things-based devices, and privacy-preserving online medical diagnosis system may be built with nonlinear SVMs, a machine learning technique. However, attackers can exploit AI and use AI-based security threats, like adversarial examples, which have been discussed later, as well to hack the system and pose more dangerous security threats to patients and practitioners alike. Nowadays, with the proliferation of ubiquitous healthcare with the need of constant automated sensor-based monitoring of patients' health in real time, the vulnerabilities can also be exposed in biomedical wearable and implantable devices (Gu et al., 2014) that contain a trove of biometric and sensor-based physiological data of the wearer, which may face cyber-security failure and attract lethal security threats to the wearer. Mobile healthcare networks are related in this regard, and security and privacy issues also plague such devices from the perspective of "quality of protection." Thus, the healthcare and biomedical sectors utilizing AI mechanisms are very vulnerable to privacy and security threats, most importantly. It is potentially dangerous as the lives of innumer-able patients are at stake.

13.3.1 ADVERSARIAL ATTACKS

Adversarial attacks (Bos et al., 2016, Konečný et al., 2016a) are very common security issues in the machine learning field. Studies have shown that "adversarial fooling of neural networks" may happen or may be caused by malicious actors, very small, almost imperceptible (to human sense) changes are made in the data to make the neural network classifier misclassify the data, that is, make the wrong prediction. In the case of biomedical AI systems, this poses a very big threat to medical practitioners and more importantly, patients; for, mere modification of some pixels in medical images (Alnemari et al., 2017) may cause the deep learning algorithm/neural network to predict a benign tumor as malignant or a malignant one as benign, both of which cases are very unfortunate and life-threatening as it can cause the wrong diagnosis and wrong treatment. Moreover, since such types of attacks are very subtle due to the imperceptible noise added to fool the network, detecting the presence of an error is hard. Celik et al. (2017) have also noted that such adversarial attacks may guide the attacker toward the susceptible locations in biomedical images, which may be distorted to cause the system to misclassify. Such identification of attacking technique to identify vulnerable pixels in images may then be applied to obtain a "susceptibility score" to alert biomedical systems and make them more resistant to such attacks.

Finlayson et al. (2018) have pointed at several factors in the specific case of medical data that makes biomedical AI systems more vulnerable to adversarial attacks, e.g., ambiguous ground truth due to disagreement among human medical specialists and the dearth of diversity in neural network architectures utilized. The authors also noted that adversarial patch attack techniques are more powerful as well as universal. IEEE Spectrum (Abouelmehdi et al., 2018) also reports that there may be a lot of "incentives" behind such attacks, e.g., existing cases of healthcare fraud and the enormous revenue generated by the global healthcare economy (Abuwardih et al., 2016) that makes the situation even more threatening. Some studies indicate that ethical hacking can come to our help in this regard as well since the expertise of ethical hackers lies in feigning attacks on the data or the system to ultimately understand how to protect the system against real attackers.

Studies have suggested that careful auditing of biomedical AI systems and several rounds of testing by cybersecurity specialists and AI experts can help detect such vulnerabilities in the system. There is also the need to develop algorithmic and infrastructural defense mechanisms to these adversarial attacks (Alnemari et al., 2017). In the long run, we must prioritize the development of robust machine learning/deep learning models that are resistant to or not susceptible to such kinds of attacks (Bose, 2016).

Other security threats may be privacy breaches—considering that the medical information of a patient is and should be confidential, the breach of privacy through various intentional means results in the vulnerability of the patients' data to be leaked into the public domain, which is not only unethical but also poses a danger to the patients' lives. Data poisoning techniques like model skewing, the weaponization of feedback, etc. (Cooper and Elstun, 2018).

13.3.2 DATA POISONING AND MODEL THEFT

Mozaffari-Kermani et al. (2015) studied systematic data poisoning (Chen et al., 2015) of machine learning systems related to healthcare, which may introduce either targeted errors or arbitrary errors. Targeted errors refer to the manipulation of the algorithm to yield results pertaining to a specific (predetermined by the malicious actor in cases of fraud) label or class, and the measures that may be taken to counter them.

Another type of attack is model theft or model stealing (Juuti et al., 2018; Tramèr et al., 2016; Wang et al., 2018). It refers to the technique of duplicating machine learning and deep learning models via membership leakage of data points to understand whether a certain data point belongs to the training set or not, or extracting information, for example, hyperparameters tuned on a confidential model, to reconstruct models has a high threat to security as it could disclose patient data or allow the attacker to gain insight into the model to be able to efficiently attack it. With respect to the data leak, appropriate leakage detection techniques also need to be devised; leak detection may be carried out along with privacy guaranteeing methods.

13.3.3 OTHER TECHNIQUES OF PRIVACY THREATS

There may be a variety of other threats, related or similar to adversarial examples or otherwise, which may be equally harmful to biomedical AI systems.

Transferability attacks have been studied by Papernot et al. (2016) and many other researchers in relation to the transferability of adversarial examples and how that can harm the model and breach privacy. Tracing attacks refers to the technique to deduce that a data point is in the targeted dataset just by gaining API access to a deep learning model learned on it, which is very dangerous.

Dataset reconstruction may be exploited in a different way

altogether, in this regard, that is, to build privacy-preserving techniques to protect sensitive biomedical data. Reconstruction attacks may also be staged to harm the model (Nasr et al., 2018).

Linkage attacks are another important type of attack in AI systems. Data linkage refers to the idea that data in the public domain can be easily related to sensitive information (say, confidential patient data) that can, thus, be accessed by malicious actors. The linkage can be of several types such as attribute linkage or table linkage (Kieseberg et al., 2014).

To counter linkage attacks and security threats of the similar sort, which is especially important in the wake of healthcare services in mobile devices that can compromise user identity and location data, recent studies have suggested private record linkage and entity resolution techniques, such as deriving unique fingerprints from genomes to preserve patient identity. On the other hand, inference attacks (Shokri et al., 2017; Nasr et al., 2018) are data mining techniques wherein sensitive and robust information can be deduced or "inferred" from trivial information in a database with high confidence value by malicious actors, hence the name. Membership inference attacks are made by adversaries by querying whether a specific example was a part of the training data or on the summary statistics to reveal the underlying distribution and other useful statistical features (Shokri et al., 2017) and, thus, exploit the predictions made by the AI model or compute some sensitive attribute of the dataset in question. The attack model may be trained with the help of what is called "shadow models," which are trained on either real or artificially generated fake data or both. The target model is made to train on such shadow models—it has been suggested in some studies that the more the number of shadow models, the better the perspective of accuracy—though there is a cost disadvantage in such situations, however, such claims have not yet been supported with valid proofs. The difference between such types of attacks and reconstruction attacks (described above) is that they work even when the specific example does not belong to the membership set.

13.4 POSSIBLE SOLUTIONS TO SECURITY AND PRIVACY ISSUES IN BIOMEDICAL AI SYSTEMS

13.4.1 GENERAL TECHNIQUES

General techniques to database security such as auditing, that is, rechecking for detection of anomalous activity, bug reporting, and authentication may be utilized for biomedical AI systems also. However, given the nature of the issue and the threat to human life and very private information of patients, newer advanced algorithms must be exploited in this regard.

Encryption mechanisms must also be studied in this regard, both conventional and advanced techniques like homomorphic encryption, which is especially used in case of ensuring privacy in cloud-based biomedical AI systems and allows computations on data without the need to decrypt it first. Sensitive patient information may be encrypted before being uploaded to the cloud service, via suitable encryption schemes. Bos et al. (2014) studied the particular case of cardiovascular patient data and encryption-based privacy-preserving methods with regard to considerable accuracy and the result that the cloud has no knowledge of the encrypted data in this regard.

Authentication techniques have also been widely researched and especially with respect to biomedical systems (He et al., 2015; Wu et al., 2018; Punithavathi et al., 2019) such as hashing techniques and Kerberos (Valdez and Ziefle, 2019). Access control methods like role-based access control and attribute-based access control are widely used conventional techniques in this regard.

13.4.2 DIFFERENTIAL PRIVACY AS A SOLUTION TO SECURE BIOMEDICAL AI MODELS

Secure AI models dealing with biomedical data can be designed using the idea of a probabilistic theory called differential privacy (Abadi et al., 2016; Dwork et al., 2015; Ji et al., 2015), which extends its general idea from the principles of cryptography and, of course, mathematics and in some cases, game theory even, and deals with the idea of quantifying the notion of privacy.

In the words of Papernot and Goodfellow, "Differential privacy is a framework for measuring the privacy guarantees provided by an algorithm." Differential privacy (DP) can be added as a constraint to any algorithm, for example, differentially private-stochastic gradient descent (SGD) or DP SGD can ensure that no information extracted from a database can be uniquely attributed to specific users. The basic SGD algorithm that is commonly used in deep learning systems for training the neural network may be modified to make it DP such as by clipping and randomizing the gradients. The differential privacy algorithm/technique provides a "privacy guarantee" and a "privacy budget" that are related by the idea that if the privacy budget is smaller, the privacy guarantee would be larger that is advantageous to us from a security point of view.

Experts have also pointed towards the composition property of this technique, meaning the privacy loss of any AI technique applied to a biomedical system can be added over subsequent queries made to the training data, making it easier to estimate the total privacy loss. These properties make differential privacy so powerful and an important benchmark in privacy-preserving experiments.

Several variations of this concept exist; other than the pure differential privacy, there exist useful variants like Rényi differential privacy (Mironov et al., 2017; Geumlek et al., 2017). Another technique of designing DP algorithms is the exponential mechanism (McSherry et al., 2007). The idea of "local differential privacy" versus centralized differential privacy, "multiparty differential privacy" and differential privacy with interactive mechanisms has also been studied in this regard. Another example of an adaptive algorithm based on differential privacy has been suggested by Alnemari et al. (2017)—where it is shown that the partitioning technique can help improve the privacy guarantee. Also, considering the sensitivity of the queries prior to making predictions is shown to be essential in this regard.

Moreover, the private aggregation of teacher ensembles algorithm is mentionable here, and it has been studied extensively with some studies suggesting variations and extensions to the original algorithm (Papernot et al., 2018; Jordon et al., 2018), based on the idea of differential privacy, which is discussed in the later sections.

Privacy concerns are everywhere and more so in AI systems but are amplified when it comes to AI systems for clinical surveillance dealing with biomedical data, because of the specific vulnerabilities of this type of data.

Biomedical AI systems such as healthcare recommender services (Valdez and Ziefle, 2019) need to use privacy-aware recommendation techniques, which are conducive to the sparsity of data commonly encountered in this regard. Valdez and Ziefle (2019) have studied the user, that is, a patient point of view in this regard, which is often neglected in contemporary research.

Differential privacy has been used successfully in the context of healthcare systems in previous studies (Lin et al., 2016; Dankar and El Emam, 2012) such as sensor data in body area networks.

However, like all other good techniques, it also has a few limitations. For example, one such disadvantage is that an adversary would be able to estimate the sensitive information that we are trying to protect if repeated queries, that is, multiple attempts are made and privacy breach may happen quite easily.

13.4.3 PRIVACY PRESERVATION AS A SOLUTION TO SECURE BIOMEDICAL AI MODELS

13.4.3.1 PRIVACY-PRESERVING DATA MINING (PPDM)

Privacy-preserving data mining (PPDM) and a similar concept privacy-preserving machine learning are important paradigms that must be understood and utilized with respect to security threats in biomedical AI systems. Privacy-preserving techniques have already been utilized in biomedical use cases and especially those involving distributed computing, such as cluster analysis of healthcare data (Fung et al., 2018).

13.4.3.1 DATA PERTURBATION

Data perturbation involves the addition of random noise, which makes it harder to attack the sensitive patient data by the adversary. These techniques include additive and multiplicative noise modeling as well as other techniques like geometric and random space perturbation.

13.4.3.2 ANONYMIZATION

Appropriate anonymization techniques (Szarvas, et al., 2007; Shin, et al., 2018), that is, those dealing with removal of private information and also the linking of information for such cases, must be applied to ensure the protection of patients' confidential information in biomedical data. Examples of such techniques include the FAST algorithm (Mohammadian et al., 2014) and identity-based anonymization (Abouelmehdi et al., 2018).

However, anonymizing data is not always sufficient and the privacy it provides quickly degrades as adversaries obtain auxiliary information about the individuals represented in the dataset. A much-cited example of such circumstances was studied by Narayanan and Shmatikov (2008) in relation to "breaking the anonymity" of a Netflix dataset. The authors experimented with the fact that privacy breaches may occur by way of the revelation of anonymized identities via linkage attacks.

A technique to achieve anonymization is anatomization, which deals with splitting the data into separate records and rendering the linkage ambiguous.

Pseudonymization—in pseudonymization, the linkage concerned with data linkage as described above, and pseudonyms (reversible or irreversible types) such as cryptographic keys, hashes, or other products of encryption techniques are used as a link between private and public data that can be reidentified only by authorized people (Kieseberg et al., 2014).

The broader idea related to this can be traced to traditional data deidentification and data reidentification processes that prohibit data leakage by employing various techniques.

Data reidentification is the technique of matching deidentified or anonymous data with public or auxiliary data to identify the owners via different techniques like generalizing, masking, etc. The likelihood of reidentification is a factor in determining the probability that the patient's information has been compromised. The probability of reidentification can also be determined by the uniqueness factor that may be estimated by several methods. Dankar et al. (2012) have studied different uniqueness estimators such as Pitman's estimator and Zayatz estimator on biomedical datasets and concluded that an appropriate "decision rule" developed by the combination of different such estimator mechanisms may be able to achieve best possible accuracy in the predictions.

However, these may also harm biomedical AI systems, as it is thought that if institutions release the data after deidentification, it may be in the public domain and susceptible to adversary attacks. Often, healthcare data in the form of electronic health records or otherwise is encouraged by organizations and government initiatives based on open data policy to be shared and disseminated publicly to help low-cost medical research and investigations. This gives rise to privacy attacks and risks based on linkage and reidentification by prospective attackers. Some studies (Loukides, et al., 2014) have experimented with dissociation to prevent wrongful data linkage in such cases, with lower privacy loss than other existing methods.

Some techniques to combat the disadvantages of reidentification and deidentification are more uniform standards related to deidentification and deidentification with appropriate privacy-preserving techniques for the same. Proper rules and legislations need to be passed to prevent reidentification from being used for malicious purposes.

However, some studies tend to dispel this notion as the risk of reidentification by adversaries may be less in most instances.

k-Anonymity (Sweeney 2002) is a concept used frequently in relation to PPDM. Related concepts are p-sensitive anonymity (Wibowo 2018; Cooper and Elstun, 2018) and l-diversity (Tu et al., 2019; Machanavajjhala et al., 2006), which is a technique based on group-based anonymization that focuses on decreasing the "granularity" of the dataset, t-closeness, which is basically an improvement over the former, and quasi-identifiers, which are attributes that may have linkage with an external dataset (Veličković et al., 2017). However, these methods are not foolproof and we often need to look at other newer techniques like differential privacy to ensure maximal privacy and security. For example, in situations when one has auxiliary information from secondary sources, these methods fail. They also tend to overfit or overgeneralize resulting in misleading predictions, which is very harmful to biomedical AI systems. A relevant study in this regard is Pycroft and Aziz (2018), where the idea of k-anonymity has been improved by introducing "semantic linkage k-anonymity" to balance the privacy loss and accuracy.

However, it is not advisable to use anonymization techniques in such situations; rather differential privacy would be a good weapon for adaptive attacks on biomedical data.

13.4.4 FEDERATED LEARNING AS A SOLUTION TO SECURE BIOMEDICAL AI MODELS

Federated learning (Konečný et al., 2016b) is a decentralized manner of learning a model in a collaborative manner, that is, by several clients in which no training data is exchanged. The original idea behind federated learning stems from the premise of this concept being utilized in mobile phones, such as predictive typing for mobile keyboards being dependent on the particular user's history and not on the global users' history.

Moreover, the concept of differential privacy, as mentioned in the previous subsection, can be integrated with the idea of federated learning, as proposed by Geyer et al. (2017). In techniques that are concerned with "differential privacy-preserving federated optimization" (Geyer et al., 2017), the breach of privacy in the data is less likely, as the loss of privacy and the model performance are balanced better.

This manner of training AI or deep learning models in a privacy-conscious fashion has brought about a paradigm shift for the way machine learning is traditionally done, which usually fails to emphasize the priority issue, and must be duly extended and employed for biomedical systems.

13.5 BEST PRACTICES TO ENSURE SECURITY AND PRIVACY IN BIOMEDICAL AI SYSTEMS AND FUTURE DIRECTIONS

To counter such security threats, many studies have suggested the adoption of a set of best practices when working with biomedical data and also to ensure the optimal use of predictive models in research, especially to discourage inadequate studies with inaccurate results that may compromise the credibility of important and valid research in the field.

Moreover, working with regulators is important and emphasizing on the aspect of regulatory affairs in biomedical AI will help in assessing the reliability of machine learning and deep learning techniques in this field.

With regard to advanced technologies such as AI finding use in biomedical systems, few important factors like regulation compliance, safety assurance and risk identification, and management are also cropping up, as is commonly done in other industries. Moreover, from the software perspective, developers must analyze the "data security lifecycle" for efficient decision-making and also to ensure the other factors central to a project like software reuse, cost-effectiveness, budget issues, etc. (Abouelmehdi et al., 2018). Moreover, due to the sensitive nature of the subject, it is essential for institutions to recognize and respect the various data protection laws that exist in the system with regard to maintaining and safeguarding legal and ethical responsibilities toward patients (such laws vary from country to country, such as HIPAA Act and Patient Safety and Quality Improvement Act in the USA and IT act in India).

It is extremely important to test AI models in real-time clinical situations, which are often complex and noisy, to further understand the fragility of such models and where their vulnerabilities can be exploited so that better security schemes can be devised to counter the problems. After all, it is always essential to understand the problem thoroughly to come up with actual and effective solutions.

13.6 CONCLUSION

Biomedical AI systems suffer from the threats of security and privacy risks in various capacities due to the complexities and vulnerabilities of clinical data as well as AI-based attacking techniques gaining advancement over normal cases of healthcare fraud. The users or patients are at stake in this regard the most, and such attacks may even prove fatal to them because it misleads AI systems to make wrong predictions regarding the diagnosis of a disease.

This problem must be looked at from several different perspectives. The most important one being the patient point of view—whether the patient is willing to have the sensitive personal medical data shared or not and whether they fear the risk of reidentification from publicly shared data records. Studies have been carried out in this regard such as the one by Ziefle and Valdez (2018) where patients' decisions regarding whether or not to share their private medical data for scientific purposes were investigated and analyzed from various perspectives like the age of the patient and the nature of the illness. Studies in this direction must be encouraged more and must be adopted with more relevance by researchers and healthcare institutions alike to understand and respect the perception of privacy by patients and users, who are most at stake in biomedical AI systems suffering from lack of security and privacy. However, patients often withhold sensitive information that does not help AI-based biomedical systems either. In this regard, some studies have suggested other techniques like privacy distillation, which gives control to patients about the amount of personal sensitive data that is fed to these systems (Celik et al., 2017).

In this chapter, we have outlined the various ranges of security and privacy threats to biomedical AI systems such as linkage attacks, inference attacks, adversarial examples, and so on. Similarly, solutions to such problems have been discussed, such as conventional techniques, like auditing, etc., and newer advancements in research, like differential privacy and federated learning.

It is to be reiterated that practically usable biomedical and healthcare systems utilizing AI techniques have a long way to go with regard to security and privacy issues and that advances in research do not necessarily translate into advancements or improvements in these systems that patients and medical practitioners make use of in real life. There are always practical challenges to apply the research directions and results to privacy-preserving real-time systems, but efforts are being made in the right direction (Sharma et al., 2018).

However, it must be understood that the idea of ideal privacy or absolute guarantee of patient privacy is impossible to be achieved; hence, the focus should be on maximizing the security and privacy guarantee, prioritizing users, that is, patients, and balancing the tradeoff between performance/accuracy and privacy loss.

Hence, not only must research in this regard are being prioritized (which is, of course, the dominant aspect since it drives us toward newer advanced algorithms to fight privacy breaches) but also a diverse and multidisciplinary range of professionals belonging to fields ranging from software, risk management, and cybersecurity to regulatory affairs, machine learning validation, and database applications must be welcomed in this regard to truly succeed in ensuring security and privacy guarantees in biomedical AI systems. However, research must be given prime importance. We are confident that suitable advancements in research in this field will result in quality solutions to achieve a true balance between performance and privacy that is conducive to users and patients in the healthcare and biomedical domains.

In conclusion, addressing the problem of security and privacy in biomedical AI systems is complex, multidisciplinary, and also involves ethical and legal perspectives. As newer and better machine learning and deep learning algorithms are devised to tackle the problems in the healthcare and medical domain, newer security threats will also emerge.

KEYWORDS

- **machine learning security**
- **privacy**
- **differential privacy**
- **adversarial examples**

REFERENCES

Abadi, M; Chu, A; Goodfellow, I; McMaha, H. B., Mironov; I., Talwar, K.; Zhang, L. Deep learning with differential privacy. In *Proceedings of ACM SIGSAC Conference on Computer and Communications Security*, **2016**, 308–318.

Abouelmehdi, K; Beni-Hessane, A.; Khaloufi, H. Big healthcare data: preserving security and privacy. *Journal of Big Data*, **2018**, 5(1), 1.

Abuwardih, L. A.; Shatnawi, W. E.; Aleroud, A. Privacy preserving data mining on published data in healthcare: A survey. In *Proceedings of the IEEE 7th International Conference on Computer Science and Information Technology (CSIT)*, **2016**, 1–6.

Alnemari, A.; Romanowski, C. J.; Raj, R. K. An adaptive differential privacy algorithm for range queries over healthcare data. In *Proceedings of IEEE International Conference on Healthcare Informatics (ICHI)*, **2017**, 397–402.

Bos, J. W.; Lauter, K.; Naehrig, M. Private predictive analysis on encrypted medical data. *Journal of Biomedical Informatics*, **2014**, 50, 234–243.

Bose, R. Intelligent technologies for managing fraud and identity theft. In *Proceedings of 3rd International Conference on Information Technology:*

New Generations (ITNG'06), **2006**, 446–451.

Celik, Z. B.; Lopez-Paz, D.; McDaniel, P. Patient-driven privacy control through generalized distillation. In *Proceedings of IEEE Symposium on Privacy-Aware Computing (PAC)*, **2017**, 1–12.

Çiçek, Ö.; Abdulkadir, A.; Lienkamp, S. S.; Brox, T.; Ronneberger, O. 3D U-Net: learning dense volumetric segmentation from sparse annotation. In *Proceedings of International Conference on Medical Image Computing and Computer-Assisted Intervention*, Springer, Cham, **2016**, 424–432 .

Cooper, N.; Elstun, A; User-controlled generalization boundaries for p-sensitive k-anonymit, **2018**.

Dankar, F. K.; El Emam, K. The application of differential privacy to health data. In *Proceedings of the 2012 Joint EDBT/ICDT Workshops*. ACM, **2012**, 158–166

Dankar, F. K; El Emam, K.; Neisa, A.; Roffey, T. Estimating the re-identification risk of clinical data sets. *BMC Medical Informatics and Decision Making*, **2012** 12(1), 66.

Dwork, C.; Roth, A. The algorithmic foundations of differential privacy. *Foundations and Trends in Theoretical Computer Science*, *9*(3–4) **2014**, 211–407.

Elmisery, A. M.; Fu, H. Privacy preserving distributed learning clustering of healthcare data using cryptography protocols. In *Proceedings of IEEE 34th Annual Computer Software and Applications Conference Workshops*. **2010**, 140–145.

Finlayson, S. G.; Chung, H. W.; Kohane, I. S.; Beam, A. L. Adversarial attacks against medical deep learning systems. *arXiv:1804.05296*, **2018**.

Fung, C.; Yoon, C. J.; Beschastnikh, I. Mitigating Sybils in federated learning poisoning. *arXiv:1808.04866*, **2018**.

Geumlek, J.; Song, S.; Chaudhuri, K. Renyi differential privacy mechanisms for posterior sampling. In *Proceedings of Advances in Neural Information Processing Systems*. **2017**, 5289–5298

Geyer, R. C; Klein, T.; Nabi, M. Differentially private federated learning: A client level perspective. *arXiv:1712.07557*, **2017.**

Goodfellow, I. J.; Shlens, J.; Szegedy, C. Explaining and harnessing adversarial examples. *arXiv:1412.6572*, **2014.**

Gu, S.; Rigazio, L. Towards deep neural network architectures robust to adversarial examples. *arXiv:1412.5068*, **2014.**

He, D.; Kumar, N.; Chen, J.; Lee, C. C.; Chilamkurti, N.; Yeo, S. S. Robust anonymous authentication protocol for health-care applications using wireless medical sensor networks. *Multimedia Systems*, 21(1), **2015**, 49–60.

Jordon, J.; Yoon, J.; van der Schaar, M. PATE-GAN: Generating Synthetic Data with Differential Privacy Guarantees, **2018**

Juuti, M.; Szyller, S.; Dmitrenko, A.; Marchal, S.; Asokan, N. PRADA: protecting against DNN Model Stealing Attacks. *arXiv:1805.02628*.**2018**

Kieseberg, P.; Hobel, H.; Schrittwieser, S.; Weippl, E.; Holzinger, A.; Protecting anonymity in the data-driven medical sciences. Interactive knowledge discovery and data mining: state-of-the-art and future challenges in biomedical informatics, *Springer Lecture Notes in Computer Science LNCS*, 8401, **2014**, 303–318.

Kohl, J.; Neuman, C. The Kerberos network authentication service (V5) (No. RFC 1510), **1993**

Konečný, J.; McMahan, H. B.; Ramage, D.; Richtárik, P. Federated optimization: Distributed machine learning for on-device intelligence. *arXiv:1610.02527*, **2016a**

Konečný, J.; McMahan, H. B.; Yu, F. X.; Richtárik; P.; Suresh, A. T.; Bacon, D. Federated learning: Strategies for improving communication efficiency. *arXiv:1610.05492*, **2016b**

Lacharité, M. S.; Minaud, B.; Paterson, K. G. Improved reconstruction attacks on encrypted data using range query leakage.

In *Proceedings of IEEE Symposium on Security and* Privacy (SP), **2018**, 297–314.

Li, N.; Li, T.; Venkatasubramanian, S. t-Closeness: Privacy beyond k-anonymity and l-diversity. In *Proceedings of IEEE 23rd International Conference on Data Engineering*, **2007**, 106–115.

Li, X.; Qin, J. Protecting privacy when releasing search results from medical document data. In *Proceedings of 51st Hawaii International Conference on System Sciences*, **2018**.

Lin, C.; Song, Z.; Song, H.; Zhou, Y.; Wang, Y.; Wu, G. Differential privacy preserving in big data analytics for connected health. *Journal of Medical Systems*, **2016**, *40*(4), 97.

Loukides, G.; Liagouris, J.; Gkoulalas-Divanis, A.; Terrovitis, M. Disassociation for electronic health record privacy. *Journal of Biomedical Informatics*, **2014**, 50, 46–61.

Lu, Y.; Sinnott, R. O.; Verspoor, K.; Parampalli, U. Privacy-preserving access control in electronic health record linkage. In *Proceedings of* the 17th IEEE International Conference On Trust, Security and Privacy/*12th IEEE International Conference on Big Data Science and Engineering (TrustCom/BigDataSE)*, **2018**, 1079–1090.

Machanavajjhala, A.; Gehrke, J.; Kifer, D.; Venkitasubramaniam, M. l-Diversity: Privacy beyond k-anonymity. In *Proceedings of IEEE 22nd International Conference on Data Engineering (ICDE'06)*, **2006**, 24–24.

McSherry, F.; Talwar, K. Mechanism design via differential privacy. In *Proceedings of Foundations of Computer Science (FOCS)*, vol. 7, **2007**, 94–103.

Mironov, I. Rényi differential privacy. In *Proceedings of IEEE 30th Computer Security Foundations Symposium (CSF)*, **2017**, 263–275.

Mohammadian, E.; Noferesti, M.; Jalili, R. FAST: fast anonymization of big data streams. In *Proceedings of ACM International Conference on Big Data Science and Computing*, **2014**, 23.

Mozaffari-Kermani, M.; Sur-Kolay, S.; Raghunathan, A.; Jha, N. K. Systematic poisoning attacks on and defenses for machine learning in healthcare. *IEEE Journal of Biomedical and Health Informatics*, 19(6), **2015**, 1893–1905.

Narayanan, A.; Shmatikov, V. Robust De-Anonymization of Large Datasets (*How to Break Anonymity of the Netflix Prize Dataset*). University of Texas at Austin, **2008**.

Nasr, M.; Shokri, R.; Houmansadr, A. comprehensive privacy analysis of deep learning: stand-alone and federated learning under passive and active white-box inference attacks. *arXiv:1812.00910*, **2018**

O'Keefe, C. M.; Westcott, M.; O'Sullivan, M.; Ickowicz, A.; Churches, T. Anonymization for outputs of population health and health services research conducted via an online data center. *Journal of the American Medical Informatics Association*, 24(3), **2016**, 544–549.

Papernot, N.; McDaniel, P.; Goodfellow, I. *Transferability in machine learning*: from phenomena to black-box attacks using adversarial samples. *arXiv:1605.07277*, **2016**

Papernot, N.; Song, S.; Mironov, I.; Raghunathan, A.; Talwar, K.; Erlingsson, Ú. Scalable private learning with PATE. *arXiv:1802.08908*.**2018**

Punithavathi, P.; Geetha, S.; Karuppiah, M; Islam, S. H.; Hassan, M. M.; Choo, K. K. R. A lightweight machine learning-based authentication framework for smart IoT devices. *Information Sciences*, *484*, **2019**, 255–268.

Pycroft, L.; Aziz, T. Z. Security of implantable medical devices with wireless connections: the dangers of cyber-attacks, **2018**

Ronneberger, O.; Fischer, P.; Brox, T. U-net: Convolutional networks for biomedical image segmentation. In *Proceedings of International Conference on Medical Image Computing and Computer-Assisted Intervention*, Springer, Cham, **2015**, 234–241.

Sevastopolsky, A. Optic disc and cup segmentation methods for glaucoma detection with modification of U-net convolutional neural network. *Pattern Recognition and Image Analysis*, *27*(3), 2017, 618–624.

Shakeel, P. M; Baskar, S.; Dhulipala, V. S.; Mishra, S.; Jaber, M. M. Maintaining security and privacy in health care system using learning based deep-Q-networks. *Journal of Medical Systems*, 42(10), **2018**, 186.

Sharma, S.; Chen, K.; Sheth, A. Toward practical privacy-preserving analytics for iot and cloud-based healthcare systems. *IEEE Internet Computing*, 22(2), **2018**, 42–51.

Shin, H. C.; Tenenholtz, N. A.; Rogers, J. K.; Schwarz, C. G.; Senjem, M. L.; Gunter, J. L.; Michalski, M. Medical image synthesis for data augmentation and anonymization using generative adversarial networks. In *Proceedings of International Workshop on Simulation and Synthesis in Medical Imaging*, 2**018**, 1–11. Springer, Cham.

Shokri, R.; Stronati, M.; Song, C.; Shmatikov, V. Membership inference attacks against machine learning models. In *Proceedings of IEEE Symposium on Security and Privacy (SP)*, **2017**, 3–18.

Sweeney, L. Achieving k-anonymity privacy protection using generalization and suppression. *International Journal of Uncertainty, Fuzziness and Knowledge-Based Systems*, 10(05), **2002**, 571–588.

Sweeney, L. k-Anonymity. A model for protecting privacy. *International Journal of Uncertainty, Fuzziness and Knowledge-Based Systems*, 10(05), **2002**, 557–570.

Szarvas, G.; Farkas, R.; Busa-Fekete, R. State-of-the-art anonymization of medical records using an iterative machine learning framework. *Journal of the American Medical Informatics Association*, 14(5), **2007**, 574–580.

Tramèr, F.; Zhang, F.; Juels, A., Reiter; M. K.; Ristenpart, T. Stealing machine learning models via prediction apis. In *Proceedings of 25th {USENIX} Security Symposium ({USENIX} Security 16)*, **2016**, 601–618.

Tu, Z.; Zhao, K.; Xu, F.; Li, Y.; Su, L.; Jin, D. Protecting trajectory from semantic attack considering anonymity, diversity, and closeness. In *Proceedings of IEEE Transactions on Network and Service Management*, 16(1), **2019**, 264–278.

Valdez, A. C.; Ziefle, M. The users' perspective on the privacy-utility trade-offs in health recommender systems. *International Journal of Human-Computer Studies*, 121, **2019**, 108–121.

Veličković, P.; Lane, N. D., Bhattacharya, S.; Chieh, A.; Bellahsen, O.; Vegreville, M. Scaling health analytics to millions without compromising privacy using deep distributed behavior models. In *Proceedings of ACM 11th EAI International Conference on Pervasive Computing Technologies for Healthcare*, **2017**, 92–100.

Wang, B.; Gong, N. Z. Stealinghyperparameters in machine learning. In *Proceedings of IEEE Symposium on Security and Privacy (SP)*, **2018**, 36–52.

Wibowo, W. C. A. Distributional model of sensitive values on p-sensitive in multiple sensitive attributes. In *Proceedings of IEEE 2nd International Conference on Informatics and Computational Sciences (ICICoS)*, **2018**, 1–5.

Wu, F.; Li, X.; Sangaiah, A. K.; Xu, L.; Kumari, S.; Wu, L.; Shen, J. A lightweight and robust two-factor authentication scheme for personalized healthcare systems using wireless medical sensor networks. *Future Generation Computer Systems*, 82, **2018**, 727–737.

Yuan, X.; He, P.; Zhu, Q.; Li, X. Adversarial examples: Attacks and defenses for deep learning. In *Proceedings of IEEE Transactions on Neural Networks and Learning Systems*.**2019.**

Zhang, K.; Yang, K.; Liang, X.; Su, Z., Shen, X.; Luo, H. H. Security and privacy for mobile healthcare networks: from a quality of protection perspective. *IEEE Wireless Communications*, 22(4), **2015**, 104–112.

Zhu, H.; Liu, X.; Lu, R.; Li, H. Efficient and privacy-preserving online medical prediagnosis framework using nonlinear SVM. *IEEE Journal of Biomedical and Health Informatics*, 21(3), **2017**, 838–850.

Ziefle, M.; Valdez, A. C. Decisions about medical data disclosure in the internet: an age perspective. In *Proceedings of International Conference on Human Aspects of IT for the Aged Population*, **2018**, 186–201, Springer, Cham.

CHAPTER 14

LIMOS—LIVE PATIENT MONITORING SYSTEM

T. ANANTH KUMAR[1*], S. ARUNMOZHI SELVI[2], R.S. RAJESH[2], P. SIVANANAINTHA PERUMAL[2], J. STALIN[2]

[1]*Department of Computer Science and Engineering, IFET college of Engineering, Tamilnadu, India*

[2]*Department of of Computer Science and Engineering, Manonmaniam Sundaranar University, Tirunelveli, India*

[*]*Corresponding author. E-mail: ananth.eec@gmail.com*

ABSTRACT

In this Hi-Tech world, the utilization of time is a very important parameter in every human's life. When it is for a physician, it is more important to save as many lives as possible. Especially, most talented resource persons are indeed very busy to get an appointment, but their services are very important to the patients who are in the intensive care struggling for their life. Of late, healthcare monitoring of patients has been the most crucial part of the medical field. In real-time, only the medical reports are translated and have been sent to the physician. In our proposed work, a complete set up for analyzing patient's health parameters has been established online and the important data are reported immediately. Finally, the report and prescription will be sent in reply to the patient. A system is developed to monitor the patient's health continuously by a medical monitor kit that includes hemodynamic, cardiac, blood glucose level, body temperature, pulse oximetry (respiration rate), and the stress of a patient. Practically nowadays, the measured information is noted down and sent to the medical physician using the Internet through Wi-Fi technology that generates harmful radiation that has electromagnetic waves. It causes health disorders to patients such as tissue damage, blindness, sterility, heating of tissues. To overcome the critical issues, Li-Fi technology is proposed to transmit the online health information to the physician without generating any vicious radiations to the patient. A general physician can get the critical health parameters of the patient from a remote place. The live patient monitoring system (LiMoS) framework is installed in the systems of the physician and ICU, where the framework interlinks the patients monitoring units with Li-Fi-enabled Arduino board from which all time live data is uploaded and updated to the other side of the monitoring unit of the LiMoS. Based on the given information, the physicians send back the report of the patient to the ICU unit. Thus, the LiMoS framework contributes good performance monitoring of the patient for this emerging medical field.

14.1 INTRODUCTION

14.1.1 OVERVIEW TO PATIENT MONITORING

The patient monitoring systems started blooming in the mid-1960s. The Technicon Medical Information System was the first and most successful system, begun in 1965 as a collaborative project between Lockheed and El Camino Hospital in Mountain View, California (Pramila et al., 2014). Patient monitoring is the examination of the medical condition of the critical organs and other vital parameters of a patient over a particular period of time. It has the facility to evaluate the condition of a patient's health by checking without letting any disease or ailment to further impediments the body. The vital health parameter includes heart rate, respiration rate, body temperature, glucose level, stress, hypertension, and blood pressure. These parameters are continuously observed by using the respective sensors and the monitored information will be transmitted to the medical practitioner using different technologies like Wi-Fi, GSM, and wireless sensor network. If any deformities or any variation from the normal value of such parameters are detected, then an alert or warning signal will be sent to the medical practitioner and nurse. So, the medical practitioner can be able to provide treatment for the health disorder in the earlier stage. Thus, the healthcare system continuously monitors the patient(s) information especially just in case of any potential irregularities, in the emergency phase, the alert system connected to the system gives an audio and video cautionary signal that the patient needs immediate attention. From (Pramila et al. 2014; Agarwal, 2013; Adivarekar et al., 2013) to learn the growth the different stages involved in patient monitoring are represented in Table 14.1.

TABLE 14.1 Stages of Patient Monitoring System

Stages	Methods of Monitoring	Type of Communication	Limitations
Stage 1	Single parameter (ECG) or multiparameter includes (ECG, respiration rate, and blood pressure) are monitored in the hospital	Only SMS is sent through mobile phones	Only a minimal amount of basic information is communicated, no complete healthcare report is given
Stage 2	In-home patient monitoring system with sensors and RF components	Gathers data from the sensors and forwards the patient's information via GPRS	More sharing of data can be generated when worked on a PC or laptop
Stage 3	In-home patient monitoring system	The necessity behind a computer is stopped by GPS-enabled smartphone	Radiation issues are not considered
Stage 4	Real-time health monitoring of multiple patients	Wireless structured system with Zigbee	Radiation issues are not considered

All the above stages work with wireless technology that uses radio frequency (RF) radiation for the patients when it is monitored. Especially, the areas like ICU cannot be entertained with RF signals. To overcome all the above issues, a radiation-free technology can be introduced in the hospitals for monitoring the patient even more securely.

14.1.2 WI-FI ROLE IN PATIENT MONITORING

The health parameters are constantly monitored and recorded. If the parameter value is observed abnormal, the alert will be sent to the medical practitioner through an alarm, SMS, or email using Wi-Fi that is an IEEE 802.11 standard that stands for "wireless fidelity." It is a popular wireless networking technology introduced by NCR Corporation/AT&T in the Netherlands (1991) (LaMonica, 1991). With the help of this technology, the collected information can be easily exchanged or shared between one or more devices. Wireless technology is required for using all home appliances such as mobiles, televisions, DVD players, digital cameras, laptops, smartphones, etc. The probability of communication with Wi-Fi shall be through the client to client communication or access point to client communication. It is an optimal option for home and business networks. The data is converted into a radio signal, which in turn transfers the data into an RF antenna for users using a computer's wireless adapter.

Previously, patient monitoring is implemented using wireless technology that uses RF waves, that is, electromagnetic waves to transmit the sensed data which is collected by various sensors. Wi-Fi commonly uses a single band (2.4 GHz) or dual band (5.8 GHz) RF that works best for light-of-sight condition. Some common materials can absorb or reflect the radio waves that restrict the range of the signal. Wi-Fi uses a half-duplex shared configuration where all stations can transmit and receive the signal on the same channel.

14.2 LI-FI TECHNOLOGY

Li-Fi is a wireless technology that uses visible light as the communication medium of standard IEEE 802.15.7. Li-Fi was proposed by Harald Haas in 2011 (Li-Fi, 2019). Li-Fi refers to an innovative wireless system of visible light communication (VLC) technology. The VLC technology can deliver bidirectional communication with high-speed data rates and networked mobile communication by using LED as the light source. The LED transmits the binary form of data in the form of light pulses and thus is an optical wireless communication (OWC) communication (Li-Fi, 2019). Li-Fi technology is also based on a visible-light wireless communication system that lies between the violet color (800 THz) and red color (400 THz). The Li-Fi uses the optical spectrum that is visible light part of the electromagnetic spectrum, whereas Wi-Fi uses RF of the electromagnetic spectrum. It uses fast strokes of LED light to transmit data, as it cannot be noticed by the normal human eye. It includes the visible light spectrum to transmit the information. VLCs features are providing wide bandwidth, that is, the optical spectrum guarantees more than 10,000 times better bandwidth than the convention of the harmful RF frequencies. The LED lights work rapidly for transmitting the binary data by switching the LED on and off because it has no interfering light frequencies like that of the radio frequencies in Wi-Fi. In Li-Fi, the LED in the transmitter is connected to the data network (Internet through the modem) and the receiver (photodetector/solar panel) on the receiving end, which obtains the data as light gesture and decrypts the information and then displays on the device connected to the receiver (An Internet of Light, 2014). In the early stage, the data transfer speed was 15 Mbps.

Later, many commercial luminaries helped to increase the speed of Li-Fi up to almost 10 Gbps, which has overcome the speed of 802.11.ad. IEEE 802.15.7 is a specific wireless network standard that defines the working of the physical and data link layer, which is a media access layer that defines the working of the mobility of optical transmission and its coexistence with the present architecture. There are many features of Li-Fi related to modulation, illumination, and dimming scheme, which is the first concern (Wikipedia, 2016). In December 2017, Velmenn introduced an advanced Li-Fi USB adapter for its use in the communication of USB components and Li-Fi-enabled LED lights (LaMonica et al., 2019). This technology that is of a light form of data transmission is highly radiation-free and plays an energetic role in the medical field. The advantages include efficiency, availability, and security (Rohner et al., 2015).

14.2.1 APPLICATIONS

Li-Fi applications are many in a broader sense due to its key features such as high data rate capability, directional lighting, high-level efficiency in energy consumption, intrinsic security, signal blockage by walls, and combined networking capability (Techopedia, 2019). Some interesting applications that can be evolved in future provide effective security, high data rate speed in dense urban environments, cellular communication, electromagnetic interference-sensitive environments on aircraft travel, augmented reality, localized advertising, especially for the abled, underwater communication, safety environment, intelligent transportation systems, and indoor navigation (Li-Fi-centre, 2019).

14.2.2 LIMITATIONS OF LI-FI TECHNOLOGY

Although there are many merits of Li-Fi technology, some demerits also exist. The Internet cannot be used without a light source. Although full operations of Li-Fi technology cannot be imposed on certain areas where Wi-Fi fails, in such places Li-Fi replaces Wi-Fi. A whole new innovative infrastructure for Li-Fi solely should be proposed for future disabilities. The area where the Li-Fi fails to operate is duplex transmission where both sending and receiving of data are in the form of light and the light may interfere. In Li-Fi, the transmitter should be able to maintain a directional link during transmission (Ramadhani et al., 2018; Classen et al., 2015). As of now, the uplink is in another mode of transmission like Wi-Fi and mobile communication. A certain area of research has not been touched like security; transmission scenario in the outdoor that has a high challenge needs to be implemented.

14.3 PROPOSED IDEA

- The main aim is to improve the health of patients by continuously monitoring them, and the monitored data has to be sent to the medical practitioner. If any abnormality is detected, the alarm signal will be given to the medical practitioner.
- The second objective of this proposal is to transmit the data without exposing any harmful radiation to the patients.

14.3.1 WORKFLOW OF THE PROPOSED IDEA

The proposed idea is pictorially represented as Figure 14.1, which consists of the workflow of the proposed Li-Fi-based system. The

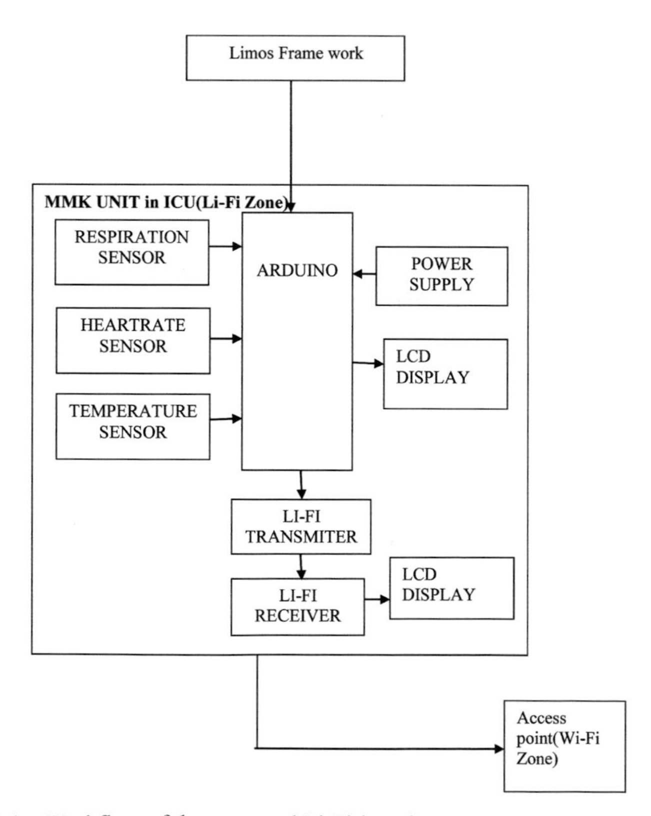

FIGURE 14.1 Workflow of the proposed Li-Fi-based system.

three phases of work have to be explored, namely, entering phase, intermediate phase, and transmission phase.

The first phase is the entering phase, which is a design interaction of the live patient monitoring system (LiMoS) framework. In this phase, two types of frameworks are designed, namely, system app and mobile app. The system app will be used by both physicians and the ICU unit section. For the second phase, the framework of the medical monitor kit (MMK) has to be designed that comprises all the instruments used for monitoring. Medical monitoring kit comprises of all the medical support given to patients for observations. The corresponding sensors are embedded with the Arduino board, namely, a respiration sensor, a heart rate sensor, and temperature sensors from where the data is generated to transfer. Finally, in the third phase, the interoperability of how the Arduino board interacts with the Li-Fi environment and the communication scenario for the Backhaul network to transfer through the Internet. The measure of its performance is compared with the real-time patient monitoring system in terms of cost by reducing the physicians' official visit for treatment, and in terms of time management for diagnostic testing procedures.

The workflow falls in three categories, such as

1. entering phase,
2. intermediate phase, and
3. transmission phase.

The entering phase is the LiMoS framework, which is the gateway for the online patient monitoring system from which the patient's information is extracted by a physician outside the hospital and the corresponding interactions are made by an internal doctor in the hospital.

The intermediate phase is the MMK, which consists of all the medical components integrated with ICU patients. This phase works in the Li-Fi zone. MMK comprises hemodynamic, cardiac, blood glucose level, body temperature, pulse oximetry (respiration rate), and the stress to a patient. The results of all these components are lively transmitted to the transmission phase.

The transmission phase is the most important part of the Li-Fi zone that transmits and receives information from the MMK kit to the Wi-Fi zone. This phase comprises of LED, transmitter, and receiver units for transmission.

14.3.2 LIMOS FRAMEWORK (LI-FI MONITORING SYSTEM): ENTERING PHASE

The terminology named for our proposed idea has the basic requirements for the process of monitoring a patient in an ICU. This framework is composed of two modules:

1. System app and
2. Mobile app.

14.3.2.1 SYSTEM APP

The System app gives the operational use of the physician and the ICU unit, hereafter mentioned as E-Doc and I-Doc/patient care taker (PCT), respectively. E-Doc mentions the physician outside the hospital, whereas the I-Doc/PCT mentions the internal physician who is inside the hospital, and PTC refers to the patient's caretaker who is inside the ICU all the time. The homepage is the open screen of the system app of the doctor who opens the LiMoS framework. Followed by which, it has doctor and admin options, where doctor refers to the doctor login portal, where the physicians have been given a separate password to know the authenticated person. The E-Doc authentication information is created by the admin. The assignment of the respective doctors for the respective ICU bed is also assigned in this phase. The E-Doc and I-Doc have separate login information that provides authentication for the security of data. Figure 14.2 shows the main source page of the LiMoS from the E-Doc side.

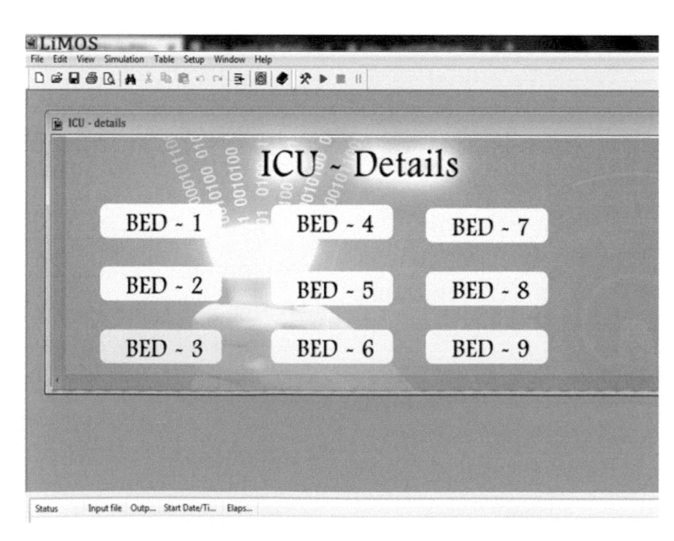

FIGURE 14.2 Main source page of the LiMoS from the E-Doc side.

Figure 14.3 shows the monitoring page of the LiMoS from E-Doc side. The main source page of the LiMoS from the E-Doc side shows all the bed information in the ICU, and the bed allocated to the particular doctor is highlighted. The E-Doc checks in the bed to view the status of the patient. The monitoring page gives information about the patient that includes hemodynamic, cardiac, blood glucose level, body temperature, pulse oximetry (respiration rate), and the stress of a patient. All this information is live updated to the physicians.

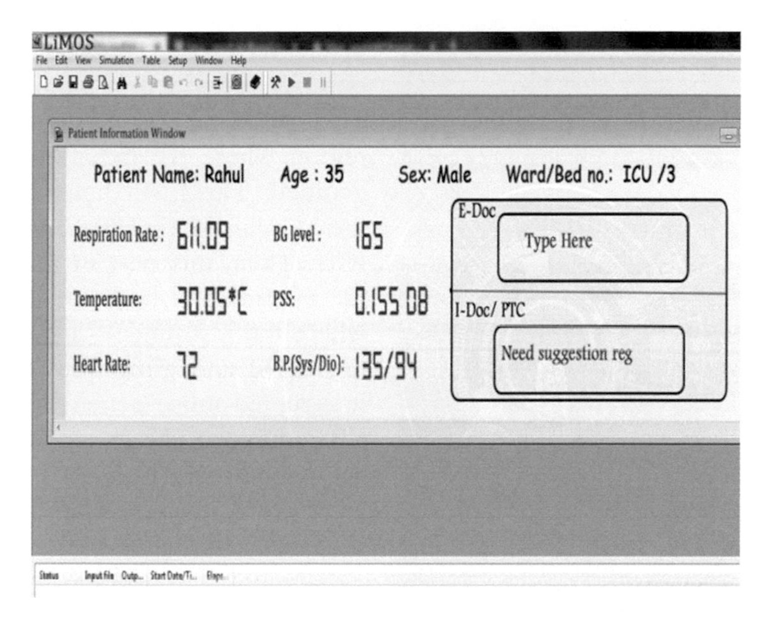

FIGURE 14.3 Monitoring page of the LiMoS from E-Doc side.

14.3.2.2 MOBILE APP

The mobile app is especially for the E-Doc who is outside the hospital. In case of any emergency even if the E-Doc is out of station or not in the hospital he can update the status through mobile app too. Certain features are not present in the mobile app. Figure 14.4 indicates the allocation of bed for the patient in LiMoS system. These features play a minor role that is avoidable when out of station. This page represents the same page as in the system APP, which is the front page of the APP for the mobile user. Figure 14.5 shows the detailed information of the patient monitored.

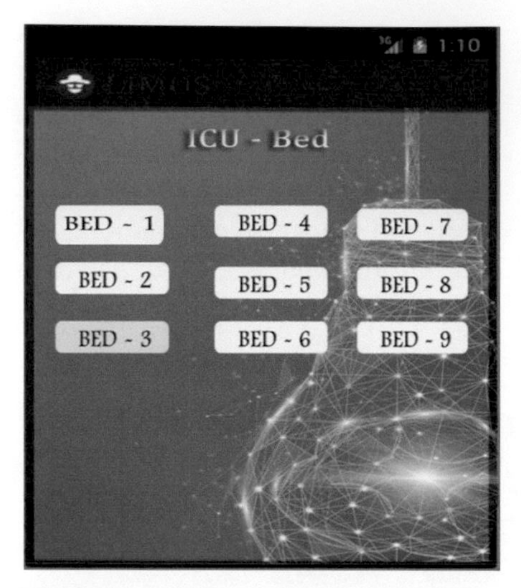

FIGURE 14.4 ICU information—LiMoS.

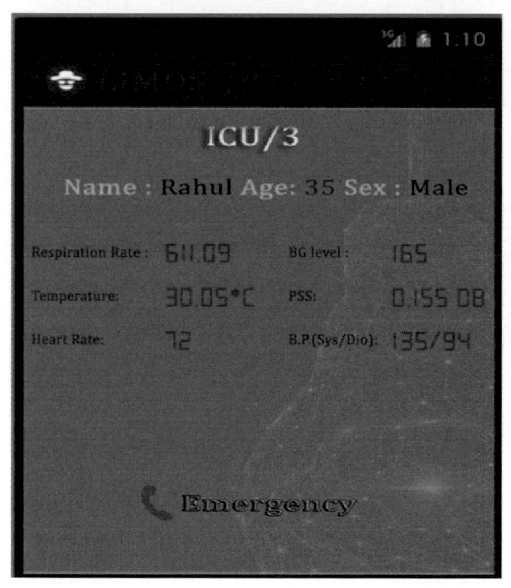

FIGURE 14.5 Patient information system—LiMoS.

14.4 MMK: INTERMEDIATE PHASE

The MMK works in the Li-Fi zone that has a common board known as Arduino. This is the motherboard for all the components to be integrated and is transmitted to the Li-Fi transmitter, which is placed near the ICU bed, from which the receiver placed in the line of sight (LOS) of the transmitter receives the data from a distance of 20 m. After that, the data is sent through the optical cable communication link to the Wi-Fi zone outside the ICU normally provided in the hospitals. This MMK is the most sensitive region where RF radiation can be avoided. The Li-Fi luminaries are the normal lighting used inside the ICU. Thus, this part of the MMK is a radiation-free environment where the instruments are no more distracted and error free from the interrupted noise generated by the RF radiation. The general setup is shown in Figure 14.6.

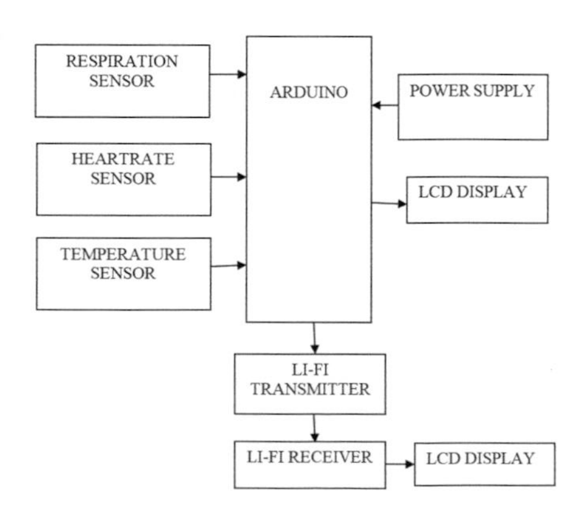

FIGURE 14.6 The block diagram of patient's MMK in ICU.

14.4.1 COMPONENTS USED

In this patient healthcare monitoring, some of the vital health parameters are continuously monitored and recorded by a device called a medical monitor. These health parameters include hemodynamic (blood pressure and blood flow), blood glucose level (blood sugar level), temperature, cardiac (electrocardiogram), and pulse oximetry (respiration rate) of a patient. These parameters are sent to the medical practitioner to improve the health condition of a patient.

14.4.2 RESPIRATION SENSOR

A respiration sensor measures the respiration rate of a human per minute. Breath rate is characterized as the number of breaths an individual takes for every moment, and it is estimated when an individual is at rest state. The rate will increase with fever, illness, and other medical conditions. The most well-known strategy for the estimation of breath rate is by physical evaluation, watching an individual's rib along with heart and sum of the number of breaths for one moment time. The limited data can be obtained using breath rate estimation as it is an actual respiration design that exposes the necessary details and other characteristics. The respiration rate differs from breathing. For the adult, the normal respiration rate is usually 12–25 breaths per minute, and if the respiration rate is above 25 or below 12, it is considered as abnormal respiration rate.

14.4.3 TEMPERATURE SENSOR

In the proposed design, thermocouple or resistance temperature detector is used to measure the body temperature. It is very essential to measure the temperature that reveals the body's metabolic rate and hormonal healthiness. It is mandatory to measure body temperature regularly. Certain diseases can be examined by measuring body temperature, and the efficiency of a treatment initiated can be analyzed by the physician. An abnormal change in body temperature can be related to fever (high temperature) or hypothermia (low temperature). These sensors are normally used in apertures, medical incubators, blood analysers, anesthesia transfer machines, sleep apnoea machines, and temperature monitoring and control for an organ transplant system.

14.4.4 HEART RATE SENSOR

A heart rate sensor is used for measuring a person's heart rate in real time and record the heart rate in a system intended for future use. Every person's fitness level can be restrained by its heart rate. The heart rate is maintained to reduce the risk of injury and mental fatigue. For measuring and displaying the heart rate continuously, a heart rate monitor device is to display the data as the number of beats per minute. The pulse rate ranges from 60 to 100 bpm, which may fluctuate and rises with exercise, illness, injury, and emotions.

14.4.5 ARDUINO UNO

Arduino is an open-source-based microcontroller that shall be easily programmable/reprogrammable at any time. It also acts as a miniature CPU just like a microcontroller by taking inputs and controlling the outputs for a variety of electronic devices that are also capable of observing and forwarding data's over the Internet through various Arduino shields.

14.4.6 LED

The most important requirement for a light source used in Li-Fi is the ability to be switched on and off repeatedly in very short intervals of time. LEDs are suitable light sources for Li-Fi due to their ability to be switched on and off quickly. The variations in rate with the dimensions of LEDs are very important in Li-Fi technology. Different sized LEDs can produce different data rates, where a micro-LED bulb can itself transmit 3.5 Gbps.

14.5 TRANSMISSION PHASE: LI-FI IN NETWORKING

Li-Fi module has two submodules: one is the transmitter, and the other is the receiver module (Figure 14.7).

14.5.1 TRANSMITTER AND RECEIVER MODULES

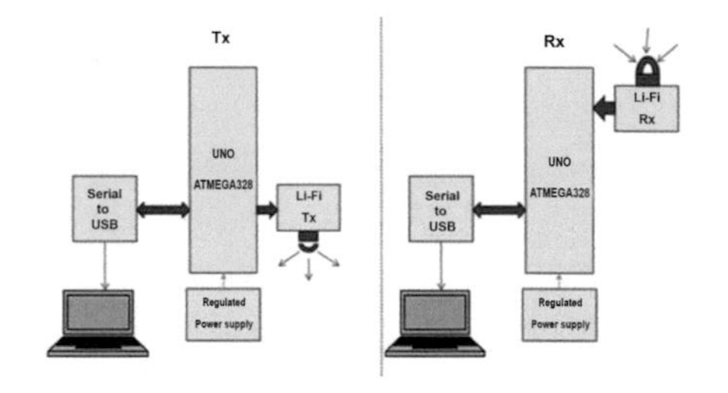

FIGURE 14.7 Block diagram of transmitter and receiver modules.

14.5.1.1 TRANSMITTER SECTION

The data received from the sensors such as heart rate, temperature, and respiration are given as input to the Arduino board, where these data are converted into a digital signal. This digital signal is then given to the Li-Fi transmitter part. The transmitter section is used to convert digital data into visible light. The general concept behind this is that the light intensity of the LED is modulated, that is, the intensity of the light corresponds to the data transmitted. The Arduino is not able to provide the right amount of current to make the light intensity strong and fast enough for transmitting the data as light. To overcome this problem, a transistor is used as a switch to turn on and off the LED, which made it possible to switch a larger current faster. Figure 14.8 shows the components used in the Li-Fi transmitter setup that consists of the Li-Fi transmitter module and LED light source.

FIGURE 14.8 Li-Fi transmitter module.

14.5.1.2 RECEIVER SECTION

The receiver module has two modules such as a demodulation circuit and a microcontroller. The transmitted optical pulse is retrieved back into an electrical signal using a photodiode that is inbuilt in the demodulation circuit. The photodiode is a semiconductor that converts the light signal into electric current. The need for the photodiode in this transmitter section is a rapid response time with spectral sensitivity in the visible spectrum and a large radiation-sensitive area. The converted electrical signal is feeble and overwhelmed by noise. Then, it undergoes demodulation through envelope detection to demodulate the data from the carrier signal. The receiver does the filtering and then amplifies the signal. After amplification, the signal will be in an analog form that stances to fed into an analog-to-digital converter, before sending it to the Arduino board. The photodiode generates the current at very low value; hence, for converting this current into a voltage, a high-value resistor is used. Further voltage is again amplified by a comparator circuit to give properly transmitted bits. The amplitude of the amplified voltage is the output of the 741 op-amp. Then, the voltage comparator transforms the signal into a digital format before feeding into the microcontroller that transmits the data serially to another device. Figure 14.9 shows the receiver module that transmits data 38,400 baud rate serially. It covers 5–15 ft distance. The coverage area can be increased by changing the LED wattage.

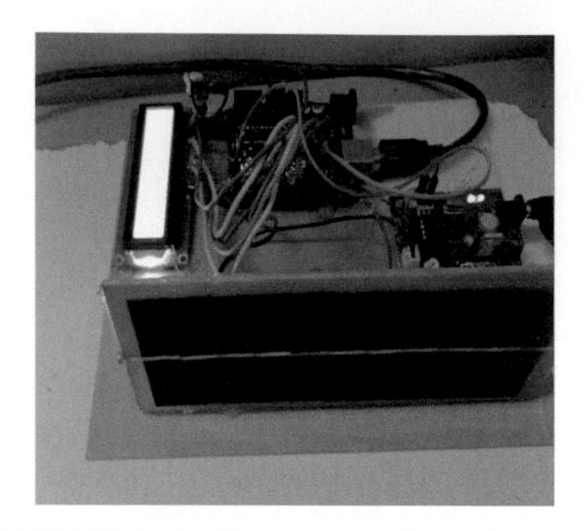

FIGURE 14.9 Receiver module.

14.5.3 LI-FI OPTICAL CHANNEL CHARACTERISTICS

The transmitter and the receiver sections communicate through an optical wireless path. The Li-Fi channel is generally called OWC, as shown in Figure 14.10 (Dimitrov et al., 2015). The optical channel is defined in Figure 14.10.

It includes LED nonlinearity, a dispersive optical wireless channel, and the AWGN. This path is the light medium where the data is transmitted much faster and in a high data rate. This type of medium follows the specific channel model for communication.

14.6 CHANNEL MODEL

OWC can be described by the following continuous-time model for a noisy communication link:

$$i(t)= x(t)\ F(j(t)) + y(t)$$

where $i(t)$ represents the received distorted replica of the transmitted signal, $j(t)$, which is subject to the nonlinear distortion function, $Fj(t)$, of the transmitter frontend. The nonlinearly distorted transmitted signal is convolved with the channel impulse response, $x(t)$, and it is distorted by AWGN, $y(t)$, at the receiver. Here, * denotes linear convolution. The generalized model of the OWC link in the time domain is illustrated in Figure 14.10. The path of the OWC is to be considered in the later stage of deep concentration. The general path LOS has to be studied for basic Li-Fi operation. LOS is known for line of sight.

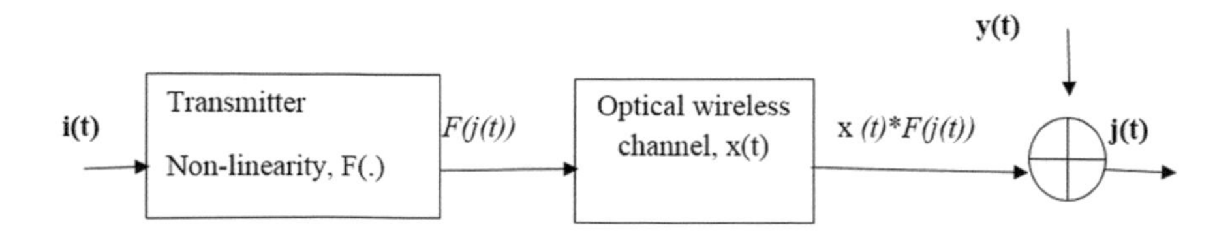

FIGURE 14.10 Generalized block diagram of the OWC link in the time domain.

Generally, the origin of research in Li-Fi is the light communication that can be transmitted through the LOS of the receiver. Later on, the research is evolved and defined as LOS and NLOS.

14.6.1 LOS—LINE OF SIGHT

This is the direct path for communication from the source to the destination easily without any distraction. If the light is distracted, the communication is lost. This type of communication will be possible in somc applications that are fixed and not moving.

14.6.2 NLOS—NON-LINE OF SIGHT

This is the redirected path for communication from the source to the destination easily with some distraction. The distraction may be the wall inside a room or an obstacle, which in turn redirects the light by reflecting on that surface. Here, the light that reflects at the obstacle from the LOS of the transmitter is known as the angle of reflection θ_{inc} and the angle from the obstacle to the destination is θ_{obs} in Figure 14.10.

The mutual orientation of the transmitter and the receiver and their orientation toward the reflecting surface are described by means of observation and incident angles with respect to the normal direction.

14.7 PATH LOSS

The modeling of the path loss of the optical wireless channel increased the significant interest after the ground-breaking work of Gfellerand and Bapst (1979). They offered an analytical model for the received optical power in LOS and single-reflection NLOS in OWC. A design of the geometry of optical wireless communication scenario is shown in Figure 14.10. Parameters describing the mutual orientation of the transmitter and the receiver, as well as their orientation toward the reflecting surface, are included. As a communal trend in literature, the channel characterization is represented by the following equation:

$$x(D) = x(o)_{\text{LOS}} + x(o)_{\text{NLOS}} + \text{noise component} \quad (14.1)$$

In Equation (14.1), the first addend represents the LOS received optical power from the direct path, while the second addend calculates the NLOS received optical power after a single reflection on the reflecting surface. Here, the received optical power is obtained after integration of the reflective surface in x and y directions and integration of the BRDF over θ and φ angles, ranging over $0 \leq \theta \leq \pi/2$ and $0 \leq \varphi \leq 2\pi$. In addition, θ_{Tx} denotes the observation angle of the transmitter on the direct path, and θ_{Rx} is the direct incident angle of the receiver. Observation and incident angles are computed with respect to the normal directions of the radiating, reflecting, or detecting elements. The distance between the transmitter and receiver on the direct path is given by *d*, and *A* is the photosensitive area of the PD. On the nondirect path, θ_{Tx} is the observation angle of the transmitter toward the reflective surface, θ_{Rx} is the incident angle of the receiver from the reflective surface, and ρ denotes the reflection coefficient of the surface. The distance between the transmitter and the reflective surface is given by *d*1, while *d*2 is the distance between the receiver and the reflective surface

$$x(o)_{LOS} = \begin{cases} \frac{(m+1)A}{2\pi d^2}\cos^m(\varphi)T_s(\Psi)g(\Psi)\cos(\Psi), 0 \le \psi < \psi_c \\ 0, o \ge \psi_c \end{cases} \tag{14.2}$$

$$x(o)_{NLOS} = \begin{cases} \frac{\rho Ah}{\pi(h^2+d^2)}T_s(\Psi)g(\Psi)\cos(\Psi), 0 \le \psi < \psi_c \\ 0, o \ge \psi_c \end{cases} \tag{14.3}$$

where

A—area of the detector
d—distance between the source and detector
—angle of incidence
H—distance above the receiver
—reflectivity
—angle of irradiance
)—optical filter gain.

Thus, *x*(*D*) represents the generalized model of the optical wireless channel. This model of communication is simulated in the transmission phase of the Li-Fi zone. The input information of the intermediate phase is communicated by the above-mentioned channel model. This communication leads to good efficiency and hassle/radiation-free communication.

14.8 RESULT ANALYZING PARAMETERS

The LiMOS framework system analysis is performed based on the basic three metrics like throughput, cost, and time.

14.8.1 THROUGHPUT

Throughput is the ratio of the actual data transferred to the receiver to the actual data sent by the transmitter. When utilized in the context of communication networks, like LAN or packet radio, output or network output is the rate of successful message delivery over a communication channel. The experimental setup includes the transmitter and receiver sections, in Figures 14.8 and 14.9, of the Li-Fi unit integrated with the patient monitoring unit. The throughput is measured when the MMK unit information is transmitted from the transmitter to the receiver. The throughput performance of existing wireless communication technologies is also measured with real-time results. Comparatively, Li-Fi achieves better performance and has advantages over other technologies. Normally, the value will fall between 0 and 1. The threshold limit is 0.5. If the throughput is above 0.5, it is considered to be a good result. Here, the value of both communication falls under the above category, but the better performance is achieved by the VLC for our LiMOS framework.

Figure 14.11 shows a comparative analysis of the patient monitoring system based on RF and Li-Fi communication. Practically, possible metrics are considered for test cases in the hospital that are taken for analysis such as cost- and time-based.

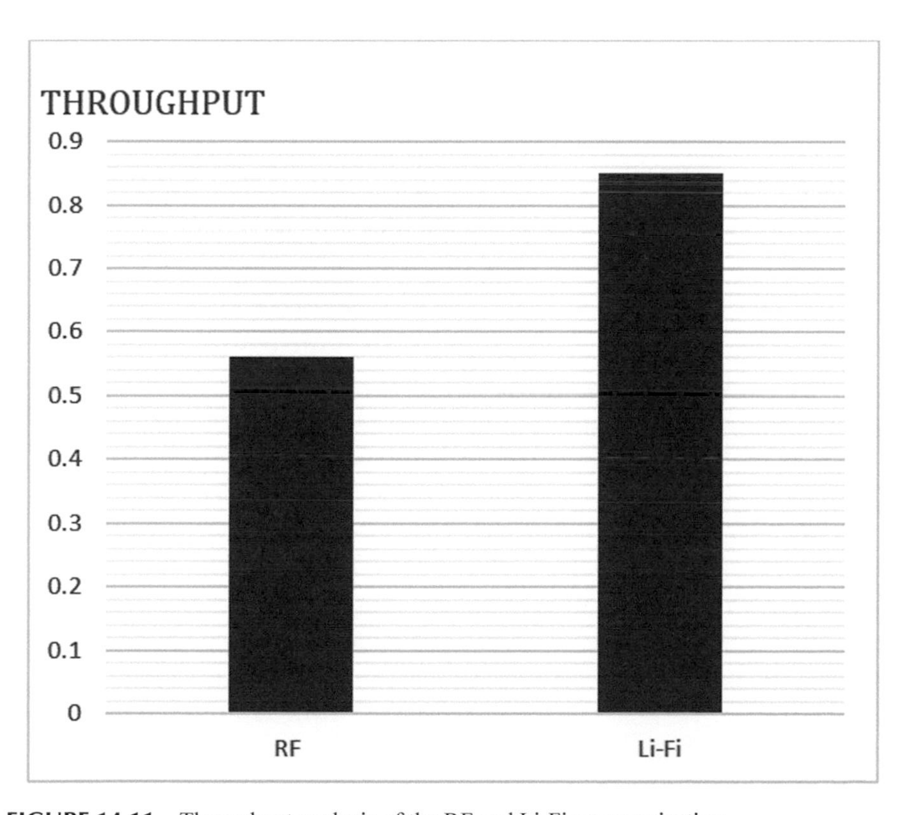

FIGURE 14.11 Throughput analysis of the RF and Li-Fi communication.

Cost: This metric is a practical parameter that is calculated based on the hospital management of the doctors such as the travel cost of the doctor in times of emergency, the special visit cost for both the hospital and the patient.

Time: This metric plays a vital role in this analysis. The time management is easily done when the doctor's visit is reduced for both the hospital and the doctor. Especially when the doctor is out of the station the time management is handled easily. The emergency cases are handled more easily in online support.

14.9 CONCLUSION

Thus, for complete monitoring of the patient, early observation of any physical deterioration in patient health, continuous health monitoring is needed, and especially radiation-free environment is very important. In ICUs, these types of setup hold good results especially in the RF distraction of equipment's leading to some noise and high error component. This methodology for ICUs in hospitals is a very useful feature. The time-consuming factors and cost-effective issues are taken into account by the hospital and the patient's side, respectively. This system is more effective for future generation medical fields who are going to run at the back of time and patients can take care of more patients.

14.10 FUTURE ENHANCEMENT

Although the area is emerging and only test cases are evolving, in near future, more research will give many ideas in this Li-Fi networking. The Li-Fi has not reached its 100%; as of now, its use complements the lags of Wi-Fi. The Li-Fi and Wi-Fi together will run the world at the back of them. This part of the work that has been concentrated in the ICU section can be enhanced for the whole hospitals one by one, especially in the area of scanning and X-ray where already full radiation is being used. Most of the demerits can be overcome by Li-Fi. In the scanning center, X-ray can also be equipped. The data transfer speed at present may be less comparatively as research gets stronger that can be overwhelmed too. In the fore coming era, artificial intelligence technique can be embedded for the betterment of the LiMoS framework in the MMK unit. In the future, this prototype can be developed on a system on chip that can be used commercially.

KEYWORDS

- **LiMoS**
- **Li-Fi**
- **IoT**
- **radiation-free device**
- **online patient monitoring**

REFERENCES

Adivarekar J. S., Chordia A. D., Baviskar H. H., Aher P. V., Gupta S., "Patient monitoring system using GSM technology," International Journal Of Mathematics and Computer Research, vol. 1, no. 2, pp. 73–78, 2013. ISSN: 2320-7167.

Agarwal T., "Project on remote patient monitoring system," Microcontroller Based Project on Patient Monitoring System. (2013) https://www.elprocus.com/microcontroller-based-project-on-patient-monitoring-system/

An IEEE Standard for Visible Light Communications, Archived 29 August 2013 at the Wayback Machine, visiblelightcomm.com, dated April 2011. It is superfast modern internet technology, accessed on 02/03/2017.

An Internet of Light: Going online with LEDs and the first Li-Fi smartphone. Archived 11 January 2014 at the Wayback Machine, Motherboard Beta, Brian Merchant.

Breton J., "Li-Fi smartphone to be presented at CES 2014." Digital Versus, 20 December 2013. Archived from the original on 8 January 2014. Retrieved on 16 January 2014.

Classen, J., Chen, J., Steinmetzer, D., Hollick, M., & Knightly, E. The spy next door: Eavesdropping on high throughput visible light communications. In Proceedings of the 2nd International Workshop on Visible Light Communications Systems (pp. 9–14), 2015.

Chow C.-W., Chen C.-Y., Chen S.-H., "Visible light communication using mobile-phone camera with data rate higher than frame rate," Optics Express, vol. 23, 26085, 2015.

Dimitrov S., Haas H., *Principles of LED Light Communications: Towards Networked Li-Fi*, Cambridge University Press, 2015.

Gfeller F. R., Bapst U., "Wireless in-house data communication via diffuse infrared radiation," Proceedings of the IEEE, vol. 67, no. 11, pp. 1474–1486, 1979.

Guha P., "Light fidelity: technical overview and its applications," International Journal of Mobile Ad-hoc and Sensor Networks, vol. 4, no. 1, 2014.

https://www.lifi-centre.com/about-li-fi/applications, accessed on 02 February 2019.

https://www.techopedia.com/7/31772/technology-trends/what-are-the-advantages-and-disadvantages-of-li-fi-technology, accessed on 02 February 2019.

http://iopscience.iop.org/article/10.1088/1757-899X/325/1/012013/pdf, accessed on 07 April 2019

https://en.wikipedia.org/wiki/Li-Fi, accessed on 02 August 2016.

Jeya A. S. , Venket S., Kumar V. L., "Data transmission by Ceaser Cipher wheel encryption using LiFi," International Journal of Advance Research, Ideas and Innovations in Technology. vol. 4, no. 2, pp. 512–517.

Kumar A., Raj A., Lokesh V., Sugacini M., "IoT enabled by Li-Fi technology," Proceedings of the National Conference on Communication and Informatics, 2016, pp. 214–243, ISSN:2320-0790

LaMonica M., "Philips creates shopping assistant with LEDs and smart phone." IEEE Spectrum. 18 February 2014.

Archived from the original on 17 February 2019.

Lee S. J., Jung S. Y., A SNR analysis of the visible light channel environment for visible light communication. Proceedings of the 18th Asia-Pacific Conference on Communication: Green and Smart Communication for IT Innovation (APCC 2012), 2012, pp. 709–12.

Liu J., Chen Y., Wang Y., Chen X., Cheng J., Yang J., "Monitoring vital signs and postures during sleep using WiFi signals," IEEE Internet of Things Journal, vol. 5, no. 3, pp. 2071–2084, June 2018.

Pottoo S. N., Wani T. M., Dar M. A., Mir S. A., "IoT enabled by Li-Fi technology," International Journal of Scientific Research in Computer Science, Engineering and Information Technology, vol. 4, no. 1, 2018, pp. 106–110. ISSN:2456-3307.

Pramila R. S., Nargunam A. S., "Secure patient monitoring system," Journal of Theoretical and Applied Information Technology, vol. 62, no. 1, April 2014.

Purwita A. A., Soltani M. D., Safari M., Harald H. Terminal orientation in OFDM-based LiFi systems, vol. 18, no. 8, pp. 4003–4016, 2018.

Ramadhani E., Mahardika G. P., IOP Conference Series: Material Science Engineering, vol. 325, 012013, 2018.

Rigg J., "Smartphone concept incorporates LiFi sensor for receiving light-based data." Engadget. 11 January 2014. Archived from the original on 15 January 2014. Retrieved on 16 January 2014.

Rohner C., Raza S., Puccinelli D., and Voigt T., "Security in visible light communication: novel challenges and opportunities," Sensors & Transducers, vol. 192, no. 9, September 2015, pp. 9–15.

Study Paper on LiFi (Light Fidelity) & Its Applications, FN Division, TEC.

LiFi Data. Lumisense Technologies, Chennai.

Sudha S., Indumathy D., Lavanya A., Nishanthi M., Sheeba D. M., Anand V., "Patient monitoring in the hospital management using Li-Fi", Proceedings of IEEE Technological Innovations in ICT for Agriculture and Rural Development (TIAR), Chennai, 2016, pp. 93–96.

Tsonev D., Sinanovic S., Haas H., "Complete modeling of nonlinear distortion in OFDM-based optical wireless communication," IEEE Journal of Lightwave Technology, vol. 31, no. 18, pp. 3064–3076, 2013.

Van Camp J., "Wysips solar charging screen could eliminate chargers and Wi-Fi." Digital Trends, 19 January 2014. Archived from the original on 7 November 2015. Retrieved on 29 November 2015.

Yadav S., Mishra P., Velapure M., Togrikar P. S., "LI-FI technology for data transmission through LED," Imperial Journal of Interdisciplinary Research, vol. 2, no. 6, pp. 21–24, 2016.

CHAPTER 15

REAL-TIME DETECTION OF FACIAL EXPRESSIONS USING K-NN, SVM, ENSEMBLE CLASSIFIER AND CONVOLUTION NEURAL NETWORKS

A. SHARMILA[1], B. BHAVYA[1], and K. V. N. KAVITHA[2*], and P. MAHALAKSHMI[1]

[1] *School of Electrical Engineering, Vellore Institute of Technology, Vellore, India*

[2]*Deloitte Consulting India Private Limited, Bengaluru, Karnataka*

[*]*Corresponding author. E-mail: kvnkavitha@yahoo.co.in*

ABSTRACT

Identifying human emotions is important in facilitating communication and interactions between individuals. They are also used as an important means for studying behavioral science and psychological changes. There are many applications that use facial expressions to evaluate human nature, their feelings, judgment, and opinions. Since recognizing human facial expressions is not a simple task because of some circumstances due to illumination, facial occlusions, face color/shape, etc., this chapter will provide the best method that can be adopted for noninvasive real-time facial expression detection. This chapter involves a comparative analysis of facial expression recognition techniques using the classic machine learning algorithms—*k*-nearest neighbor, support vector machine, ensemble classifiers, and the most advanced deep learning technique using convolutional neural networks. Successful and satisfactory results have been obtained giving the future researchers in this field an insight into which technique could be used to get the desired results.

15.1 INTRODUCTION

Facial expressions of a person are more meaningful than the words spoken by the person because they express all his feelings or a person can easily understand unspoken words by facial expressions. In many diseases such as stroke and paralysis, facial motion disorders are an early symptom. The patients with stroke might have swallowing disorders in the initial stage. In the case of paralysis, resulted from the accident or illness, the facial muscle could be the last failed motor unit. Many prosthetic devices can control facial emotions. Hence, facial emotion recognition can provide valuable reference in medical diagnosis and biomedical applications. Keeping in mind the advancements in the field of digital image processing, facial expression recognition (FER) is one the most important applications that is being utilized in today's world. In the current decade, a lot of progressive improvements have been made in domains like face recognition, face tracking, face retrieval, and FER. An FER system involves different measures. The basic framework of an FER system is shown in Figure 15.1.

FIGURE 15.1 Basic framework of an FER system.

- Face detection is the first phase in FER.
- The next step is to extract features from the detected face; then, from these features, the best features are selected and irrelevant features are eliminated in the feature selection phase (Sharmila and Geethanjali, 2016a, b).
- Finally, we use it for classification where different expressions are classified into seven different emotions like as angry, happy, sad, surprise, neutral, fear, and disgust.

The FER was introduced in 1978 and it creates the main point for face detection with feature extraction, image alignment, normalization, and categorization. FER is employed in various kinds of applications of human–computer interaction like face image processing, automatic recognition of expressions, video surveillance systems, artificial intelligence, and different challenging tasks during this decade. Researchers Ekman and Friesen represent six basic facial expressions (emotions) such as happy, surprise, disgust, sad, angry, and fear. As per the Meharabian work, 55% communicative cues can be judged by facial expressions; hence, recognition of facial expressions becomes a major modality. In this chapter, we have developed an FER system based on different classifiers and neural networks.

15.1.1 CONVOLUTION NEURAL NETWORKS

A neural network is a system of interconnected artificial "neurons" that exchange messages between each other. The connections have numeric weights that are tuned during the training process so that a properly trained network will respond correctly when presented with an image or pattern to recognize (Shan et al., 2017). The network consists of multiple layers of feature-detecting "neurons." Each layer has many neurons that respond to different combinations of inputs from the previous layers. As shown in Figure 15.1, the layers are built up so that the first layer detects a set of primitive patterns in the input, the second layer detects patterns of patterns, the third layer detects patterns of those patterns, and so on. Typical CNNs use 5–25 distinct layers of pattern recognition. A CNN is a special case of the neural network described above. A CNN consists of one or more convolutional layers, often with a subsampling layer, which are followed by one or more fully connected layers as in a standard neural network. These networks have been some of the most influential innovations in the field of computer vision, decreasing the classification error record from 26% to 15%, an astounding improvement at the time.

15.1.2 OBJECTIVE

- The objective of this work is to generate an FER system that could be used to detect the different expressions of the human face, namely, happy, sad, anger, neutral, disgust, fear, and surprise in real time using the concept of machine learning and deep learning.
- FER) is one the most important applications that is being utilized in today's world.
- Looking at how interactive the human world and computers have become, the FER system is used in artificial intelligence to create a better human–computer interaction. For example,

the FER is used in robots so that they can understand the emotions of the human they are interacting with and react accordingly.

- In studying psychological and behavioral sciences, the FER systems has been used. Whatever be the face shape, color, skin texture, age of the person, the FER systems detect the emotion represented by the face.
- Since the proposed method (CNN model) can be used to detect multiple faces in a single frame, this work can be used by teachers to check the students' emotions in class while teaching and improve their methods accordingly.
- FER systems when hardware interfaced can also be used for further applications like in ATMs to prevent ATM robberies; if the person withdrawing is scared, the system will not dispense cash.

15.1.3 BACKGROUND

Pantic and Rothkrantz (2004) developed an algorithm that recognized facial expressions in frontal and profile views. They took a sample of 25 subjects in an MMI database. They used a rule-based classifier that had an accuracy of 86%. They proposed a way to do automatic coding in profile images but not in real time. This involved the extraction of frontal and profile facial points. Bucio and Pitas (2004) performed principal component analysis (PCA) for a comparison purpose. Local non-negative matrix factorization (LNMF) outperformed both PCA and non-negative matrix factorization (NMF), whereas NMF performed the poorest. They discovered that the cumulative learning system (CSM) classifier is more reliable than Matthews correlation coefficient (MCC) and gives better recognition. They used the Cohn–Kanade database that had a sample size of 164 samples and the JAFFE database that had 150 samples.

The Cohn–Kanade database gave the highest accuracy of 81.4%, whereas the accuracy of the JAFFE database ranged from 55% to 68% for all the three methods. They applied a nearest-neighbor classifier using CSM and MCC. Their feature extraction included image representation using NMF and LNMF.

Pantic and Patras (2005) developed an algorithm that recognized 27 AUs, which was invariant to occlusions like glasses and facial hair. It was shown to give a better performance than the AFA system. It had an overall average recognition of 90%. They used Cohn–Kanade and MMI databases. The Cohn–Kanade database had 90 images, whereas the MMI database had 45 images. They tracked a set of 20 facial fiducial points using temporal rules. Zheng et al. (2006) used KCCA to recognize facial expressions. The singularity problem of the Gram matrix has been tackled using an improved KCCA algorithm. Their accuracy on JAFFE database using semantic info with leave-one-image-out cross-validation was 85.79% and with leave-one-subject-out (LOSO) cross validation was 74.32% and on Ekman's database was 78.13%. They used JAFFE and Ekman's pictures of affect database. Their JAFFE database consisted of 183 images, and Ekman's database had 96 images. Neutral expressions were not chosen from either database. The correlation is used to estimate the semantic expression vector, which is then used for classification. They converted 34 landmark points into a labeled graph using the Gabor wavelet transform. Then, a semantic expression vector is built for each training face.

Tian et al. (2001) developed an algorithm that recognized posed expressions. It was a real-time system. They used Cohn–Kanade and Ekman–Hager Facial Action Exemplars. They used 50 upper face samples from 14 subjects performing 7 AUs and 63 lower face sample sequences from 32 subjects performing 11 AUs. The accuracy of recognition of upper face AUs was 96.4%, and the accuracy of lower face AUs was 96.7%. The permanent features extracted were optical flow, Gabor wavelets, and multistate models, and the transient feature extracted was canny edge detection. Bourel et al. (2001) developed an algorithm that deals with recognizing facial expressions in the presence of occlusions. It also proposed the use of modular classifiers instead of monolithic classifiers. Classification is done locally, and then the classifier output is fused. The Cohn–Kanade database was used, and there were 30 subjects. A total of 25 sequences for 4 expressions (a total of 100 video sequences) were taken. Local spatio-temporal vectors were obtained from the Kanade lucas tomasi tracker algorithm. They used the modular classifier with data fusion. Local classifiers are rank-weighted KNN classifiers.

Pardas and Bonafonte (2002) developed an algorithm for automatic extraction of MPEG-4 FAPs. This proves that FAPs convey the necessary information that is required to extract the emotions. An overall efficiency of 84% was observed across six prototypic expressions. They used the whole Cohn–Kanade database and HMM classifier. MPEG-4 FAPs extracted using an improved active contour algorithm and motion estimation. Cohen et al. (2003) developed a real-time system. It suggests use of HMMs to automatically segment a video into different expression segments. They used Cohn–Kanade and their own database. They took 53 subjects under Cohn–Kanade, and 5 subjects under their own database. They used NB, TAN, and ML-HMM classifiers. They extracted a vector of extracted motion units using the piecewise Bézier volume deformation model (PBVD) tracker.

Cohen et al. (2003) made a real-time system that used semi-supervised learning to work with some labeled data and large amount of unlabeled data. They used the Cohn–Kanade database and the Chen–Huang database. They also extracted a vector of extracted motion units using the PBVD tracker.

Sebe et al. (2007) developed an algorithm that recognizes spontaneous expressions. They created an authentic DB where subjects are showing their natural facial expressions. They used spontaneous emotions database and also the Cohn–Kanade database. The sample for the database consisted of 28 subjects showing mostly neutral, joy, surprise, and delight whereas, the Cohn–Kanade database consisted of 53 subjects. They used Bayesian net classifier, SVM, and decision tree. They extracted MUs generated from the PBVD tracker.

The algorithm of Kotsia and Pitas (2007) recognizes either six basic facial expressions or a set of chosen AUs. Very high recognition rates have been shown. They used the whole Cohn–Kanade database. It had an accuracy of 99.7% for FER and 95.1% for FER based on AU detection. They used multiclass SVM for expression recognition and six classes of SVM, one for each expression. The feature extracted was the geometric displacement of Candide nodes.

Wang and Yin (2007) proposed a topographic modeling approach in which the gray-scale image is treated as a 3D surface. They analyzed the robustness of detected face region and the different intensities of facial expressions. They used the Cohn–Kanade database and the MMI database. They took 53 subjects, and 4 images per subject were taken for each expression, which made a total of 864 images. They used QDC, LDA, SVC, and NB classifiers that extracted topographic context expression descriptors. It had an accuracy of 92.78% with QDC, 93.33% with LDA, and 85.56% with NB on the Cohn–Kanade database.

Dornaika and Davoine (2008) proposed a framework for simultaneous face tracking and expression recognition. Two AR models per expression gave better mouth tracking and in turn better performance. The video sequences contained posed expressions. They created their own database and used several video sequences. Also, they created a challenge 1600 frame test video, where subjects were allowed to display any expression in any order for any duration. Results have been spread across different graphs and charts. First head pose is determined using online appearance models and then expressions are recognized using a stochastic approach. They extracted the Candide face model to track features.

15.2 DESCRIPTION AND GOALS

The work focuses on developing a FER system based on machine learning and deep learning algorithms. The entire work is be divided into two parts: FER systems based on classifiers and FER systems based on CNNs.

15.2.1 FER BASED ON CLASSIFIERS

Using the concept of machines learning, I created a train dataset that contains five different file classes, happy, sad, angry, disgust, surprise, fear, and neutral. This was created by dividing the the JAFFE image database into seven class files. Using the feature extraction technique, the features from each emotion class were extracted. Different classifiers were trained on the features extracted, and the validation accuracy was generated. Based on the trained classifier one chooses to export to the model, the real-time emotion is recognized in comparison to the amount of accuracy achieved. The block diagram of the proposed method is shown in Figure 15.2.

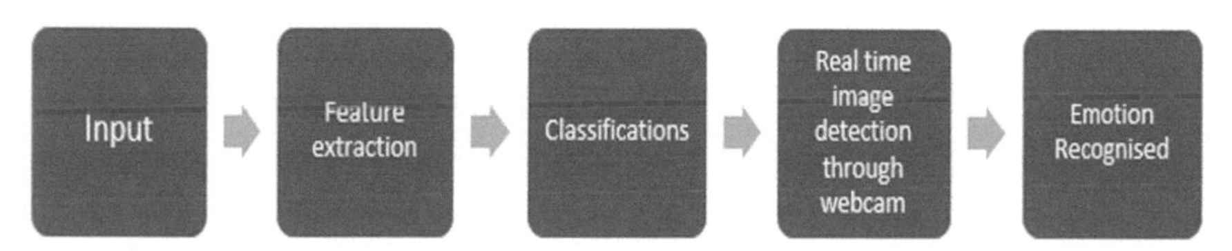

FIGURE 15.2 System block diagram followed by the classifier-based FER systems.

15.2.2 FER BASED ON CNN

The Kaggle dataset has been used for expression recognition that has 48×48 pixel labeled images (supervised learning) divided into training (28,709 images) and validation (3589 images) datasets. Using Keras lib in Python, we have built the required CNN model. The trained CNN model and the weights are saved and loaded into the model we will work with. This image is loaded into the model that has been loaded with pretrained weights that were imported from the Keras model, and finally, this unlabeled image whose class needs to be detected is passed through the model making the predictions that are displayed. The block diagram of the proposed method is shown in Figure 15.3.

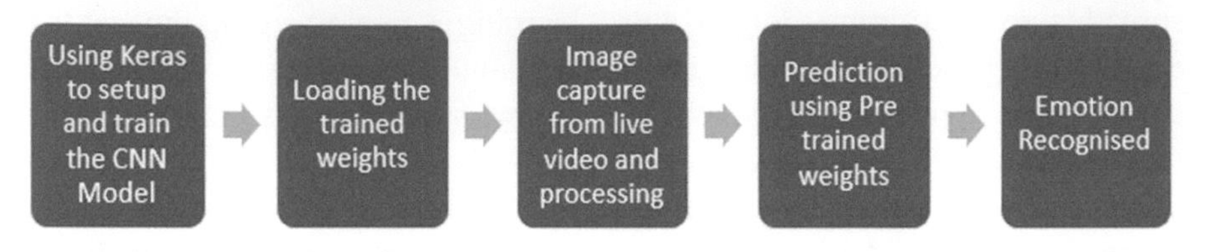

FIGURE 15.3 System block diagram followed by the CNN based FER systems.

15.3 DATASET SPECIFICATION

The JAFFE image dataset comprises 216 images of different emotion classes that are well labeled, as supervised learning requires labeled images for training the classifier. The FER2013 image dataset from Kaggle comprises of 48×48 pixel labeled images (supervised learning) divided into training (28,709 images) and validation (3589 images) datasets.

In this work, all classifiers were used with the help of the classification learner on MATLAB and the holdout validation was kept 10%. This means that 10% of training data is used by the classifier for validation purpose for the classifier to check and improve its accuracy

15.3.1 CONVOLUTION NEURAL NETWORKS

- In a CNN, convolution layers play the role of a feature extractor. However, they are not hand designed. Convolution filter kernel weights are decided as part of the training process. Convolutional layers are able to extract the local features because they restrict the receptive fields of the hidden layers to be local.
- Following are the technical specifications of the CNN model given in Table 15.1.

TABLE 15.1 Technical Specifications of the CNN Model

Batch size	100
Epochs	20
Optimizer	Adam
Loss function	Categorical cross entropy

- For real-time detection, the image is converted into gray scale and 48×48 pixels as the CNN model takes in only images of the specified size.

In this work, the CNN was built using Keras, and for validation, 3853 images of the training data are used by the CNN for validation purpose for the CNN to check and improve its accuracy.

15.4 DESIGN APPROACH AND DETAIL

15.4.1 DESIGN APPROACH: CLASSIFIER-BASED MODEL

15.4.1.1 INPUT DATA

Using the concept of machine learning, we created a train dataset for which we used the JAFFE image database that comprises 213 images of different emotion classes, namely, angry, sad, happy, surprise, disgust, and neutral. Figure 15.4 represents the sample of the facial expressions from the JAFFE image database.

15.4.1.2 FEATURE EXTRACTION

Feature extraction is the method of extracting certain characteristics from the signal. Using the SURF feature extraction technique, the features from each emotion class were extracted and a total of 500 features were used among all the features extracted.

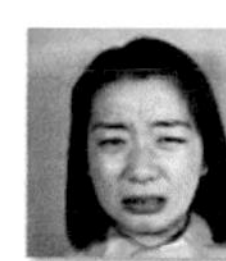

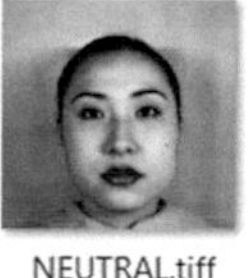

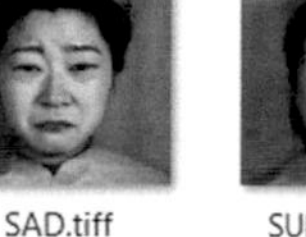
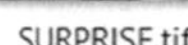

FIGURE 15.4 Sample of the JAFFE image database.

15.4.1.3 FEATURES LOADED TO CLASSIFIER

Features were converted from an array into a table, and the table was loaded into the classifier. This was done using the help of a classification learner toolbox of MATLAB. The holdout validation was kept as 10%, that is, 10% of the data was kept for the validation purpose of the classifier post the training.

15.4.1.4 TRAINING THE CLASSIFIER AND GENERATING THE VALIDATION ACCURACY

A classifier is a set pattern recognition algorithm that is used to define whether or not the test data belongs to a certain class based on the training set and the labels given in the training set. The classifiers used in this work are elaborately explained in the technical specifications. The different classifiers were trained on the features extracted, and validation accuracy was generated on each classifier. The scatter plot, ROC curve, and confusion matrix can also be generated.

15.4.1.5 IDENTIFYING IMAGE IN REAL TIME

Based on the trained classifier one chooses to export, the real-time emotion is recognized in comparison to the amount of accuracy achieved.

15.4.1.6 RESULT

The results obtained and the accuracy achieved are discussed in detail in the following sections. Also, the results were tabulated.

15.4.2 DESIGN APPROACH: CNN-BASED MODEL

15.4.2.1 INPUT DATA

The Kaggle dataset, a sample for which is shown in Figure 15.5, has been used for expression recognition that has 48×48 pixel labeled images (supervised learning) divided into training (28,709 images) and validation (3589 images) datasets.

15.4.2.2 CONSTRUCTING THE CNN MODEL

Using Keras lib in Python, we have built the required CNN model. The CNN model used in this work is shown in Figure 15.6.

FIGURE 15.5 Sample from the Kaggle database.

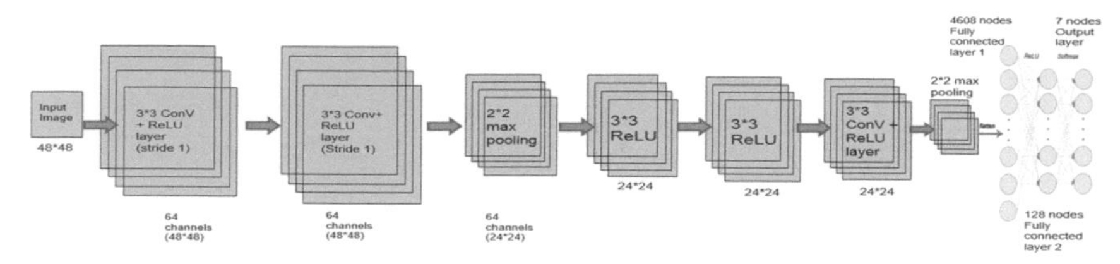

FIGURE 15.6 Architecture of the CNN model.

15.4.2.3 TRAINING THE CNN MODEL

The CNN model is trained with all the training images in every epoch. The optimizer used is Adam optimizer, and the loss function used is categorical cross entropy. A batch size of 100 was given for each epoch, and the final trained CNN model was saved along with the trained weights.

15.4.2.4 IMAGE CAPTURE FROM LIVE VIDEO AND PROCESSING

The camera is accessed to capture the image from live video and converted into gray scale. The face is detected using Harr cascade capable of detecting multiple faces in the single frame. The image is converted into 48 × 48 pixels as the CNN model takes in only images of the following size.

15.4.2.5 EMOTION RECOGNITION

The image is loaded into the model that has been loaded with pretrained weights that are imported from the Keras model, and finally, this unlabeled image whose class needs to be detected is passed through the model making the predictions that are displayed.

15.5 VARIOUS MODEL RESULTS AND DISCUSSION

15.5.1 KNN MODEL

The *KNN model* while making prediction on new data (unlabeled) takes into account the nearest *K* different classes and the majority of the classes in the neighborhood will decide the class of the data (Wang et al., 2015). This KNN classifier gives an accuracy of 87% on the JAFFE image database. Figure 15.7 represents the confusion matrix of the KNN classifier. Moreover, the ROC curves of the respective emotions are represented in Figure 15.8(a)–(f). It was observed that the ROC curves of the emotions were a perfect right angle for emotions that represented 100% accuracy.

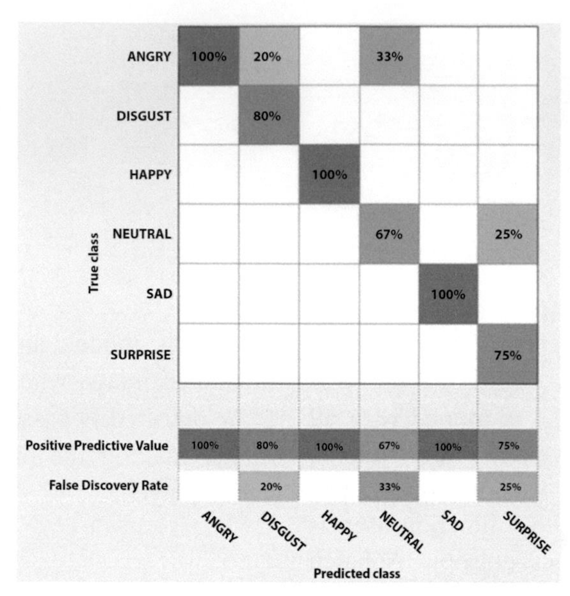

FIGURE 15.7 Confusion matrix—KNN classifier.

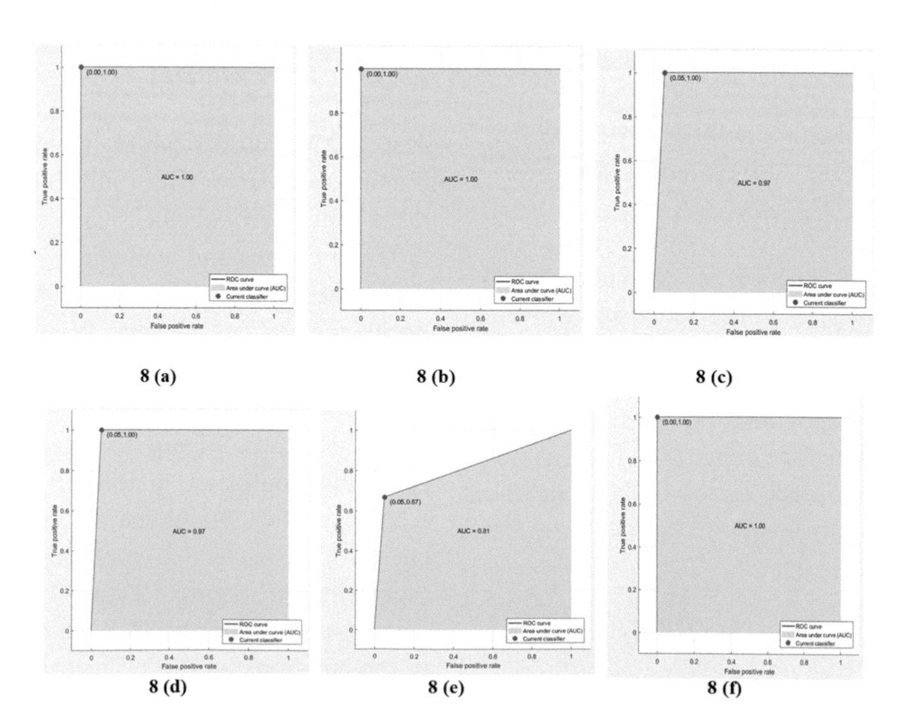

FIGURE 15.8 (a) ROC of happy, (b) ROC of sad, (c) ROC of disgust, (d) ROC of surprise, (e) ROC of neutral, and (f) ROC of anger.

15.5.2 SVM MODEL

The *SVM model* searches for the closest points, which it calls the "support vectors," draws a line connecting them, and then declares the optimal hyperplane to be a plane that bisects and is perpendicular to the connecting line dividing the classes. When a new test data is fed as input, it declares which class it belongs to, as the division has already been made (Kumbhar et al. 2012). Not all SVM problems are linearly separable; at times, the data distribution is such that on applying simple linear SVM the accuracy will be very low. So, it is very important to select the correct kernel function that gives better accuracy to segregate the data in the train dataset. Table 15.2 shows the accuracy result of the SVM classifier when different kernels are used.

TABLE 15.2 Accuracy Results When Different SVM Kernels Were Chosen

Classifier Type	Accuracy (%)
SVM classifier (linear)	73.9
SVM classifier (quadratic)	87
SVM classifier (cubic)	91.3

Thus, from Table 15.2, it can be concluded that the cubic SVM is the best fit, and hence, the cubic SVM model is exported into the model for real-time emotion detection. The confusion matrix of the model is represented in Figure 15.9. Also, the ROC curves of the respective emotions are represented in Figure 15.10(a)–(f). It was observed that the ROC curves of the emotions were a perfect right angle for emotions that represented 100% accuracy.

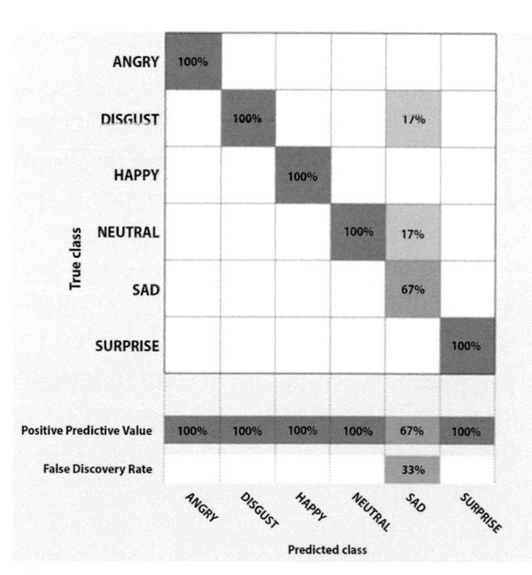

FIGURE 15.9 Confusion matrix—cubic SVM model.

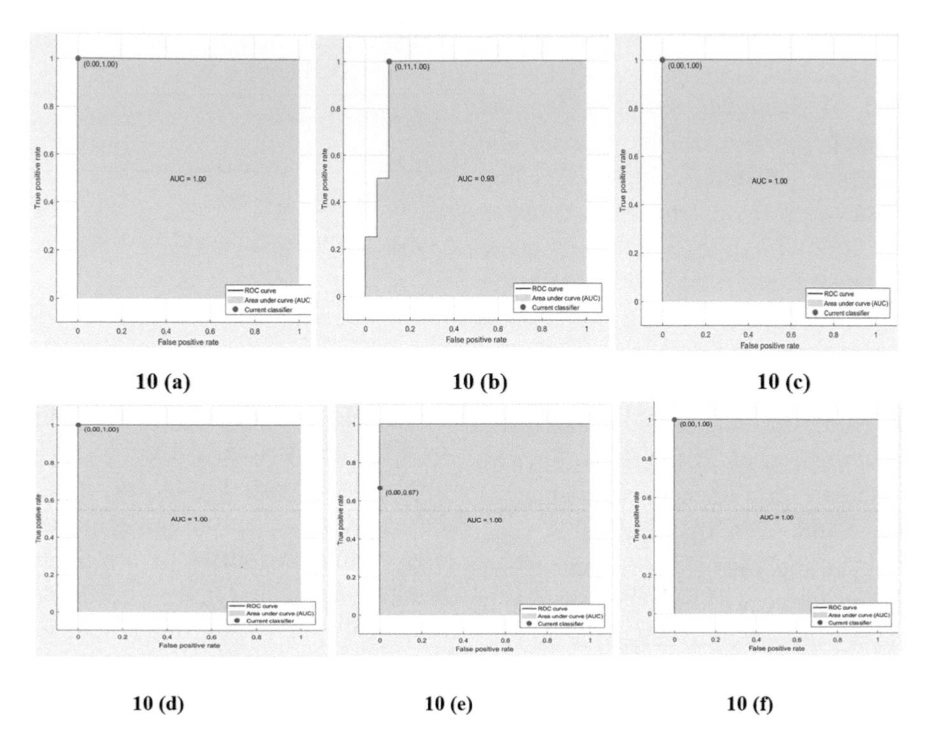

FIGURE 15.10 (a) ROC of happy, (b) ROC of sad, (c) ROC of disgust, (d) ROC of surprise, (e) ROC of neutral, and (f) ROC of anger.

15.5.3 ENSEMBLE SPACE KNN MODEL

The e*nsemble-based subspace KNN model* makes use of prediction of individual models that train random parts of the training dataset and generate results and by comparing the prediction results of these individual models. The ensemble subspace model classifies the real-time new image that is given for prediction. In our case, the individual model was the KNN model. The accuracy achieved was 95.7%. The confusion matrix of the ensemble space KNN model is represented in Figure 15.11. Moreover, the ROC curves of the respective emotions are represented in Figure 15.12(a)–(f). It was observed that the ROC curves of the emotions were a perfect right angle for emotions that represented 100% accuracy.

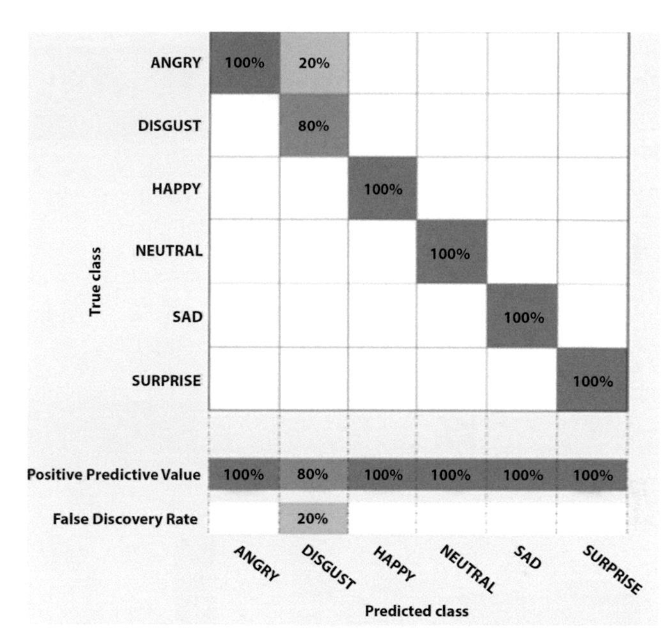

FIGURE 15.11 Confusion matrix—ensemble-based subspace KNN model.

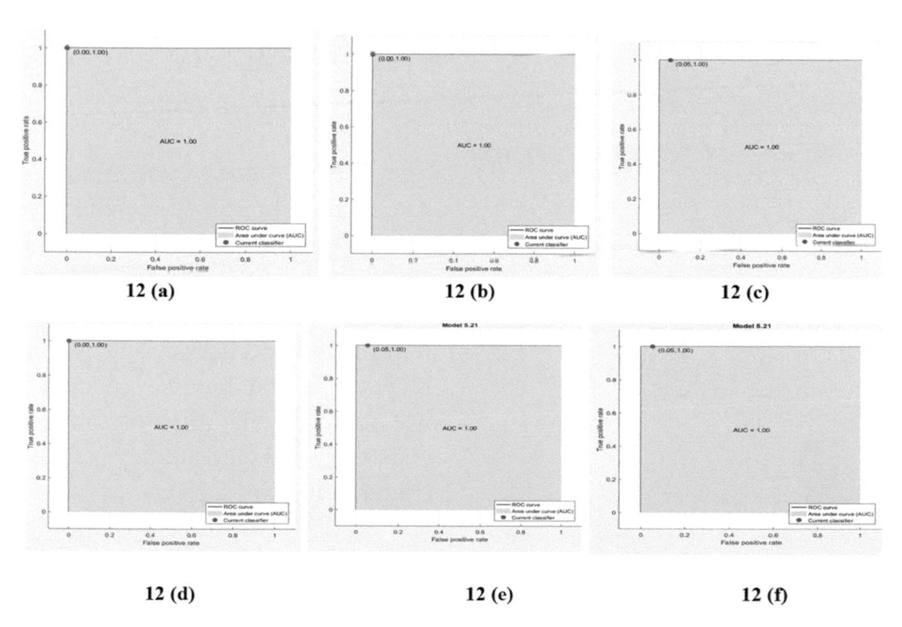

FIGURE 15.12 (a) ROC of happy, (b) ROC of sad, (c) ROC of disgust, (d) ROC of surprise, (e) ROC of neutral, and (f) ROC of anger.

The simulation results of detection of different real-time emotions like happy, sad, angry, etc. are summarized in Figures 15.13–15.17.

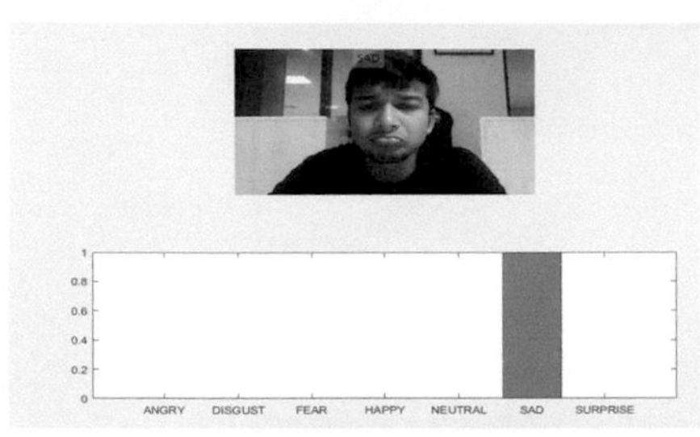

FIGURE 15.13 Real-time emotion recognized—sad.

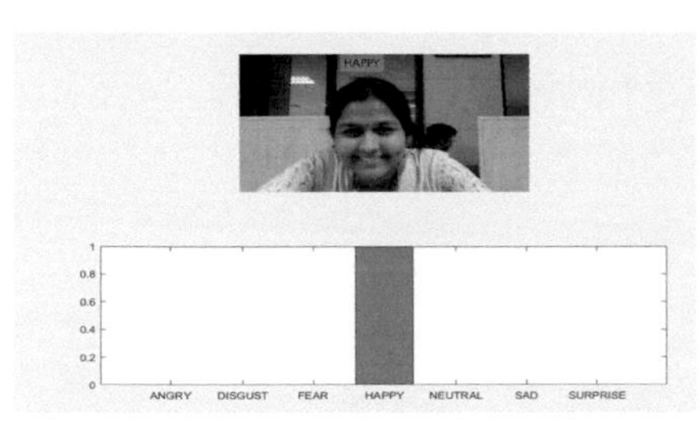

FIGURE 15.14 Real-time emotion recognized—happy.

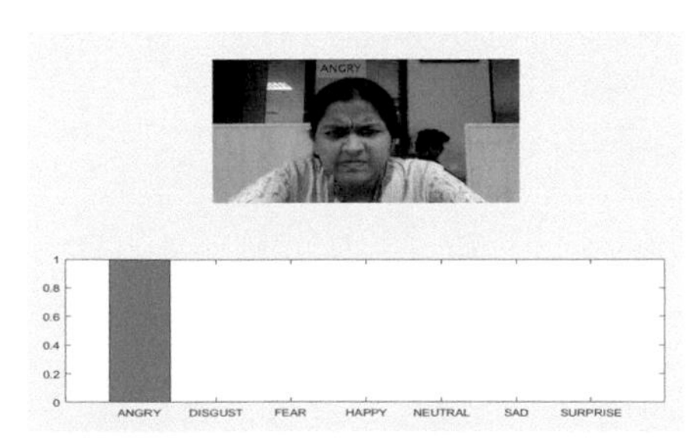

FIGURE 15.15 Real-time emotion recognized—anger.

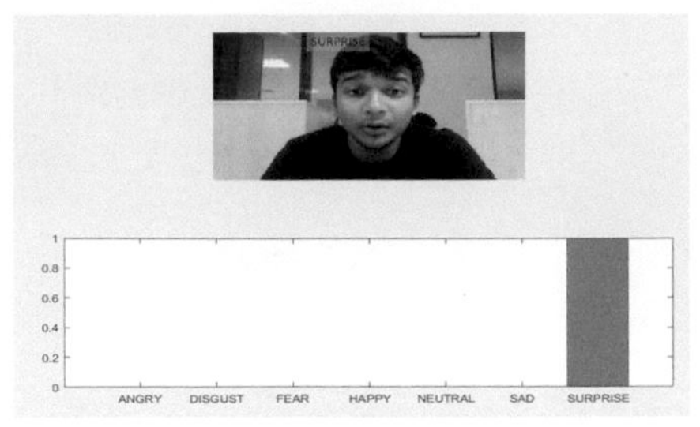

FIGURE 15.16 Real-time emotion recognized—surprise.

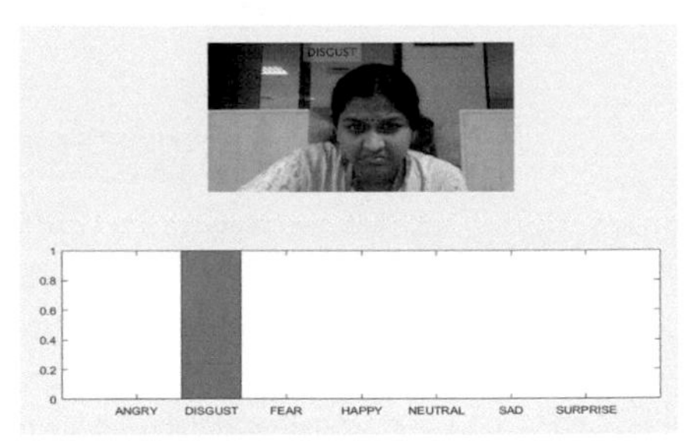

FIGURE 15.17 Real-time emotion recognized—disgust.

15.5.4 CNN MODEL

The CNN model was made on Keras, and the trained weights were loaded onto our FER model. The CNN model has greater prediction accuracy compared to the classifiers mentioned above, and with an increase in the number of epochs, one can see that the loss reduces constantly and the accuracy increases (Figures 15.18–15.20).

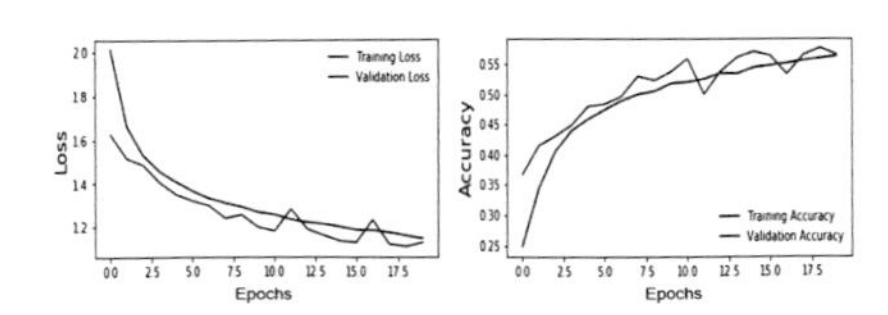

FIGURE 15.18 Loss vs epoch graph (left) and accuracy vs epoch graph (right).

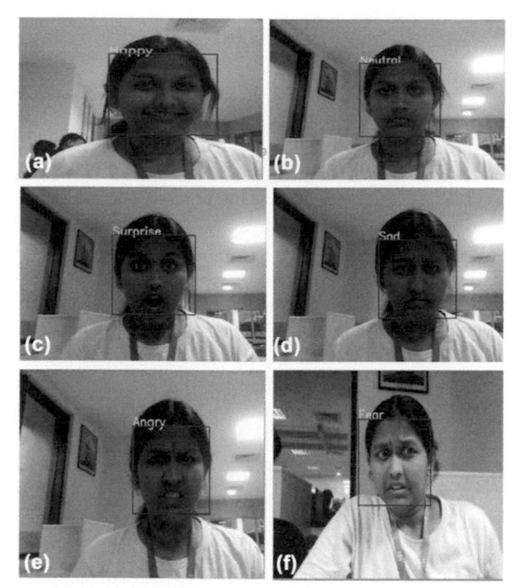

FIGURE 15.19 (a) Emotion recognized—happy, (b) emotion recognized—neutral, (c) emotion recognized—surprise, (d) emotion recognized—sad, (e) emotion recognized—angry, and (f) emotion recognized—fear.

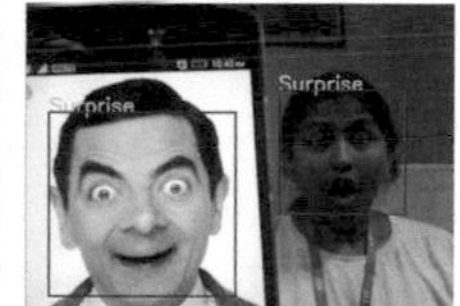

FIGURE 15.20 Emotion detection of multiple images.

15.5.5 COMPARISON OF THE ACCURACY RESULTS

It was observed that the CNN model gave the best prediction results (not validation accuracy) compared to the classifier models. Table 15.3 represents the accuracies of different models. The accuracy values do not match with the said conclusion because the classifiers are being implemented on the JAFFE image database that contains only 213 images, whereas the CNN model works on approximately 29,000 images and the obtained results are only on 20 epochs that will improve when the number of epochs is increased.

TABLE 15.3 Comparison of Various Models

Classifier	Accuracy (%)
KNN classifier	87
SVM classifier (linear)	73.9
SVM classifier (quadratic)	87
SVM classifier (cubic)	91.3
Ensemble subspace KNN	95.7
CNN model	60 (20 epochs)

Table 15.4 shows the comparison of research work conducted by other researchers. It can be seen that the JAFFE image database on the whole achieves very low accuracy as the number of images is not sufficient to train the model. So, on the whole, achieving a good accuracy on JAFFE is difficult, which our model has achieved up to a certain extent. Moreover the CNN model that we use gives considerable accuracy at 20 epochs, which is comparable to the accuracy that the neural network model has obtained. Although the CNN models different from the ones listed below are observed to show an accuracy of around 90%, which even

the model proposed in our work can achieve once the number of epochs is increased, which will be our future work.

TABLE 15.4 Comparison of Various Methods Adopted by Researchers on FER

Researcher	Year	Methods	Dataset	Accuracy (%)
Pantic et al.	2004	Rule-based classifier on frontal and profile views	MMI dataset	86
Bucio et al.	2004	Nearest neighbor classifier using CSM and MCC	JAFFE	55–68
Sebe et al.	2007	Bayesian nets classifiers, SVMs, decision trees	Cohn Kanade	72.46-93.06
Kotsia and Pitas	2007	Multiclass SVM based on AU detection	Cohn Kanade	95.1
Dornaika and Davoine	2008	First, the head pose is determined using online appearance models, and then, expressions are recognized using a stochastic approach	Own Database	56.3
Neeta Sarode et al.	2010	2D appearance-based model radial symmetry transform	JAFFE	81
Samad et al.	2011	Gabor wavelet, PCA, multiclass SVM	FEEDTUM database	81.7 (avg)
Kumbhar. et al.	2012	Neural network	JAFFE	60 (avg)
Wang et al.	2015	SVM KNN	JAFFE	85.74 83.91
Alexandru Savoiu et al.	2017	SVM classifier baseline with CNN (VGG-16 and ResNet50) ensemble-based classifier baseline	Kaggle dataset	31.8 67.2
Ke Shan	2017	KNN classifier CNN	JAFFE	65.1163 76.7
This work	2018	KNN classifier SVM (cubic) classifier Ensemble-based KNN classifier Convolution neural-network-based FER	JAFFE JAFFE JAFFE Kaggle	87 91.3 95.7 60 (20 epochs)

15.6 CONCLUSIONS

It has been found that the real-time FER model based on machine learning algorithms are successful in detecting emotions. The proposed FER system was trained on different classifiers—KNN, SVM, and ensemble-based subspace KNN models, and it has been found out that the ensemble-based subspace KNN model gives a maximum accuracy of 95.7% compared to the KNN model that gives an accuracy of 87% and the SVM model (cubic) that gave an accuracy of 91.3%. In real-time detection, it is observed that the model was getting confused between disgust, neutral, and sad. This is because the JAFFE database on which the classifier is trained has very similar images in these three emotion classes, which is why even the accuracy of the classifiers faced a decrease, as clearly represented by the confusion matrix. So, in future, more images once added into the model for training will give a clear distinction while detection. (Apart from Japanese women, different other faces need to be added.) Comparing it to the already existing research in this field, our model gives maximum accuracy on the JAFFE image database.

It is observed in our research that the convolution neural network model's performance was way better than the classifiers, the emotion detection was more accurate, and the results were more reliable. Also, multiple faces can be detected and emotions can be recognized. Although working on the CNN model has been more strenuous and time-consuming due to the bulky nature of the dataset that it requires for training, the results are really satisfactory. For instance, the accuracy obtained was only 60% at 20 epochs, which can be increased by increasing the number of epochs, but the prediction at 60% is a way better than the classifiers because the CNN model trains on large number of data, and unlike the classifiers are that trained on JAFFE, this model is trained on the Kaggle dataset that has a variety of facial expressions. Therefore, we can conclude that the deep-learning-based CNN model works better than the machine-learning-based FER system.

KEYWORDS

- **facial expression**
- ***k*-nearest neighbor**
- **support vector machine and deep learning technique**

REFERENCES

Bourel, F., C.C. Chibelushi, and A.A. Low, "Recognition of Facial Expressions in the Presence of Occlusion," *Proc. of the 12th*

British Machine Vision Conference, vol. 1, pp. 213–222, 2001.

Buciu, I. and I. Pitas, "Application of Non-Negative and Local Non Negative Matrix Factorization to Facial Expression Recognition." *Proc. of the ICPR*, pp. *288*–291, Cambridge, UK, August 23–26, 2004.

Cohen, I., N. Sebe, F. Cozman, M. Cirelo, and T. Huang, "Learning Bayesian Network Classifiers for Facial Expression Recognition Using Both Labeled and Unlabeled Data." *Proc. of the IEEE Conf. Computer Vision and Pattern Recognition (CVPR),* vol. 1, pp. I-595–I-604, 2003.

Cohen, I. N. Sebe, A. Garg, L.S. Chen, and T.S. Huang, "Facial Expression Recognition From Video Sequences: Temporal and Static Modeling." *Comput. Vis. Image Understand.*, vol. 91, pp. 160–187, 2003.

Dornaika F. and F. Davoine, "Simultaneous Facial Action Tracking and Expression Recognition in the Presence of Head Motion." *Int. J. Comput. Vision*, vol. 76, no. 3, pp. 257–281, 2008.

Ekman, P. and W.V. Friesen, "Constants across Cultures in the Face and Emotion." *J. Pers. Soc. Psychol.*, vol. 17, no. 2, pp. 124–129, 1971.

Kotsia I. and I. Pitas, "Facial Expression Recognition in Image Sequences Using Geometric Deformation Features and Support Vector Machines." *IEEE Trans. Image Process.*, vol. 16, no. 1, pp. 172–187, 2007.

Kumbhar, M., A. Jadhav, and M. Patil, "Facial Expression Recognition Based on Image Feature." *Int. J. Comput. Commun. Eng.*, vol. 1, pp. 117–119, 2012.

Mehrabian. A., "Communication without Words." *Psychol. Today*, vol. 2, no. 4, pp. 53–56, 1968.

Pantic, M. and I. Patras, "Detecting Facial Actions and Their Temporal Segments in Nearly Frontal-View Face Image Sequences." *Proc. IEEE Conf. Syst., Man Cybern.*, vol. 4, pp. 3358–3363, October 2005.

Pantic M. and J.M. Rothkrantz, "Facial Action Recognition for Facial Expression Analysis from Static Face Images." *IEEE Trans. Systems, Man Cybernet., Part B,* vol. 34, no. 3, pp. 1449–1461, 2004.

Pardas, M. and A. Bonafonte, "Facial animation parameters extraction and expression recognition using Hidden Markov Models." *Signal Process.: Image Commun.*, vol. 17, pp. 675–688, 2002.

Samad, R. and H. Sawada. "Extraction of the Minimum Number of Gabor Wavelet Parameters for the Recognition of Natural Facial Expressions." *Artif. Life Robot.*, vol. 16, no. 1, pp. 21–31, 2001.

Sarode, N. and S. Bhatia. "Facial Expression Recognition." *Int. J. Comput. Sci. Eng.*, vol. 2, no. 5, pp. 1552–1557, 2010.

Savoiu, Alexandru, and James Wong. "Recognizing facial expressions using deep learning." (2017).

Sebe, N., M.S. Lew, Y. Sun, I. Cohen, T. Gevers, and T.S. Huang, "Authentic Facial Expression Analysis." *Image Vis. Comput.*, vol. 25, pp. 1856–1863, 2007.

Shan, K., J. Guo, W. You, D. Lu, and R. Bie, "Automatic Facial Expression Recognition Based on a Deep Convolutional-Neural-Network Structure." *Proc. of the IEEE 15th International Conference on Software Engineering Research, Management and Applications (SERA)*, London, pp. 123–128, 2017.

Sharmila A. and P. Geethanjali, "Detection of Epileptic Seizure from EEG Based on Feature Ranking and Best Feature Subset Using Mutual Information Estimation." *J. Med. Imag. Health Informat.*, vol. 6, 1850–1864, 2016a.

Sharmila A. and P. Geethanjali, "DWT Based Epileptic Seizure Detection from EEG Signals Using Naïve Bayes and KNN Classifiers." *IEEE Access*, vol. 4, 7716–7727, 2016b.

Shih, F.Y., Chuang C.F., and Wang P.S.P., "Performance comparisons of facial expression recognition in JAFFE database," *Int. J. Pattern Recognit. Artificial Intell.*, vol. 22, no. 3, pp. 445–459, 2008.

Shrivastava, D. and L. Bhambu, "Data Classification Using Support Vector Machine." *J. Theor. Appl. Inf. Technol.*, vol. 12, no. 1, pp. 1–7, 2010.

Tian, Y., T. Kanade, and J. Cohn, "Recognizing Action Units for Facial Expression Analysis." *IEEE Trans. Pattern Anal. Mach. Intell.*, vol. 23, no. 2, pp. 97–115, 2001.

Wang J. and L. Yin, "Static Topographic Modeling for Facial Expression Recognition and Analysis." *Comput. Vis. Image Understand.*, vol. 108, pp. 19–34, 2007.

Wang, X.-H., A. Liu, and S.-Q. Zhang. "New Facial Expression Recognition Based on FSVM and KNN." *Optik*, vol. 126, pp. 3132–3134, 2015.

Zheng, W., X. Zhou, C. Zou, and L. Zhao, "Facial Expression Recognition Using Kernel Canonical Correlation Analysis (KCCA)." *IEEE Trans. Neural Netw.*, vol. 17, no. 1, pp. 233–238, January 2006.

CHAPTER 16

ANALYSIS AND INTERPRETATION OF UTERINE CONTRACTION SIGNALS USING ARTIFICIAL INTELLIGENCE

P. MAHALAKSHMI and S. SUJA PRIYADHARSINI*

Department of Electronics and Communication Engineering, Anna University Regional campus-Tirunelveli, Tamilnadu, Tirunelveli, India.

**Corresponding author. E-mail: sujapriya_moni@yahoo.co.in*

ABSTRACT

Electrohysterography (EHG) is the technique used to monitor the activity of the uterine signals. The EHG signals are acquired from the abdominal surface of pregnant women, and the readings used to study the electrical activity produced by the uterus. The electrical signal obtained from the abdominal surface helps to differentiate the true labor and false labor pain. EHG signals are recorded from the three channels. The objective of the proposed work is to differentiate the true labor and false labor pain from the EHG signal of the pregnant woman. The proposed work employs EHG signal available in the physionet database. Three channels are used to extract the EHG signal. The dataset consists of 300 records, in which each record consists of three signals recorded using three channels from the abdomen of a pregnant woman.

The dataset contains 160 true labor signals and 140 false labor signals. Then obtained signal is filtered by using the five pole order Butter worth band pass filter. The cut-off frequency applied for the butter worth band pass filter is 0.3–1 Hz.

The proposed work uses the features such as mean, median, maximum frequency, median frequency, kurtosis, skewness, energy and entropy extracted from the signals for identifying true labor/false labor pain signals. This work employs different classifiers individually for classifying the signals into true labor pain/false labor pain signal based on the values of the features extracted.

This chapter compares the performance of different machine learning algorithms such as support vector machine (SVM), extreme learning machine (ELM), *k*-nearest neighbor (KNN), artificial neural network (ANN), radial basis function neural network (RBNN) and random forest (RF) classifiers in identifying true and false labor pain signals. The performance of each classifiers, are evaluated individually. The performance of SVM classifier is evaluated with different kernel functions like linear, polynomial, radial basis function (RBF) and multilayer perceptron (MLP). SVM yields an accuracy of 58%, 55%, 57%, and 55% for different kernel functions linear, polynomial, RBF and MLP respectively. The Performance of KNN classifier is evaluated with kernel function 1 norm. KNN classifier yields an accuracy of 77% for kernel function 1 norm. The Accuracy for different machine learning algorithms such as ANN, RBNN, RF, ELM are 96%, 97%, 82% and 98%, respectively. The highest accuracy of 98% is obtained for ELM classifier. Hence ELM classifier outperforms other classifiers.

This chapter proposes the possibility of employing artificial intelligence technique with EHG signals for differentiating true labor and false labor pain signals. Early diagnosis of premature delivery helps to delay the delivery by proper in-time treatment and thus it helps to prevent the premature baby and its associated health issues and risk of death.

16.1 INTRODUCTION

EHG or uterine electromyography (EMG) records electrical activity signals responsible for the involuntary contractions of the uterus. EHG signals are recorded from the abdominal surface of pregnant women, and the readings used to study the electrical activity produced by the uterus. EHG is a measure of the electrical potential generated by the uterine muscles during pregnancy and labor. The technique consists of placing electrodes on the maternal abdomen and recording electrical activity. Monitoring uterine signals is critical during pregnancy to determine whether the onset of labor pains is indicative of true or false labor.

EHG signals make it possible to detect uterine activity related to contractions during both gestation and active labor. EHG signals are mostly used to predict true/false labor pains and prevent premature birth. The aim of the proposed work is to demonstrate the efficacy of the machine learning algorithm in classifying EHG signals as indicators of true or false labor pains. In the proposed work, the EHG signal dataset from the PhysioNet database is used for evaluation.

Delivery prior to the completion of 37 weeks of gestation is referred to as preterm and is currently a challenge in obstetrics. A preterm delivery with its associated complications is a major cause of infant deaths, with studies reporting thousands of infant deaths daily (Zorz et al., 2008).

Efforts have been made to alleviate the effects of preterm births. Consequently, it is important to predict or distinguish between true and false labor pains. EHG signals occur as a result of the propagation of electrical activity between the muscular cells of uterus walls, and reveal potential differences between electrodes. HG signal studies help establish an ambulatory monitoring system for risky pregnancies, and provide alerts if a premature pregnancy threat occurs (Fergus et al., 2015).

The objective of prenatal care is to sustain the health of both mother and fetus, and retain the fetus in the uterus till healthy birth. Monitoring uterine contractility is crucial during pregnancy in order to differentiate normal contractions from those causing early stretching of the cervix (Fergus et al., 2015). To obtain a better performance, the three-channel EHG signals are band pass-filtered using a fifth-order Butterworth band pass filter before analysis.

Methods presently used in obstetrics are not accurate enough to detect the risk of labor early. Therefore, a more consistent method is needed for early recognition and prevention of false labor threats (Zorz et al., 2008).

The EHG signal recorded superficially represents the internal uterine electrical activity. Though the EHG represents the uterine electrical activity it is contaminated by the noise that distorts the signal. Hence preprocessing is step required to eliminate the noise before further processing the signal.

In the proposed work, features such as the mean, median, kurtosis, skewness, peak frequency, median frequency, energy, and entropy are extracted from EHG signals. The extracted features are applied to different machine learning algorithms individually to determine whether the signals herald true or false labor pains.

This work aims to evaluate the features extracted from EHG signals, in conjunction with several advanced artificial intelligence algorithms, to assess their ability to distinguish between true and false labor pains.

16.2 A REVIEW OF LITERATURE

In this section, brief analysis of some significant contributions of the existing literature in EHG signal processing and artificial intelligence are presented.

Radomski et al. (2008) proposed a method based on a nonlinear feature analysis of EHG signals. Monitoring uterine electromyographic signals during pregnancy is critical to clinical medicine. This work evaluated the possibility of a nonlinear analysis of electrohysterographical signals to assess uterine contractile activity during pregnancy. The analysis was based on sample entropy statistics, and the initial results confirmed that the method could provide clinically useful information for obstetrical care.

Hassan et al. (2010) developed a method to distinguish active labor from normal pregnancy contractions. Labor prediction using the EHG has far-reaching clinical applications. Different linear methods, as in, for instance, classic spectral analysis, fail to offer significantly beneficial clinical results. This work presented two useful methods such as one linear and two non-linear methods. The linear method is based on the mean power frequency, and the two nonlinear methods on approximate entropy and time reversibility. The comparisons demonstrate that time reversibility is an excellent method for classifying pregnancy and labor signals. The results show that time reversibility is a very promising tool for distinguishing between labor and physiological contractions during pregnancy.

Ivancevic et al. (2008) proposed an analysis of EHG signals to assess modern nonlinearity methods used in preterm birth analysis. A nonlinear analysis of uterine contraction signals furnishes information on the physiological changes undergone during the menstrual cycle and pregnancy, which can be used for both preterm birth predictions and preterm labor control.

Baghamoradi et al. (2011) proposed a method to predict preterm labor and evaluated the application of cepstral analysis for the classification of both term and preterm labor. In all, 20 EHG records of term delivery (pregnancy duration ≥37 weeks) and preterm delivery (pregnancy duration ≤37 weeks) were analyzed. A multilayer perception (MLP) neural network is used to classify the two groups. An improved classification accuracy of 72.73% is obtained using the sequential forward feature selection scheme.

Arora and Garg (2012) proposed a discrete wavelet transform, based on a pyramid set of rules method, to decompose EHG signals and obtain the final feature vector matrix. EHG signals are classified into two, term and preterm. Classification is carried out with the SVM, dividing data into testing and training sets. It is validated on a standard database from

PhysioNet. The experimental results illustrate that the technique gives an accuracy of 97.8% and can be a constructive tool for investigating the risk of preterm labor.

Hassan et al. (2012) analyzed the propagation of uterine EMG signals using the nonlinear correlation coefficient. EMG signals from 49 women (36 during pregnancy and 13 in labor) at different gestational ages were recorded by placing a 4 × 4 matrix of electrodes on the abdomen. Receiving operating characteristics curves evaluate the various methods in differentiating between contractions recorded during pregnancy and labor. The results indicate that the nonlinear correlation analysis performs better than classical frequency parameters in distinguishing labor contractions from normal pregnancy contractions. The paper concludes that analyzing the propagation of uterine electrical activity using the nonlinear correlation coefficient underscores the usefulness of uterine EMG signals for clinical purposes, such as monitoring pregnancy, detecting labor, and predicting preterm labor.

Li et al. (2013) proposed a method for EEG signal recognition using empirical mode decomposition (EMD) and an SVM. Automatic seizure detection is vital to monitoring epilepsy, in addition to its diagnosis and treatment. This paper proposed a new method for EEG feature extraction and pattern recognition is proposed, based on empirical mode decomposition (EMD) and the SVM. EEG signals are decomposed into intrinsic mode functions (IMFs) using the EMD, followed by the extraction of features including the coefficient of variation and fluctuation index of IMFs. The task of recognizing EEG signals is undertaken by classifying features using the SVM classifier. The experimental results report that the algorithm delivers 97.00% sensitivity and 96.25% specificity for multiracial EEGs, and 98.00% sensitivity and 99.40% specificity for normal EEGs on the Bonn datasets.

Hussian (2013) proposed a method to predict the preterm deliveries in pregnant women using immune algorithm. Identifying the preterm deliveries and treatment to the preterm infants paves a way to the chance of survival. This work employed an immune algorithm which helps in diagnosing and classifying the EHG signal in true labor and to predict the preterm delivery. This work pays attention to determine the delivery as whether term and preterm delivery. The machine learning classifier produces the overall accuracy of 90%.

Shulgin and Shepel (2014) proposed a method to detect and characterize uterine activity. The fetal heart rate (FHR) and uterine contraction activity (UA) during pregnancy and labor are normally monitored using external tocography and ultrasound respectively. Given that these methods, however, are not accurate and sensitive enough, more precise methods are called for. Abdominal electrocardiography and EHG are safe and noninvasive techniques that monitor the FHR and uterine contractions during pregnancy. Signal processing methods are developed for three algorithms, based on the amplitude demodulation, spectrogram, and root mean square. The algorithms extract uterine activity signals from multichannel abdominal signals.

Far et al. (2015) proposed a method that predicts preterm labor using statistical and non-linear features. Predicting preterm labor plays a key role in decreasing neonatal deaths. Statistical and non-linear features extracted from EHG signals are classified into term and preterm labor signals using the SVM. A dataset comprising 26 records from term delivery (pregnancy duration $\geq$ 37 weeks) and 26 records from preterm delivery (pregnancy duration $<$37 weeks) was used. The results show that the highest accuracy can be attained by using the four statistical features of mean, standard deviation, median and zero crossing from channel 1.

Fatima et al. (2017) proposed a method for EHG signal classification for true and false pregnancy analysis. A baby born on forty week of pregnancy is called as normal and healthy baby. A premature baby was born within the period of after 20th week and before 37 weeks of pregnancy. In this work, linear features (mean, root mean square) and nonlinear features (entropy and cepstrum) were extracted and the EHG signal is classified into term and preterm pregnancy using SVM classifier.

16.3 METHODOLOGY

This section describes the method adopted for the classification of EHG signals into true and false labor pains, using artificial intelligence. Figure 16.1 describes the proposed work in detail.

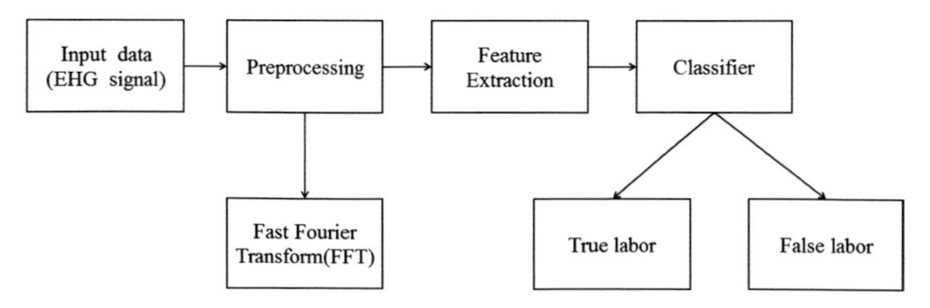

FIGURE 16.1 Flow diagram of the proposed work.

16.3.1 RAW DATA COLLECTION

In this work, EHG signals used for the analysis were obtained from PhysioNet. Signals were recorded from pregnant women by placing electrodes on the abdominal surface at horizontal and vertical distances, between 2.5 and 7 cm apart. The EHG signals were preprocessed to denoise them.

16.3.2 PREPROCESSING

EHG signals are tainted, owing to contact with different sources, during recording. The recorded, raw EHG signals contain noise, and are preprocessed to denoise them. This is done with the digital fifth-order Butterworth band pass filter with cut-off frequencies ranging from 0.34 to 1 Hz. The sampling rate of each signal is 20 Hz.

16.3.3 FEATURE EXTRACTION

Features extracted from the EHG signals, and used in the present work, are described below. Energy, entropy, kurtosis, median frequency, mean, median, peak frequency and skewness are extracted from the power spectrum of the EHG signals. The power spectrum is computed using the fast Fourier transform.

Energy:

The energy of a signal is defined as

$$E = \sum \left[x(n)^2 \right] \quad (16.1)$$

where $x(n)$ represents the number of samples.

Kurtosis:

Kurtosis is a measure of the peakness and tailedness of the probability distribution (Zorz et al., 2008)

$$k = \sum (x - \bar{x}\sigma) \backslash n - 3 \quad (16.2)$$

where $\bar{x}$ represents σ the mean, the standard deviation, and n the number of samples.

Mean:

The mean is the average value of a signal, indicated by μ. All the samples are added together and divided by N (Far et al., 2015)

$$\mu = \frac{1}{N} \sum_{i=0}^{N-1} xi \quad (16.3)$$

where xi is the sum of the values of the signal.

Median:

The median is a value separating the higher half of a data sample from the lower half. The median is the value such that a number is equally likely to fall above or below it (Far et al., 2015)

$$M = \frac{(n+1)}{2} \quad (16.4)$$

where n is the number of samples in the set.

Skewness:

It estimates the symmetry of a distribution, that is, the relative frequency of positiveand negative extreme values (Zorz et al., 2008)

$$s = (x - \mu / \sigma)^3 \quad (16.5)$$

where σ represents the standard deviation and μ is the mean.

16.4 CLASSIFIERS

In the proposed work, classifiers such as the SVM, ELM, KNNs, RF, ANN, and RBFNN are individually used to categorize the EHG signals into true or false labor pains. The performance of each classifier is individually evaluated and a comparison drawn.

16.4.1 K-NEAREST NEIGHBORS

The KNNs is a nonparametric method used for classification and regression. The input consists of the "*k*" closest training set in the feature space and the output depends on whether the *k*-nearest neighbor is used for classification. For classification, a skilful technique can be assigned and weighted to the contributions of the neighbors so that the nearest neighbors contribute more to the average than distant ones. The KNNs is a simple algorithm that stores all available cases and classifies new ones based on a similarity measure (https//datascience.com).

16.4.2 SUPPORT VECTOR MACHINES

An SVM is a discriminative classifier strictly defined by sorting out a hyperplane. Support vector machines are supervised learning models that analyze the data used for classification and regression analysis. A set of training samples is marked as belonging to one or two categories, and the SVM efficiently performs linear and nonlinear classification using a kernel function. In a linear function, if the training data is linearly separable, the two parallel hyper planes that separate the two classes of data are selected. By varying the kernel functions, accuracy can be improved (Li et al., 2013).

16.4.3 EXTREME LEARNING MACHINES

ELMs are feedforward neural networks used for classification and regression analysis. A feedforward neural network such as back-propagation is slow. Training using a gradient-based algorithm requires many iterative steps for an enhanced performance. The ELM, which is designed to overcome these issues, is a single hidden layer feedforward neural network. The ELM randomly chooses input weights and hidden unit biases. Training the network by finding the least square solution to the linear system, the ELM analytically determines the output weight and provides the best reductive performance at an extremely fast learning speed (Chen et al., 2017).

16.4.4 ARTIFICIAL NEURAL NETWORKS

An ANN is based on a collection of connected units or nodes called artificial neurons. Each connection between artificial neurons can transmit a signal from one node to another node. The artificial neuron network receives the signal and processes it. Artificial neurons and connections typically have a weight that adjusts as learning proceeds. The weight increases or decreases, depending on the strength of the signal at a connection (Fergus et al., 2015).

16.4.5 RADIAL BASIS FUNCTION NEURAL NETWORKS (RBFNN)

The RBF network is an ANN that uses RBFs as activation functions. The output of the network is a linear combination of the RBFs of inputs and neuron parameters. The RBFNN typically has three layers: an input layer, a hidden layer with a nonlinear RBFNN activation function, and a linear output layer (https//datascience.com).

16.4.6 RANDOM FOREST (RF)

The RF is a flexible, easy-to-use machine learning algorithm that produces good results, and is used for both classification and regression. It is a supervised learning algorithm, largely used because of its simplicity. It builds multiple decision trees, merges them together to obtain stable predictions, and adds additional randomness while growing the trees. Rather than search for the most important features, it looks for the best feature among a random subset of features (https//datascience.com).

16.5 RESULTS AND DISCUSSION

This section presents the results and analyses the performance of different classifiers in terms of grouping signals into true and false labor pains.

16.5.1 PERFORMANCE METRICS

Evaluating artificial intelligence algorithms for the classification of EHG signals into true and false labor pains is essential. The performance is evaluated using the three measures of accuracy, sensitivity, and specificity.

16.5.1.1 CLASSIFICATION ACCURACY

$$\text{Accuracy} = \frac{(TP + TN)}{TP + FP + FN + TN} \times 100 \quad (16.6)$$

where

TP—True positive represents the correct classification, that is, true labor signals classified as true labor,

TN—True negative represents the correct classification, that is, false labor signals classified as false labor,

FP—False positive represents misclassification, that is, false labor signals misclassified as true labor, and

FN—False negative represents misclassification, that is, true labor signals misclassified as false labor.

16.5.1.2 SENSITIVITY

A measure of the performance of a binary classification test, sensitivity is also called True Positive Rate. It measures the proportion of actual positives that are correctly identified as such

$$\text{Sensitivity} = \frac{(TP)}{(TP + FN)} \times 100 \quad (16.7)$$

16.5.1.3 SPECIFICITY

It is a measure of the performance of a binary classification test, and is also called the True Negative Rate. It measures the proportion of actual negatives that are correctly identified as such

$$\text{Specificity} = \frac{(FP)}{FP + TN} \times 100 \quad (16.8)$$

16.5.2 DATA COLLECTION FOR UTERINE CONTRACTION

Raw EHG signals, obtained from the PhysioNet database, were recorded using four bipolar electrodes. Each signal was either recorded early, at 26 weeks (at around 23 weeks of gestation) or later, at 26 weeks (at around 31weeks). Within the dataset, three signals per record were obtained simultaneously by recording them through three different channels (Zorz et al., 2008).

In the proposed work, EHG signals from the PhysioNet database were used for analysis. Raw EHG signals obtained from the PhysioNet database were recorded using four bipolar electrodes, stuck to the abdominal surface and spaced at horizontal and vertical distances between 2.5 and 7 cm apart. The database contains a total of 300 signals, of which 160 are indicative of true labor and 140 of false labor. The raw EHG signals were preprocessed for denoising.

16.5.3 RESULTS

Figure 16.2 depicts the raw EHG uterine signals containing the output of three channels, with each channel comprising three channel signals, respectively. Therefore, the EHG signals constitute a total of nine.

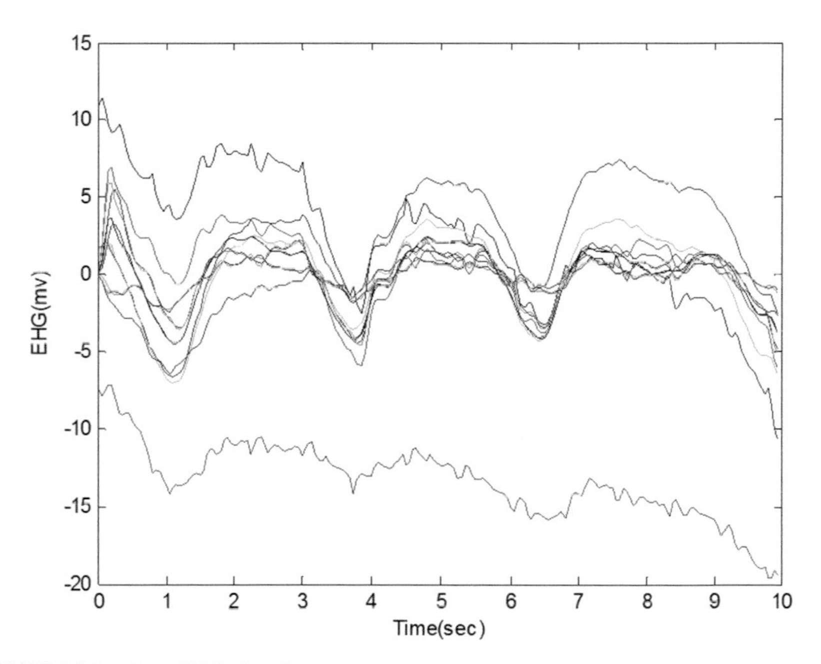

FIGURE 16.2 Raw EHG signals.

Figure 16.3 shows the filtered response of the EHG signals.

The signals are filtered using the five-order Butterworth band pass filter, with a cut-off frequency ranging between 0.34 and 1 Hz.

Figure 16.4 represents the power spectral density plot of the EHG signals. By applying the fast Fourier transform to the signals, their power spectral density is obtained.

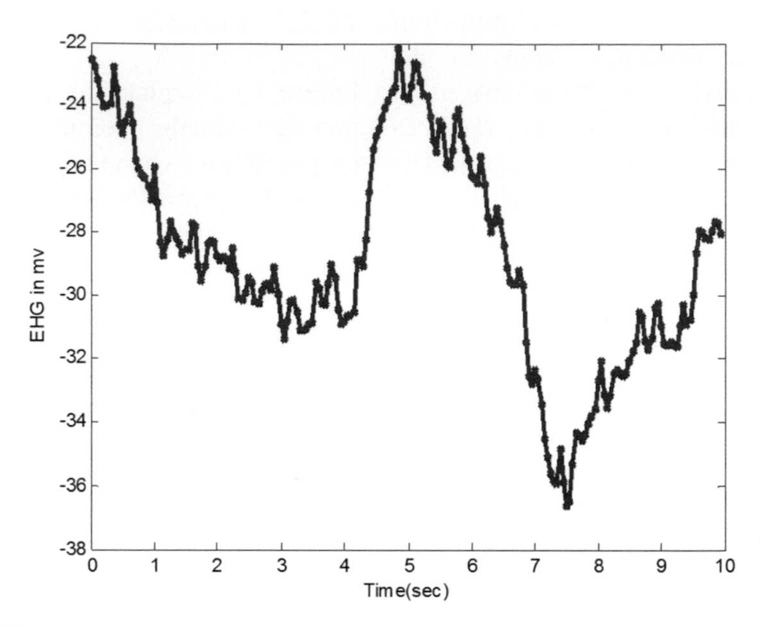

FIGURE 16.3 Filtered response of the EHG signals.

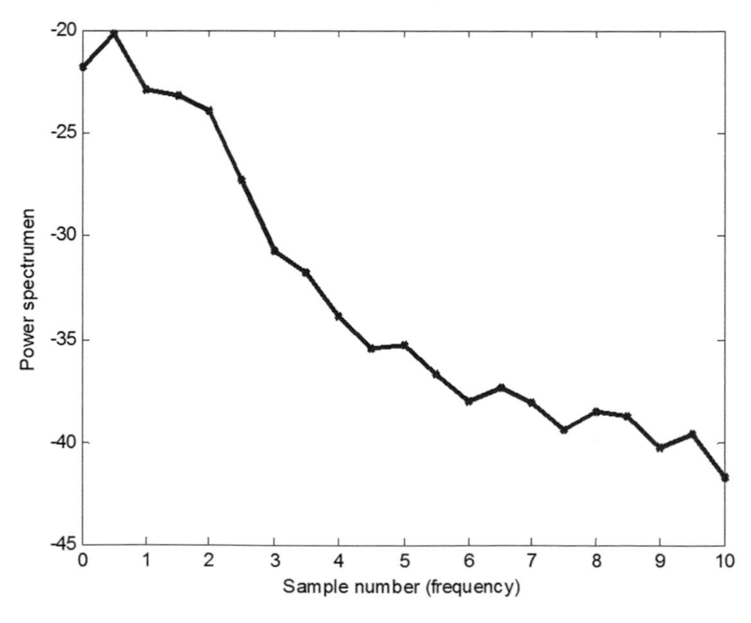

FIGURE 16.4 Power spectral density of the EHG signal.

In this work, the performance of classifiers such as the SVM, ELM, KNN, RF, ANN and RBFNN is evaluated individually, in terms of accuracy, to classify the EHG signals into true and false labor pains. EHG records from the Term–Preterm Electrohysterogram Database of PhysioNet were used for evaluation. It comprises 300 records of pregnant women, of which 262 are full-term pregnancies and 38 are premature births. Here, 162 records were taken before the 26th week of pregnancy and 138 records after, collected from 1997 until 2005 at the Department of Obstetrics and Gynaecology, Ljubljana University Medical Centre Slovenia. The records were acquired from the general population, in addition to patients admitted to the hospital with a diagnosis of anticipated preterm labor One record per pregnancy was recorded with a sampling frequency (F_s) of 20 Hz.

The records were gathered from the abdominal surface using four $AgCl_2$ electrodes, placed in two horizontal rows, equally under and above the umbilicus, spaced 7 cm apart. Three channels were made using the four electrodes. The first channel acquired a signal combining electrodes E2-E1, the second acquired a second signal combining electrodes E2–E3, and the third acquired a third signal combining electrodes E4–E3 (Zorz et al., 2008).

The proposed work employs a dataset of 300 signals, in which 160 and 140 signals belong to true and false labor pains respectively. The classifier is trained to classify the signals into true and false labor pains.

The performance of different classifiers such as the SVM, ELM, KNN, RF, ANN, and RBFNN is evaluated for different features. SVM and ELM classifiers are evaluated using the different kernel functions. The SVM is evaluated using four kernel function linear, polynomial, RBF and multilayer perceptions (MLP). Moreover, ELM is evaluated using kernel functions like sin, sigmoid, radial basis function (RADBAS), and triangular basis function (TRIBAS).

A performance analysis of different classifiers in classifying the EHG signals recorded into true and false labor, with different feature sets, is shown in Table 16.1.

16.5.4 DISCUSSIONS

From the tabulation, it is evident that the ELM classifier outperforms others in classifying EHG signals for all kernel functions, except sigmoid and sin, for energy and entropy features. Next to the ELM, the RBNN classifier excels in classifying the EHG signals for every feature set. For the different features

TABLE 16.1 A Performance Analysis of Different Classifiers

Features	SVM				ELM				RBFNN	ANN	KNN	RF
	Kernel functions				Kernel functions							
	Linear	Polynomial	RBF	MLP	Sigmoid	sin	TRIBAS	RADBAS				
Maximum frequency Median frequency	52%	55%	54%	48%	98%	98%	98%	98%	97%	96%	76%	81%
Mean median	58%	50%	57%	55%	98%	98%	98%	98%	96%	94%	77%	82%
Kurtosis skewness	49%	49%	49%	44%	98%	98%	98%	98%	85%	72%	70%	68%
Energy entropy	49%	50%	49%	52%	87%	89%	98%	98%	91%	91%	71%	81%

set ANN classifier produces a good accuracy except for kurtosis and skewness features set. In the RF classifier yields a highest accuracy than SVM classifier. In the case of the SVM classifier, of the different kernel functions and features, the mean and median features, along with the linear kernel function, yield the higher accuracy in comparison with other kernel functions and feature sets. In the proposed work SVM classifier is inferior than other classifiers in classifying the EHG signals into true \ false labor pain.

16.6 CONCLUSION

This chapter offers insights into the development of an intelligent system for the early diagnosis of premature birth by correctly identifying true/false labor pains. Raw term–preterm electrohysterography signals from PhysioNet were analyzed in this work. The database contains a total of 300 signals. In this study, raw EHG signals are preprocessed before analysis and filtered using the five-order Butterworth band pass filter. The raw EHG signals obtained from the three channels comprise nine signals that are filtered with cut-off frequencies of between 0.34 and 1 Hz. The power spectrum of the signals is computed by applying the fast Fourier transform.

Features are extracted from the preprocessed EHG signals, and energy, entropy, kurtosis, mean, median, maximum frequency, median frequency and skewness are calculated After the features are extracted, the classifier is trained and tested using a different feature set The performance of classifiers such as the SVM, ELM, KNN, ANN, RBNN, and RF is individually evaluated in terms of classifying EHG signals into true labor and false labor pains, and an analysis is carried out. The accuracy obtained from the different classifiers is 58%, 98%, 77%, 96%, 97%, and 82% respectively. It is evident from the results that the ELM classifier outperforms others in classifying EHG signals into true and false labor pains.

16.7 FUTURE WORK

In the present work, EHG signals are classified using different classifiers. In future, an embedded system can be developed by integrating sensors for signal acquisition and an artificial intelligence algorithm using suitable software to differentiate between true and false labor pains using EHG signals.

KEYWORDS

- **electrohysterography (EHG)**
- **support vector machine (SVM)**
- **extreme learning machine (ELM)**

REFERENCES

Acharya RU, Sudarshan KV, Rong QS, Tan Z, Min CL, Koh EWJ, Nayak S, Bhandary VS. Automated detection of premature delivery using empirical mode and wavelet packet decomposition techniques with uterine electromyogram signals, *Computers in Biology and Medicine,* **2017,** 85, 33–42.

Arora S and Garg G. A Novel scheme to classify EHG signal for term and preterm pregnancy analysis, *International Journal of Computer Application,* **2012**, 51(18), 37–41.

Askar A, Jumeily AD, Jager F, Fergus P. Dynamic neural network architecture inspired by the immune algorithm to predict preterm deliveries in pregnant women, *Neurcomputing*, **2015**, 151, 963–974.

Baghamoradi SMS, Naji M and Aryadoost H. Evaluation of cepstral analysis of EHG signals to prediction of preterm labor, *Iranian Conference of Biomedical Engineering,* **2011.**

Batra. A., Chandra A, Matoria V. Cardiotocography Analysis Using Conjunction of Machine Learning Algorithms, *International Conference on Machine Vision and Information Technology* **2017**.

Chen L and Hao1 Y. Feature extraction and classification of EHG between pregnancy and labour group using Hilbert–Huang transform and extreme learning machine, *Computational and Mathematical Methods in Medicine*, **2017,** 2017, 1–9.

Chudacek V, Spilka J, Bursa M, Janku P, Hruban L, Huptych M and Lhotska L. *Open Access Intrapartum CTG Database BMC Pregnancy Childbirth,* **2014**.

Far TD, Beiranvand. M and Shahbakthi M. Prediction of preterm labor from EHG signals using statistical and non-linear features, *Biomedical Engineering International Conference,* **2015.**

Fatima U, Goskula. T. EHG signal classification for true and false pregnancy analysis, *International Journal on Recent and Innovation Trends in Computing and Communication,* **2017,** 5 (6), 811–814.

Fergus P, Dowu. I, Hussain. A, Dobbins. C. Advanced artificial neural network classification for detecting preterm births using EHG records, *Neurocomputing* **2015,** 188, 42–49.

Hassan Terrien, Alexanderson, Marque, Karlsson. Nonlinearity of EHG signals used to distinguish active labor from normal pregnancy contractions, *32nd Annual International Conference of the IEEE,* **2010**.

Hassan M, Terrien. J, Muszynski. C, Alexanderson. A, Marque. C and Karrlonson. B. Better pregnancy monitoring using non linear correlation analysis of external uterine electromyography, *IEEE Transaction in Biomedical Engineering,* **2013,** 60(4), 1160–1166.

Huang ML, Hsu YY. Fetal distress prediction using discriminant analysis, decision tree, and artificial neural network, *Journal of Biomedical Science and Engineering*, 2012, 5, 526–533.

Ivancevic T, Jain CL, Pattison EJ, Hariz A. Preterm birth analysis using nonlinear methods, *Recent Patents on Biomedical Engineering*, **2008**, 160–170.

Jezewski M, Czabanski R, Wrobel J and Horoba K. Analysis of extracted cardiotocographic signal features to improve automated prediction of fetal outcome, *Biocybernetics and Biomedical Engineering*, **2010,** 30(4), 29–47.

Li S, Zhou W, Yuan Q, Geng S, Cai D, Feature extraction and recognition of ictal EEG using EMD and SVM, *Computers in Biology and Medicine,* **2013,** 43(7), 807–816.

Murray M. Antepartal and intrapartal fetal monitoring. 3rd ed. Springer, **2006.**

Radomski D, Grzanka A, Graczyk S, and Przelaskowski A. Assessment of uterine contractile activity during a pregnancy based on a nonlinear analysis of the uterine electromyographic signal, *Information Technologies in Biomedicine Springer*, **2008**, 47, 325–331.

Shulgin V and Shepel O, Electrohysterographic signals processing for uterine activity detection and characterization, *International Scientific Conference Electronics and Nanotechnology*, **2014.**

Sundar C, Chitradevi M, Geetharamani G. Classification of cardiotocogram data using neural network based machine learning technique, *International Journal of Computer Applications,* **2012**, 47(14), 19–25.

Zorz F.G, Kavsek G, Novak Z, Franc Jager. A, comparison of various linear and non linear signal processing techniques to separate uterine EMG records of term and preterm delivery groups. *Medical & Biological Engineering & Computing,* **2008**, 46(9), 911–922.

CHAPTER 17

ENHANCED CLASSIFICATION PERFORMANCE OF CARDIOTOCOGRAM DATA FOR FETAL STATE ANTICIPATION USING EVOLUTIONARY FEATURE REDUCTION TECHNIQUES

SUBHA VELAPPAN,[1*] MANIVANNA BOOPATHI ARUMUGAM,[2] and ZAFER COMERT[3]

[1]*Department of Computer Science & Engineering, Manonmaniam Sundaranar University, Tirunelveli, India*

[2]*Instrumentation & Chemicals Division, Bahrain Training Institute, Kingdom of Bahrain*

[3]*Department of Software Engineering, Samsun University, Turkey*

[*]Corresponding author. E-mail: subha_velappan@msuniv.ac.in

ABSTRACT

Role of computers became inevitable in healthcare sector and computers with information and communication technologies are found to be widely used for assessment, patient monitoring, documentation, and telemedicine. Data mining is a field which helps to obtain knowledge from massive amount of data from any industry or organization. Cardiotocography (CTG) is a test that is done during the third trimester of pregnancy to measure the heart rate and movements of fetus and helps to monitor the contractions in the uterus and thereby for monitoring the signs of any distress, before the delivery of baby and during the labour. The physical interpretation of information from CTG is found to be a challenging task, and any contradictory interpretation will lead to erroneous diagnosis on fetal condition, which may lead to the extent of fetal death. Feature selection is the process in which an optimal subset of features is selected based on some defined criterion which helps to considerably improve the performance of classification in terms of learning speed, accuracy of prediction, simplicity of rules, etc. Also, the reduction in size of feature subset helps to remove noise and irrelevant features. Several approaches have been introduced for improving the performance of computerized classification of CTG data which leads to an improved diagnosis of fetal status. In this chapter, Filter and Wrapper feature selection techniques are applied to CTG dataset available in UCI machine learning repository. Evolutionary algorithms such as genetic algorithm, firefly algorithm, and a hybrid technique incorporating information gain and opposition-based firefly algorithm have been used to improve the classification performance of CTG dataset. The results of simulations show that the proposed methodologies are highly promising when compared to the other existing methods. To assess the performance of these proposed methodologies, various performance measures namely accuracy, sensitivity (or) recall, specificity, precision (or) positive predictive value, negative predictive value, geometric mean, *F*-measure, and area under ROC have been used and the hybrid model incorporating information gain and ppposition-based firefly algorithm proves better performance than the other techniques.

17.1 INTRODUCTION

Healthcare is one of the major sectors which exploits computers and modern techniques of information technology for efficient patient information storage, management, retrieval, documetation, diagnosis, etc. Data mining techniques are employed in clinical decision support systems (CDSS) for efficiently handling these huge amount of healthcare data in order to assist the this industry in identifying good practices of patient monitoring, hospital administration, diagnosis, treatment and documentation. This eventually brings the cost down by almost 30% (HealthCatalyst, 2019). However, identifying and employing efficient data mining techniques for this purpose remains still a challenge because of the complex nature of healthcare data and inability to adapt to new technologies. The knowledge-based CDSSs make use of if-then rules in the knowledge base with an inference system and a communication system in order to obtain the inferences by combining the if-then rules with the patient data. The CDSSs which do not rely on knowledge base, utilize the machine learning techniques such as support vector machines (SVM), artificial neural networks (ANNs), etc. (Wagholikar et al., 2012) instead of prewritten knowledge base.

Cardiotography (CTG) is a popular CDSS used for monitoring the well being of fetus in the mother's womb. A CTG recording containts the information of heart rate (HR) of fetus and uterine contraction (UC). Fetal hypoxia is an abnormality in fetal condition which results due to the scarcity of oxygen for the fetus (Chudáček et al., 2010). If not diagnosed well and treated promptly, this torment of the fetus may lead to severe neurological disorder or death (CÖMERT et al., 2018b).

Manual interpretation of information from CTG signals is a difficult task for the physicians. The poor and inconsistant intrepretation of CTG signals will lead to poor diagnosis and treatment and eventually to fetal death. Using computers with machine learning capabilities on the attributes of CTG data the fetal condition can be classified more efficiently and accurately. The preformance of this classification task can be improved by feature selection (FS) process by which the appropriate attributes of CTG data are identified and selected.

Attribute reduction performed by the FS process aids to improve the learning speed and accuracy of prediction using simple rules, ability to visualize the data for selection of model. Further, it also helps to reduce the dimensionality and remove the noise present in the data.

The FS process also assists the knowledge discovery process which is performed in three stages namely preprocessing of data, data mining and the postprocessing. The FS process ensures that a good quality of data is supplied by pre-processing stage to the mining stage, a powerful and systematic process of mining and a meaningful knowledge being delivered by the post-processing stage. Excessive amount of features present in the data generally leads to a relatively ineffective mining results and hence the number of features is reduced to the possible extent without compromising the quality of mining.

Even though there are many methodologies available for interpretation of CTG data, their prediction accuracy are still not up to the mark (Liu and Motoda, 2000). Hence, it is still a challenge to develop an efficient and effective methodology with excellent accuracy of prediction.

In this chapter, three new and efficient FS techniques based on evolutionary methodologies such as Firefly Algorithm (FA), Opposition-Based Firefly Algorithm (OBFA) and OBFA melded with Information Gain (IG-OBFA) are presented in detail. The first two methods employ the wrapper method and the third one employs both filter and wrapper methods. These FS techniques are used to find an optimal feature subset and then combined with SVMs in order to enhance the accuracy of classification done by SVM. The CTG data that is widely used by researchers (UCI Machine Learning Repository, 2019) has been used for the experiments.

Various performance measures such as accuracy, sensitivity, specificity, positive predictive value, negative predictive value, geometric mean, F1-measure, and area under ROC have been evaluated to assess the performance of these methods. Figure 17.1 shows the classification of classifiers used for classification of CTG data set.

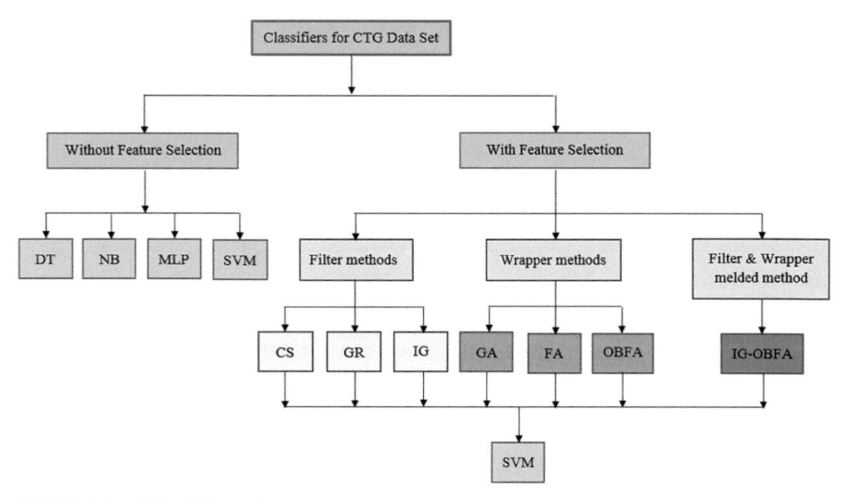

FIGURE 17.1 Classifiers for CTG data set.

Most of the existing attribute reduction techniques exploit the benefits of soft computing techniques such as ANNs, fuzzy logic, genetic algorithm (GA), and the combination of these such as neuro-fuzzy, genetic-neuro, etc.

Among the techniques employing ANNs, neuro-fuzzy system to recognize the accelerative and decelerative patterns of fetal heart rate signal (Romero et al., 2002), neural-network-based classifier with eliminated potential outliers (Chitradevi et al., 2013; Tang et al., 2018), machine learning technique (Sahin and Subasi, 2015), modular neural network models (Jadhav et al., 2011), models using discriminant analysis, decision tree (DT) and ANN (Huang and Hsu, 2012), particle swarm optimization (PSO) and GA-aided BPNNs (Hongbiao and Genwang, 2012), optimal neural network (ONN) (Bryan at al., 2012), ANN-based clustering adaptive resonance theory 2 (ART2) with fuzzy decision trees (Ping et al., 2012) and adaptive neuro fuzzy inference system (ANFIS) (Hasan and Ertunc, 2013) are found to be appreciable in certain aspects of classification performance.

Further, the methods based on genetic algorithms such as CDSS using Improved adaptive genetic algorithm (IAGA) and extreme learning machine (ELM) (Sindhu at al., 2015), SVM classifier with GA (Hasan, 2013; Subha et al., 2015), intelligent heart disease decision support system (Ratnakar et al., 2013), enhanced heart disease prediction system using GA (Anbarasi at al., 2010), decision support system using SVM and integer-coded genetic algorithm (ICGA) (Bhatia et al., 2008), GA and ANNA-based algorithm (ElAlami, 2009), two stage optimisation using GA (Huang et al., 2007), GA with linear and nonlinear Great Deluge algorithm (Jaddi and Abdullah, 2013), Mining and FS using GA (Sikora and Piramuthu, 2007), *K*-nearest neighbor (KNN) with GA algorithm (Deekshatulu and Chandra, 2013), differential evolution (DE) with GA (Bharathi and Subashini, 2014), Wrapper method using GA and SVM (Zhuo et al., 2008), integer and binary coded GA-based SVMs (Nithya et al., 2013; DİKER et al., 2018), intrusion detection systems using GA (Aziz et al., 2013), Correlation-based GA method (Tiwari and Singh, 2010) and ANN trained by back propagation algorithm combined with GA (Venkatesan and Premalatha, 2012) have contributed remarkable performance improvement.

The contributions to improve the performance of FS for CTG classification to predict the fetal well being using other optimization techniques include ant colony optimization (ACO) technique with SVM (Abd-Alsabour and Randall, 2010; Al-Ani, 2005), complementary particle swarm optimization algorithm (Chuang et al., 2013), Gray-Wolf Optimization technique (Emary et al., 2015), Complementary binary particle swarm optimization algorithm (Chuang et al., 2013), multi-objective algorithms (Xue et al., 2012a), multiobjective binary PSO (Xue et al., 2012b), combination of ACO and GA for SVM (Imani et al., 2012), bat algorithm and optimum-path forest (Rodrigues et al., 2014), hybridized PSO, PSO-based relative reduct and PSO-based Quick Reduct (Inbarani et al., 2014), artificial bee colony (ABC) (Schiezaro and Pedrini, 2013), bat algorithm for attribute reduction (Taha and Tang, 2013), combination of binary bat algorithm with the Optimum-Path Forest classifier (Nakamura et al., 2012), binary cuckoo search (Rodrigues et al., 2013), binary particle swarm optimisation (Xue et al., 2012c), modified multi-swarm PSO (Liu et al., 2011), FA with rough set theory (Banati and Bajaj, 2011) and modified FA optimization (Emary et al., 2015).

In addition to the above-mentioned methods, which use the soft computing techniques, there are the other methodologies with remarkable contribution to the improvement of classification performance namely information gain and adaptive models , evolutionary neural network FS algorithm (Chudacek et al., 2008), adaptive boosting decision trees and machine learning algorithm (Karabulut and Ibrikci, 2014), knowledge discovery in databases in machine learning, statistics and databases (Fayyad et al., 1996), Wrapper around random forest classification feature selection algorithm using Borutapackage (Kursa and Rudnicki, 2010), Mutual information-based greedy feature selection (Hoque et al., 2014), emprical mode decomposition-based approach (Krupa et al., 2011; CÖMERT et al., 2018a), Combined system identifcation and meashine learning methods (Warrick et al., 2010), Random Forest, REPTree and linear discrimination analysis-based algorithms (Tomáš et al., 2013), FS methods for naïve Bayes (NB) classifier (Menai et al., 2013), continuous CTG monitoring using electronic fetal monitoring (Alfirevic et al., 2006), clustering, detection of outlier and classification, by random tree and Quinlan's C4.5 algorithm (Jacob and Ramani, 2012), supervised ANN (Sundar et al., 2013), medical decision support systems using NB, machine learning (Sontakke et al, 2019), multilayer perceptron and C4.5 (Aftarczuk, 2007), fetal heart rate (FHR) baseline estimation algorithm (Nidhal et al., 2010), algorithm using memory-less fading statistics (Rodrigues et al., 2011), non-stress test-based algorithm (Ergun et al., 2012), Data mining techniques for heart disease prediction (Bhatla and Jyoti, 2012), SVM for comparative genomic hybridized data (Liu et al., 2008), Two stage decision support system diabetes disease (Ambica et al., 2013), Chi Square and T-Tests (Jeyachidra and Punithavalli, 2013), Fisher Ratio and Mutual Information-based method (Vidyavathi and Ravikumar, 2008), separability index matrix (Han et al., 2013), margin-based feature selection (Bachrach et al., 2004), multiobjective genetic algorithm (MOGA) and multiobjective version of forward sequential selection (Pappa et al., 2002), dynamic mutual information-based algorithm (Liu et al., 2009), multiclass SVM using GA (Agarwal and Bala, 2007), association rule mining with decision tree algorithm (Rajendran and Madheswaran, 2010), cAnt-Miner2 and Max-Relevance-based algorithm and

Min-redundancy feature selection algorithms (Michelakos et al., 2010), Levenberg–Marquardt algorithm (Buck and Zhang, 2006), Wavelet transform (CÖMERT and Kocamaz, 2019), prognostic model (CÖMERT et al., 2018c) and hybrid data mining model (Ha and Joo, 2010).

17.2 CARDIOTOCOGRAPHY (CTG)

To monitor the UC and FHR and thereby the condition of the fetus such as fetal hypoxia, cardiotocography (CTG) is used (Patient.info, 2019). It is obtained from the abdomen of pregnant woman using two external probes of cardiotocograph instrument during the last trimester and also at intrapartum. Interpretation of CTG signal is done based on the parameters of the structure "DR C BRaVADO" which is listed in Table 17.1 (Geeky.medics, 2019). Figure 17.2 shows the typical CTG recorded signal containing FHR and UC (Healthnetconnections, 2019).

TABLE 17.1 Parameters of CTG Signal

Parameter	Description
DR	Define risk
C	Contractions
BRa	Baseline rate
V	Variability
A	Accelerations
D	Decelerations
O	Overall impression

(*Source:* Adapted from Geeky.medics, 2019.)

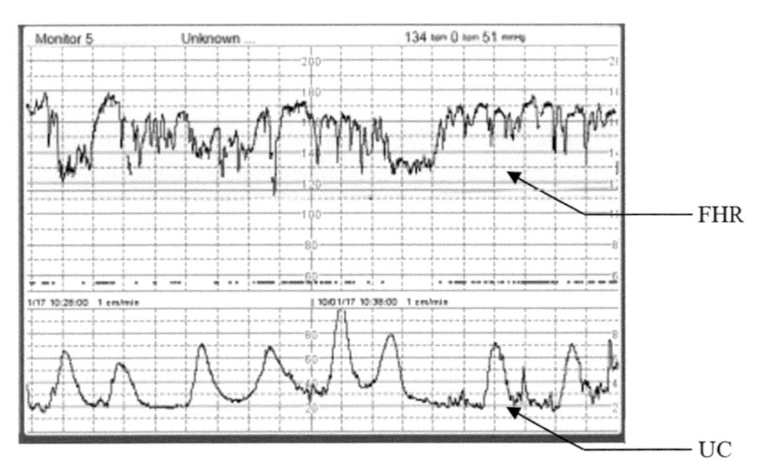

FIGURE 17.2 Example of a CTG recorded signal.

(*Source:* Adapted from Healthnetconnections, 2019.)

Sometimes, internal monitoring of CTG is done when the external monitoring is difficult or not possible. During internal CTG monitoring, the probe is inserted in to the mother's womb to touch the scalp of the fetus to get the FHR. The UC is measured using Catheter, a flexible tube inserted in to the uterus.

17.2.1 UCI CTG DATASET

The Centre for Machine Learning and Intelligent Systems, Bren School of Information and Computer Science, University of California at Irvine, USA maintains a free to access data bank named UCI Machine Learning Repository (UCI Machine Learning Repository, 2019). The CTG data available in this data bank is one of the widely referred data sets. There are 2126 fetal CTG recordings classified into three classes namely Normal (N), Suspect (S), and Pathologic (P). There are 21 attributes for each CTG with 1 attribute to represent the class. More details of these attributes and classes can be found in (UCI Machine Learning Repository, 2019).

17.3 CLASSIFIERS AND PERFORMANCE

Classifiers are the models used to classify the given input data in to a class, by fitting the data with the label of class based on the relationship between the atributes and label. Some of the popular classifiers are SVM, NB classifiers, decision tree classifiers, rule-based classifiers, neural networks, etc. Usually, the two stages of classification process are, learning the model from the training data set which has class labels and applying the model to the test set. The table containing the results of classification is called as "Confusuion matrix." The confusion matrix shows the number of times the classes are predicted correctly and wrongly, which gives the accuracy of classification. From the confusion matrix, there are a number of performance measures such as accuracy, error rate, sensitivity, specificity, negative predictive value, geometric mean, precision (or) positive predictive value, *F*-measure, area under ROC, etc., are evaluated in order to specify the effectiveness of the classifier. Following are the expressions for these performance metrics measured from the confusion matrix given in Table 17.2.

TABLE 17.2 Confusion Matrix

		Predicted	
		Class A	***Class B***
ACTUAL	***Class A***	TRUE POSITIVE (TP)	FALSE NEGATIVE (FN)
	Class B	FALSE POSITIVE (FP)	TRUE NEGATIVE (TN)

$$\text{Accuracy} = \left[\frac{TP+TN}{TP+TN+FP+FN}\right] = \frac{\text{Number of correct predictions}}{\text{Total number of predictions}} \tag{17.1}$$

$$\text{Error rate} = \left[\frac{FP + FN}{TP + TN + FP + FN}\right] = \frac{\text{Number of wrong predictions}}{\text{Total number of predictions}} \quad (17.2)$$

$$\text{Negative predictive value: } NPV = \left[\frac{TN}{TN + FN}\right] \quad (17.3)$$

$$\text{Precision (or) positive predictive value: } PPV = \left[\frac{TP}{TP + FP}\right] \quad (17.4)$$

$$\text{Sensitivity (or) Recall} = \left[\frac{TP}{TP + FN}\right] \quad (17.5)$$

$$\text{Specificity} = \left[\frac{TN}{TN + FP}\right] \quad (17.6)$$

$$\text{Geometric mean: } Gmean = \sqrt{\text{specificity} \times \text{sensitivity}} \quad (17.7)$$

$$\text{F-measure} = 2 \times \left[\frac{\text{precision} \times \text{sensitivity}}{\text{precision} + \text{sensitivity}}\right] \quad (17.8)$$

$$\text{Area under ROC} = \frac{\text{Sensitivity} + \text{Specificity}}{2} \quad (17.9)$$

17.4 FEATURE SELECTION

The FS or variable selection or attribute selection, eliminates the features which are not relevant, unnecessary and containing noise and as a result creates a reduced feature subset. The reduced size of the feature subset helps to improve the performance of classification. FS is done either by scalar selection or by vector selection (Dua and Du, 2011). Since the scalar selection invloves selection of features individually, it is simple but least efficient method. On the other hand, the vector selection establishes a relationship across the features based on a wrapper or a filter and hence it is relative complex and efficient. The filter-based selection of features is done based on the statsitical properties of the features (John et al., 1994). Since, it does not employ any learning algorithms for this task, it is very simple and fast. However, because it does not consider the dependencies of any feature with other features, it may result in a poor classification. The wrapper-based feature selection is done using repeated learnings which looks for the optimal or near optimal subset then followed by cross-validation. As a result, it is more computationally expense but more accurate too.

Methods like information gain, chi-square test, Fisher score, correlation coefficient and variance threshold, gain ratio attribute evaluator (Kantardzic, 2013), information gain attribute evaluator (Liu and Motoda, 2008), etc., are the popular filter-based methods. Genetic algorithms, multiclass SVM classifiers (Ahuja and Yadav, 2012), recursive elimination, sequential selection algorithms and, etc., are the examples for wrapper-based selection methods.

17.5 FEATURE SELECTION USING FIREFLY ALGORITHM

17.5.1 FIREFLY ALGORITHM

Artificial intelligence (AI) techniques based on the behavior of swarms such as bees, birds, fishes, ants, etc., are called as swarm intelligence (SI). These swarms possess an organized behavior due to the interactivity among the individuals. One of the latest SI methods is based on the behavior of Fireflies, called as FA. It was introduced by Yang (Yang, 2010) as meta-heuristic and stochastic algorithm to solve the optimization problems. The flashing behavior of fireflies is considered as randomization for searching the optimum solutions. The FA has an advantage that it prevents the searching processes from being confined in a local optima and the candidate solution is improved by the local search process untill the algorithm attains the optimum solution even in fastly changing and noisy environments. It is found that FA, its modified versions, and hybridized versions are widely used in varieties of single and multiobjective engineering optimization problems (Yang, 2010; Fister at al., 2013; Manivanna Boopathi and Abudhahir, 2015; Mohamed Ali et al., 2018). It is also found that the FAs are efficiently employed for the problems of classification (Subha and Murugan, 2014).

The fireflies emit a flashing light from their body in an orderly manner (Yang, 2010), which is the result of a bioluminscent reaction which happens in their body. They are capable of producing the light as a high intesity and disjunct flashes. These beetles use the flashing light to attract their partners for copulation and also as a cautioning message to other fireflies when required. It is an interesting fact that upon seeing the flashing light from the male firefly, the female partner generates a response light flash comprising information about its identity and gender.

When the distance from where the flashing light of a firefly is seen decreases the the intensity of light being seen increases and vice versa. Hence, the attractiveness is directly proportional to the intensity. The female fireflies are more attracted by the male firefly's light of more intensity. However, it is to be noted that the female fireflies are not capable of differentiating the larger intensity of light from longer distance and smaller intensity from shorter distance. In addition to the above facts being considered for developing the Firefly Algorithm, three facts are assumed; (i) All fireflies are genderless, (ii) The light intensity of each firefly has the direct relationship with its attractiveness, (iii) The nature of fitness function has the influence in the light intensity of a firefly.

The overall idea of FA is on the relationship that the light intensity (I) is in square relationship with the distance (x). It means that the light is seen much brighter than the actual intensity at the source (I_s), when the distance from which it is seen decreases. This relationship can be mathematically written as

$$I(x) = \frac{I_s}{x^2} \tag{17.10}$$

Hence, the fitness or objective function of FA is evaluated in such a way that the solutions are represented by the light intensity of each firefly which are directly proportional to the value of the fitness or objective function.

In FA, a random initial population is initialized with the defined values of parameters such as randomization parameter (r), attractiveness (a) and coefficient of absorption (c). With these arrangements, the solutions are searched by determining the fitness values continuously for the given number of iterations.

The light intensity of a firefly seen by another firefly in a medium varies with distance (x) as

$$I = I_s e^{-cx} \tag{17.11}$$

where the I_s is the intensity of light at source.

Hence, it can be written from the above two equations of light intensity that

$$I(x) = I_s e^{-cx2} \tag{17.12}$$

However, the attractivness equation relating it with the light intensity can be written using the attractiveness at zero distance (a_s) as

$$a = a_s e^{-cx^2} \tag{17.13}$$

The Euclidean distance between two fireflies namely p_y and p_z can be written by representing the mth component of spatial coordinate, p_y as $p_{y,m}$ and p_z as $p_{z,m}$;

$$x_{yz} = \left\| p_y - p_z \right\| = \sqrt{\sum_{m=1}^{n} \left(p_{y,m} - p_{z,m} \right)^2} \tag{17.14}$$

The attraction of firefly y by another firefly z can be written using ψ_y as a vector containing Gaussion ditributed random numbers in the space [0,1] as

$$p_y = p_y + a_s e^{-cx_{yz}^2} \left(p_z - p_y \right) + r\psi_y \tag{17.15}$$

The present state of yth firefly, attraction of yth firefly by another firefly and the movement of yth firefly in a random manner are the three factors which are considered to represent the firefly's movement.

Therefore, these three parameters have to be adjusted in order to improve the performance of FA. The whole sequence of FA for finding the optimum solution in the given number of iterations is presented in the form of pseudocode in Figure 17.3 (Manivanna Boopathi and Abudhahir, 2015; Mohamed Ali et al., 2018; Subha and Murugan, 2014).

```
Begin
Define a, r and c
Objective function f(p): f(p1,p2,..., px )^T
Initial population of fireflies pi(i=1, 2, .... n)
Evaluate Intensity of light Ii for pi using f(pi)
g=1;
while (g< Gmax)
do
 for i=1 : n all n fireflies do
   for j=1 : i all n fireflies do
if (Ij > Ii) then
  move firefly i towards j using equation 5.6
  end
if (a varies with x as in equation 5.3) then
Find new solutions and update I
end for j
end for i
find current best firefly
g++
end while
end begin
```

FIGURE 17.3 Firefly algorithm — pseudocode

17.5.2 FEATURE SUBSET SELECTION USING FIREFLY ALGORITHM

In order to improve the classification accuracy and other performance measures, FS is employed. The process of FS results in a feature subset containing *g* number of features from the whole feature set containing *h* number of features. As mentioned earlier, the FS methods can be based on either wrapper or filter techniques.

Feature subsets are found using various optimization techniques such as ABC (Uzer et al., 2013; Schiezaro and Pedrini, 2013), ACO, PSO, etc. The feature selection method using SVM with FA is presented here which results in a better classification performance compared to the above mentioned methods.

All features present in the UCI CTG data set are represented either by 0 or by 1 to represent the presence or absence of a particular feature, respectively. For the FA to find the optimum feature subset, the objective function is taken as the accuracy of classification of SVM.

The FA is developed and run with the initial population of 25 for 100 iterations which is set as the stopping criteria for the algorithm. The objective function of FA is to maximize the light intesity of firefly, which in turn maximizes the classification accuracy. The other parameters of FA such as randomization (r), attractiveness (a), and coefficient of absorption (c) are selected as 0.5, 0.2 and 1, respectively. At the end of 100 iterations, an optimal feature subset is found by FA which improves the accuracy of classification of SVM. Totally 25 trials were done using FA and the best results of these 25 runs are presented.

17.5.3 RESULTS OF SIMULATION EXPERIMENTS

The results of simulation experiments performed in MATLAB with both data sets with the actual feature set without using FA-based FS and reduced feature set by using FA-based FS are presnted in Table 17.3.

To substatiate the better performance of SVM with FA-based FS, other performance measures namely Sensitivity, Specificity, Positive Predictive Value and Negative Predictive Value are also evaluated and presented in Table 17.4.

TABLE 17.3 Average Accuracy of SVM With and Without FA-based FS

Class	Accuracy (%)	
	Actual Full Feature Set (Without FS)	**Reduced Optimal Feature Set (With FA-based FS)**
Normal	94.44	95.64
Suspect	66.77	77.62
Pathologic	72.15	81.25
Average	88.75	91.92

TABLE 17.4 Other Performance Metrics of SVM with and Without FA-based FS

Performance Metrics (%)	Without FS	With FA-based FS
Sensitivity	77.79	84.83
Specificity	90.22	93.78
PPV	78.29	83.14
NPV	90.70	93.26

TABLE 17.4 *(Continued)*

Performance Metrics (%)	Without FS	With FA-based FS
G-Mean	83.77	89.19
F-Measure	78.08	83.94
Area under ROC	84.00	89.30

Tables 17.3 and 17.4 show that the FA-based FS with SVM exhibits an appreciable improvement of performance in all aspects. The measures of these performance metrics are also presented graphically in Figures 17.4 and 17.5.

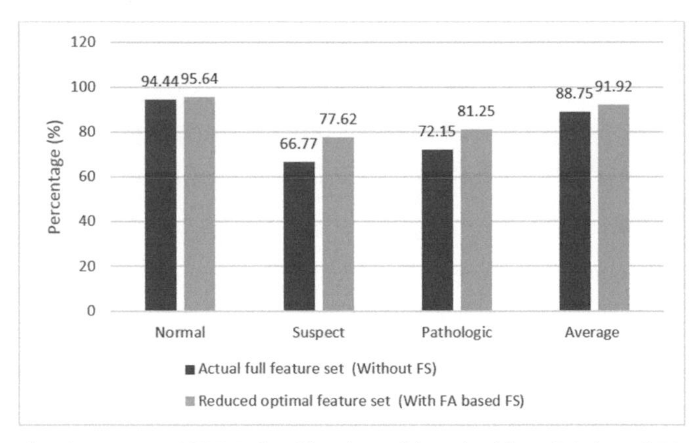

FIGURE 17.4 Accuracy of SVM classification with and without FA-based FS.

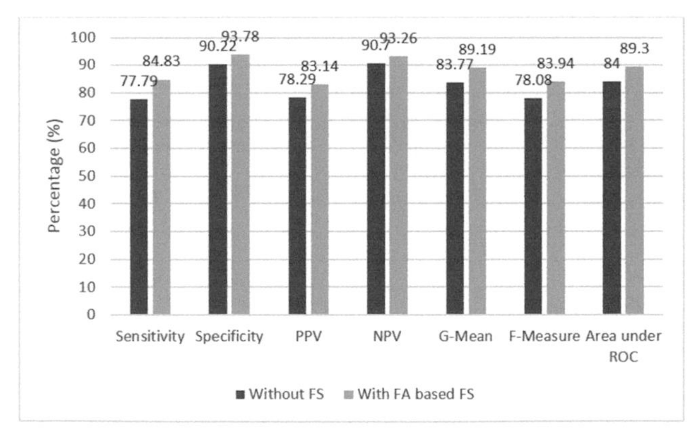

FIGURE 17.5 Other performance metrics of SVM with and without FA-based FS.

The percentage increases in the performance of SVM using FA-based FS for each of the performance matrics are consolidated in Table 17.5 and graphically presented in Figure 17.6.

TABLE 17.5 Percentage Increase in Performance of SVM using FA-based FS

Sl. No.	Performance Metric	Improvement (%)
1.	Average accuracy	3.57
2.	Sensitivity	9.05
3.	Specificity	3.95
4.	PPV	6.19
5.	NPV	2.82
6.	*G*-Mean	6.47
7.	*F*-Measure	7.51
8.	Area under ROC	6.31

From the results shown above, it is clear that using FA-based FS has improved the performance of SVM classifier by minimum of 2.82% increase in NPV and maximum of 9.05 % increase in sensitivity.

17.6 FEATURE SELECTION USING OPPOSITION-BASED FIREFLY ALGORITHM

17.6.1 OPPOSITION-BASED FIREFLY ALGORITHM

It is encouraging that the performance of simple FA is much better than the other evolutionary algorithms. However, it is still a challenge in FA to prevent the premature convergence before reaching the optimum solution.

To get rid of this challenge and expedite the convergence, the FA has been modified with added features. The popular modified FAs are Fuzzy FA, Lévy-flight FA, Jumper FA, chaotic FA and self-adaptive step FA (Uzer et al., 2013). Another efficient modified FA is named opposition-based FA which uses opposition-based learning (OBL) (Schiezaro and Pedrini, 2013; Draa et al., 2015; Subha and Murugan, 2016; Tizhoosh, 2006; Xu et al., 2011; Yu et al., 2015). The other optimization algorithms such as GA, ACO, PSO, bio-geography optimization,

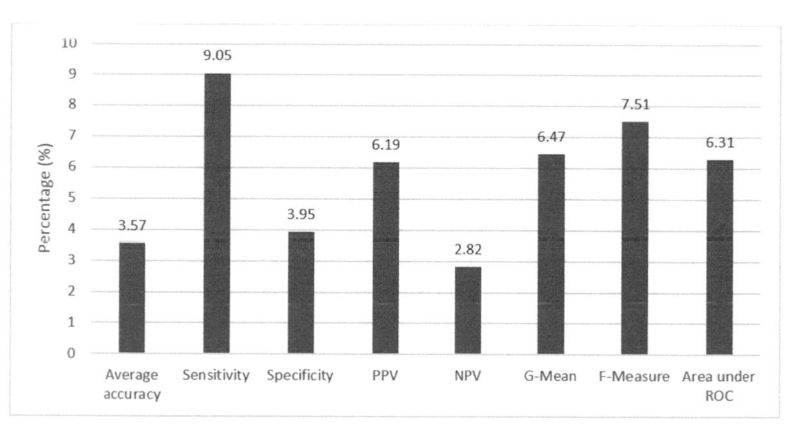

FIGURE 17.6 Percentage increase in performance of SVM using FA-based FS.

differential evolution algorithm, gravitational search algorithm, and simulated annealing have also been modified using OBL for improving their performance.

The OBL is developed by taking both a solution and its opposite for consideration. In simple words, OBL-based FA uses the opposite of a worst firefly as a new firefly to replace the worst firefly. This process expels the worst firefly from its actual path so that it go out from the local optima.

17.6.2 OPPOSITION-BASED LEARNING

As mentioned earlier, the OBL simultaneously uses a solution and also its equivalent and opposite solution.

For instance, if a real number is denoted as $a \in [b, c]$, then its opposite number can be denoted as $a` = b + c - a$. Extending this idea to large dimension can be as follows. For a vector $A(a_1, a_2, \ldots\ldots a_n)$ of dimension $n$ such that $a_i \in [b_i, c_i]$ with $i = 1, 2, \ldots, n$, the opposite vector can be obtained as $A`(a`_1, a`_2, \ldots, a`_n)$ with $a`_i = b_i + c_i - a_i$.

In the OBL-based FA developed for Feature Selection, the OBL has been used in during population initialization and creating new generations as two stages. In the OBFA firstly an initial population of size n is created with number of their opposite positions resulting a total of $2n$ fireflies. From this $2n$ fireflies n number of fittest fireflies are identified based on their fitness values. The overall OBFA sequence is given as pseudocode in Figure 17.7.

The OBL is used to update the positions of fireflies in every iteration by replacing the fireflies with worst fitness by their opposite ones. During the initial stage of optimization, the number of worst fireflies (w) is kept large in order to perform a productive search globally. However, as the number of iterations increases, the number of worst fireflies (w) being considered is gradually reduced to ensure the local exploitation. The way in which the number of worst fireflies (w) are considered as the iteration progresses is given by an equation using g as the present generation, G_{max} as the maximum limit of generations and the function Round ( ) to round the number to its nearest integer.

$$w = \text{Round}\left(\frac{0.33n(G_{max} - g)}{G_{max}}\right) \quad (17.16)$$

17.6.3 FEATURE SUBSET SELECTION USING OPPOSITION-BASED FIREFLY ALGORITHM

To perform feature selection, the OBFA is used with the SVM classifier on the UCI CTG data set. The

```
Begin
Define a, r and c
Objective function f(p): f(p1,p2, ..., px)^T
Initial population of fireflies pi(i=1, 2, .... n)
Opposite population of fireflies po-i(i=1, 2, .... n)
Get fittest population from pi and po-i and set as
initial population
Evaluate Intensity of light Ii for pi using f(pi)
g=1;
while (g< Gmax)
do
  for i=1 : n all n fireflies do
    for j=1 : i all n fireflies do
if (Ij > Ii) then
  move firefly i towards j using equation 5.6
  end
if (a varies with x as in equation 5.3) then
Find new solutions and update I
Replace the worst fireflies found using equation
(6.1) by their opposite ones
end for j
end for i
find current best firefly
g++
end while
end begin
```

FIGURE 17.7 Opposition-based firefly algorithm—pseudocode.

presence and absence of a feature in the data set is represented by 1 and 0, respectively. The data set is divided into two parts in such a way that the three-fourth of the data set is used for training the classifier and the remaining one-fourth is used to test it. The FA parameters such as objective function, initial population, randomization, attractiveness, coefficient of absorption, and number of iterations, are taken as same as that of the standard FA used earlier. Also, 25 trials of simulations were performed using OBFA and the best of these 25 runs are presented.

17.6.4 RESULTS OF SIMULATION EXPERIMENTS

Two SVM classification experiments are performed using the full data set and reduced data set by OBFA. The accuracy of these two classifications are presented in Table 17.6. Further, the other measured performance metrics for all these three classifiers are consolidated in Table 17.7.

TABLE 17.6 Average Accuracy of SVM With and Without OBFA-based FS

Data set	Average accuracy (%)
Actual full feature set (without FS)	88.75
Reduced optimal feature set (with OBFA-based FS)	92.85

TABLE 17.7 Other Performance Metrics of SVM With and Without OBFA-based FS

Performance Metrics (%)	Without FS	With OBFA-based FS
Sensitivity	77.79	83.81
Specificity	90.22	93.72
PPV	78.29	85.45
NPV	90.70	95.02
G-mean	83.77	88.62
F-measure	78.08	84.62
Area under ROC	84.00	88.76

It is found that the average accuracy is 88.75% with full feature set and the same is achieved as 91.92% with optimal feature set produced by FA and as 92.85% with optimal feature set produced by OBFA.

Figures 17.8 and 17.9 are the graphical presentations of the results of OBFA-based SVM classifier for UCI CTG data set.

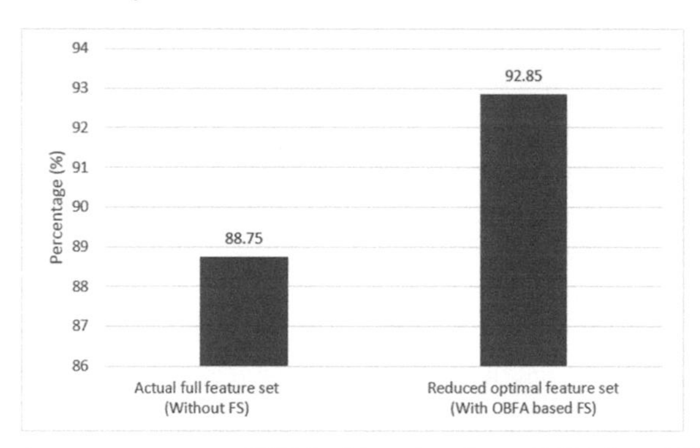

FIGURE 17.8 Average accuracy of SVM with and without OBFA-based FS.

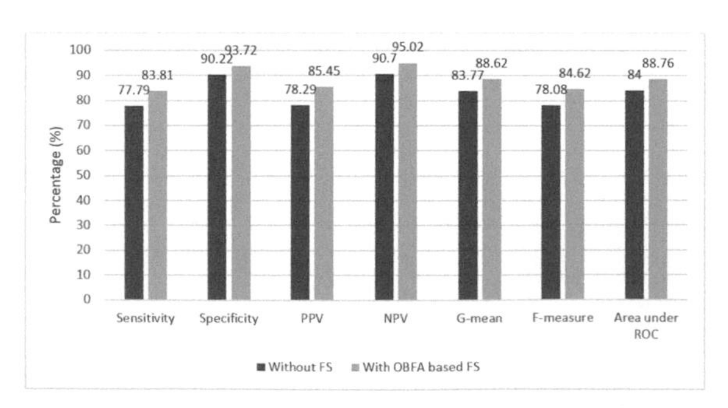

FIGURE 17.9 Other performance metrics of SVM with and without OBFA-based FS.

As presented in FA-based FS, the increase in performance of OBFA-based SVM is given in terms of the performance metrics, in Table 17.8 and Figure 17.10.

TABLE 17.8 Percentage Increase in Performance of SVM Using OBFA-based FS

Sl. No.	Performance Metric	Improvement (%)
1.	Average accuracy	4.62
2.	Sensitivity	7.74
3.	Specificity	3.88
4.	PPV	9.15
5.	NPV	4.76
6.	*G*-Mean	5.79
7.	*F*-Measure	8.38
8.	Area under ROC	5.67

The OBFA-based FS has resulted a maximum of 8.8% increase in PPV and the minimum of 3.9% increase in specificity. Hence, the OBFA performs well on feature selection.

17.7 FEATURE SELECTION USING OPPOSITION-BASED FIREFLY ALGORITHM MELDED WITH INFORMATION GAIN

17.7.1 IG-OBFA-BASED FEATURE SELECTION

It is always important that any feature in a data set which contains useful information about it should not be ignored during feature selection process. Removing such an apposite feature will lead to poor classification and thereby poor prediction too. Hence, it is a good practice to employ some techniques to evaluate the relevancy of each feature to the data set before ignoring it for feature reduction (Zhang et al., 2018). One of the successful filter-based techniques for assessing the relevance of

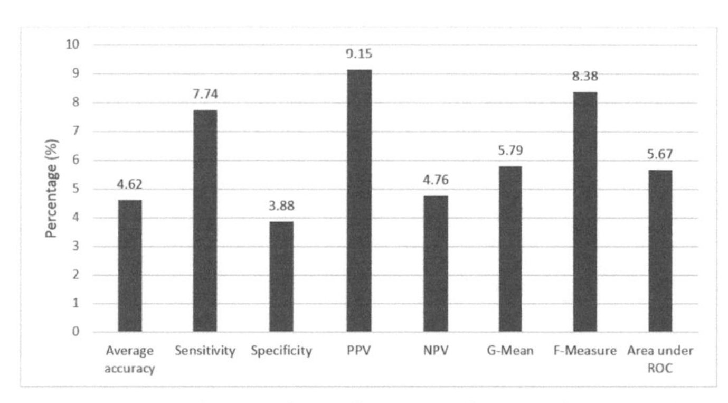

FIGURE 17.10 Percentage increase in performance of SVM using OBFA-based FS.

feature to its associated data set is IG (Sui, 2013; Azhagusundari and Thanamani, 2013; Mitchell, 1997; Porkodi, 2014; Subha et al., 2017).

In order to improve the classification performance of OBFA-based SVM classifier further, a new melded method is presented here which employs IG with the OBFA for SVM classifier to classify the UCI CTG data set.

In the IG-OBFA-based feature selection process, the IG of all features in the data set are determined and these features are arranged in descending order based on their IG values. Then, the top 15 features are taken as the reduced feature set and presented as initial population for the OBFA with 1's and 0's representing the presence and absence of a feature in the data set, respectively, in order to produce the optimum feature set. As performed in FA and OBFA, 25 trials were done using IG-OBFA too and the best of the results are presented here.

17.7.2 RESULTS OF SIMULATION EXPERIMENTS

As done in the previous experiments, the training and testing data are selected as 75:25 ratio of the full data set. Classification experiments are performed with the full feature set, reduced feature set produced by IG only and the reduced optimum feature set produced by IG-OBFA and the results are peresented below.

TABLE 17.9 Average Accuracy of SVM Without FS, with IG and with IG-OBFA-based FS

Data Set	Average Accuracy (%)
Actual full feature set (without FS)	88.75
Reduced feature set (with IG-based FS)	89.47
Reduced feature set (with IG-OBFA-based FS)	96.24

Table 17.9 shows that using IG alone for feature selection has slightly increased the average accuracy than using the full feature set. However, there has been a great improvement in average accuracy from 88.75% to 96.24% when using IG-OBFA-based FS instead of using a full feature set. The other performance metrics such as Specificity, Sensitivity, PPV, NPV, *G*-mean, *F*-measure and area under ROC are also measured for these classifications and presented in Table 17.10.

Figures 17.11 and 17.12 are the graphical presentations of the results of IG and IG-OBFA-based SVM classifiers for the classification of UCI CTG data set.

TABLE 17.10 Other Performance Metrics of SVM Without FS, with IG and with IG-OBFA-based FS

Performance Metrics (%)	Without FS	With IG-based FS	With IG-OBFA-based FS
Sensitivity	77.79	81.07	96.26
Specificity	90.22	91.14	91.92
PPV	78.29	78.48	93.33
NPV	90.70	91.29	97.44
G-mean	83.77	85.96	94.06
F-measure	78.08	79.75	92.61
Area under ROC	84.00	86.11	94.09

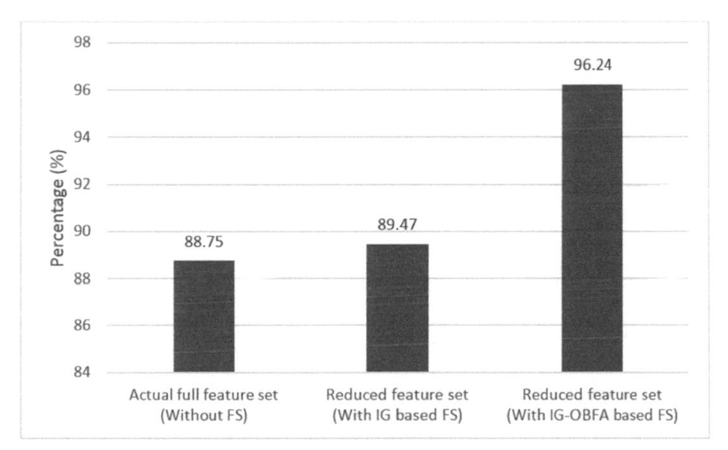

FIGURE 17.11 Average accuracy of SVM without FS, with IG and with IG-OBFA-based FS.

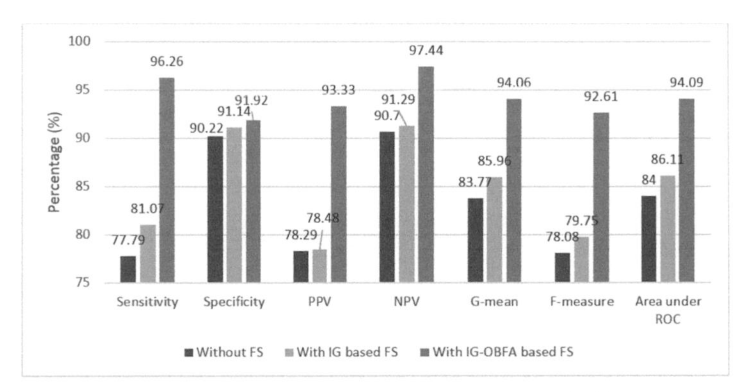

FIGURE 17.12 Other performance metrics of SVM without FS, with IG and with IG-OBFA-based FS.

TABLE 17.11 Performance Measures of Classification Using Various Feasture Selection Methods for UCI CTG Data Set

Performance Metrics (%)	Without FS				With FS						
					Filter Techniques			Wrapper Techniques			
	DT	MLP	NB	SVM	Chi-Squared	Gain Ratio	IG	GA	FA	OBFA	IG-OBFA
Average Accuracy	88.15	83.45	79.69	88.75	87.40	86.46	89.47	91.35	91.92	92.85	96.24
Sensitivity	78.60	72.21	69.74	77.79	74.77	74.00	81.07	80.71	84.83	83.81	91.92
Specificity	90.99	90.12	86.83	90.22	89.33	89.2	91.14	92.50	93.78	93.72	96.26
PPV	75.94	71.60	62.38	78.29	74.92	72.44	78.48	83.06	83.14	85.45	93.33
NPV	90.09	86.43	83.11	90.70	89.78	88.89	91.29	93.77	93.26	95.02	97.44
G-Mean	84.12	79.93	77.34	83.77	80.90	80.46	85.96	85.92	89.19	88.62	94.06
F-Measure	77.14	71.45	65.28	78.08	74.78	73.03	79.75	81.87	83.94	84.62	92.61
Area under ROC	84.79	81.17	78.29	84.00	82.05	81.60	86.11	86.61	89.30	88.76	94.09

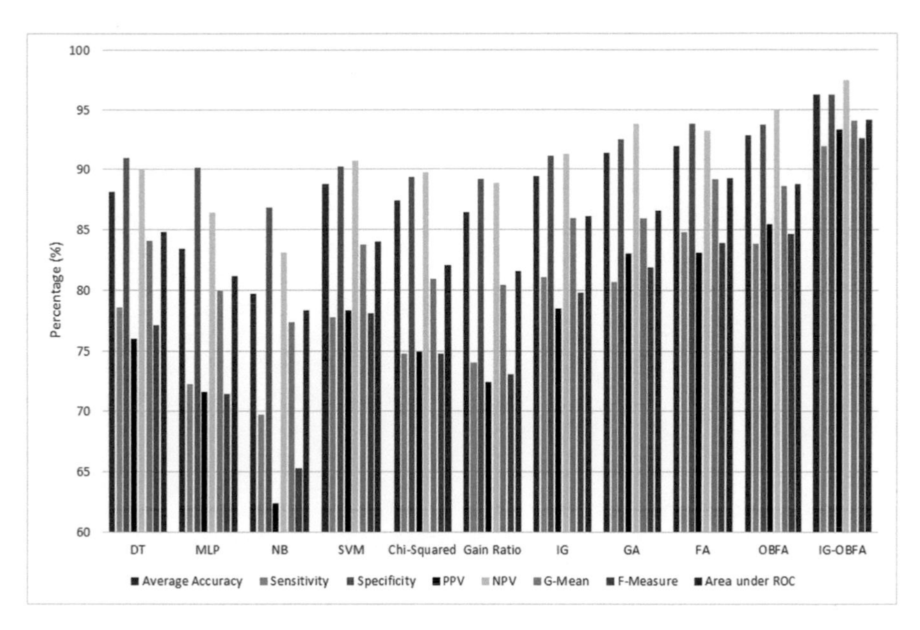

FIGURE 17.13 Performance measures of classification using various feature selection methods for UCI CTG data.

17.8 CONCLUSION

This chapter has briefly presented the feature selection techniques to improve the performance of SVM classification of UCI CTG data set. By performing feature selection, the size of the feature set is reduced and therby the task of classification is made less computationally expensive and more efficient. Optimization techniques such as FA, OBFA, and IG-OBFA have been used to find the optimum feature set containing only the most infulential features of the whole feature set. These three feature selection techniques and their performance when combined with SVM classifier were presented in this chapter.

The overall comparison of performances of various classification methods have been experimented. These methods include classifiers without using any feature selection methods such as DT, multilayer perceptron (MLP), NB, and SVM, classifiers using filter techniques for feature selection such as chi-squared, gain ratio and IG methods and classifiers using wrapper techniques such as GA, FA, OBFA, and IG-OBFA methods. Various performance measures namely average accuracy, sensitivity, specificity, positive predictive value (PPV), negative predictive value (NPV), *G*-mean, *F*-measure and area under ROC are evaluated for all these classifiers. The results of all these classifications for UCI CTG data set are summarized in Table 17.11. This table shows that the classification performance is highly improved when employing IG-OBFA.

KEYWORDS

- **cardiotocogram (CTG)**
- **feature selection**
- **classification**
- **performance metrics**
- **fetal heart rate**
- **uterine contraction**
- **clinical decision support system**
- **firefly algorithm**
- **opposition-based learning**
- **information gain**
- **support vector machines**

REFERENCES

Abd-Alsabour, N.; Randall, M. Feature selection for classification using an ant colony system. *Sixth IEEE International Conference on e-Science Workshops*, **2010**, 86–91.

Aftarczuk, K. Evaluation of selected data mining algorithms implemented in Medical Decision Support Systems. *Master Thesis, Blekinge Institute of Technology,* Sweden, **2007**.

Agarwal, RK.; Bala, R. A hybrid approach for selection of relevant features for microarray datasets. *International Journal of Computer and Information Engineering.* **2007**, 1(2), 1319–1325.

Ahmed, Al-Ani. Feature subset selection using ant colony optimization. *International Journal of Computational Intelligence*. **2005**, 2(1), 53–58.

Ahuja, Y.; Yadav, SK. Multiclass classification and support vector machine. *Global Journal of Computer Science and Technology Interdisciplinary*, **2012**, 12(11), 14–20.

Alfirevic, Z.; Devane, D.; Gyte, GM.; Cuthbert, A. Continuous cardiotocography (CTG) as a form of electronic fetal monitoring (EFM) for fetal assessment during labour. *Cochrane Database Systematic Reviews*, **2006**, 2(2). DOI: 10.1002/14651858.CD006066.pub3

Ambica, A.; Gandi, S.; Kothalanka, A. An efficient expert system for diabetes by naive Bayesian classifier. *International Journal of Engineering Trends and Technology*. **2013**, 4(10), 4634–4639.

Anbarasi, M.; Anupriya, E.; Iyengar, N. Enhanced prediction of heart disease with feature subset selection using genetic algorithm. *International Journal of Engineering Science and Technology*. **2010**, 2(10), 5370–5376.

Arumugam, Manivanna Boopathi.; A, Abudhahir. Firefly algorithm tuned fuzzy set-point weighted PID controller for antilock braking systems. *Journal of Engineering Research*. **2015**, 3(2), 79–94.

Azhagusundari, B.; Thanamani, Antony Selvadoss. Feature selection based on information gain. *International Journal of Innovative Technology and Exploring Engineering*, **2013**, 2(2), 18–21.

Aziz, ASA.; Azar, AT.; Salama, MA.; Hassanien, AE.; Hanafy, SEO. Genetic algorithm with different feature selection techniques for anomaly detectors generation. *Federated Conference on*

Computer Science and Information Systems (FedCSIS), **2013**, 769–774.

Bachrach, RG.; Navot, A,; Tishby, N. Margin based feature selection-theory and algorithms. *Proceedings of the Twenty-First International Conference on Machine Learning*. **2004**, 43–51.

Banati, H.; Bajaj, M. Firefly based feature selection approach. *International Journal of Computer Science Issues*. **2011**, 8(4), 473–480.

Bharathi. PT; Subashini, P. Differential evolution and genetic algorithm based feature subset selection for recognition of river ice types. *Journal of Theoretical & Applied Information Technology*. **2014**, 67(1), 254–262.

Bhatia, S.; Prakash, P.; Pillai, GN. SVM based decision support system for heart disease classification with integer-coded genetic algorithm to select critical features. *Proceedings of the World Congress on Engineering and Computer Science (WCECS)*, **2008**, 22–24.

Bhatla, N.; Jyoti, K. An analysis of heart disease prediction using different data mining techniques. *International Journal of Engineering*. **2012**, 1(8), 1–4.

Buck, TE.; Zhang, B. SVM kernel optimization: An example in yeast protein subcellular localization prediction. *Project Report, School of Computer Science, Carnegie Mellon University,* Pittsburgh, USA, **2006**.

Chitradevi, Muthusamy.; Sundar, Chinnasamy.; Geetharamani, Gopal. An outlier based Bi-level neural network classification system for improved classification of cardiotocogram data. *Life Science Journal*, **2013**, 10(1), 244–251.

Chuang, LY.; Jhang, HF.; Yang, CH. Feature selection using complementary particle swarm optimization for DNA microarray data. *Proceedings of International Conference of Engineers and Computer Scientists,* Hong Kong, **2013**.

Chudáček, V.; Spilka, J.; Huptych, M.; Georgoulas, G.; Lhotská, L.; Stylios, C.; Koucky, M.; Janku, P. Linear and non-linear features for intrapartum cardiotocography evaluation. *Computing in Cardiology*, **2010**, 37, 999–1002.

Chudacek, V.; Spilka, J.; Rubackova, B.; Koucky, M.; Georgoulas, G.; Lhotska, L.; Stylios, C. Evaluation of feature subsets for classification of cardiotocographic recordings. *Computers in Cardiology*, **2008**, 845–848.

Cömert, Zafer.; Kocamaz, AF.; Subha, Velappan. Prognostic model based on image-based time-frequency features and genetic algorithm for fetal hypoxia assessment. *Computers in Biology and Medicine*, **2018**, 99, 85–97.

Cömert, Z.; Kocamaz, AF. Using wavelet transform for cardiotocography signals classification, *25th Signal Processing and Communications Applications Conference (SIU)*, Turkey, 2017

CÖMERT, Zafer.; Yang, Zhan.; Subha, Velappan.; Kocamaz, Adnan Fatih.; Manivanna Boopathi, Arumugam. Performance evaluation of empirical mode decomposition and discrete wavelet transform for computerized hypoxia detection and prediction. *26th IEEE Signal Processing and Communication Applications (SIU) Conference,* Turkey, 2018a.

CÖMERT, Zafer.; Yang, Zhan.; Subha, Velappan.; Kocamaz, Adnan Fatih.; Manivanna Boopathi, Arumugam. The influences of different window functions and lengths on image-based time-frequency features of fetal heart rate signals. *26th IEEE Signal Processing and Communication Applications (SIU) Conference,* Turkey, 2018b.

Deekshatulu, BL.; Chandra, P. Classification of heart disease using k-nearest neighbor and genetic algorithm. *Procedia Technology*. **2013**, 10, 85–94.

DİKER, Aykut.; CÖMERT, Zafer.; Subha, Velappan.; AVCI, Engin. Intelligent system based on genetic algorithm and support vector machine for detection of myocardial infarction from ECG signals. *26th IEEE Signal Processing and Communication Applications (SIU) Conference,* Turkey, 2018.

Draa, Amer.; Benayad, Zeyneb.; Djenna, Fatima Zahra. An opposition-based firefly algorithm for medical image contrast enhancement. *International Journal of Information and Communication Technology*. **2015**, 7(4/5), 385–405.

ElAlami, ME. A filter model for feature subset selection based on genetic algorithm. *Knowledge-Based Systems*. **2009**, 22(5), 356–362.

Emary, E.; Zawbaa, HM.; Ghany, KKA.; Hassanien, AE.; Parv, B. Firefly optimization algorithm for feature selection. *Proceedings of the 7th Balkan Conference on Informatics Conference*, **2015**, 26.

Emary, E.; Zawbaa, HM.; Grosan, C.; Hassenian, AE. Feature subset selection approach by gray-wolf optimization. *Afro-European Conference for Industrial Advancement*. **2015**, 1–13.

Ergun, B.; Sen, S.; Kilic, Y.; Kuru, O.; Ozsurmeli, M. The role of non-stress test to decision making procedure in pregnant women with Cesarean delivery "outcomes of our clinic and literature review". *Turkish Journal of Obstetrics and Gynecology*. **2012**, 9(1), 59–64.

Fayyad, U.; Shapiro, GP.; Smyth, P. From data mining to knowledge discovery in databases. *AI Magazine*, **1996**, 17(3), 37–54.

Fontenla, Romero.; Guijarro, Berdiñas.; Alonso, Betanzos. Symbolic, neural and neuro-fuzzy approaches for pattern recognition in cardiotocograms. *Advances in Computational Intelligence and Learning*, **2002**, 18, 489–500.

Geeky.medics. http://geekymedics.com/how-to-read-a-ctg/ (accessed June 21, 2019)

Ha, SH.; Joo, SH. A hybrid data mining method for the medical classification of chest pain. *International Journal of Computer and Information Engineering*. **2010**, 4(1), 33–38.

Han, JS.; Lee, SW.; Bien, Z. Feature subset selection using separability index matrix. *Information Sciences*. **2013**, 223, 102–118.

HealthCatalyst. https://www.healthcatalyst.com/data-mining-in-healthcare (accessed June 21, 2019)

Healthnetconnections. http://www.hnc.net/products/trium-ctg-monitoring/ (accessed June 21, 2019)

Hongbiao, Zhou.; Ying, Genwang. Identification of CTG based on BP neural network optimized by PSO. *11th International Symposium on Distributed Computing and Applications to Business, Engineering & Science (DCABES)*, **2012**, 108–111.

Hoque, N.; Bhattacharyya, DK.; Kalita, JK. MIFS-ND: A mutual information-based feature selection method. *Expert Systems with Applications*. **2014**, 41(14), 6371–6385.

Huan, Liu.; Hiroshi, Motoda. Computational methods of feature selection, *Data Mining and Knowledge Discovery Series.* Chapman & Hall/CRC, **2008**.

Huan, Liu.; Hiroshi, Motoda. Feature Selection for Knowledge Discovery and Data Mining. Springer Science & Business Media, New York, **2000**.

Huang, J.; Cai, Y.; Xu, X. A hybrid genetic algorithm for feature selection wrapper based on mutual information. *Pattern Recognition Letters*. **2007**, 28(13), 1825–1844.

Huang, Mei-Ling.; Yung-Yan, Hsu. Fetal distress prediction using discriminant analysis, decision tree, and artificial neural network. *Journal of Biomedical Science and Engineering*, 2012, 5(9), 526.

Huang, Yo-Ping.; Shin-Liang, Lai.; Frode Eika, Sandnes.; Shen-Ing, Liu. Improving classifications of medical data based on fuzzy ART 2 decision trees, *International Journal of Fuzzy Systems*. **2012**, 14(3), 444–453.

Imani, MB.; Pourhabibi, T.; Keyvanpour, MR.; Azmi, R. A new feature selection method based on ant colony and genetic algorithm on Persian font recognition. *International Journal of Machine Learning and Computing*. **2012**, 2(3), 278–282.

Inbarani, HH.; Azar, AT.; Jothi, G. Supervised hybrid feature selection based on PSO and rough sets for medical diagnosis. *Computer Methods and Programs in Biomedicine*. **2014**, 113(1), 175–185.

Iztok, Fister.; Iztok, Fister Jr.; Xin-She, Yang.; JanezBrest. A comprehensive review of firefly algorithms. *Swarm and Evolutionary Computation*. **2013**, 13(1), 34–46.

Jacob, SG.; Ramani, RG. Evolving efficient classification rules from cardiotocography data through data mining methods and techniques. *European Journal of Scientific Research*. **2012**, 78(3), 468–480.

Jaddi, NS.; Abdullah, S. Hybrid of genetic algorithm and great deluge algorithm for rough set attribute reduction. *Turkish Journal of Electrical Engineering & Computer Sciences*. **2013**, 21(6), 1737–1750.

Jadhav, S.; Nalbalwar, S.; Ghatol, A. Modular neural network model based foetal state classification. *IEEE International Conference on Bioinformatics and Biomedicine Workshops (BIBMW)*, **2011**, 915–917.

Jeyachidra, J.; Punithavalli, M. A study on statistical based feature selection methods for classification of gene microarray dataset. *Journal of Theoretical and Applied Information Technology*. **2013**, 53(1), 107–114.

John, GH.; Kohavi, Ron.; Pfleger, Karl. Irrelevant features and the subset selection problem. machine learning: *Proceedings of 11th International Conference*, **1994**, 121–129.

Johnson, Bryan.; Alex, Bennett.; Myungjae, Kwak.; Anthony, Choi. Automated evaluation of fetal cardiotocograms using neural network. *IEEE International Conference on Systems, Man, and Cybernetics (SMC)*. **2012**, 408–413.

Karabulut, EM.; Ibrikci, T. Analysis of cardiotocogram data for fetal distress determination by decision tree based adaptive boosting approach. *Journal of Computer and Communications*. **2014**, 2(9), 32–37.

Krupa, N.; Ali, M.; Zahedi, E.; Ahmed, S.; Hassan, FM. Antepartum fetal heart rate feature extraction and classification using empirical mode decomposition and support vector machine. *Biomedical Engineering Online*, **2011**, 10(1), 1–15.

Kursa, MB.; Rudnicki, WR. Feature selection with the Boruta package, *Journal of Statistical Software*, **2010**, 36(11), 1–13.

Liu, H.; Sun, J.; Liu, L.; Zhang, H. Feature selection with dynamic mutual information. *Pattern Recognition*. **2009**, 42(7), 1330–1339.

Liu, J.; Ranka, S.; Kahveci, T. Classification and feature selection algorithms for multi-class CGH data. *Bioinformatics*. **2008**, 24(13), i86-i95.

Liu, Y.; Wang, G.; Chen, H.; Dong, H.; Zhu, X.; Wang, S. An improved particle swarm optimization for feature selection. *Journal of Bionic Engineering*. **2011**, 8(2), 191–200.

Mehmed, Kantardzic. Data Mining-Concepts, Models, Methods, and Algorithms. John Wiley & Sons, **2011**.

Menai, MEB.; Mohder, FJ.; Al-mutairi, F. Influence of feature selection on naive Bayes classifier for recognizing patterns in cardiotocograms. *Journal of Medical and Bioengineering*, **2013**, 2(1), 66–70.

Michelakos, I.; Papageorgiou, E.; Vasilakopoulos, M. A hybrid classification algorithm evaluated on medical data. *19th IEEE International Workshop on Enabling Technologies: Infrastructures for Collaborative Enterprises (WETICE)*. **2010**, 98–103.

Mitchell, T. Machine Learning. McGraw-Hill, New York, **1997**.

Mohamed Ali, EA.; Abudhahir, A.; Manivanna Boopathi, A. Firefly Algorithm optimized PI controller for pressure regulation in PEM Fuel Cells. *Journal of Computational and Theoretical Nanoscience*, **2018**, 15(1), 1–9.

Nakamura, RYM.; Pereira, LAM.; Costa, KA.; Rodrigues, D.; Papa, JP.; XS, Yang. BBA: A binary bat algorithm for feature selection. *25th SIBGRAPI Conference on Graphics, Patterns and Images*, **2012**, 291–297.

Nidhal, S.; Ali, MAM.; Najah, H. A novel cardiotocography fetal heart rate baseline estimation algorithm. *Scientific Research and Essays*. **2010**, 5(24), 4002–4010.

Nithya, D.; Suganya, V.; RSI, Mary. Feature selection using integer and binary coded genetic algorithm to improve the performance of SVM classifier. *Journal of Computer Applications*. **2013**, 6(3), 57–61.

Ocak, Hasan. A medical decision support system based on support vector machines and the genetic algorithm for the evaluation of fetal well-being. *Journal of medical systems*. **2013**, 37(2), 1–9.

Ocak, Hasan.; Huseyin Metin, Ertunc. Prediction of fetal state from the cardiotocogram recordings using adaptive neuro-fuzzy inference systems. *Neural Computing and Applications*. **2013**, 23(6), 1583–1589.

Pappa, GL.; Freitas, AA.; Kaestner, CAA. A multiobjective genetic algorithm for attribute selection. *Proceedings of 4th International Conference on Recent Advances in Soft Computing (RASC-2002)*. **2002**, 116–121.

Patient.info. http://patient.info/in/health/cardiotocography (accessed June 21, 2019)

Porkodi, R. Comparison of filter based feature selection algorithms an overview. *International journal of Innovative Research in Technology & Science*. **2014**, 2(2), 108–113.

Rajendran, P.; Madheswaran, M. Hybrid medical image classification using association rule mining with decision tree algorithm. *Journal of Computing*. **2010**, 2(1), 127–136.

Ravindran, Sindhu.; Asral Bahari, Jambek.; Hariharan, Muthusamy.; Siew-Chin, Neoh. A novel clinical decision support system using improved adaptive genetic algorithm for the assessment of fetal well-being. *Computational and Mathematical Methods in Medicine*. **2015**, 2015, 283532.

Rodrigues, D.; Pereira, LAM.; Almeida, TNS.; Papa, JP.; Souza, AN.; CCO, Ramos.; XS, Yang. BCS: A binary cuckoo search algorithm for feature selection. *IEEE International Symposium on Circuits and Systems (ISCAS2013)*, **2013**, 465–468.

Rodrigues, D.; Pereira, LAM.; Nakamura, RYM.; Costa, KAP.; Yang, XS.; Souza, AN.; Papa, JP. A wrapper approach for feature selection based on bat algorithm and optimum-path forest. *Expert Systems with Applications*. **2014**, 41(5), 2250–2258.

Rodrigues, PP.; Sebastiao, R.; Santos, CC. Improving cardiotocography monitoring: a memory-less stream learning approach. *LEMEDS'11 Learning from Medical Data Streams*. **2011**, 12.

Sahin, H.; Subasi, A. Classification of the cardiotocogram data for anticipation of fetal risks using machine learning techniques. *Applied Soft Computing*, **2015**, 33, 231–238.

Schiezaro, M.; Pedrini, H. Data feature selection based on artificial bee colony algorithm. *EURASIP Journal on Image and Video Processing*. **2013**, 47(1), 1–8.

Shruti, Ratnakar.; K, Rajeshwari.; Rose, Jacob. Prediction of heart disease using genetic algorithm for selection of optimal reduced set of attributes. *International Journal of Advanced Computational Engineering and Networking*, **2013**, 1(2), 2106–2320.

Sikora, R.; Piramuthu, S. Framework for efficient feature selection in genetic algorithm based data mining. *European Journal of Operational Research*. **2007**, 180(2), 723–737.

Sontakke, S.; Lohokare, J.: Dani, R.; Shivagaje, P. Classification of Cardiotocography Signals Using Machine Learning. *Proceedings of the 2018 Intelligent Systems Conference* (IntelliSys), Volume 2, 2019.

Subha, V.; Murugan, D. Foetal state determination using support vector machine and firefly optimisation. *International Journal of Knowledge Based Computer System*, **2014**, 2(2), 7–12.

Subha, V.; Murugan, D. Opposition-based firefly algorithm optimized feature subset selection approach for fetal risk anticipation. *Machine Learning and Applications: An International Journal*. **2016**, 3(2), 55–64.

Subha, V.; Murugan, D.; Manivanna Boopathi, A. A hybrid filter-wrapper attribute reduction approach for fetal risk anticipation. *Asian Journal of Research in Social Sciences and Humanities*. **2017**, 7(2), 1094–1106.

Subha, Velappan.; Murugan, D.; Prabha, S; Manivanna Boopathi, Arumugam. Genetic algorithm based feature subset selection for fetal state classification. *Journal of Communications Technology, Electronics and Computer Science*, **2015**, 2, 13–17.

Sui, Bangsheng. Information gain feature selection based on feature interactions. M.S. thesis, University of Houston, **2013**.

Sumeet, Dua.; Xian, Du. Data Mining and Machine Learning in Cybersecurity. CRC Press, **2011.**

Sundar, C.; Chitradevi, M.; Geetharamani, G. An overview of research challenges for classification of cardiotocogram data. *Journal of Computer Science*. **2013**, 9(2), 198–206.

Taha, AM.; Tang, AYC. Bat algorithm for rough set attribute reduction. *Journal of Theoretical and Applied Information Technology*. **2013**, 51(1), 1–8.

Tang, H.; Wang, T.; Li, M.; Yang, X. The design and implementation of cardiotocography signals classification algorithm based on neural network. Computational and Mathematical Methods in Medicine, **2018**, 2018, 12.

Tiwari, R.; Singh, MP. Correlation-based attribute selection using genetic algorithm. *International Journal of Computer Applications*. **2010**, 4(8), 28–34.

Tizhoosh, HR. Opposition-based learning: a new scheme for machine intelligence. *International Conference on Computational Intelligence for Modelling, Control and Automation Jointly with International Conference on Intelligence Agents, Web Technologies and Internet Commerce (CIMCA-IAWTIC'05)*, **2006**, 1, 695–701.

Tomáš, P.; Krohova. J.; Dohnalek, P.; Gajdoš, P. Classification of cardiotocography records by random forest. *36th International Conference on Telecommunications and Signal Processing (TSP)*, **2013**, 620–923.

UCI Machine Learning Repository. http://archive.ics.uci.edu/ml/index.php (accessed June 21, 2019)

Uzer, MS.; Yilmaz, Nihat.; Inan, Onur. Feature selection method based on artificial bee colony algorithm and support vector machines for medical datasets classification. *The Scientific World Journal*. **2013**, 2013, 419187.

Venkatesan, P.; Premalatha, V. Genetic-neuro approach for disease classification. *International Journal of Science and Technology*. **2012**, 2(7), 473–478.

Vidyavathi, BM.; Ravikumar, CN. A novel hybrid filter feature selection method for data mining, *Ubiquitous Computing and Communication Journal*. **2008**, 3(3), 118–121.

Wagholikar, Kavishwar.; V, Sundararajan.; Ashok, Deshpande. Modeling paradigms for medical diagnostic decision support: a survey and future directions. *Journal of Medical Systems. Journal of Medical Systems*, **2012**, 36(5), 3029–3049.

Warrick, PA.; Hamilton, EF.; Kearney, RE.; Precup, D. Classification of normal and hypoxic fetuses using system identification from intrapartum cardiotocography. *IEEE Transactions on Biomedical Engineering*, **2010**, 57(4), 771–779.

Xin-She, Yang. Nature-Inspired Metaheuristic Algorithms. 2nd edition, Luniver Press, UK, **2010**.

Xue, B.; Cervante, L.; Shang, L.; Browne, WN.; Zhang, M. A multi-objective particle swarm optimisation for filter-based feature selection in classification problems, *Connection Science*, **2012**, 24(2–3), 91–116.

Xue, B.; Zhang, M.; Browne, WN. Multi-objective particle swarm optimisation (PSO) for feature selection. *Proceedings of the 14th Annual Conference on Genetic and Evolutionary Computation*, **2012**, 81–88.

Xue, B.; Zhang, M.; Browne, WN. New fitness functions in binary particle swarm optimisation for feature selection. *IEEE Congress on Evolutionary Computation*, **2012**, 1–8.

Xu, Q.; Wang, L.; Baomin, H.; Wang, N. Modified opposition-based differential evolution for function optimization. *Journal of Computational Information Systems*, **2011**, 7(5), 1582–1591.

Yu, Shuhao.; Zhu, Shenglong.; Ma, Yan.; Mao, Demei. Enhancing firefly algorithm using generalized opposition-based learning. *Computing*. **2015**, 97(7), 741–754.

Zhang, Zhongheng.; Trevino, Victor.; Hoseini, Sayed Shahabuddin.; Belciug, Smaranda.; Manivanna Boopathi, Arumugam.; Gorunescu, Florin.; Subha, Velappan. Variable selection in logistic regression model with genetic algorithm. *Annals of Translational Medicine*. **2018**, 6(3):45.

Zhuo, L.; Zheng, L.; Li, X.; Wang, F.; Ai, B.; Qian, J. A genetic algorithm based wrapper feature selection method for classification of hyperspectral images using support vector machine. *Geoinformatics and Joint Conference on GIS and Built Environment: Classification of Remote Sensing Images*, **2008**, 71471, 71471J.

CHAPTER 18

DEPLOYMENT OF SUPERVISED MACHINE LEARNING AND DEEP LEARNING ALGORITHMS IN BIOMEDICAL TEXT CLASSIFICATION

G. KUMARAVELAN* and BICHITRANANDA BEHERA

Department of Computer Science, Pondicherry University, Karaikal, India

Corresponding author. E-mail: gkumaravelanpu@gmail.com

ABSTRACT

Document classification is a prevalent task in natural language processing with broad applications in the biomedical domain, including biomedical literature indexing, automatic diagnosis codes assignment, tweets classification for public health topics, patient safety reports classification, etc. In recent years, the categorization of biomedical literature plays a vital role in biomedical engineering. Nevertheless, manual classification of biomedical papers published in every year into predefined categories becomes a cumbersome task. Hence, building an effective automatic document classification for biomedical databases emerges as a significant task among the scientific community. Hence, this chapter investigates the deployment of the state-of-the-art machine learning (ML) algorithms like decision tree, *k*-nearest neighborhood, Rocchio, ridge, passive–aggressive, multinomial naïve Bayes (NB), Bernoulli NB, support vector machine, and artificial neural network classifiers such as perceptron, random gradient descent, BPN in automatic classification of biomedical text documents on benchmark datasets like BioCreative Corpus III (BC3), Farm Ads, and TREC 2006 genetics Track. Finally, the performance of all the said constitutional classifiers are compared and evaluated by means of the well-defined metrics like accuracy, error rate, precision, recall, and *f*-measure.

18.1 INTRODUCTION

Biomedical engineering introduces different innovative techniques and materials in medicine and healthcare for the development of novel biomedical tools. In the era of Internet-connected devices in every minute, a tremendous amount of biomedical data is generated with high throughput. More specifically, biomedical research publishes numerous scientific articles in electronic text form, which focuses on innovative research. In this case, the manual classification of these biomedical documents leads to a cumbersome task. Hence, building an automatic classifier model for these biomedical documents plays an important area of research.

In general, an automatic document classification algorithm assigns a predefined label to the instances of the text documents (test data set) based on the classifier model developed using the machine learning (ML) algorithm. ML is a subfield of artificial intelligence which disseminates intelligence to the classifier model from the training data set. So that the build-in classifier model captures the inherent patterns and relationship based on the corresponding labels assigned to the given text documents (training data set).

Depending on the usage of the ML algorithm, automatic document classification task is often classified into three broad classes specifically supervised document classification, unsupervised document classification, and semisupervised document classification. In supervised document classification, some external mechanism is needed manually to the classifier model, which contributes information related to the precise document classification. In unsupervised document classification, there is no scope of having an external mechanism to provide information to the classification model to the correct document classification. In semisupervised document classification, a partial amount of the documents are labeled by an external mechanism. This chapter focuses on the deployment of state-of-the-art supervised ML algorithms for biomedical text classification.

The classifier model build using supervised ML algorithms are broadly divided into two forms, namely multiclass and multilabel classification. The multiclass classification is the one where a single class label out of many is assigned to one instance. Decision tree (DT) classifier, *k*-nearest neighborhood (*k*-NN) classifier, Rocchio classifier (RC), ridge classifier, passive–aggressive (PA) classifier, multinomial naïve Bayes (M_NB) classifier, Bernoulli naïve Bayes (B_NB) classifier, support vector machine (SVM) classifier, artificial neural network

(ANN) classifier including perceptron (PPN), stochastic gradient descent (SGD), BPN are the most prominent classifier found in the literature of supervised ML community. However, multilabel classification assigns more than one class labels among the instances, and it is considered to be more complex classification than multiclass classification. Specifically, multilabel classification falls into two main categories, namely problem adaption and algorithm adaption. Problem adaptation method transforms the multilabel problem into a single-label or multiclass problem(s). The main aim of this type of transformation is to fit the data to the multiclass algorithm.

This chapter provides an overview of the deployment of the state-of-the-art supervised ML in biomedical text document classification. The performance of the built-in classifiers is compared and empirically evaluated using well-defined metrics such as accuracy, error rate, precision, recall, and f-measure on publicly available benchmark biomedical data sets like BioCreative Corpus III (BC3), Farm Ads, and TREC 2006 Genomics Track. Except for farm Ads, data set remaining BC3 and TREC 2006 Genomics Track datasets consist of documents which are extracted from PubMed central digital repository.

In the literature, only a few analysis works have been meted out that examine the progressive ML algorithms to the benchmark biomedical set in one platform. Therefore, the primary aim of this book chapter is to perform an end-to-end performance analysis of all the distinguished supervised ML algorithms for automatic document classification in the biomedical domain.

The organization of this chapter is as follows: Section 18.2 elaborates the background details for text document classification process including preprocessing along with document representation, document classification which includes mathematic formulation for document classification, and literature review for biomedical text document classification using ML algorithms. Section 18.3 depicts, in a nutshell, the various ML algorithms used in this book chapter for document classification purpose. Section 18.4 describes the novel experiments conducted towards the deployment of ML solutions in document classification. Section 18.5 gives the conclusion and suggests topics for further research.

18.2 BACKGROUND

18.2.1 TEXT CLASSIFICATION PROCESS

Biomedical text classification deals with unstructured text documents from different biomedical repositories like PubMed and MedLine, web blogs, e-newspapers, medical reports, and social media. The major aim of the text classification process is to predict a class label of the given test document with the prior knowledge of trained dataset. In general, text classification process involves three important steps: text preprocessing, text classification, and postprocessing. Figure 18.1 shows the various steps involved in building an automatic document classification model.

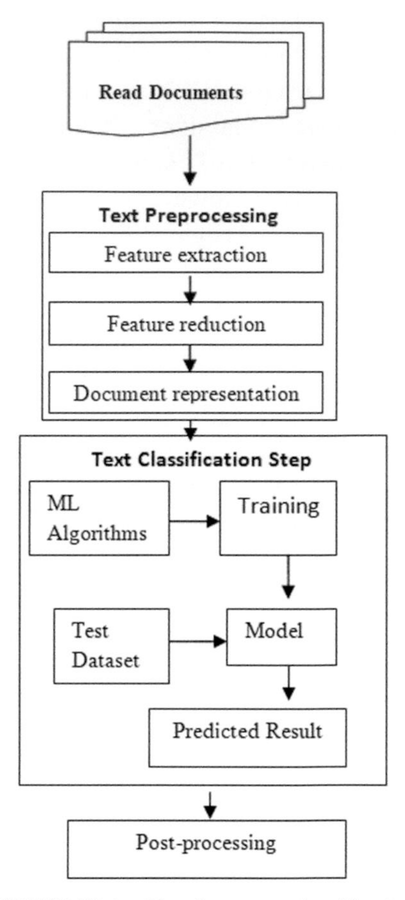

FIGURE 18.1 Text document classification process using ML algorithm.

18.2.1.1 TEXT PREPROCESSING

Generally, in document classification model development, the first and important key component is text preprocessing, which has a great impact on classification performance. It normally consists of three tasks, namely *feature extraction*, *feature reduction*, and *document representation*.

Feature Extraction: It includes many activities such as tokenization, filtering or stop-word removal, lemmatization, and stemming of words to scale down the document complexity and to present the classification method in an accessible manner.

- *Tokenization:* The input for tokenization activity is the raw text data or text document. It breaks the sequence of strings from the given raw text data into small character pieces that can be a distinctive word, phrases, or keywords known as tokens (Webster & Kit, 2010).

- *Filtering:* It removes unwanted words from the documents so that more focus is given to special words in the document. *Stop-Words removal* is a well-known filtering method in which those words that are often used without meaningful content is get removed (Saif et al., 2014; Silva, 2003). Examples of such stop-words are prepositions, conjunctions, and determiners.
- *Lemmatization:* In documents, there is varied inflected sort of words whose meaning are almost in identical nature. In such a situation, lemmatization is a kind of task which performs grouping of those words having similar meaning into one word by using vocabulary and morphological analysis of those words in that cluster of words.
- *Stemming:* It is the task of reducing derived words to their base or root form. Otherwise, it is like a crude chopping of affixes. For example, words like "running"and "runs" will be reduced to their base form like "run." Several stemming algorithms have been developed with time. In the field of Text Mining, Porter Stemmer is the mostly used stemming technique (Porter, 1980; Hull, 1996).

Feature Reduction: Normally, in a text document, the numbers of words otherwise called *features* are incredibly large, and those words play a vital role in document representation. Therefore, it is necessary to use the feature reduction methods to make an effective representation of the given text documents without changing the meaning of text data. Feature reduction methods are loosely divided into two categories, namely, *feature selection* and *feature transformation*.

- *Feature selection:* It involves the selection of a subset of features that can equivalently represent the original physical meaning with a better understanding of data which leads to the elegant learning process (Liu & Motoda, 1998). The major goal of the feature selection method is to reduce the curse of dimensionality to make the training dataset in the smaller size that can lead to lesser computational time. The advantage of reducing the curse of dimensionality is to increase the classification accuracy and to decrease the over-fitting problem. There are different types of feature selection methods available text mining literature namely term frequency (TF), mutual information, information gain, Chi-square statistic and term strength (Yang & Pederson,1995).

- ***Feature Transformation:*** It generates a new and smaller set of features by transforming or mapping the original set of features. Some well-known feature transformations methods are Latent Semantic Indexing (LSI) (Deerwester et al., 1990), PLSI (Hofmann, 1999), linear discriminant analysis (Fisher, 1936; Chakrabarti et al., 2003) and generalized singular value decomposition methods (Howland et al., 2003, 2004)

Document Representation: Once the features are extracted from the raw text data, all the given documents are normalized to unit length to perform classification in an economical manner. Basically, there are three most used models on the market within the literature for document representation namely, vector space method (VSM) (Salton et al., 1975), probabilistic models (Manning et al., 2008), and the inference network model (Turtle & Croft, 1989). Among the three models, VSM is the most used model, and the following section describes briefly about VSM.

It initially used for indexing and information retrieval (IR). It converts documents into numerical vectors with the document set *D*; vocabulary set *V* and the term vector $\vec{t}_d$ for document *d*. Set $D=\{d_1,d_2,...,d_D\}$ is a collection of Documents, the set $V=\{w_1,w_2,...,w_v\}$ is a set of unique words or terms in the set *D* and the term vector $\vec{t}_d=(f_d(w_1),f_d(w_2),...,f_d(w_v))$ where $f_d(w)$ represents the frequency of term $w\in V$ in the document $d\in D$ and $f_D(w)$ represents a number of documents contain the word *w*.

In VSM, the Boolean model and TF-IDF are the two-term weight schemes are used to calculate the weight of each feature. The Boolean model assigns $w_{ij}>0$ to each term w_i if $w_i\in d_j$ and assigns $w_{ij}=0$ $w_i\notin d_j$ if. However, the TF-IDF scheme calculates the term weight of each word $w\in d$ as follows:

$$q(w)=f_d(w)*\log\frac{|D|}{f_D(w)} \tag{18.1}$$

where $|D|$ is the number of documents in the set *D*.

18.2.1.2 TEXT CLASSIFICATION STEP

Mathematically, the text classification problem wants three sets to outline. First one is the training document set $D=\{d_1,d_2,...,d_n\}$, the second one is the category label set $C=\{c_1,c_2,...,c_n\}$ and third one is the test document set $T=\{d_1, d_2, ..., d_n\}$. Every document d_i of the training document set D is labeled with a category label c_i from the category label set C; however, every document of the test document set T has not been labeled. The most aim of text classification is to construct a text classification model, that is, a text classifier from the training document set by relating the features within the text documents to one of the target class labels. When the classification model is trained, it will predict the category labels of the test document set. Mathematical formula of text classification algorithm both for training and testing is given as

$$f: D \rightarrow C \; f(d)=c \tag{18.2}$$

In Equation 18.2, classifier assigns the proper class label to new document d (test instance). If a class label is assigned to the test instance, then this sort of classification is termed hard or multiclass classification, and on the other hand, classification is termed soft if a probability value is assigned to the test instance. In multilabel classification, multiple class labels are allotted to a test instance.

18.2.1.3 POSTPROCESSING STEP

In postprocessing step evaluation of the classifier is performed. The evaluation of the classification models is performed through various elegant performance measures like accuracy, precision, recall, and $F-1$ scores.

18.2.2 LITERATURE REVIEW

The organization and access to biomedical information are in great demand nowadays because of the exponential growth of biomedical documents evolved from different biomedical research publications and clinical trials. Sebastiani surveys concerning the various types of text document classification, application of text document classification, and mentioned the role of ML algorithms in automatic text document classification (Sebastiani, 2002) thoroughly.

Cohen developed a replacement classification algorithm by assembling SVM with rejection sampling and chi-square feature selection technique for automatic document classification (Cohen, 2006). The TREC 2005 genomics track biomedical dataset was used to compare the classification performance of the classifier with a different variant of SVM classifier. Almeida et al. conferred supervised ML approaches like NB, SVM, and provision Model Trees to perform text classification of PubMed abstracts, to support the triage of documents (Almeida et al., 2014).

García et al. developed a bag-of-concepts representation of documents and applied ML algorithm like SVM for biomedical document classification (García et al., 2015). Nguyen et al. proposed an improved feature weighting technique for document representation and SVM as a classifier (Nguyen et al., 2016). The proposed document representation technique provides the best classification performance compared to the documents represented in bag-of-words or TF-IDF document representation.

Samal et al. measured the performance of most of the supervised classifiers for sentiment analysis using movie review dataset and concluded that SVM classifiers performed best among all classifiers for large movie review datasets (Samal et al., 2017).

Mishu et al. analyzed the performance of various supervised ML algorithms such as multinomial NB, B_NB, logistic regression, stochastic gradient descent, SVM, BPN for classification on Reuters corpus, brown corpus and movie review corpus and concludes that BPN is best among them (Mishu et al., 2016).

Jiang et al. applied NB and random forest (RF) for classifying biomedical publication documents associated with mouse gene expression database (Jiang et al., 2017).

18.3 SUPERVISED ML ALGORITHMS FOR TEXT DOCUMENT CLASSIFICATION

18.3.1 DECISION TREE (DT) CLASSIFIER

In the DT classification model, the instances are the documents and attributes of every document are itself a bag of words or terms. The DT classifier (Li & Jain, 1998) performs hierarchical decomposition of text documents of training dataset by labeling its internal nodes with names of the text documents, branches of the tree with the test condition on terms and leaves of the tree with categories (labels). The test condition on terms could also be of two varieties supported the document representation model.

The first category of the test is to test whether or not a selected term out there within the documents or not. The second kind of test is to look at the weight of the terms within the text document. The primary class of the test is used if document representation is of the shape of the binary or Boolean model and also the second category of the test will be used if document representation is of the form of TF-IDF model. During the training phase, the DT is made from the training dataset, whereas making the DT from the training data set, totally different splitting criteria are used, and most of the DT classifiers use single attribute split combination wherever the one attribute is employed to perform the division (Aggarwal, 2012). The attribute or term whose information gain is high is considered as a base node, and also the procedure is continual consequently for choosing the remaining nodes. Meanwhile within the testing phase, to predict the category label of a new untagged document, the DT classifier tests the terms of the against the DT ranging from the root node (base node) to until it reaches a leaf node and assigns the category label of the leaf node.

18.3.2 NAÏVE BAYES (NB) CLASSIFIER

NB classifier is a probabilistic classifier based on Bayesian posterior probability distribution. It holds the restriction with the independent relationship among the attributes through conditional probability. There is two variant of NB classifier, namely *the multivariate Bernoulli model (B_NB)* and *multinomial model (M_NB)* (McCallum & Nigam). The multivariate Bernoulli naive Bayes model works only on binary data. Hence, in document preprocessing steps, each attributes corresponding to the list of documents in VSM must be either one or zero depending on the presence or absence of that particular attribute in that document (Lewis, 1998). However, the multinomial model works on the frequencies of attributes available in the VSM representation of the documents (McCallum, 1998). If the vocabulary size is small, then the Bernoulli model performs better than the multinomial model.

18.3.3 K-NEAREST NEIGHBORHOOD CLASSIFIER (K-NN)

Most of the classifiers within the literature pay longer in the training part for building the classification model are considered as an *eager learner*. However, *k*-NN classifier spends longer within the testing part for predicting the category label of the new untagged test document. Hence, it is known as a *lazy learner*.

In the training section of the model construction, *k*-NN classifier stores all the training documents together with their target class. Meanwhile, in testing phase, once any new test document comes for classification whose target class is unknown, k-nearest-neighborhood classifier finds the distance of the test document from all the training documents and assigns the category label of the training documents that is nearest or most like the unknown document (Sebastiani, 2002; Han et al., 2001). For this reason, *k*-NN classifier is thought of as an instant-based learning algorithm (Han et al., 2001). Euclidian distance and cosine similarity are the foremost oftentimes used approaches for measurement similarity quotient to find the NN.

18.3.4 SUPPORT VECTOR MACHINE (SVM)

SVM is a kind of classifier has the potential to classify each linear and nonlinear data (Cortes & Vapnik, 1995). The core plan behind the SVM classifier is that it first non-linearly maps the initial training data into sufficiently higher dimension let be *n*, so the data within the higher dimension is separated simply by *n-1* dimension decision surface known as *hyperplanes*. Out of all hyperplanes, the SVM classifier determines the simplest hyperplane that has most margins from the *support vectors*. Thanks to non-linearity mapping, SVM classifier works expeditiously on an oversized data set and has been with success applied in text classification (Drucker,1999).

18.3.5 ARTIFICIAL NEURAL NETWORK (ANN)

ANN is a reasonably a data processing nonlinear model cherish the structure of the brain, and it will learn from the prevailing training data to perform tasks like categorization, prediction or forecast, decision-making, visualization, and others. It consists of a compilation of nodes otherwise known as neurons that are the middle of data processing in ANN. With context to the problem statement, these neurons are organized into three different layers, specifically the input layer, an output layer, and hidden layer. Within the context of text classification, the quantity of words or terms outlines the neuron numbers within the input layer, and therefore the classes (class label) of documents define the number of neurons in the output layer. ANN will have a minimum of one input layer and one output layer; however, it's going to have several hidden layers relying upon the chosen drawback. All links from the input layer to the output layer through hidden layers are appointed with some weights that represent the dependence relation between the nodes. Once the neurons get weighted data, it calculates the weighted sum, and a well-known activation function processes it. The output value from the activation function is fed forward to all the neurons within the input layer to map the proper neuron in the output layer. Some examples of well-known activation functions are Binary step, Sigmoid, TanH, Softmax, and

Rectifier linear unit functions. ANN can be additional versatile and more powerful by employing additional hidden layers. In particular, PPN, SGD neural network, and BPN are the three widespread neural networks primarily based classifiers that extensively used for text classification.

18.3.6 ROCCHIO CLASSIFIER (RC)

Rocchio classification algorithm is outlined on the conception of relevance feedback theory established within the field of IR (Rocchio, 1971). It uses the properties of centroid and similarity measure computations among the documents within the training and testing phase of model construction and usage, respectively. If $D=<d_1,d2,...,d_n>$ represents Document set which holds all the training documents and If $C=<c_1,c_2,...,c_m>$ represents class set which have all the distinct class labels. For each class $c_1 \in C, D_{c_i}$ represent all the documents of D *the set* which belong to class c_i and $\vec{v_d}$ represents the VSM document representation for each document. In the training phase, the Rocchio classifier computes the centroid $\vec{\mu}(c_i)$ for each class from the relevant documents and establishes the centroid of each class as its representative. The RC computes the centroid $\vec{\mu}(c_i)$ for the class c_i using the equation

$$\vec{\mu}(c_i) = \frac{1}{|D_{c_i}|} \sum_{d \in D_{c_i}} \vec{v_d} \tag{18.3}$$

In testing phase to predict the category label $c_i \in C$ of an untagged test document $d \notin D$, Rocchio classifier calculates its Euclidean distance from the centroid of every class $\vec{\mu}(c_i)$ and assigns that class label which has a minimum distance from untagged test document using the following equation:

$$\text{dist} = \arg_c \min \left\| \vec{\mu}(c_i) - \vec{v_d} \right\| \tag{18.4}$$

18.3.7 RIDGE CLASSIFIER (RIDGE)

The Ridge classification algorithm relies on subspace assumption, which states that samples of a specific class lie on a linear subspace and a new test sample to a category will be described as a linear combination of training samples of the relevant class (He et al., 2014). The ridge classification algorithm is presented in Figure 18.2.

18.3.8 PASSIVE–AGGRESSIVE (PA) CLASSIFIER

The PA classifiers belong to the family of a large-scale learning

Input: Data matrix X holds training dataset and data matrix X-test holds the test dataset
Procedure:

Step-1: For each test data $x \in$ X-test, calculate the regression parameter vector $\hat{\alpha}$ as

$$\hat{\alpha} = \arg\min_{\alpha_i} \left\| x - X_i \alpha_i \right\|_2^2 + \lambda \left\| \alpha_i \right\|_2^2$$

where, λ the regularization parameter and irepresent each class.

Step-2: Perform projection of new test sample x onto the subspace of each class i using $\hat{\alpha}_i$ as follows

$$\vec{x}_i = X_i \hat{\alpha}_i$$

Step-3: Calculated the distance between the test sample x and the class-specific sub-space $\vec{x}_i$.

Step-4: The test sample x is assigned to that class whose distance is minimum.

FIGURE 18.2 Ridge classification algorithm.

algorithm (Crammer et al., 2006). The working principle of this kind of classifier is similar to that of *Perceptron* classifier; meanwhile, they do not require a learning rate. However, it includes a regularization parameter *c*. Figure 18.3 shows the pseudocode description of the passive aggressive classifier.

Algorithm: Passive-Aggressive (PA) classifier for multi-class classification
Input:

- D=<X, Y> is the dataset where X holds training instances and Y holds class labels
- Cost function $\rho(y, \bar{y})$

Initialize: Weight vector $w_1 = (0, \ldots, 0)$

1. **for** $i = 1, 2, \ldots$
2. Consider, instance $x_i \in X$ and its corresponding label $y_i \in Y$
3. **if** PA method == prediction-based(PB)
4. return $\bar{y}_i = \arg\max_{y \in Y} (w_i . \phi(x_i, y))$
5. **if** PA method==Max-loss(ML)
6. return $\bar{y} = \arg\max_{r \in Y} (w_i . \phi(x_i, r) - w_i . \phi(x_i, y_i) + \sqrt{\rho(y_i, r)})$
7. Loss function: $\ell_i = w_i . \phi(x_i, \bar{y}_i) - w_i . \phi(x_i, \bar{y}_i) + \sqrt{\rho(y_i, \bar{y}_i)}$
8. Compute: $\tau_i = \dfrac{\ell_i}{\left\| \phi(x_i, y_i) - \phi(x_i, \bar{y}_i) \right\|^2}$
9. Update weight $w_{i+1} = w_i + \tau_i (\phi(x_i, y_i) - \phi(x_i, \bar{y}_i))$

FIGURE 18.3 Pseudocode for PA classifier.

18.3.9 RANDOM FOREST (RF)

RF classifier is a *bagging* type ensemble-learning algorithm. Figure 18.4 shows the overall architecture of the random forest classifier. In the training phase, it builds several DT classifiers from the random subsample of documents. In the testing phase, each DT performs prediction for a new test document and assigns that class label, which is mostly predicted by all of the DT classifiers. The main advantage of random forest over the DT is it eliminates the problem of over-fitting and increases the classification accuracy.

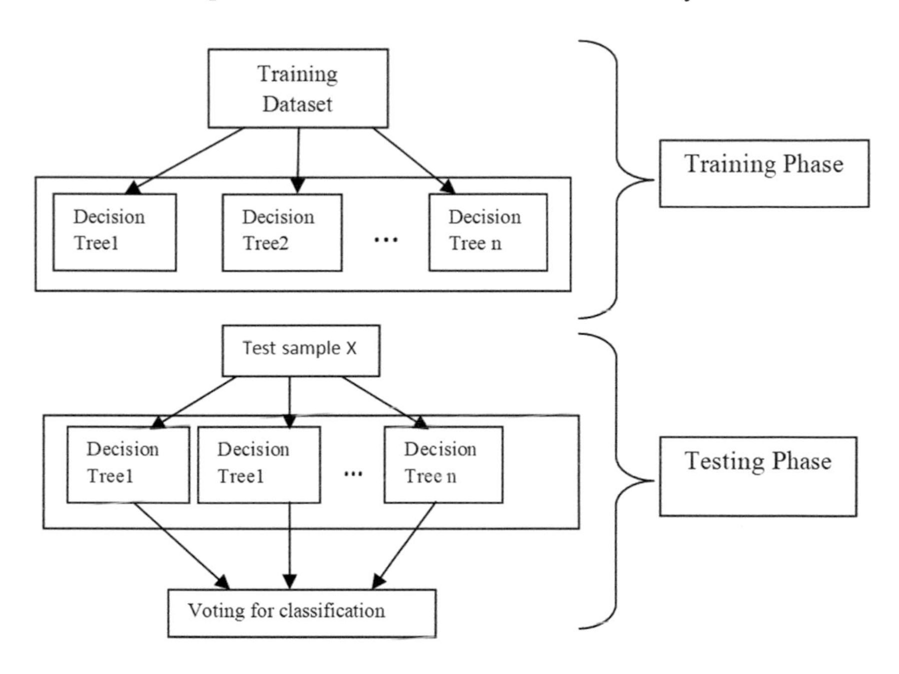

FIGURE 18.4 Random forest classification.

18.4 RESULTS AND DISCUSSION

18.4.1 EXPERIMENTAL SETUP

ML solutions for document classification has been implemented in python 3.6.7, and the experimentation is performed on a machine having Intel® Pentium® CPU 3825U processor 1.90 GHz with 4.00 GB of RAM. Four benchmark biomedical dataset namely; BioCreative Corpus III (BC3), Farm dataset and *TREC 2006 Genomics Track* have been used to perform an empirical evaluation of various ML algorithms mentioned in Section 18.3. The summary of these datasets are presented in Table 18.1, and their descriptions are detailed below:

- **BioCreative Corpus III (BC3):** The BC3 dataset has been created by the BioCreative III interactive task of the BioCreative workshop that was conducted in 2010. The BC3 dataset is divided into BC3-part 1 and BC3-part 2 datasets. Both BC3-part 1 and BC3-part 2 datasets are originally in XML format and have size 32.5 MB and 46.5 MB, respectively. For document classification, all the abstract and respective class label of each document is extracted from the XML file and represent in a CSV file. For BC3-part 1, the CSV file is of size 3.12 MB and represents 2280 article abstract with the class label of each abstract. Similarly, BC3-part 2 CSV file has size 5.73 MB and holds 4000 article abstract with their class label. This dataset is available at https://biocreative.bioinformatics.udel.edu/ resources/corpora/biocreative-iii-corpus/
- **Farm Ads dataset:** This dataset contains 4143 farm ads texts documents that represent various topics of farm animal. This is a binary classification problem where each of the documents or content either approves the ads or not. This dataset has size 12.4 MB and is available at the UCI ML repository (Lichman, 2013).
- ***TREC 2006 Genomics Track dataset*:** This dataset is the collection of biomedical full-text HTML documents from 49 journals in the area of Genomics Track. In this experiment, 1077 biomedical article abstract or document is collected from five journals. The number of document collections from each of the five journals is presented in Table 18.2.

TABLE 18.1 Summary of Four Biomedical Text Datasets

Dataset	Classes	Number of Documents
BC3—part 1	2	2280
BC3—part 2	2	4000
Farm Ads	2	4143
TREC 2006 Genomics Track	5	1077

TABLE 18.2 TREC 2006 Genomics Track Dataset

Journal Name	No. of Documents
Cerebral Cortex CC	201
Glycobiology GLY	203
Alcohol and Alcoholism AA	202
International Journal of Epidemiology IJE	206
International Immunology II	265

Extensive experimentation was carried out with eighty percentage of dataset contemplated for training and the remaining twenty percentage of dataset intended for testing, respectively. Using python *Scikit-learn* ML library (Pedregosa et al., 2011), and *TfidfVectorizer* envisage all the text preprocessing routines to build a dictionary and finally to transform all the documents to VSM representation. Subsequently, classification is performed with different ML algorithms, and finally, the classifiers are evaluated using the well-established performance measures.

18.4.2 PERFORMANCE MEASURE

Measures such as *Accuracy*, *Error Rate*, *Precision*, *Recall*, and *F1-Score* are used to evaluate the performance of the classifier (Sokolova & Lapalme, 2009). Aforementioned measures are defined by means of the following features, which defines the properties of the *confusion matrix*, as shown in Table 18.3.

TABLE 18.3 Confusion Matrix for Class C_i

Total Documents		Predicted Class	
		C_i	Not C_i
Actual Class	C_i	*True positive (TP)*	*False negative (FN)*
	Not C_i	*False positive (FP)*	*True negative (TN)*

a. True Positive (tp_1): The documents, which belong to class C_i,are correctly predicted to class $C_i$by the classifier.
b. True Negative (tn_1): The documents, which do not belong to the class C_i, are correctly predicted to other class rather than class C_i.
c. False Positive (fp_1): The documents, which do not belong to the class C_i, are wrongly predicted to the class C_i.
d. False Negative (fn_1): The documents, which belong to the class C_i, are wrongly predicted to different class rather than class C_i.

Now the performance measures are defined as follows

- Accuracy: *It is the average of per class ratio of correctly classified documents to the total documents.*

$$\frac{\sum_{i=1}^{n} \frac{tp_i + tn_i}{tp_i + fp_i + fn_i + tn_i}}{n} \quad (18.5)$$

- Error Rate: *It is the average of per class ratio of incorrectly classified documents to the total documents.*

$$\frac{\sum_{i=1}^{n} \frac{fp_i + fn_i}{tp_i + fp_i + fn_i + tn_i}}{n} \quad (18.6)$$

- Precision: *It is the average of per class ratio of true positive prediction to total positive prediction.*

$$\frac{\sum_{i=1}^{n}\frac{tp_i}{tp_i + fp_i}}{n} \quad (18.7)$$

- Recall: *It is the average of per class ratio of true positive prediction to a total number of actual positive documents in the test set.*

$$\frac{\sum_{i=1}^{n}\frac{tp_i}{tp_i + fn_i}}{n} \quad (18.8)$$

- $F1$-Score:

$$\frac{2(\text{Precision} \times \text{Recall})}{\text{Precision} + \text{Recall}} \quad (18.9)$$

In all the above cases, n is the no of classes or labels in the dataset.

18.4.3 HYPER-PARAMETERS FOR DIFFERENT CLASSIFIERS

The initialization of the input parameters among the different classifiers has a great impact on the classification performance measurements. Table 18.4 highlights the respective parameter setting procedures adapted in the experimental process of building the corresponding classifier.

TABLE 18.4 Hyper-parameters Settings of Different Classifiers

Classifiers	Parameters		
DT	**Splitting="Gini"**	**splitter="best"**	**min_samples_split=2**
M_NB	alpha=0.01	fit_prior=True	class_prior=None
B_NB	alpha=0.01	binarize=0.0	fit_prior=True
K-NN	K=10	metric="minkowski"	weights="uniform"
SVM	penalty factor="l2"	tolerance (tol)="1e-4"	loss="hinge"
PPN	max_iter="50",	tolerance(tol)="1e-3"	n_iter_no_change=5
SGD	alpha="0.0001"	Maximum iteration="50",	loss="hinge"
Ridge	solver="sag"	tolerance(tol)="1e-2"	max_iter=None
RC	metric="Euclidean"	shrink_threshold="None"	
PA	max_iter="50",	tolerance(tol)="1e-3"	loss="hinge"
RF	n_estimator="100"	Splitting="Gini"	min_samples_split=2
BPN	max_iter=200	Hidden layer size="1000"	activation function=relu

18.4.4 PERFORMANCE ANALYSIS

The extensive experiment is conducted on different ML solutions or algorithms such as DT, M_NB, B_NB, K-NN, SVM, PPN, SGD, Ridge, RC, PA, RF and BPN algorithms for biomedical benchmark dataset like BC3-part 1, BC3-part 2, Farm ads and TREC 2006 Genomics Track.

The Execution time of different algorithms is provided in Table 18.5. Execution time is the sum of training and testing time of the classification algorithm. Execution time plays a great role along with performance measures for comparing different classification algorithms.

TABLE 18.5 Performance of Classifiers with Respect to Execution Time

Algorithms	Execution Time in Seconds for Each Dataset			
	BC3-p1	BC3-p2	Farm	TREC
DT	1.674046	2.537979	2.484233	0.212883
M_NB	0.005981	0.021965	0.013005	0.006006
B_NB	0.010983	0.033979	0.034966	0.008998
K-NN	0.144941	0.413833	0.471888	0.041975
SVM	0.075956	0.232876	0.295832	0.114936
PPN	0.013019	0.048980	0.042972	0.019991
SGD	0.349801	0.819924	0.458874	0.314822
Ridge	0.130532	0.386032	0.338842	0.243861
RC	0.028429	0.033968	0.052979	0.009999
PA	0.025990	0.089968	0.102929	0.031979
RF	2.736432	4.748942	6.632810	1.309248
BPN	186.247590	271.775795	918.549152	92.036547

The performance of all the ML solutions is evaluated using different performance measures like accuracy, error rate, precision, recall, and $F1$ score. These performance measurements will provide a general overview of each ML solution performance from a different perspective. Table 18.6 shows the performance measurements of ML solutions on various benchmark biomedical dataset for automatic document classification.

The result for the BC3-part 1 dataset in Table 18.6 shows that RC classifier performs best among all the classifiers with respect to all the classification performance measures. The classification accuracy of the RC classifiers is 62.94.

TABLE 18.6 Performance of Classifiers (in %) using a Different Dataset

Dataset	Performance Measure	Classification Algorithms											
		DT	M_NB	B_NB	K-NN	SVM	PPN	SGD	Ridge	RC	PA	RF	BPN
BC3-p 1	Accuracy	53.73	57.68	62.28	57.02	58.11	57.68	59.43	60.31	62.94	57.46	60.31	57.56
	Error Rate	46.27	42.32	37.72	42.98	41.89	42.32	40.57	39.69	37.06	42.54	39.69	42.54
	Precision	53.83	57.66	62.27	57.26	58.10	57.65	59.42	60.32	62.93	57.43	60.36	57.45
	Recall	53.80	57.66	62.25	57.14	58.04	57.61	59.41	60.31	62.89	57.39	60.34	57.37
	*F*1-measure	53.68	57.66	62.25	56.88	57.99	57.59	59.41	60.21	62.89	57.36	60.30	57.30
BC3-p 2	Accuracy	77.28	85.02	84.64	83.65	86.89	84.89	86.89	86.39	79.40	84.89	83.65	84.89
	Error rate	22.72	14.98	15.36	13.35	13.11	15.11	13.11	13.61	20.60	15.11	16.35	15.11
	Precision	78.24	83.89	84.42	80.03	85.52	84.09	85.50	85.22	85.10	83.71	83.59	82.95
	Recall	77.28	85.02	84.64	83.65	86.89	84.89	86.89	86.39	79.40	84.89	83.65	84.89
	*F*1-measure	77.74	84.31	84.53	78.57	85.12	84.43	85.41	83.67	81.24	84.15	76.89	83.34
Farm ads	Accuracy	85.04	91.44	86.25	85.28	90.71	89.63	90.35	90.59	86.13	90.47	89.51	89.14
	Error rate	14.96	8.56	13.25	14.72	9.29	10.37	9.65	9.41	13.87	9.53	10.49	10.86
	Precision	85.28	91.61	86.72	85.26	90.71	89.61	90.35	90.60	86.51	90.46	89.52	89.13
	Recall	85.28	91.44	86.25	85.28	90.71	89.63	90.35	90.59	86.13	90.47	89.51	89.14
	*F*1-measure	85.09	91.37	86.30	85.25	90.71	89.62	90.33	90.56	85.95	90.46	89.47	89.13
TREC 2006 Genomics Track	Accuracy	84.26	96.30	94.44	21.30	97.69	96.30	93.98	97.69	95.83	97.22	95.37	98.15
	Error rate	15.74	3.70	5.56	78.70	2.31	3.70	6.02	2.31	4.17	2.78	4.63	1.85
	Precision	84.36	96.48	95.21	64.34	97.74	96.30	94.02	97.74	96.00	97.25	95.75	98.21
	Recall	84.26	96.30	94.44	21.30	97.69	96.30	93.98	97.69	95.83	97.22	95.37	98.15
	*F*1-measure	84.29	96.23	94.36	10.62	97.69	96.27	93.96	97.69	95.86	97.23	95.39	98.12

The B_NB classifier performs well next to RC classifier. After B_NB, the Ridge and RF classifiers have the same classification accuracy of 60.31, but if both are compared with respect to precision, recall, and *F*1 score, then Ridge classifiers show better performance. However, DT^^ classifier yields the lowest classification performance among all the classifiers for BC3-part 1 dataset. Meanwhile, the remaining classifiers provide an average classification performance.

For BC3-part 2 dataset, SVM and SGD classifiers stand top among all the classifiers with the same 86.89 percentage classification accuracy, 13.11 percentage error rate, and 86.89 percentage recall. But SVM works better than SGD with respect to precision; on the other hand, SGD outperforms SVM with respect to *F*1 score. Next to SVM and SGD, Ridge and M_NB classifier perform well. Meanwhile, PA, PPN, and BPN classifier have an equal classification accuracy of 84.89%. However, if precision and *F*1 score are taken for ranking the classifiers, then PPN classifier performs better than PA and BPN classifier. Usually for BC3-part 2 dataset DT classifier generates lowest classification performance.

The M_NB classifier shows the best classification performance among all the classifiers for Farm ads dataset. The classification accuracy of M_NB classifier is 91.44%. Meanwhile, after Ridge classifiers SVM shows good performance with classification accuracy in percentage is 90.71. Next to SVM, Ridge, PA, and SGD classifiers provide good classification performance.

For TREC 2006 Genomics Track dataset, BPN classifiers estimate more than 98.12 percentages of classification accuracy, precision, recall, and *F*1 score. SVM, Ridge and PA classifiers have good classification performance next to BPN. For TREC 2006 Genomics Track dataset KNN classifier shows the least performance among all the classifiers.

Thus from Table 18.6, it is clear that no one classifier is best for all the benchmarking dataset. From dataset to dataset, the performance of the classifier varies. But, among all the classifiers BPN, SGD, Ridge, PA, M_NB, B_NB, and SVM provide good classification performances and they have approximately possess the same classification performances. In particular, BPN classifier provides good classification accuracy, but it consumes more execution time.

The comparison of different algorithms with respect to accuracy and execution time is shown in Figure 18.5. In Figure 18.5, the accuracy values are in the range of 0 to 1 and the execution time of various ML algorithms on each dataset are normalized by dividing execution time of each algorithm by maximum execution time of any algorithm for concern dataset. The main purpose of having a normalized execution time is to compare the accuracy of the respective algorithms.

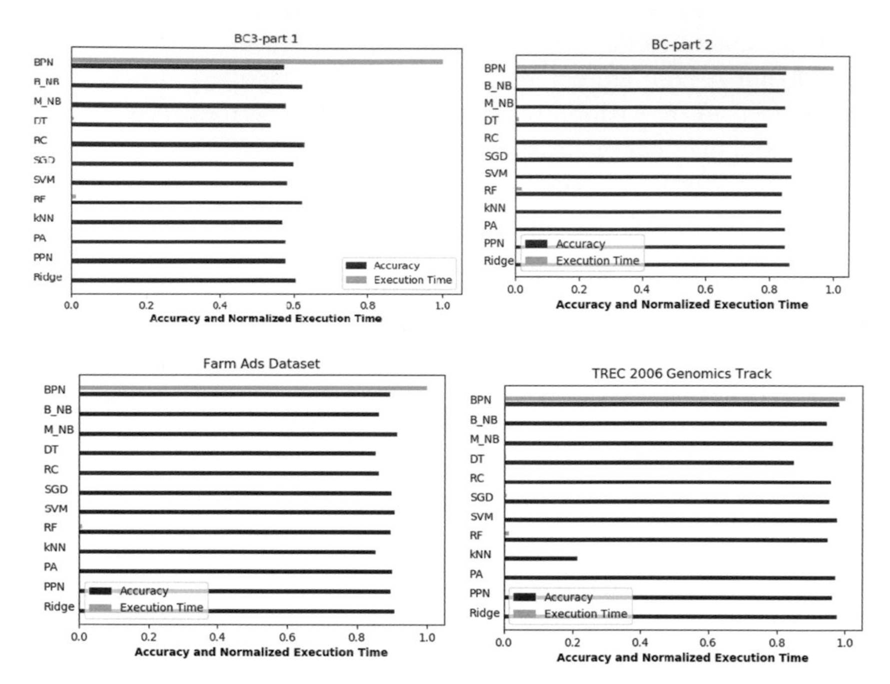

FIGURE 18.5 Performance comparison in accuracy and execution time.

18.5 CONCLUSION AND FUTURE SCOPE

Medical document classification is a multidisciplinary field of research in biomedical engineering. Many supervised ML algorithms have been successfully applied for automatic classification biomedical literature. But only a few authors addressed the performance measurements of all the classifiers in one platform. Hence, this book chapter summarizes in detail the procedures involved automatic document classification process, exemplifies the working logic of the state-of-the-art supervised ML algorithms and empirically evaluates how all the ML algorithms which are constituted to act as a classifier to the benchmark biomedical dataset. Particularly, classifiers like SGD, Ridge, PA, BPN, and SVM provides good results on the given dataset compared to the other classifiers. However, the performance of KNN and Decision Tree classifiers has shown poor results for the chosen dataset compared to other classifiers.

Meanwhile, other classifiers have an average classification performance. The future scope is to make an improvement among those classifiers to adapt well in connection to the large-scale dataset. As a result, application of deep learning-based models like multilayer feedforward neural networks, convolution neural networks, recurrent neural networks and ensemble deep learning models becomes an evitable avenue of further research.

KEYWORDS

- **text mining**
- **machine learning**
- **documents classification**
- **information retrieval**
- **information extraction**

REFERENCES

Aggarwal, C.C. ; Zhai, C. X. Mining Text Data, Springer. 2012.

Almeida, H.; Meurs, M. J.; Kosseim, L.; Butler, G.; Tsang, A. Machine learning for biomedical literature triage, PLoS One. 2014, 9(12).

Chakrabarti, S.; Roy, S.; Soundalgekar, M. Fast and accurate text classification via multiple linear discriminant projections, VLDB Journal. 2003, 12(2), 172–185.

Cohen, AM. An effective general purpose approach for automated biomedical document classification. AMIA Annual Symposium Proceedings. 2006,161–165.

Cortes, C. ; Vapnik, V.; Support-vector networks. Machine Learning. 1995, 20, 273–297.

Crammer, K.; Dekel, O.; Keshet, J.; Shalev-Shwartz, S.; Singer, Y. Online passive aggressive algorithms, Journal of Machine Learning Research. 2006, 7, 551–585.

Deerwester, S.; Dumais, S.; Landauer, T.; Furnas, G.; Harshman, R. Indexing by latent semantic analysis. JASIS. 1990, 41(6), 391–407.

Drucker, H.; Wu, D.; Vapnik, V. Support vector machines for spam categorization, IEEE Transactions on Neural Networks. 1999, 10(5), 1048–1054.

Pedregosa, F. et al. Scikit-learn: Machine learning in Python, Journal of Machine Learning Research. 2011, 12, 2825–2830.

Fisher, R. The use of multiple measurements in taxonomic problems. Annals of Eugenics. 1936, 7, 179–188.

García, M.A.M.; Rodríguez, R.P.; Rifón, L.E.A. Biomedical literature classification using encyclopedic knowledge: a Wikipedia-based bag-of-concepts approach. PeerJ. 2015, 3, e1279.

Han, E.S.; Karypis, G.; Kumar, V. Text categorization using weight adjusted *k*-nearest neighbor classification. Springer. 2001

He, J.; Ding, L.; Jiang, L.; Ma, L. Kernel ridge regression classification. Proceedings of the International Joint Conference on Neural Networks. 2014, 2263–2267.

Hofmann, T. Probabilistic latent semantic indexing. ACM SIGIR Conference, 1999.

Howland, P.; Jeon, M.; Park, H. Structure preserving dimension reduction for clustered text data based on the generalized singular value decomposition. SIAM Journal of Matrix Analysis and Applications. 2003, 25(1), 165–179.

Howland, P.; Park, H. Generalizing discriminant analysis using the generalized singular value decomposition, IEEE Transactions on Pattern Analysis and Machine Intelligence. 2004, 26(8), 995–1006.

Hull, D.A. Stemming algorithms: A case study for detailed evaluation. JASIS 47. 1996, 1, 70–84

Jiang, X.; Ringwald, M.; Blake, J.; Shatkay, H. Effective biomedical document classification for identifying publications relevant to the mouse Gene Expression Database (GXD). 2017.

Lewis, D.D. Naive (Bayes) at forty: The independence assumption in information retrieval. In Machine learning: ECML-98, Springer. 1998, 4–15.

Li, Y. ; Jain, A. Classification of text documents. The Computer Journal. 1998, 41(8), 537–546.

Lichman, M.. UCI Machine Learning Repository Irvine, CA: University of California, School of Information and Computer Science. https://archive.ics.uci.edu/ml/datasets.html, 2013.

Liu, H.; Motoda H. Feature Extraction, construction, and selection: A Data Mining Perpective. Boston, Massachusetts, MA, USA: Kluwer Academic Publishers,1998.

Manning, C.D.; Raghavan, P.; Schütze, H. Introduction to information retrieval. Cambridge University Press Cambridge. 2008,1.

Mishu, S. Z.; Rafiuddin, S. M. Performance analysis of supervised machine learning algorithms for text classification, 19th Int. Conf. Comput. Inf. Technol. 2016, 409–413.

Nguyen, D.B.; Shenify, M.; Al-Mubaid, H. Biomedical Text Classification with Improved Feature Weighting Method. BICOB 2016, April 4–6 2016, Las Vegas, Nevada, USA. 2016.

Porter, M.F. An algorithm for suffix stripping. Program: Electronic Library and Information Systems. 1980, 14, 3, 130–137.

Rocchio, J.J. "Relevance Feedback in Information Retrieval" The SMART Retrieval System. 1971, 313–323.

Saif, H.; Fernández, M.; He, Y.; Alani, H. On stopwords, filtering and data sparsity for sentiment analysis of twitter. In Proceedings of the Ninth International Conference on Language Resources and Evaluation (LREC 2014), Reykjavik, Iceland, 26–31 May 2014.

Salton, G.; Wong, A.; Yang, C.S. A vector space model for automatic indexing. Commun. ACM 18. 1975, 11, 613–620.

Samal, B.R.; Behera, A.K.; Panda, M. Performance analysis of supervised machine learning techniques for sentiment analysis. Proceedings of the 1st ICRIL International Conference on Sensing, Signal Processing and Security (ICSSS). Piscataway, IEEE. 2017, 128–133.

Sebastiani, F. Machine learning in automated text categorization, ACM Computing Surveys. 2002,34(1).

Silva, C.; Ribeiro, B. The importance of stop word removal on recall values in text categorization. In Proceedings of the International Joint Conference on Neural Networks. Portland, OR, USA. 2003, 3, 1661–1666.

Sokolova, M.; Lapalme, G. A systematic analysis of performance measures for classification tasks, Information and Processing and Management. 2009, 45(4), 427–437.

Turtle, H.; Croft, W.B. Inference networks for document retrieval. In Proceedings of the 13th annual international ACM SIGIR Conference on Research and Development in Information Retrieval. ACM. 1989, 1–24.

Webster, J.J.; Kit, C. Tokenization as the initial phase in NLP. In Proceedings of the 14th Conference on Computational Linguistics. Association for Computational Linguistics. 2010, 4, 1106–1110.

Yang Y.; Pederson, J.O. A comparative study on feature selection in text categorization, ACM SIGIR Conference, 1995.

CHAPTER 19

ENERGY EFFICIENT OPTIMUM CLUSTER HEAD ESTIMATION FOR BODY AREA NETWORKS

P. SUNDARESWARAN* and R.S. RAJESH

Department of Computer Science and Engineering, Manonmaniam Sundaranar University, Tirunelveli, India

Corresponding author. E-mail: psundareswaran@msuniv.ac.in

ABSTRACT

Wireless sensor networks (WSNs) are becoming increasingly familiar since they are an inevitable part of human-centric applications. The WSNs are also used in health applications. It has small, low energy sensing devices that can sense the data as required and send it to the collecting base station. Since the sensor nodes have limited energy and in most of the applications sensors are not replaceable and rechargeable, energy conservation of sensor nodes is the primary goal over the design of WSNs. Although a lot of techniques are available for energy conservation, clustering is the important method used for preserving energy. Body area networks (BANs) are one of the specific applications of WSNs. This type of network is normally used to monitor the health of the human and functionalities of various organs of the human body. The sensors are generally implanted or positioned in the human body. Therefore, the BANs are helping the medical attendants to access the patient's conditions regularly and keep track of the medical data of the person. The BANs consist of different type of sensors, they can sense patients pulse rate, sugar level, blood pressure, etc. Similarly, wearable devices are also having sensors to watch the behaviour of the organs of the body. These sensors sense the data and send them to the base station, where they will be analyzed and processed by the medical experts at any time. If any emergency occurs, the system will alarm the medical assistants and rapid actions will be taken. As the sensors are implanted, it is not possible to replace them. Since they are energy limited, the conservation of energy is very important in this type of applications. In this work, the sensors implanted and used in wearable are considered as reactive sensors. The reactive sensors trigger to send the sensed data when it is beyond the hard threshold value or the difference between the two consecutive sensed values is greater than the soft threshold value. These sensors are grouped under several clusters so that they can send the data to the cluster heads instead of the base stations to preserve energy. The cluster heads in turn send the received data to the base station after aggregation is performed. The triggered nodes or otherwise called as active nodes are the only busy nodes at the corresponding round and other sensors are kept under inoperative state. Hence, the active sensors are considered for the calculation of the optimum number of cluster heads. The optimum number of clusters plays a crucial role in deciding the clusters in each round of operation as they can also save the energy consumption of sensor nodes (cluster heads). The numbers of active nodes present in the earlier round and current round are determined and the change in the total network energy between two consecutive rounds is also computed. The ratio of active nodes among two consecutive rounds and the ratio of total network energy between two consecutive rounds are calculated. These values play a significant role in computing the suitable cluster heads for each round. After the cluster heads count is computed, the LEACH method is used to elect the cluster heads. If the elected cluster heads are below the optimum number, the balance cluster heads are selected among the remaining eligible nodes. If the cluster heads elected are greater than the optimum number, the cluster heads above the optimum numbers with least energy are converted as normal nodes. Simulations are performed and the results are compared with the existing protocols. The experimental results show that the proposed protocol outperforms the existing protocols in terms of network lifetime and throughput.

19.1 INTRODUCTION

A wireless sensor network (WSN) is a centralized, distributed network having a lot of tiny, self-directed, and little powered devices called sensor nodes. Each sensor node is having a RF transceiver, multiple types of memory, a power supply, accommodates different types of sensors and actuators with processing capability. The nodes are deployed in a random or an organized manner depending upon the application in which they are being used. These nodes can exchange data among themselves through the wireless mode and are self-organized after beginning to function. The sensor devices are energy efficient and include the potential for multifunctionality. WSNs are classified into homogeneous and heterogeneous WSNs. Sensor devices within the homogeneous WSNs have an equivalent amount of initial energy. Heterogeneous sensor network nodes are provided with dissimilar initial energy. Widespread industrial applications (Akyildiz et al., 2002) use the services of the wireless sensors. The group of sensors collect information from the environment to carry out and process them to meet the particular application objectives. Wireless sensors are commonly used in industrial, social, and commercial applications because of their advances in processing power, communication ability, and capacity to utilize minimum power. Sensor nodes are used to sense environmental metrics like heat, pressure, humidity, noise, vibration, etc. They also have the capability of sensing the air particles (carbon dioxide, etc.) and underwater components. The important application areas of sensor nodes include industry (Lin et al., 2004; Gungor et al., 2009), military (Hussain et al., 2009), environment monitoring (Yick et al., 2008), and transport and agriculture (Feng et al., 2008; Coates et al., 2013; Wang et al., 2006). Sensors are used in hazardous environments, where the maintenance and replacement of components by the human beings is tricky and dangerous. Since the sensor devices are provided with limited energy, energy saving is an important criteria as far as the design of a WSN or BAN is concerned. During the last decade, considerable amount of research in the area of BANs has focused on issues associated with wireless sensor design, size of the sensor devices, power-aware sensor circuit design, cost-effective sensors, signal synthesizing methods, and design of communications protocols.

In the earlier works, it is reported that the sensor devices transmit the acquired data directly to the sink. Therefore, these devices would become dead rapidly that leads to the death of the total network. Alternatively, with the introduction of clustering techniques, the sensor nodes transmit the data directly to their respective cluster head rather than sending them to the far away from the base station. This approach thus reduces a considerable quantity of energy utilization that affects the increase of network lifetime. Threshold sensitive energy efficient network (TEEN) protocol (Manjeshwar et al., 2001) is the significant cluster-based mechanism designed for reactive sensor networks. In reactive sensor networks, the sensors are kept idle. The sensors are triggered to activate, sense, and send the information based on the sudden changes in the sensed attributes, thus preserving the energy. The computation of the number of clusters for each round of function has been an important issue in the reactive sensor networks or BANs using clustering protocols. In equally distributed cluster head methodology (DECH; Sundareswaran et al., 2015), the cluster heads are distributed equally within the clusters when the cluster heads are placed close to each other. In this work, in addition to the DECH, we have computed the change in the total energy of the network between successive rounds and the change in the number of active nodes between successive rounds. An inverse exponential function has been employed to compute the optimum cluster heads for each round of operations that leads to improvement in a network lifetime and stability period. As wireless body area network (WBAN) is one of the applications of the WSN, it is assumed that this research is based on the WSN. All the sensor nodes are considered as wireless sensor nodes. It is assumed that the nodes are having the characteristics of a WSN node. The rest of this chapter is organized as follows: Section 19.2 deals with BANs, Section 19.3 explains the related works, and Section 19.4 describes the concept of the proposed methodology. Section 19.5 discusses the results and concluding remarks are provided in Section 19.6.

19.2 BODY AREA NETWORKS

The incorporation of new technologies with WSNs leads to the development of BANs. A BAN, also referred to as a WBAN or otherwise called a body sensor network, is a wireless network having wearable computing devices and sensor devices implanted within the human body. The BAN is one of the significant applications of WSNs, normally used for monitoring the human physical condition (Bao et al., 2006). The most noticeable application of WBANs is in the medical field, human body care, and patient caring. The BAN devices may be implanted inside the body or otherwise surface-mounted on the body at a permanent spot (Darwish et al., 2011). These sensors can be used for continuously watching the movement of a person, reading the essential factors like heart rate, ECG, EEG, blood pressure, etc. and sensing the neighboring atmosphere (Chen et al., 2011). Even though many of the existing healthcare systems work based on the wired connections, BAN can be a very effective solution in a healthcare system where a patient needs to be observed constantly and needs mobility. BAN differs from other WSN by few significant aspects. Mobile sensors are mostly used in BANs. The patient wearing the sensor devices can move either within the predefined environment or through different environments. The BAN nodes support low energy usage. The cost is also cheaper when compare with the WSN nodes. Considering the aspects of reliability, node complexity and density, BAN nodes are however traditional. In battlefield situation with large number of soldiers, it is essential to watch the signs of soldiers and amount of stress induced by temperature or similar factors, to read the physical and psychological performance of the troops (Jovanov et al., 2003). In order to save the energy of the sensors in the WBANs, the reactive sensor nodes can be used. The reactive sensors transmit the acquired information to the sink when the data sensed are deviated from the threshold values.

19.2.1 CHARACTERISTICS OF BODY AREA NETWORK

A node of the WBANis defined (Movassaghi et al., 2014) as a physical entity that has the capacity of communication with others and possesses limited ability to process the data. The components present in a typical WBANs includes the following:*Personnel device*: The responsibility of this device is to acquire the data transmitted from the sensors and actuators. It is also used as an interface to communicate with other users. It can be otherwise called body control unit or sink.

Sensors: Sensors are used in WBAN to measure the given properties internally or externally. Based on the physical stimuli, the sensors read the information, process, and transmit them to the personnel device. These sensors are classified as physiological sensors, ambient sensors, or biokinetics.

Actuators: The role of the actuator is to cooperate with the user upon getting data from the sensor nodes. Based on the sensed data, it gives feedback to the network.

Implant nodes: These tiny sensor nodes are kept inside the human body.

Body surface nodes: These nodes are mounted on the surface of the human body or few centimeters away from the body.

External nodes: These nodes will not be in contact with the patient's body. They are kept away from the body by more than 5 cm but within 5 m.

The other components present in the WBANs, which are mainly used for the communication purpose are as follows:

The coordinator: The function of the coordinator node is to act as a gateway to the rest of the world, similar to WBAN, a security-based trust center. In general, the coordinator of a WBAN is the access point or a PDA device, which can be used for entire communication among the sensor nodes.

Relay: These nodes are transitional nodes used for transferring the messages. The relay nodes have parent and child nodes.

End nodes: The end nodes are developed such that they can be used for the specific application only. They do not have the capability of passing the data.

In a BAN with a large number of nodes as in war places, the topology used is dynamic. If a WBAN has a limited number of sensor nodes, the IEEE 802.15.6 standard recommends the star topology to operate among the nodes. There are two types of the star topology, namely, the single-hop star topology and double-hop star topology. The star topology uses two types of communications and they are beacon mode and nonbeacon mode. Using the beacon mode, periodic beacons are broadcasted by the network coordinator to define the starting and the finishing of a frame to permit network association control and synchronization of the devices. In the nonbeacon mode, carrier sense multiple access with collision avoidance is used by the nodes to send data to the coordinator.

19.2.2 BODY AREA NETWORK ARCHITECTURE

The BAN is divided into three tiers of devices based on function and communication. Figure 19.1 illustrates the architecture of a WBAN. Tier 1 consists of body sensor nodes. The location of these nodes may be planted on the body or within the body otherwise near, and there are two types of sensors. Among the body sensors in tier 1, certain sensors set to react and gather data based on the physical stimuli of the one, and then process and transmit the quantity to the portable device. The second one is actuators that perform patient's medical supervision. The sensors within the same network send the information to the actuator, or the user can interact with the actuator. Two methods of communication are performed in tier 1 of the BANs. One communication is between the body sensor nodes available in tier 1. The other is between sensors and a portable personal server or device (PS/PD). In tier 2, the PD or access points are used to collect the data sent by the body sensors. The PD is the device used in tier 2. The information sensed by the sensors and acquired by the actuators are collected at the

PD and sent through access points to outside networks wirelessly. The communication standards used in this tier include bluetooth/bluetooth low energy, ZigBee, ultra-wide band (UWB), cellular, and WLAN. The outside users can communicate with the BAN using a gateway. Therefore, the medical supervisors remotely attending the patients can get the information immediately through wireless communication or the Internet. Therefore, the state of the patient is clearly monitored, and on emergency conditions, the ambulance that is connected to the outside WBAN is informed. The important design areas in the WBAN architecture are (a) sensors, energy or power, and network hardware components that helps in the design of body sensors nodes, position and locates the sensor nodes in WBAN, signal processing, data storage and feedback mechanism, power source, energy harvesting technologies, dynamic control, and antenna design. (b) Protocol stack for radio and wireless transmissions, channel modeling, interfaces with other wireless communication standards, interference, efficient medium access control (MAC) protocols, error correction methods, and cross-layer techniques. (c) Position and mobility deal with the position and movable property of the sensor nodes. (d) Security issues related to integrity, confidentiality, authentication, and secured communication.

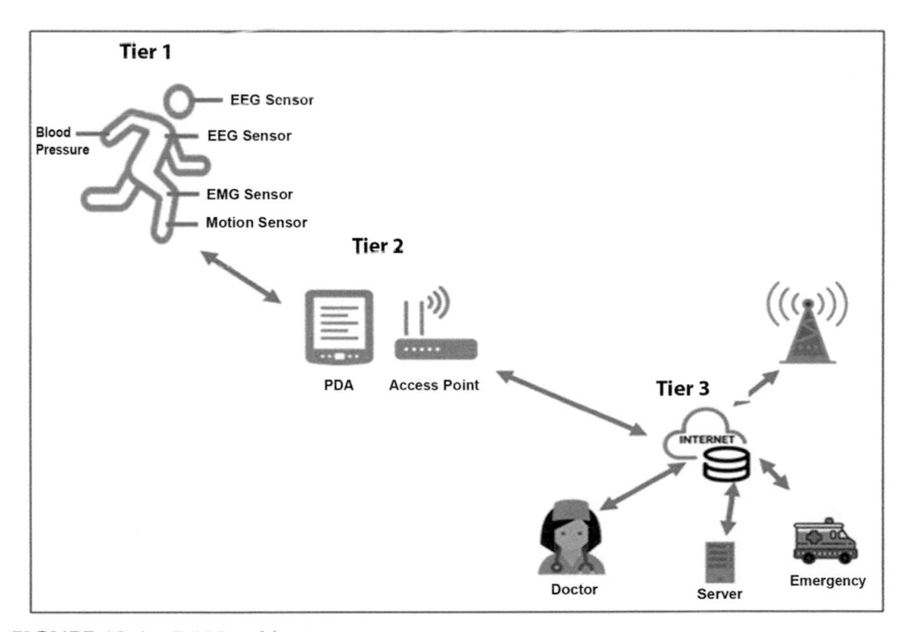

FIGURE 19.1 BAN architecture.
(*Source:* Adapted from Movassaghi et al., 2014.)

19.2.3 BODY AREA NETWORK APPLICATIONS

BANs have a huge potential to change the future of health care monitoring and patient health information by identifying critical health conditions and providing real-time tracking of patient health. The applications of BAN is categorized into medical and nonmedical applications. The medical applications includes assessing soldier fatigue and battle readiness, aiding professional and amateur sports guidance, sleep staging, diabetes and asthma control, and patient monitoring. The nonmedical applications are real-time streaming, emergency (nonmedical), entertainment applications, emotion measure, and secure authentication.

19.2.4 BODY AREA NETWORK LAYERS

The communication used in a BAN environment should be in such a way that it has high reliability, low complexity, low price, ultra-low energy consumption, and short-range communication. Since the existing layers of the IEEE standards are not meeting the requirements of the WBANs exactly, the new physical (PHY) and MAC layers exclusive for BANs has been defined by IEEE 802.15.6 (WBAN) working group.

Physical layer: The PHY layer of IEEE 802.15.6 designed for the BANs performs the following tasks: activation and deactivation of the radio transceiver built in the device, clear channel assessment (CCA) within the current channel and data transmission and reception. Depending upon the application, the physical layer is selected. Three types of physical layers have been specified by the IEEE 802.15.6 and they are classified as human body communication (HBC), narrow band (NB), and UWB. NB PHY is specifically meant for data communication by a WBAN node, activation or deactivation of the radio transceiver within the node and CCA in the existing channel. HBC PHY issues the electrostatic field communication requirements that encapsulate modulation, preamble/start frame delimiter, and packet structure. The UWB physical layer is mainly focussed on communication between on-body devices and transmission/reception between both on the body and off-body devices. Since the HBC PHY has been defined in different bandwidths for different countries, the finalization of physical layer frequency bands was the key issue in the development of the IEEE 802.15.6 standard.

MAC layer: The IEEE 802.15.6 standard working group places the MAC layer above the PHY layer so that channel access can be easily controlled. The whole channel is divided into a series of superframes for the purpose of time referenced resource allocations. This is performed by the hub or otherwise called the coordinator. The coordinator is responsible for channel access coordination. The coordinator performs this through any one of the following access modes: beacon mode with beacon period superframe boundaries, non-beacon mode with superframe boundaries, and nonbeacon mode without superframe boundaries. The responsibilities of the MAC layer in IEEE 802.15.6 are the same as with other wireless communication standards and topping with additional responsibilities.

19.2.5 ROUTING IN BAN

A lot of routing protocols are available for ad-hoc networks and WSNs. Even though BANs have identical functionalities of WSN and ad-hoc networks, it has its own distinct characteristics. As BANs have more stringent energy restrictions in terms of energy transmission compared to a traditional sensor and ad-hoc networks, the node replacements specifically for implant nodes (nodes within the human body) is not a viable one and might require sometimes a surgery in particular scenarios. Therefore, to avoid the charging of power source and replacing the batteries, the network lifetime of the BAN should be improved. An important issue in communication among the BAN devices is the reliability of the transmission since the monitored data should be properly reached to the medical professionals.

The routing protocols in BANs are classified as follows:

(a) *Cluster-based algorithms:* These types of algorithms allot nodes in WBANs into separate groups named clusters, and a cluster head is allotted for every single group. These cluster heads are used to collect the data from the nodes to the sink. Therefore, the multihop transmission is avoided and the amount of direct communications from the nodes to the sink is also decreased that save a considerable amount of energy. The direct transmission of data from the sensor node to a faraway base station requires more energy than sending data by the sensor node to nearby cluster heads. However, the huge overhead and delay related to cluster and cluster head selection are the main drawbacks of these protocols.

(b) *Probabilistic algorithms:* These types of algorithms

use a cost function to establish a path between nodes. Protocols using this type of algorithm use link-state information to renew the cost function. These algorithms construct the path among routes between nodes with minimum cost. The disadvantage of these algorithms is that a lot of transmissions are required to update the link-state information.

c) *Cross-layer algorithms:* These algorithms combine the network layer interface difficulties with neighboring layers. The advantages of this type of algorithms are minimum energy utilization, good throughput, and constant end-to-end delay. A network with high path loss and mobility will perform poorly while using these algorithms.

d) *Temperature-based algorithms:* The wireless communications generate radio signals that in turn produce electromagnetic fields. These electromagnetic fields are passed into the human body that causes a rise in human body temperature. This will cause a decrease in blood flow and damage the sensitive organs due to the rise in heat. Therefore, the primary concern of all thermal-based algorithms for BANs is to stay away from routing to hot-spots. Tissue temperature of the human body is also varied due to the electromagnetic fields. Heavy data traffic is one of the main causes of tissue heating. Preventive mechanisms for tissue heating are the effective implementation of traffic control mechanisms and decreasing the power of transceivers.

e) *QoS-based routing algorithms:* The final classification among the routing algorithms is QoS routing protocols. A modular method is followed here by providing individual modules for each QoS parameter. These modules are functioning cooperatively and sharing resources among them. The modules used in QoS-based routing algorithms are the reliability-sensitive module, the power efficiency module, the neighbor manager, and the delay-sensitive module. Hence, these algorithms provide higher reliability, lower end-to-end delay, and better throughput.

WSNs have a great number of clustering algorithms while few

algorithms are addressed for BANs. This research points towards the clustering algorithms to study the existing clustering methods and how to improve the performance of these algorithms with respect to network lifetime, throughput, and packet delivery ratio on specific environmental conditions.

19.3 RELATED WORKS

A lot of methods have been developed toward energy conservation of nodes in the WSNs. Diverse mechanisms have been suggested in the literature for conserving the energy in WSNs. Duty cycling and data-driven approaches (Giuseppe et al., 2009; Rezaei et al., 2012) are mainly applied in sensor nodes. The duty cycling approach switches off the transceiver to the sleep mode when data are not transmitted and makes the nodes ready to receive the information as soon as available. The time duration of the nodes in the active state is called a duty cycle. A collection of energy-efficient MAC protocols is evolved (Pei et al., 2013; Demirkol et al., 2006; Naik., 2004; Batra., 2016) to preserve the energy. Mobility (Sara et al., 2014) also plays a role in energy conservation in sensor networks. As the traffic around the sink in a network is always greater than that in the rest of the area, the nodes around the base station die soon. Therefore, the mobile base stations can be used to collect the data. Another approach to conserve the energy is by using the clustering approach in which the information read by the sensor nodes is directly routed to the cluster heads. A considerable amount of energy will be spent to transmit the data depending on the distance between the sensor node and the sink. Therefore, the far-away nodes would drain earlier. This will affect the performance of the network at the initial rounds. The clustering technique perhaps avoids this situation by selecting the higher energy nodes as cluster heads and the remaining nodes send the data to the nearby cluster heads. The cluster head in turn collects and aggregates the data and sends it to the base station. Therefore, minimum energy is required to transmit the data to the nearest cluster head. Data aggregation (Nakamura et al., 2007) is a useful method performed by the cluster heads so that the redundant data are eliminated instead of being sent to the sink.

Many of the clustering algorithms (Heinzelman et al., 2002; Manjeshwar et al., 2002; Younis et al., 2004; Qing et al., 2006) focus on the selection of the cluster heads between the sensor devices. LEACH (Heinzelman et al., 2002) is the pioneer in protocols used for clustering the WSN nodes. In the

LEACH protocol, the nodes are selected as cluster heads depending on the probability value. Each node is assigned with a probability Pi(t) at time *t*. A sensor will be elected as cluster head only when it has not been a cluster head in most recent rounds (r mod (N/k)), and which presumably has higher energy than the other sensors. The probability to become a cluster head is thus calculated as:

$$P_i(t) = \begin{cases} \dfrac{k}{N - k_0^* \left(r \bmod \dfrac{N}{k_{otherwise}} \right)} C_i(t) = 1 \end{cases} \quad (19.1)$$

where N is the nodes present in a WSN, r is the current round and k is the likely number of cluster heads at round r. Variants of LEACH (Salim et al., 2014; Batra et al., 2016; Arumugam et al., 2015) are developed in later stages with improved performances. A coordinator-based cluster head election method was proposed (Wu et al., 2011) and the network performed in a better way. Facility location theory is used to resolve the incapacitated facility location problem (Jain et al., 2011) and the clustering model saved the energy to a specific extent. Another distributed clustering scheme HEED (Younis et al., 2004) selects cluster heads from the deployed sensors based on a hybrid of communication cost and energy. In HEED, each sensor is directly connected with only one cluster head. Ding et al. (2005) have devised another algorithm DWEHC to overcome the drawback of HEED. Every node finds its weight after identifying its neighbor nodes in neighboring vicinity. The weight is the collection of energy and closeness to the neighbors. A node having a larger weight among the others will be elected as the cluster head. Even though HEED and DWEHC look similar, the cluster heads are evenly distributed in DWEHC than that in HEED. WSNs with nodes having different energy levels at the beginning are called heterogeneous WSNs. Two classes of sensors with dissimilar energy levels are employed in Stable Election Protoco (Smaragdakis et al., 2004) and they are called *normal* nodes and *advanced* nodes. The *advanced* nodes have ($1 + \alpha$) times the energy of the *normal* nodes. The threshold value for finding the eligible cluster heads is calculated based on the weighted probabilities. Another distributed energy-efficient clustering protocol has been developed for heterogeneous WSNs, called distributed energy efficient clustering (DEEC) (Qing et al., 2006). In DEEC, a probability ratio between the residual energy of each node in the network and the average energy of the network is used to select the cluster head. A node having more initial and residual energy will be having a better chance of becoming a cluster head among the nodes.

Threshold-sensitive energy efficient sensor network (TEEN), an example of the homogeneous clustering method used in WSNs, uses two threshold values, namely, *hard threshold* and *soft threshold*. If the acquired value is far away from the *hard threshold*, the sensor immediately gets activated and transmits the sensed value. *The soft threshold* has the highest deviation among the two acquired values, beyond which the node triggers to transmit. The numbers of clusters are calculated based on the probability function. In most of the applications using TEEN, the nodes will be triggered to active state only if the sensed value exceeds the soft and hard threshold values. Otherwise, the nodes are kept in an idle state. In a real-time deployment of thousands of sensor nodes, only a selected quantity of nodes at a specific location may be activated. For example, like sensor networks monitoring temperature, at a place where the temperature exceeds the threshold value, the nodes around that area gets activated. TEEN protocol considers all nodes for cluster head computation, whereas the TEEN-DECH approach computes the closely placed cluster heads and distributes them equally. A system-level energy utilization model related to communication distance and communication speed is designed (Yi et al., 2015) for on-body wireless communication. In this research work, an attempt is made to analyze the role of network energy and active nodes for cluster head computation. Therefore, the ratio of active nodes present in the successive rounds and network energy is considered to decide the (optimum) number of clusters for the round.

19.4 PROPOSED METHODOLOGY

The suggested technique uses the first order radio model for experimental purposes. This work proposes a new methodology called adaptive energy efficient cluster head estimation (ACHE) methodology to compute the optimum amount of cluster heads at each round. The subsequent section discusses the radio network model followed by the description of the proposed TEEN-ACHE methodology.

19.4.1 NETWORK MODEL

It is understood that all the sensors in the WSN will be having the same initial energy, that is, the nodes are homogeneous. First-order radio model (Heinzelman et al., 2000) is considered for the simulation study. Depending upon the distance among the source and destination, these radio models are classified into *the multipath* and *free space* model. The radio signals propagate from the source and reach the receiving antenna over two or more paths. This method is known as multipath propagation. Causes for multipath occurrence are due to ionospheric reflection and refraction, atmospheric ducting, and reflection from water bodies and terrestrial objects like mountains and buildings. Phase shifting of the signal and constructive and destructive interference are the effects of multipath. The multipath signals are received in a terrestrial environment, that is, where different types of propagation are present and the signals reach at the receiving station via different ways of paths. Therefore, multipath interference occurs here and causes multipath fading. Transmission antenna and receiving antenna are kept in an obstacle-free environment to have a free space propagation model. The absorbing obstacles and reflecting surfaces are not considered in the free space propagation model. The distance d_0 is calculated as,

$$d_0 = \sqrt{\frac{E_{fs}}{E_{mp}}} \quad (19.2)$$

where E_{fs} is the energy needed to send data within the free space, E_{mp} is the energy needed for sending the data in multipath networks. Therefore, it is assumed that E_{fs} and E_{mp} are the amplifier types in the respective media. The distance d_0 is considered as a threshold value for selecting the media.

Energy used for sending the data:

Let E_T be the power required to transmit a packet of size P at a distance d. The d and d_0 are used to select the media for sending the packet. If $d \leq d_0$, the free space amplifier type is used for sending the data; otherwise, the multipath amplifier is considered.

If $d \leq d_0$

$$E_T = (E_{el} * P) + (E_{fs} * P * d^2) \quad (19.3)$$

else

$$E_T = (E_{el} * P) + (E_{mp} * P * d^4) \quad (19.4)$$

where $E_{el} = E_t + E_{AG}$. E_{AG} is the power used for performing aggregation. In the case of cluster heads, only E_{AG} will be considered, and for the normal sensors, this value would be nothing. The cluster heads collect the data from the nodes within this cluster, process, and aggregate; and a single data is transmitted to the sink instead of sending complete data received from each node. This will reduce energy while transmitting the data. E_t is the energy needed for transmitting 1 bit/m^2.

Energy requirement for receiving the data:

Let E_R be the energy spent on receiving a packet of size P over the distance d

$$E_R = (E_r \times P) \quad (19.5)$$

where E_r is the power required to receive a bit/m^2. Since receiving a bit of message is consuming a significant amount of energy, the protocols focus on minimizing the transmission distances and number of transmission/reception operations for each message. The radio model used here is assumed as symmetric so the energy needed for transmitting a packet from a sensor to another sensor is identical in both directions.

19.4.2 ADAPTIVE ENERGY EFFICIENT CUSTER HEAD ESTIMATION METHODOLOGY

In real-time applications, sensor networks have thousands of sensor nodes. When considering the reactive WSNs, the nodes are stimulated during the sensed value exceeds the threshold. Therefore, the assumption is made that only part of the sensor devices is getting activated at a time. For example, applications like forest fire detection or temperature monitoring, only sensor nodes around the affected area may be activated and the remaining nodes are under an idle state. In a similar way, a WBAN having thousands of nodes is implanted within the soldiers in a war zone; these sensor nodes implanted in the body of the soldiers are considered as reactive sensors. In the earlier works and also in TEEN, the cluster count is calculated based on the probability ratio and the nodes with higher energy that are not being selected in the recent rounds [r mod(N/k)]. The nodes in these cases mean the normal nodes that are periodically sensing and broadcasting the information to the sink. In this work, the nodes are not always sending the data. They are always under idle mode that saves energy consumption. When the sensed information is beyond the defined threshold value, these nodes get activated and start sending the information to the base station or sink or otherwise called the controller. Hence, every round of operation has active nodes which are less than the total quantity of alive nodes. The nodes that are inactive but idle at the corresponding round are called normal nodes. In the earlier works, the normal nodes are also considered for finding the optimum number of clusters. As this work focuses on reactive sensor nodes, the normal nodes have not been considered for the computation of the optimum cluster number. Therefore, the number of clusters found using the earlier method had not been optimized, when less number of nodes are activated. If the cluster heads are placed close with each other, the energy dissipation becomes uneven, which leads to the shortfall of the network lifetime. The TEEN-DECH method has been used to resolve the above drawback.

In this chapter, a novel methodology is suggested to compute the optimum amount of cluster heads needed for each round. The number of cluster heads for each round is calculated by the ratio of the total energy of all the nodes in a WSN at the current round to that at the previous round. The ratio of the number of active nodes at the current round divided by that at the previous round is also taken into consideration. If E_i is the energy of the *i*th node in the WSN, the total energy ($E_{tot}(r)$) of the WSN at round *r* is calculated as

$$E_{tot}(r) = \sum_{i=1}^{n} E_i \quad (19.6)$$

The change in the total energy level (E_c) of the sensor network between two consecutive rounds can be found as:

$$E_c = \frac{(E_{tot}(r))}{(E_{tot}(r-1))} \quad (19.7)$$

where (Etot(r)) is the total energy of the network in round *r* and (Etot(r–1)) is the total energy of the network in the previous round. Let M_t be the ratio of the active nodes between the current and previous rounds

$$M_t = \frac{M_r}{M_{r-1}} \quad (19.8)$$

where Mr has been the total amount of active sensors in round *r* and Mr–1 has been the total amount of active sensors in the round *r*–1. The rate of change of the above-said parameters can be computed as

$$X = [E_c M_t \times 10] \quad (19.9)$$

Now, the given expression is used to find the optimum cluster count

$$Y = 1 - e^{(-\alpha X)}; 0 < \alpha < 1 \quad (19.10)$$

The optimum cluster count is computed as

$$CH_{opt} = [(p * n_a * Y)] \quad (19.11)$$

where *p* is the probability value and n_a is the total number of alive sensor nodes in the network. CH_{opt} gives the total number of optimum cluster heads to be selected in the respective round. If the number of clusters already computed is less than the optimal amount of cluster head (CH_{opt}), the balance cluster heads will be elected from the sensors with more energy and not being elected as cluster head in the recent *r* mod (*N*/*k*) rounds are selected as cluster heads. Otherwise, if the clusters already computed are greater than the optimum cluster count, the numbers of excess cluster heads are found and among the existing cluster heads, cluster heads with least energy are converted as normal nodes. After the cluster heads are selected, the

TEEN-DECH method is applied to find the closely placed cluster heads. If these cluster heads have been closely placed, these closely placed cluster heads are replaced by the members with in the same cluster having minimum link cost.

19.5 SIMULATION RESULTS

Modifications have been made on the existing TEEN and TEEN-DECH protocols and the simulations have been performed using MATLAB R2013a. A number of simulations with different parameters were executed to evaluate the performance of TEEN and TEEN-DECH protocols with the proposed methodology. Experimental results show that the proposed methodology outperforms the existing protocols in terms of a network lifetime and throughput. A field dimension of 100 × 100 has been considered as a simulation environment for the experiment. The nodes used for this simulation are 100. Various network parameters assumed for this simulation are given in Table 19.1.

19.5.1 SIMULATION ENVIRONMENT

Every researcher wants to have an easy, reliable, flexible, and error-free simulation tool for new prototype development, modification, and testing. This tool should incorporate appropriate analysis on output data, integrate various models, finite mathematical functions, and statistical accuracy of the simulation results. The tool used for simulating the WSN for performing the developed protocol is MATLAB R2013a. The required simulation procedures for transmitting nodes, modeling the communication channel, and receiving node architecture are available here.

This tool is an easy-to-use environment for beginners where problems and solutions are expressed in recognizable mathematical expressions. Due to this, this tool is recognized as one of the benchmark network simulations environment and has a significant number of users, including students, researchers, so on so forth.

TABLE 19.1 Network Parameters

Parameters	Values
Base station position	50 × 50
E_{fs} (amplifier used in the free space model)	10×10^{-12} J
E_{mp} (amplifier used in multipath fading model)	0.0013×10^{-12} J
E_{AG} (aggregation energy)	5×10^{-9} J
E_o (initial energy)	0.01 J
Size of packet	2000 bits
Size of control packet	100 bits
P_o (cluster head election probability)	0.1
Number of rounds	500
Number of sensor nodes (n)	100

19.5.2 PERFORMANCE METRICS

The objectives of the performance analysis are to learn and analyze the performance of the work, identify performance problems, identify the factors that should influence expectations, and understand the relationship between expected output and performance analysis. Performance metrics measure and analyze the actions that lead to the result, and this is the key data everyone needs to have to make proper decisions and directions on their research. The given below metrics have been used to analyze the efficiency of the proposed TEEN-ACHE technique with the available TEEN-DECH and TEEN protocols.

1. *Network lifetime*: It is defined as the time until all nodes are drained of their energy. Network lifetime is defined in another way as the time at which the first network node drains out its energy to send a packet.
2. *Throughput*: This is the cumulative quantity in packets reached the base station or controller sent by the sensor devices in a WBAN.
3. *Stability period*: It is the time duration starting from the functioning of a BAN to the demise of the first sensor node.

19.5.3 RESULT ANALYSIS

Several rounds of iterations are performed in this experiment with α ranging from 0.1 to 0.9 in the equation $1-e^{-\alpha x}$ to get the best result for finding the optimal amount of cluster. Experimental results make it clear that the proposed technique performs better while $\alpha = 0.6$. Figure 19.2 illustrates the amount of cluster heads for each round of operation used by proposed method and the existing protocols. According to the algorithm, first the cluster heads have been computed using Equation (19.1) then the algorithm finds the optimal quantity of cluster heads. If the number of cluster heads already computed is greater than the optimum number of cluster heads, the difference numbers of cluster heads with least energy are changed as normal nodes. When the optimal quantity of cluster is greater than the already computed cluster heads, the balance clusters are regenerated using Equation (19.1). The TEEN and TEEN-DECH calculate the cluster heads based on the probability value, but the TEEN-ACHE method computes the cluster headcount using the ratio of active nodes and change in network energy between two successive rounds. The active nodes M_r at round r and M_{r-1} at round $r-1$ are definitely not equal to total number of nodes. Therefore, the active nodes are significant in the computation of cluster count.

The outcome of the suggested TEEN-ACHE methodology is observed using the metrics given in Section

19.5.2. In addition, the performance of the proposed method is compared with that of the existing TEEN-DECH and TEEN protocols and the graphs are shown in Figures 19.2–19.5. Figure 19.2 illustrates the amount of cluster heads in each iteration using TEEN-ACHE and the existing TEEN-DECH and TEEN protocols. It indicates that the TEEN-ACHE methodology optimally computes the cluster heads so that the energy dissipation is minimized that leads to an increase in life of the BAN. Figure 19.3 illustrates the network lifetime of the BAN. Since the optimum amount of cluster heads has been computed and used for the simulation, the unnecessary usage of nodes as cluster heads are averted. This avoids the energy loss due to nodes being cluster heads and improves the network lifetime. Similarly, a small amount of cluster heads inside a network makes a rapid drain of the energy of these cluster heads due to heavy traffic. Therefore, an energy hole has been created inside the network and decreases the network lifetime. It is observed that up to round 320; the TEEN-ACHE performs better than the other protocols, and at the last stages, the TEEN-DECH has little more live nodes.

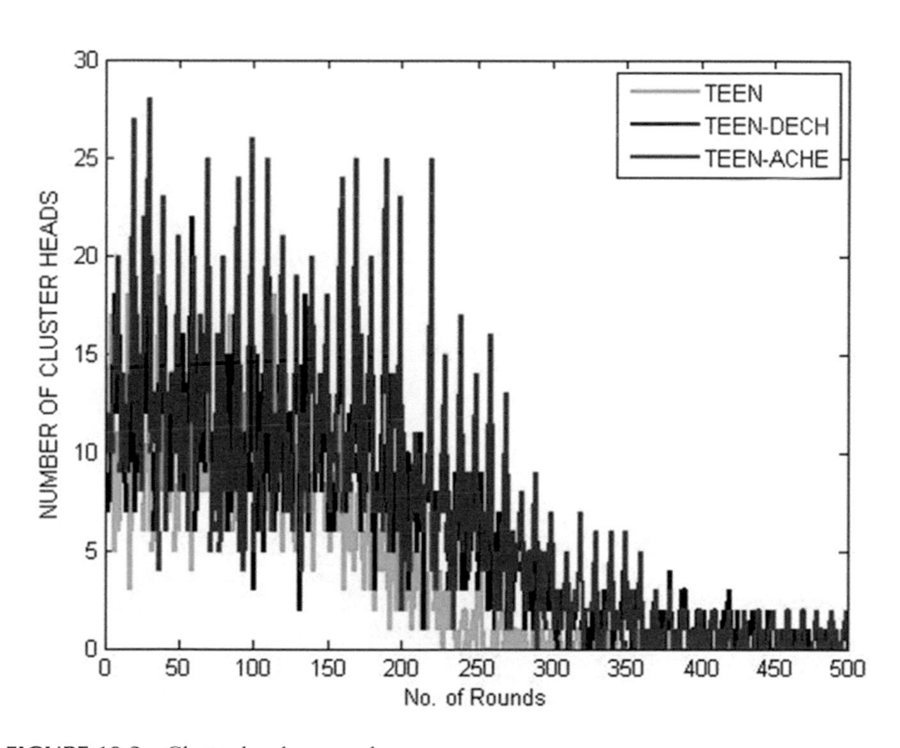

FIGURE 19.2 Cluster head vs rounds.

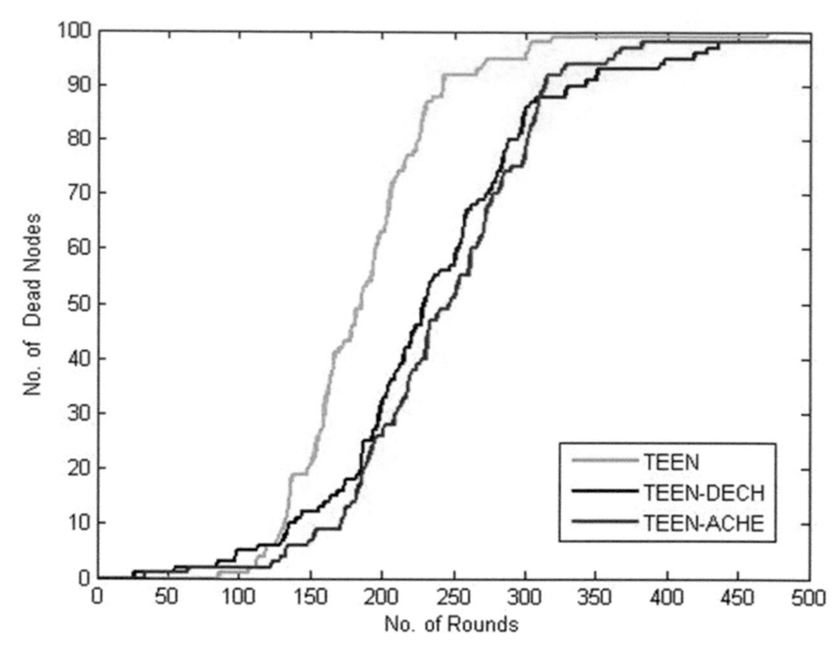

FIGURE 19.3 Network life time.

Figure 19.4 illustrates the amount of data packets received by the base station or sink from the sensor devices available in the BAN. It is inferred that the sink receives much amount of data than that received using TEEN and TEEN-DECH. Since the network lifetime is improved using TEEN-DECH, the nodes are alive till round 350. Therefore, the total quantity of packet received by the base station is increased. The number of packets received by the cluster heads from the respective member sensors for each round is shown in Figure 9.5. This states that the cluster heads receive more data from the sensor nodes when the TEEN-ACHE methodology is applied. Since optimal cluster heads are selected, the network lifetime has been enhanced so that more data are transmitted. Using the optimal cluster count, usage of either more number of clusters or less number of cluster is prevented. This avoids unnecessary use of clusters. Therefore, energy loss due to the cluster is avoided. Similarly, less number of cluster causes heavy traffic at cluster heads that leads to rapid death of cluster heads. Therefore, it is understood that the TEEN-ACHE algorithm functions superior to TEEN-DECH and TEEN with respect to throughput and network lifetime.

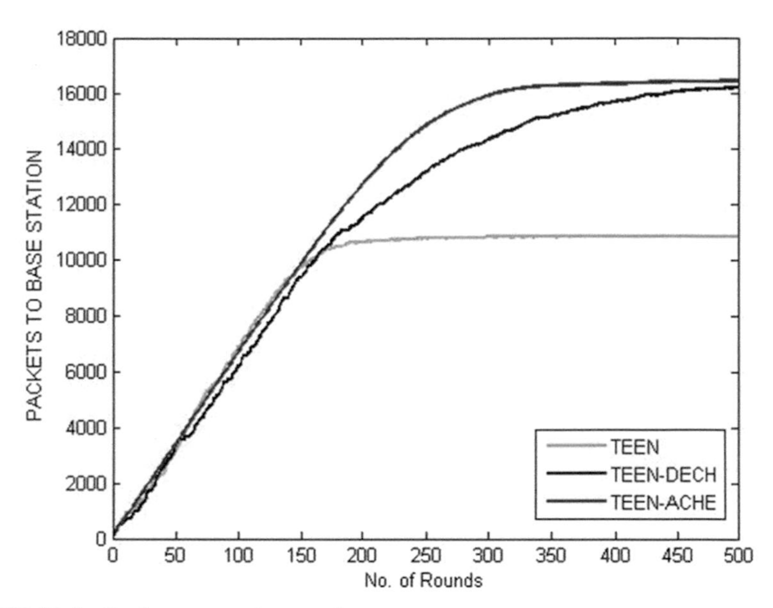

FIGURE 19.4 Packets sent to base station.

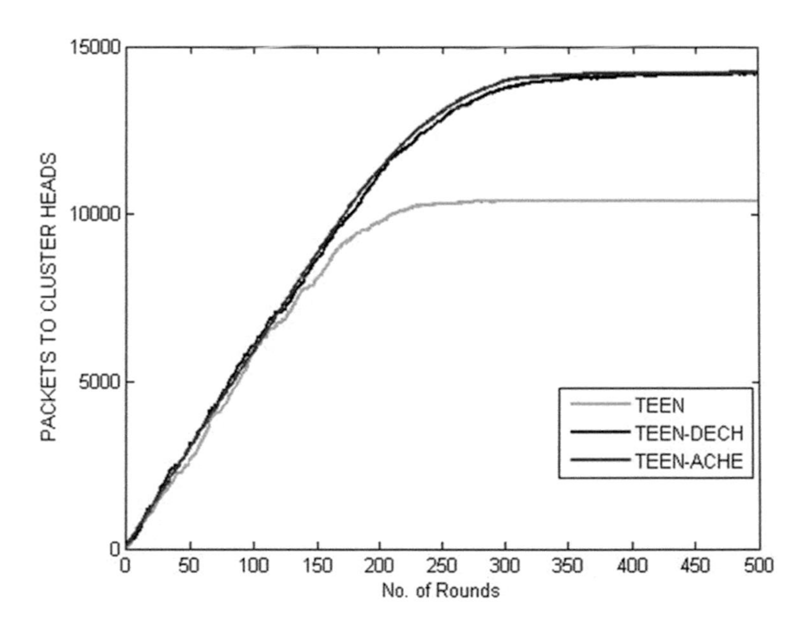

FIGURE 19.5 Packets sent to cluster head.

19.6 CONCLUSION

BANs are striving for finding novel methods to minimize the energy utilization of the nodes. Although numerous methods have been used for the conservation of energy, clustering is a premier mechanism used in WSN and BAN. Clustering techniques reduce energy consumption, packet drop rate, and extend scalability and network lifetime. The requirement is there to identify or compute the correct number of clusters to nullify network overhead that is a demanding task for energy-efficient BAN. There is a difficulty in choosing the most advantageous number of clusters for each round of operation manually. Finding the appropriate clusters to reduce energy utilization is an important challenge in BANs. In this work, a novel methodology is suggested to find the best possible cluster heads for a single round of operations in reactive BANs. The change in total network energy between consecutive rounds and the change in the number of active nodes among consecutive rounds of operations play a major role in the calculation of the optimal cluster headcount. The inverse exponential function is used to optimally find the cluster count for each round. The outcome of the experiments makes clear that during the final rounds of operations the nodes become dead quickly and sustained its level with TEEN-DECH at the last stage. The experimental results depending upon the topology and network metrics illustrate that the network's functionality is enhanced with respect to network lifetime and throughput when using the proposed methodology. For the future extension of this work, the fuzzy logic will be used in selecting the cluster head. Since the heterogeneous devices are connected in a BAN, the algorithm used for electing the cluster head and finding the optimum number should support the heterogeneity of the BAN.

KEYWORDS

- **clustering**
- **sensor networks**
- **energy efficiency**
- **body area networks**
- **TEEN**

REFERENCES

Akyildiz, Ian F., et al. "A survey on sensor networks." IEEE Communications Magazine 40.8 (2002): 102–114.

Anastasi, Giuseppe, et al. "Energy conservation in wireless sensor networks: A survey." Ad Hoc networks 7.3 (2009): 537–568.

Arumugam, Gopi Saminathan, and Thirumurugan Ponnuchamy. "EE-LEACH: Development of energy-efficient LEACH Protocol for data gathering in

WSN." EURASIP Journal on Wireless Communications and Networking 2015.1 (2015): 1–9.

Bao, Shu-Di, Yuan-Ting Zhang, and Lian-Feng Shen. "Physiological signal based entity authentication for body area sensor networks and mobile healthcare systems." *Proceedings of the 2005 27th Annual Conference on IEEE Engineering in Medicine and Biology*. IEEE, 2006.

Batra, Payal Khurana, and Krishna Kant. "A clustering algorithm with reduced cluster head variations in LEACH protocol." International Journal of Systems, Control and Communications 7(4) (2016): 321–336.

Batra, Payal Khurana, and Krishna Kant. "LEACH-MAC: A new cluster head selection algorithm for wireless sensor networks." Wireless Networks 22(1) (2016): 49–60.

Chen M, Gonzalez S, Vasilakos A, Cao H, Leung VC. Body area networks: A survey. Mobile Networks and Applications 16(2) (2011):171–93.

Coates, Robert W., et al. "Wireless sensor network with irrigation valve control." Computers and Electronics in Agriculture 96 (2013): 13–22.

Darwish, Ashraf, and Aboul Ella Hassanien. "Wearable and implantable wireless sensor network solutions for healthcare monitoring." Sensors 11(6) (2011): 5561–5595.

Demirkol, Ilker, Cem Ersoy, and Fatih Alagoz. "MAC protocols for wireless sensor networks: A survey." IEEE Communications Magazine 44(4) (2006): 115–121.

Ding, Ping, JoAnne Holliday, and Aslihan Celik. "Distributed energy-efficient hierarchical clustering for wireless sensor networks." International Conference on Distributed Computing in Sensor Systems. Springer, Berlin, Heidelberg, 2005.

Feng, Gao, et al. "Preliminary study on crop precision irrigation system based on wireless sensor networks for stem diameter microvariation." Transactions of the Chinese Society of Agricultural Engineering 2008(11) (2008): 7–12.

Gungor, Vehbi C., and Gerhard P. Hancke. "Industrial wireless sensor networks: Challenges, design principles, and technical approaches." IEEE Transactions on Industrial Electronics 56(10) (2009): 4258–4265.

Heinzelman, Wendi B., Anantha P. Chandrakasan, and Hari Balakrishnan. "An application-specific protocol architecture for wireless microsensor networks." IEEE Transactions on Wireless Communications 1(4) (2002): 660–670.

Heinzelman, Wendi Rabiner, Anantha Chandrakasan, and Hari Balakrishnan. "Energy-efficient communication protocol for wireless microsensor networks.". Proceedings of the 33rd Annual Hawaii International Conference on System Sciences. IEEE, 2000.

Huang, Pei, et al. "The evolution of MAC protocols in wireless sensor networks: A survey." IEEE Communications Surveys & Tutorials 15(1) (2013): 101–120.

Hussain, Md Asdaque, and Kwak kyung Sup. "WSN research activities for military application.". *Proceedings of the 11th International Conference on Advanced Communication Technology, ICACT'2009*. Vol. 1. IEEE, 2009.

Jain, Kamal, and Vijay V. Vazirani. "Approximation algorithms for metric facility location and k-median problems using the primal-dual schema and Lagrangian relaxation." Journal of the ACM 48(2) (2001): 274–296.

Jovanov, Emil, et al. "Stress monitoring using a distributed wireless intelligent sensor system." IEEE Engineering in Medicine and Biology Magazine 22(3) (2003): 49–55.

Lin, Ruizhong, Zhi Wang, and Youxian Sun. "Wireless sensor networks solutions for real time monitoring of nuclear power plant." *Proceedings of the 5th World Congress on Intelligent Control and Automation, WCICA'2004*. Vol. 4. IEEE, 2004.

Manjeshwar, Arati, and Dharma P. Agrawal. "APTEEN: A hybrid protocol for efficient routing and comprehensive information retrieval in wireless sensor networks." *Proceedings of the International Parallel and Distributed Processing Symposium, IPDPS*. Vol. 2. 2002.

Manjeshwar, Arati, and Dharma P. Agrawal. "TEEN: A routing protocol for enhanced efficiency in wireless sensor networks." *Proceedings of the International Parallel and Distributed Processing Symposium, IPDPS*. Vol. 1. 2001.

Movassaghi S, Abolhasan M, Lipman J, Smith D, Jamalipour A. Wireless body area networks: A survey. IEEE Communications Surveys & Tutorials 16(3) (2014): 1658–86.

Naik, Piyush, and Krishna M. Sivalingam. "A survey of MAC protocols for sensor networks." *Wireless Sensor Networks*. Springer, New York, NY, USA, 2004. pp. 93–107.

Nakamura, Eduardo F., Antonio AF Loureiro, and Alejandro C. Frery. "Information fusion for wireless sensor networks: Methods, models, and classifications." ACM Computing Surveys 39(3) (2007): 9.

Qing, Li, Qingxin Zhu, and Mingwen Wang. "Design of a distributed energy-efficient clustering algorithm for heterogeneous wireless sensor networks." Computer Communications 29(12) (2006): 2230–2237.

Rezaei, Zahra, and Shima Mobininejad. "Energy saving in wireless sensor networks." International Journal of Computer Science and Engineering Survey 3(1) (2012): 23.

Salim, Ahmed, Walid Osamy, and Ahmed M. Khedr. "IBLEACH: Intra-balanced LEACH protocol for wireless sensor networks." Wireless Networks 20(6) (2014): 1515–1525.

Sara, Getsy S., and D. Sridharan. "Routing in mobile wireless sensor network: A survey." Telecommunication Systems 57(1) (2014): 51–79.

Smaragdakis, Georgios, Ibrahim Matta, and Azer Bestavros. SEP: A Stable Election Protocol for Clustered Heterogeneous Wireless Sensor Networks. Boston University Computer Science Department, 2004.

Sundareswaran, P., K. N. Vardharajulu, and R. S. Rajesh. "DECH: Equally distributed cluster heads technique for clustering protocols in WSNs." Wireless Personal Communications 84(1) (2015): 137–151.

Wang, Ning, Naiqian Zhang, and Maohua Wang. "Wireless sensors in agriculture and food industry—Recent development and future perspective." Computers and Electronics in Agriculture 50(1) (2006): 1–14.

Wu, Shan-Hung, Chung-Min Chen, and Ming-Syan Chen. "Collaborative wakeup in clustered ad hoc networks." IEEE Journal on Selected Areas in Communications 29(8) (2011): 1585–1594.

Yi, Chenfu, Lili Wang, and Ye Li. "Energy efficient transmission approach for WBAN based on threshold distance." IEEE Sensors Journal 15(9) (2015): 5133–5141.

Yick, Jennifer, Biswanath Mukherjee, and Dipak Ghosal. "Wireless sensor network survey." Computer Networks 52(12) (2008): 2292–2330.

Younis, Ossama, and Sonia Fahmy. "HEED: A hybrid, energy-efficient, distributed clustering approach for ad hoc sensor networks." IEEE Transactions on Mobile Computing. 4 (2004): 366–379.

CHAPTER 20

SEGMENTATION AND CLASSIFICATION OF TUMOUR REGIONS FROM BRAIN MAGNETIC RESONANCE IMAGES BY NEURAL NETWORK-BASED TECHNIQUE

J. V. BIBAL BENIFA[1*] and G. VENIFA MINI[2]

[1]*Department of Computer Science and Engineering, Indian Institute of Information Technology, Kottayam, India*

[2]*Department of Computer Science and Engineering, Noorul Islam Centre for Higher Education, Kumaracoil, India*

[*]*Corresponding author. E-mail: benifa.john@gmail.com*

ABSTRACT

Brain tumor imaging and interpretation is a challenging problem in the field of medical sciences. In general, healthy brain as well as the tumor tissues appears similar without any differences in the brain images while captured with magnetic resonance technology. It makes several complications in the course of diagnosis and treatment of affected people. Presently, image processing segmentation algorithms are extensively employed to detect the tumor regions in MRI, however it is incompetent to differentiate tumor tissue from the healthy tissues. In this chapter, a practical solution is proposed through a novel algorithm by segmenting the tumor part and healthy part that works based on image segmentation and self-organizing neural networks (NNs). The segmentation algorithm identifies the tumor regions and further boundary parameters will be gathered from the segmented images to feed into the neural network system. The classification efficiency of proposed algorithm is also improved with the aid of neural networks. NN-based tumor segmentation algorithm is implemented using MATLAB 12b and it has been tested with the dataset consists of real images. The results indicate that the proposed neural network-based algorithm to find brain tumors is a promising candidate with intensive practical application capabilities.

20.1 INTRODUCTION

Cancer is a life-threatening syndrome that occurs because of uncontrolled growth of cells causing a lump called a tumor (Autier, 2015). Typically, the tumor region grows and spreads in an uncontrollable manner that leads to causality. Brain tumor or neoplasm within the cranium happens when anomalous cell grows inside the brain region (Gonzales, 2012). Tumours are broadly classified into two categories, namely malignant tumors and benign tumors. Further, cancerous or malignant tumor in brain is categorized into primary tumors and secondary tumors. Here, primary tumors originate within the brain while secondary tumor (metastasis) spreads from any other location to brain. The symptoms of brain tumors differ based on the affected regions of the brain and may comprise of headaches, seizures, vision related issues, and mental disorders. Other typical symptoms include complexity in walking, utterance, and frequent unconsciousness. Hence, detection and segmentation of tumor regions obtained from Magnetic resonance images through advanced image processing techniques are essential for the present scenario.

Magnetic resonance imaging (MRI) technology helps for acquiring the image of brain and it differentiates healthy and tumor tissue from the acquired image. Tissue level differentiation is always challenging due to the unavailability of accurate image classification techniques (Nabizadeh and Kubat, 2015; Breiter et al., 1996). It occurs because of variation in size, location, and image intensities exist in the real images and tumor regions cannot be easily detected by computer segmentation algorithms. Conversely, manual segmentation is observed to be a complex and time-consuming process. MRI is a common technique used for observing and collecting detailed images of the brain. Segmentation of tumor regions from MR images plays a vital role on diagnosis and tumor cell growth rate forecasting as well as to plan effective treatment strategies. Tumors such as meningiomas can be easily segmented, while gliomas and glioblastomas are more complicated to be localized on MRI. Further, these tumors are always hard to segment because the MR images are always diffused and in low contrast with an extended tentacle-like structures. The shape of tumor is always irregular and the size varies at regular intervals that pose additional complications. Hence, the voxel values for the images obtained through MR cannot be standardized like X-ray computed tomography (CT) scans. The type of MR machine employed (tesla intensity level) and the acquisition protocol (voxel resolution, gradient strength, field view magnitude and b0 value) influence the picture resolution. For instance, different MR machines offer drastically different grayscale values for the same type of tumor (examined at different cancer diagnosis centers). Healthy brains comprise three types of tissues that include: (i) white matter, (ii) gray matter, and the (iii) cerebrospinal fluid. The objective of brain tumor segmentation is to perceive the tumor regions known as active tumorous tissue, necrotic tissue, and edema (swelling area that is adjacent to tumor). This is done by subtracting the normal tissues from the abnormal areas through MRI. However, the borders of infiltrative tumors like glioblastomas are always fuzzy and complex in nature to discern from normal tissues (Hall et al., 1992).

Multifaceted MRI modality offers better brain tumor segmentation through various (e.g., T1 (spin-lattice relaxation), T2 (spin-spin relaxation), and diffusion MRI (DMRI) pulse sequences). The dissimilarity among these values (modalities) yields a sole signature to every class of tissues. Most recent automated brain tumor segmentation techniques employ tailored features using classical machine learning (ML) pipeline. As said by these methods, image features are initially extracted and then introduced into a classifier algorithm and the training practice would not influence the characteristics of those extracted features (Bengio et al., 2013). Feature description through a specific optional method should be selected based on the competency to learn intricate features openly from in-domain data. For brain tumor region segmentation as well as to merge data across MRI modalities, deep neural network (DNN) is employed to learn feature hierarchies (Menze et al., 2015). Recently, convolution neural networks (CNN) became a foundation for the computer vision researchers because of its excellent performance in the ImageNet Large-Scale Visual Recognition Challenge. CNN is productively used for segmentation problems in nonmedical research areas as well. In this chapter, numerous CNN-based tumor segmentation architectures are presented which includes Maxout hidden units and Dropout regularization (Havaei et al., 2017).

The drawback associated with few machine learning methods is that they carryout pixel classification by not including the local dependencies of labels. To overcome this problem, structured output methods such are usually employed and however these methods are computationally expensive as compared with other methods. Conditional random fields (CRFs) are a typical example for one of the structured output methods. Alternatively, label dependencies can be modeled by considering the pixel-wise probability approximation of an initial CNN as additional input to specific layers of a second DNN and it forms a cascaded architecture. Since, convolutions have the potential to execute competent process it would be significantly faster than CRF (He and Garcia, 2009).

Neural networks (NN) execute classification process by learning from data and exclusive of using rule sets. NN generalizes the variables using previous data and learns from past experience and it performs well on complex, multivariate, and noisy domains such as brain tissue segmentation. Self-organizing map (SOM) is a typical NN that uses an unendorsed competitive learning algorithm and SOM involuntarily classifies itself based on the input data using a similarity factor such as Euclidean distance. Studies show that the employment of SOM is essential to group the output relevant to the topology because it has additional output neurons beyond the types of tissue to be segmented. Topological relations in SOM are preserved in the input and contiguous inputs are connected to adjoining neurons. Clustering the analogous output neurons is generally done by an extra NN that exploits weight vectors as the input parameter (Specht, 1991).

20.2 RELATED WORK

Demirhan et al. (2010) performed a study to detach brain tumor region with SOM where image segmentation procedure is done by separating the images into segments known as classes or subsets according to the features or background subtraction. Subsequently, the images were segmented by means of SOM networks and gray level co-occurrence matrices (GLCM). Previously, the performance of SOM networks and GLCM methods on image segmentation were evaluated by many researchers and it has shown more than 90% success rate on image segmentation applications.

Kaus et al. (1999) described a novel iterative technique for the programmed segmentation of MRI images of brain. This iterative technique integrates a statistical taxonomy scheme and anatomical understanding from an aligned digital atlas. For justification, the iterative method was applied to 10 tumor cases at various places of the brain including meningiomas and astrocytomas (grade 1–3). The brain and tumor segmentation outputs were evaluated with physical segmentations done by four autonomous medical experts. Then, it is confirmed that the algorithm offers better accuracy than the manual segmentation process in a short period.

Reddick et al. (1997) proposed a classification method that does not necessitate any former understanding of anatomic structures and it simply uses the subpixel precision in the zone of interest. They have evaluated the potential of this novel algorithm in the course of segmentation of anatomic structures on a simulated as well as actual brain MR images. Further, the CFM was evaluated to the level-set-based methodology in segmenting objects in a range of brain MR images. The experimental results through MRI signify that the CFM algorithm attains excellent segmentation results for brain and tumor applications.

Song et al. (2007) employed SOM by means of weighted probabilistic NN to desperately fragment the T1 and T2 MR images. The fractional contributions of every direction vector to various target classes are estimated using the training sets and posteriori probabilities are estimated using Bayesian theorem. Their parametric technique assumes that a probability density function of the tissues which is not precise and it does not match with real data distribution. Subsequently, Iftekharuddin (2005) employed feedforward NN combined with automated Bayesian regularization method for classification and further used SOM for clustering.

Low signal-to-noise ratio or contrast-to-noise ratio is found to reduce the correct segmentation ratio irrespective of the technique used. Filtering methods such as space invariant low-pass filtering techniques can be utilized to images to resolve this problem. The limitations of conventional filtering techniques are blurred object boundaries and vital features along with repression of fine structural points such as small lesions in the images. This constraint is set on by using the space-variant filters with feature-dependent techniques. Gerrig and Murphy (1992) investigated the performance of anisotropic diffusion filter that is proposed by Perona and Malik and compared the results with a wide range of filters used to get rid of the arbitrary noise of the MRI. It is observed that anisotropic diffusion filter smudge homogeneous region, boosts the ratio of signal-to-noise and sharpens the object borders. This type of filter diminishes the noise and reduces partial volume effects. It is evident that accurate segmentation of the images relies on the automated feature extraction methods that decide the best features to differentiate dissimilar tissues (Gerrig and Murphy, 1992).

Wavelet transform method is generally employed in feature extraction for brain MR image segmentation because it offers efficient localization in both spectral and spatial domains. The drawback of this method is translation-variant characteristic and it generates different features from the same two images with a minor realignment.

From the literature, brain tumor segmentation methods for MRI are broadly divided into two categories: (i) generative models and (ii) discriminative models (Gerrig and Murphy, 1992). Generative models require domain-specific prior information about the healthy and tumorous tissues of brain images and it was discussed extensively by Prastawa et al. (2003). The appearance of tissues is complex to characterize using generative models and generally it identifies the tumor as a shape or an indication that is different from a typical brain image (Clark et al., 1998).

Specifically, brain tumor segmentation methods depend on anatomical models that are acquired subsequent to the alignment of 3D MR image on an atlas estimated from several healthy brain's images (Doyle et al., 2013). The method specified by International Consortium for Brain Mapping atlas computes the later probabilities of healthy tissues including white matter, gray matter and cerebrospinal fluid. Then, the Tumorous regions are identified by localizing the voxels where the posterior probability estimation is under a definite threshold. Subsequently, postprocessing action is employed to guarantee an excellent spatial regularity in the results (Gerrig and Murphy, 1992). Prastawa et al. (2003) registered brain images on top of an atlas to obtain likelihood map for deformity and hence an active curve is initialized on the map. Subsequently, the above-mentioned process is iterated until posterior probability goes under a threshold.

Several active contour methods rely on left–right brain symmetry features or alignment-oriented features (Khotanlou, 2009; Cobzas et al., 2007). Apart from this, alternatively brain tumor segmentation can be done by employing discriminative models. These methods need comparatively less understanding about the brain's anatomy and necessitate extracting the image features during postprocessing. Subsequently, the relationship among these extracted features will be modeled and labeled. The features include values of original input pixels in local histograms along with surface features that include Gabor filter banks, or alignment-oriented features. Here, the considered alignment-oriented features are inter-image gradient, region shape distinction, and symmetry analysis (Tustison, 2013).

The discriminative models use hand-designed features and the classifier is specifically trained to differentiate healthy from nonhealthy tissues. It is also assumed that the key features have elevated discriminative power because the behaviour of classifier algorithm is self-regulating from the characteristics of those extracted features. The complexity of these hand-designed features requires the computation of huge number of features for ensuring accuracy when employed with conventional ML methods. Efficient techniques always use less features thereby employing dimensionality reduction or feature identification methods for better accuracy. Preliminary analysis has shown that Deep CNNs for brain tumor image segmentation is a proficient technique (Havaei et al., 2017). Hinton et al. (2012) proposed a CNN-based method that consists of a series of convolutional layers for feature detection. Similarly, Nabizadeh and Kubat (2015) investigated various classification methods for tumor segmentation in brain images and suggested that neural network-based methods offer best results as compared to other techniques for the above-stated problem.

20.3 PROPOSED WORK

The brain tumor regions from MR images are differentiated from the healthy tissues by image segmentation and thereby training the NN to obtain the classification efficiency. In order to enhance the accuracy of differentiation, a NN-based classification is proposed and its classification efficiency is computed for brain tumor images. The results obtained through the proposed method are more robust and it supports in accurate segmentation of brain tumor regions.

The sequence of operations involved in the brain tumor segmentation process is presented in Figure 20.1. The input images are selected from standard cancer image databases that are commonly known as MR images. Here, MRI of the brain is balanced by converting it into grey scale image and resized to a standard resolution. The real two-dimensional (2D) MRI scan image of a brain from the DICOM data is obtained from the 64-slice CT scan machine. The sequences of images are taken from different projections by rotating the gantry. This image provides the complete view of a brain and it is used for the subsequent analysis purpose. In the course of analysis, the brain image is segmented for measuring the cancerous area thoroughly inside brain. The regions except the cancerous area are considered to be noise and it has to be filtered.

The input MRI images were normalized, preprocessed and registered initially. Subsequently, the noise in the image, such as the strip of the skull or any other clinically unwanted portion is removed. Once the irrelevant part is removed an anisotropic diffusion filtering is applied all over the image (Nair et al., 2019). Anisotropic diffusion filters outperform isotropic related to few applications such as de-noising of highly degraded edges. Anisotropic diffusion filters typically employ spatial regularization strategies by considering the modulus of the edge detector, ∇u_σ and its direction. The orthonormal system of eigenvectors u_1, u_2 of the diffusion tensor is estimated by $u_1 \parallel \nabla u_\sigma$. In order to achieve smoothing along the edge, Galic et al. (2008) considered to select the relevant eigenvalues λ_1 and λ_2 as

Please check equation something missing

$$\lambda_1(\nabla u_\sigma) := g\left(\left|\nabla u_\sigma\right|^2\right) \qquad (20.1)$$

In mathematical morphology, dilation and erosion are the operations usually performed with binary images to expand the boundaries consist of foreground pixels (Chen and Haralick, 1995). Hence, the regions of foreground pixels increase in size as the holes inside those areas happen to be smaller in size. The input to the dilation operator is an image to be dilated and the coordinate points are called as a kernel. A kernel is a structuring element that decides the accurate consequence of the dilation operator on the input MRI.

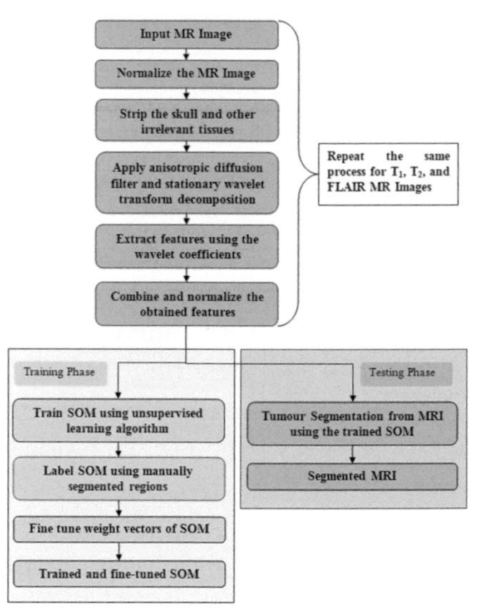

FIGURE 20.1 Block diagram of the proposed system.

In a mathematical insight, Gaussian function is applied to minimize the noise exists in the MRI images (Chaddad, 2015). When an image is blurred by the influence of Gaussian function then it is called as Gaussian blur or Gaussian smoothing. The visual outcome of Gaussian blurring method is a smooth blur that is similar to the visualization of a MRI through a transparent screen. Gaussian smoothing can also be exploited in the preprocessing phase of computer vision-based algorithms to improve the image structures. Further, it minimizes the high-frequency components of an input image. One of the essential characteristics of Gaussian filters is to have zero overshoot to a step function input as the rise and fall duration is minimized. A Gaussian filter whose impulse response is a Gaussian function and it transforms the input signal by convolution and this transformation is called as the Weierstrass transform (Kavitha and Chellamuthu, 2013).

The background pixels of a binary input image along with its structuring element are used for computing the dilation as well as the erosion factors. The structuring element lies on apex of an input image is placed over such that the origin of an element matches with the corresponding input pixel positions at the background. If at any rate one pixel in the structuring element matches with a foreground pixel in a base image, then the input pixel will be set as the corresponding forefront value (Chen and Haralick, 1995). Similarly, if the entire corresponding pixels in the base image lie at the backdrop, then the input pixel is fixed as the background value. The structuring element should be assigned as a small binary image or in a unique matrix format for postprocessing. In general, the dilation and erosion operations are performed prior to the feature extraction and image segmentation processes. Here, the shapes and different objects exist in the images are analysed along with the boundary properties.

20.3.1 FEATURE EXTRACTION THROUGH DISCRETE WAVELET TRANSFORM (DWT)

The DWT is a potential mathematical means of feature extraction, and it has been employed to filter the wavelet coefficient from MR images (Arizmendi, 2012). Typically, Wavelets are localized base functions derived from mother wavelets in which they provide localized frequency information about the function of a signal. A continuous wavelet transform (w_ψ) of a signal $x(t)$, as compared to a real-valued wavelet, (t) is expressed as

$$w_\psi(a,b) = \int_{-\infty}^{\infty} f(x) * \psi_{a,b}(t)\,dx \quad (20.2)$$

where, "a" is the Dilation factor, "b" is the translation parameter

$$\psi_{a,b}(t) = \frac{1}{|a|} \quad (20.3)$$

The wavelet ($\psi_{a,b}$) is estimated from the mother wavelet (ψ) by the process of translation and dilation. It is assumed that the mother wavelet (ψ) satisfies the condition of zero mean and the equation can be discredited by preventing "*a*" and "*b*" from a discrete lattice to represent the DWT. The DWT is a linear transformation method that functions on a data vector whose dimension is the numeral power of two and it transforms various vectors of the equivalent length (Arizmendi, 2012). Using DWT, the data is divided into multiple frequency components and it can be expressed as

$$x(n)=\begin{cases} d_{j,k}=\sum x(n)\,h^*j(n-2jk) \\ d_{j,k}=\sum x(n)\,g^*j(n-2jk) \end{cases} \quad (20.4)$$

The coefficients $d_{j,k}$, are the key elements in signal $x(n)$ and correspond to the wavelet function. The $h(n)$ and $g(n)$ in Equation (20.4) signify the coefficients of the high-pass and low-pass filters while the other two factors j and k corresponds to wavelet scale and conversion factors. The key aspect of DWT is multidimensional expression of function and it can be analyzed at different range of resolution through wavelets. The input image is analyzed along the x and y axes by $h(n)$ and $g(n)$ functions that is the row wise demonstration of the actual image. The output of these transformations is summarized as sub-bands such as LL, LH, HH, and HL.

Texture feature extraction using GLCM simply distinguishes normal and abnormal (malignant) tissues (Joshi et al., 2010). In first-order numerical texture analysis, texture data is filtered from the histogram that represents the image intensity. GLCM method determines the occurrence of a specific gray-level at an arbitrary image position and does not include the correlations among pixels. The geometric features from MRI are acquired through gray level spatial dependence matrix. GLCM approach is a 2D histogram where (*i*, *j*)th node is the occurrence of event "*I*" that coincides with "*j*." It has a function with distance $d = 1$, angle at 0° (for horizontal), 45° (along the +ve transverse), 90° (for normal) and 135° (along the –ve transverse) to calculate pixel intensity. GLCM approach has been formulated to capture the texture features including contrast, correlation, energy, homogeneity and entropy that can be obtained from LH and HL sub-bands.

20.4 IMAGE SEGMENTATION

Segmentation method isolates an image into numerous parts and it is exercised to recognize an object or other applicable information exists in digital images. Image segmentation can be performed through multiple strategies that include (i). Thresholding method (Marquez, 2016), (ii). Color-based segmentation methods (Marquez, 2016), and (iii). transform methods (Arizmendi, 2012). The segmentation results obtained through Otsu's thresholding method, color-based segmentation such as *K*-means clustering (Selvakumar, 2012) and transform methods are presented in Figure 20.2(a), (b), and (c), respectively.

(a) Thresholding methods such as Otsu's method

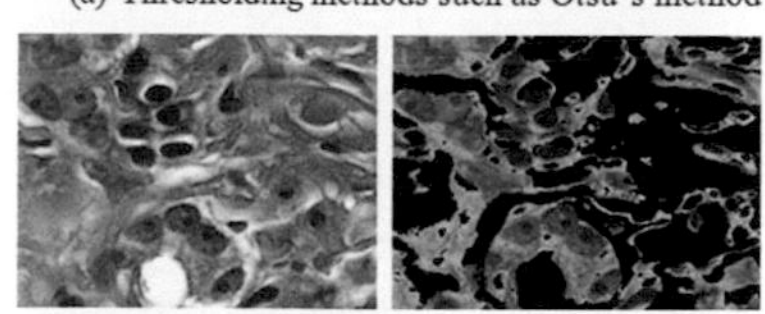

(b) Colour-based Segmentation such as K-means clustering

(c) Transform methods such as watershed segmentation

FIGURE 20.2 Sample results for image segmentation techniques.

In the presented results (Figure 20.2), the boundary features along with their values are acquired and fed into a NN classification tool. Primarily, SOM is trained to plot the input image against the relevant tissue regions based on their features by taking into account of their usual clusters in the matrix (Kohonen, 1998). SOM reduces the dimension and clusters linked regions jointly for understanding large image datasets. SOM consists of two layers, namely (i) input nodes (in layer 1), and (ii) output nodes (in layer 2). Here, output nodes are presented in the form of 2D grid and there are adaptable weights exist between every output. The map characterizes various features with finest accuracy using a constrained set of clusters that is confirmed by multidimensional observation. At the finale training stage, the clusters turn into organized grid so that similar clusters are close to each other and unlike clusters are isolated from each other. At the end of training, confusion matrix is plotted in order to assess the efficiency of NN classifiers.

The classifier employed in this chapter is a back-propagation based NN classifier (Nekovei, 1995). Each neuron admits a signal from the neurons exists in the preceding layer, and the signals are multiplied with an assigned weight value. The weighted values are aggregated and conceded by a limiting function that scales the result to a finite range of values. The relationship among the neurons has an exclusive weighting value and the inputs from previous neurons are independently weighted and summed subsequently. The successive results are nonlinearly scaled between the values 0 and +1, meanwhile the output is moved to the neurons in the subsequent layer. The distinctiveness of the NN subsists in the values of the assigned weights among the neurons. Hence, a mechanism is required for regulating the assignment of weights to aforementioned problems. Typically, backpropagation (BP) network learning algorithm is employed for such problems and it learns by few input examples (training dataset) and the correctly known output for each case.

The BP learning process functions in a repetitive manner and sample cases are applied to the NN, that generates the output related to the present state of the assigned weights. This output is validated with a sample output and a mean-squared error signal will be determined. Subsequently, the determined error value is then back propagated through the NN, and the weights in each layer are regulated to the minimum level. The weight differences are estimated to minimize the error signal and the entire process is repeated for each sample cases. This cyclic process is done until the net error magnitude drops below the encoded threshold. At this stage, the NN has learned the function and it asymptotically advances towards the ideal function.

20.5 EXPERIMENTAL RESULTS AND DISCUSSION

The objective of this chapter is to identify the brain tumor tissue in an MRI image by differentiating the healthy surrounding tissues. To achieve the proposed objective, the following evaluations are done through the sequence of operations mentioned in Figure 20.1.

1. Identify tumorous parts by differentiating them from the healthy tissue region in the preprocessing phase by MRI segmentation.
2. To obtain the boundary parameters and coefficients of the tumor area from the MRI data.
3. To train the NN with past classification data and obtain performance parameters.
4. To analyze the overall efficiency and performance parameters of the proposed system to compare the performance with the state-of-the-art methods.

The proposed work has been simulated using the MATLAB environment and the results are presented in the subsequent sections with discussion. The results are classified into three categories, namely preprocessing, segmentation and NN performance evaluation results.

20.5.1 RESULTS FOR PREPROCESSING PHASE

In the preprocessing phase, the training images have been collected from the National Cancer Archives of United States of America database and used as the input for this research work (Kinahan, 2019). The image dataset for the proposed work is presented in Figure 20.3.

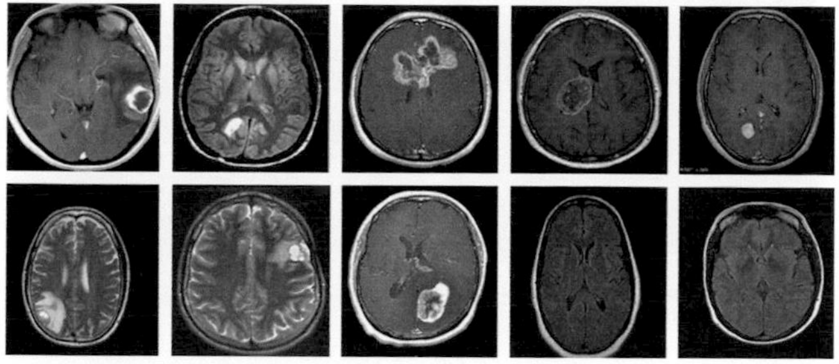

FIGURE 20.3 Dataset used in the proposed research work.
(*Source:* Used with permission from Kinahan et al., 2019.)

The input MRI of the brain is taken using a MR scanner which is considered to be the test input for the proposed work. The test input MRI is presented Figure 20.4(a) and the corresponding Gray-scale converted image is displayed in Figure 20.4(b). Here, the conversion is essential to process input image easily by the upcoming blocks. Subsequently, a filtering operation is performed to remove unwanted noise present in the image. Once the grey scale conversion is completed and the unwanted noise is removed using a Sobel edge detector. These edge detectors will identify the edges present in the image using Sobel algorithm (Rulaningtyas, 2009). The output image obtained from edge detector block is presented in Figure 20.5(a). Once the edge detection is completed, the output will be utilized for creating the dilated image with gradient mask as highlighted in Figure 20.5(b).

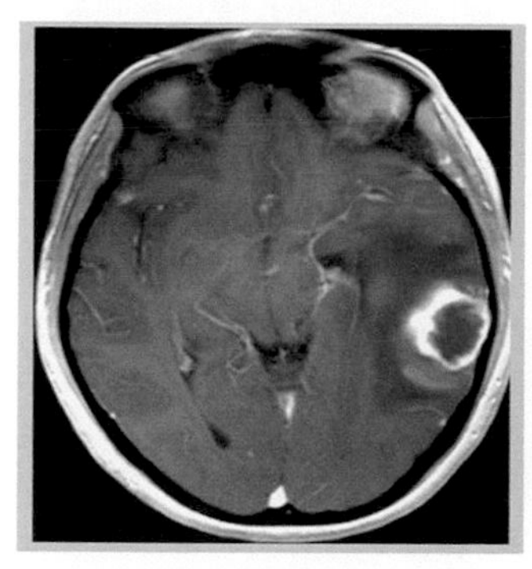

(a) Input MRI of Brain

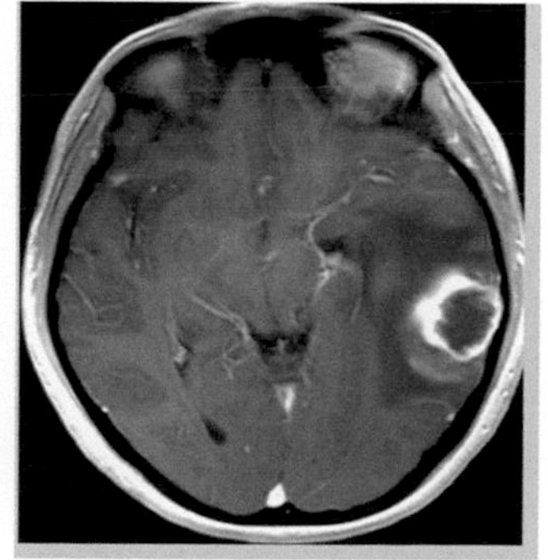

(b) Image after Grey scale conversion

FIGURE 20.4 Input MRI of brain and converted images.

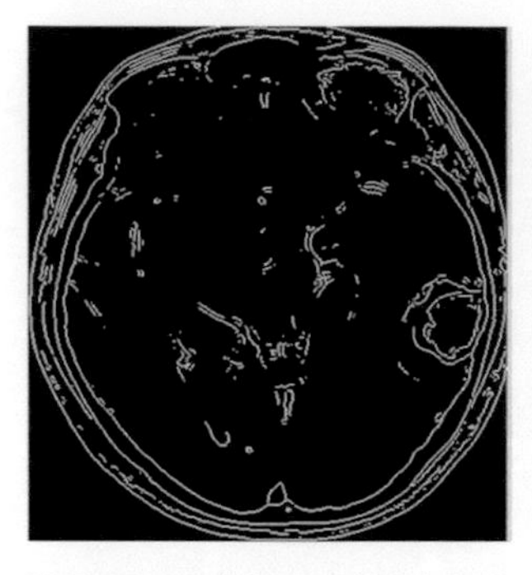

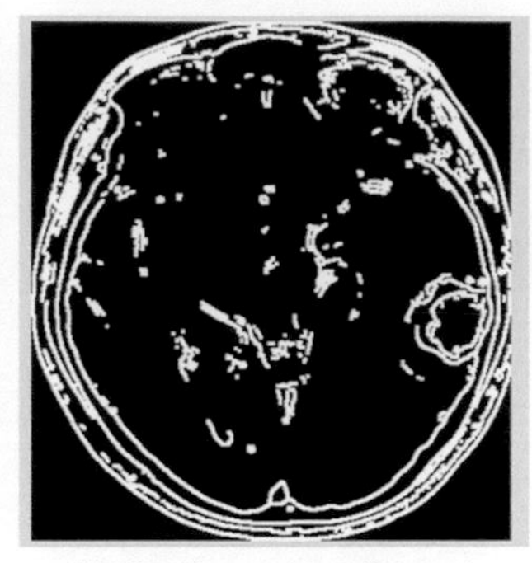

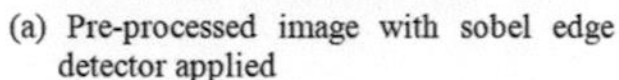

(a) Pre-processed image with sobel edge detector applied (b) Dilated image with gradient mask

FIGURE 20.5 Preprocessed and dilated images for edge detector block.

20.5.2 RESULTS FOR SEGMENTATION PHASE

Image dilation is attributed for identifying neighbourly connected components of an object present in an image. The edge detected image is dilated using a gradient mask in order to identify the neighbourhood connected components. After first level dilation is completed, stage-2 dilation is performed to thicken the connected components and the corresponding output is presented in Figure 20.6(a). After image dilation process is completed, a binary gradient mask is applied to the second level dilated image (Zacharaki et al., 2009). The output image after binary gradient mask applied is presented in Figure 20.6(b).

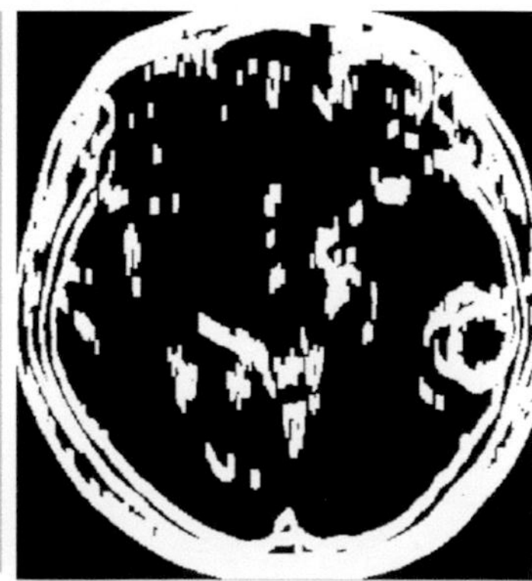

(a) Second stage dilated image with gradient mask (b) Image after binary gradient mask applied

FIGURE 20.6 Dilated images with gradient masks.

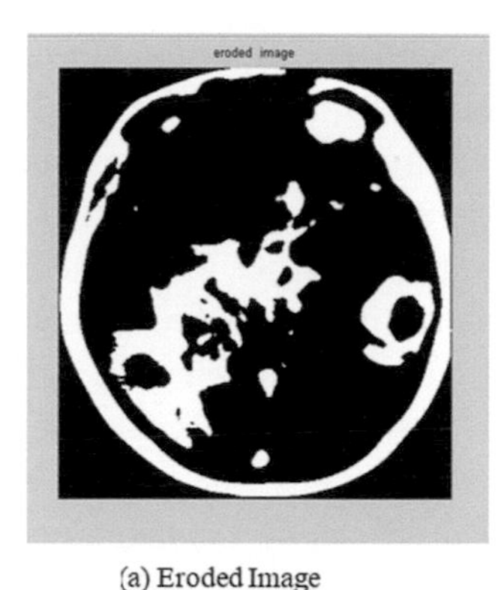

(a) Eroded Image

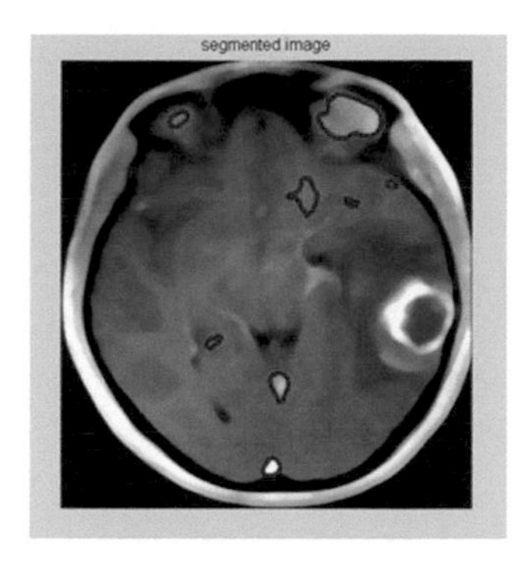

(b) Final Marked Image

FIGURE 20.7 Eroded image and final marked image after segmentation process.

An image erosion procedure is applied after the binary gradeint mask phase is completed. The abnormal peaks of intensities are vaded away from the dilated image in the course of erosion process and the output is shown in Figure 20.7(a). The eroded image is segmented for abnormal features and the tumerous tissues are marked. Once the segmentation and marking process is completed, the object wise features such as contours, area and their properties are extracted using region properties algorithm (Hazen, 1988). The feature properties are obtained for the entire image in dataset that will be further fed into the NN classifier.

20.5.3 NEURAL NETWORK CLASSIFIER PERFORMANCE EVALUATION

The object properties obtained for all the images are assigned as input to the NN classifier. The extracted object features include white, grey and Edema along with their structural and intensity features (Nabizadeh and Kubat, 2015).

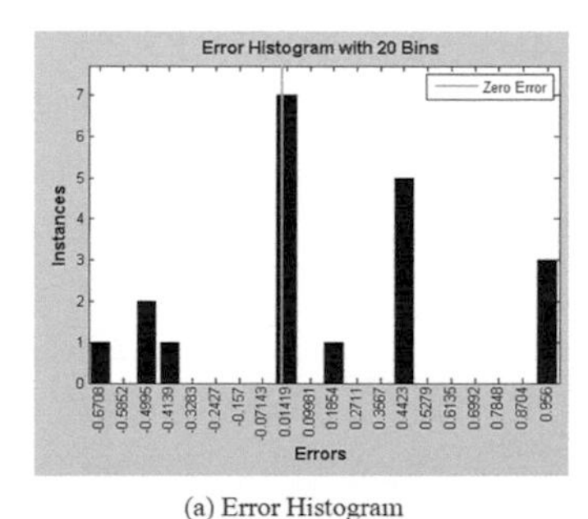

(a) Error Histogram

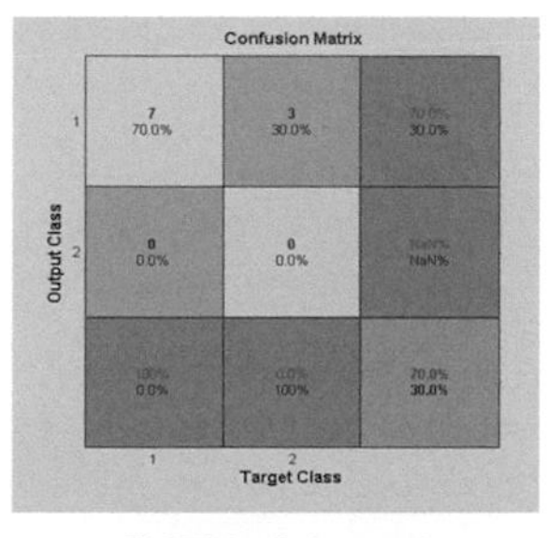

(b) NN Confusion matrix

FIGURE 20.8 Error histogram and confusion matrix of NN.

A histogram is an approximation of probability distribution of a continuous variable and generally it is represented using graphical distribution of numerical data. To create a histogram, initially "bin" the range of values and subsequently identify the values that fall into each time step. The bins are commonly specified as successive, nonoverlapping time step of a variable and the intervals should be uniformly located at equidistance. In this work, the classification error occurred at each interval is plotted in the histogram as displayed in Figure 20.8(a). In particular, at the error interval from 0.01419 to 0.956, there are totally 8 instances that occur with a probability of zero error. This is known as the optimum interval of the classification using NN classifier.

The NN confusion matrix is vital for performance evaluation of the NN based on classification as highlighted in Figure 20.8(b). It is observed that 70% tumor images were successfully classified by the NN based classifier through BP algorithm based on the image features. Except the 70%, there are three images which is true positive but it could not classified correctly by the classifier. From the confusion matrix plot, it is found that the classifier was competent to achieve 0% error while classifying true negative images. Thus, the overall classification efficiency is maintained at 70% and the false classification is about 30% during the first instance of training process. The level of classification is promising one which can be improved substantially with a larger dataset with additional training.

The training state statistics shows that the NN classifier achieves an optimal minumum value at the point 10 and the state is maximum at the points 4 and 5. The gradient peak achieved without error is about 0.011789 and the training state statistics is presented in Figure 20.9. The overall performance analysis curve is given in Figure 20.10. It shows that the validation curve achieves its

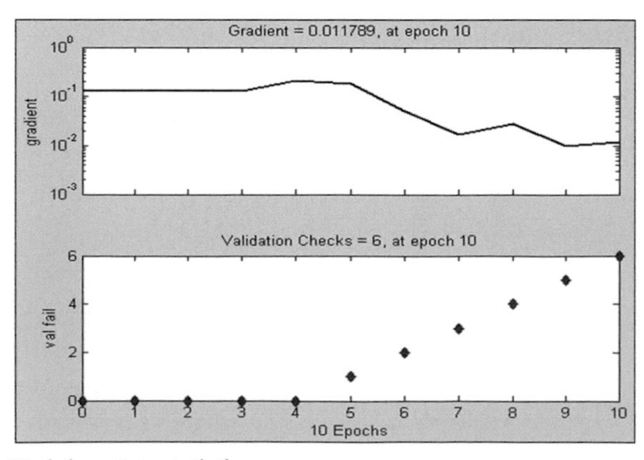

FIGURE 20.9 Training state statistics.

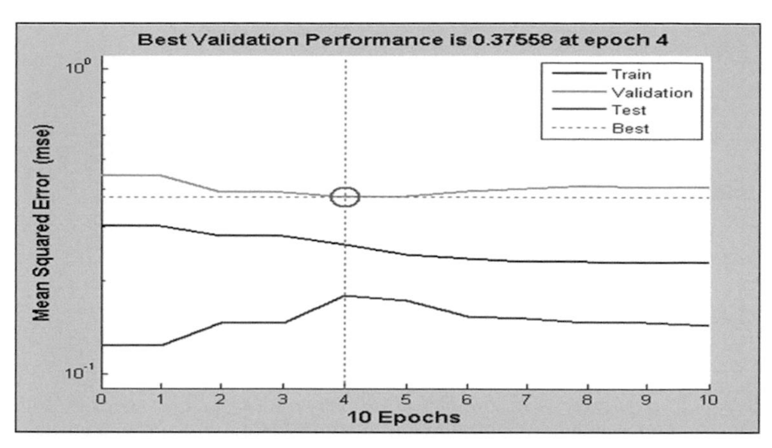

FIGURE 20.10 Overall performance of the proposed research work.

peak without error at the epoch 4. The peak best performance is occurred at epoch 4 with a score of 0.37558. From the training state statistics, it is observed that the overall validation efficiency is constantly above the required level. The state of testing and training is exponentially varying until the epoch 4, and it is stabilized afterwards.

20.5.4 COMPARISON WITH THE STATE-OF-THE-ART METHODS

Four supervised robust classification techniques are applied for the comparison and validation of the present work. The techniques such as SVM, KNN, NSC, and *k*-means clustering are compared with the NN-based method used for brain tumor segmentation process (Nabizadeh and Kubat, 2015). The number of healthy frames is more than the number of tumor frames that formulates the training set fully unbalanced. To minimize the difficulties caused while handling unbalanced training datasets, equal number of healthy and tumor frames were preferred for the experimental analysis.

The accuracy of different classification methods was compared with the proposed work using statistical features with noise reduction as presented in Figure 20.11. In addition, the number of features along with the recognition rate for SVM and NN based methods are presented in Figure 20.12. The results show that the proposed NN based segmentation and classification methods are efficient for determining the tumorous tissues from the brain MRI and classifying them precisely.

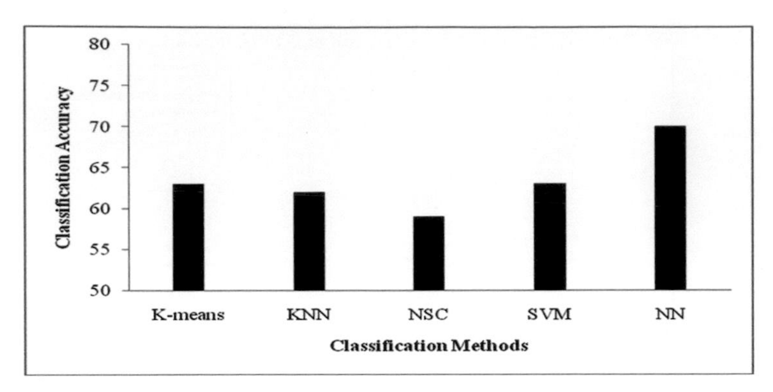

FIGURE 20.11 Accuracy of various classification methods through statistical features with noise reduction.

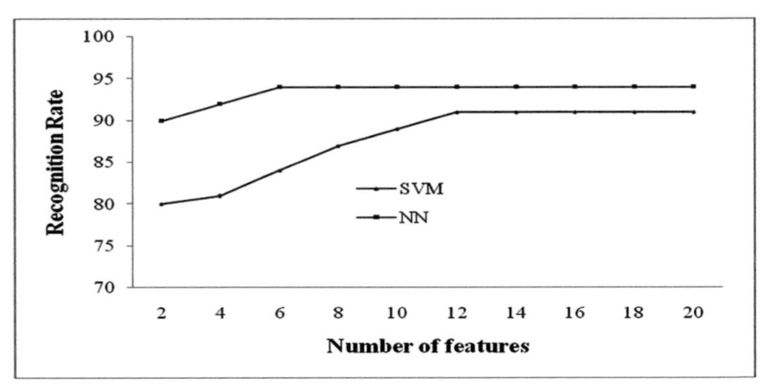

FIGURE 20.12 Feature recognition rate.

20.6 CONCLUSION

The brain tumor detection is done by image segmentation process and NN-based classification. The input brain MR images are preprocessed by filtering and noise removal followed by removal of unwanted data present in the image such as skull strip. Here, the NN based classification method is proficient for segmentation and it differentiates the brain tumor area from the healthy tissue region. The proposed algorithm is able to obtain the object features present in the image and subsequently it can be fed into the NN classifier. The BP based NN classifier trains itself with the input dataset and achieves subsequent learning with feed forward learning. Once the learning is completed, the performance efficiency of the system is evaluated with confusion matrix plot and performance curves. This system is known to be outperforming than the other existing techniques in terms of accuracy and performance efficiency. The present chapter shall be extended with a larger dataset by including prior knowledge about shape and model features to the tumor segmentation. In addition, it can be extended further by including the morphological structure related data of the input brain MRI to train the NN. Hence, such a detailed input data helps the NN tool to perceive greater information from the MRI for extensive medical applications.

KEYWORDS

- **brain tumor**
- **segmentation**
- **discrete wavelet transform**
- **neural networks**

REFERENCES

Autier, P. Risk factors and biomarkers of life-threatening cancers. *Ecancer Medical Science*. 2015, 9:596, 1–8.

Ahmad Chaddad. Automated feature extraction in brain tumor by magnetic resonance imaging using Gaussian mixture models. *International Journal of Biomedical Imaging*. 2015, 2015, 1–11.

Breiter, Hans C., Scott L. Rauch, Kenneth K. Kwong, John R. Baker, Robert M. Weisskoff, David N. Kennedy, Adair D. Kendrick et al. Functional magnetic resonance imaging of symptom provocation in obsessive-compulsive disorder. *Archives of General Psychiatry*. 1996, 53:7, 595–606.

Bengio, Yoshua, Aaron Courville, and Pascal Vincent. Representation learning: A review and new perspectives. *IEEE Transactions on Pattern Analysis and Machine Intelligence*. 2013, 35:8, 1798–1828.

Clark, M.C., Hall, L.O., Goldgof, D.B., Velthuizen, R., Murtagh, F.R., Silbiger, M.S., Automatic tumor-segmentation using knowledge-based techniques. *IEEE Transactions on Medical Imaging*. 1998, 117, 187–201.

Cobzas, D., Birkbeck, N., Schmidt, M., Jagersand, M and Murtha, A. 3D variational brain tumor segmentation using a high dimensional feature set. *Mathematical Methods in Biomedical Image Analysis (MMBIA 2007)*. 2007, 1–8.

Carlos Arizmendi, Alfredo Vellido and Enrique Romero, Classification of human brain tumours from MRS data using discrete wavelet transform and Bayesian neural networks. *Expert Systems with Applications*. 2012, 39:5, 5223–5232.

Chen S. and Haralick R. M. Recursive erosion, dilation, opening, and closing transforms, *IEEE Transactions on Image Processing*, 1995, 4:3, 335–345.

Demirhan, Ayse, Memduh Kaymaz, Raşit Ahıska, and Inan Guler. A survey on application of quantitative methods on analysis of brain parameters changing with temperature. *Journal of Medical Systems*. 2010, 34:6, 1059–1071.

Doyle S., Vasseur F., Dojat M., and Forbes F. Fully automatic brain tumor segmentation from multiple MR sequences using hidden Markov fields and variational EM. *Procs. NCI-MICCAI BraTS*. 2013, 18–22.

Galic, I., Weickert, J., Welk, M., Bruhn, A., Belyaev, A., Seidel, H.P. Image compression with anisotropic diffusion. *Journal of Mathematical Imaging and Vision*. 2008, 31:2–3, 255–269.

Gerrig, Richard J., and Gregory L. Murphy. Contextual influences on the comprehension of complex concepts.

Language and Cognitive Processes. 1992, 7:3–4, 205–230.

Hall, Lawrence O., Amine M. Bensaid, Laurence P. Clarke, Robert P. Velthuizen, Martin S. Silbiger, and James C. Bezdek. A comparison of neural network and fuzzy clustering techniques in segmenting magnetic resonance images of the brain. *IEEE Transactions on Neural Networks*. 1992, 3:5, 672–682.

Havaei, Mohammad, Axel Davy, David Warde-Farley, Antoine Biard, Aaron Courville, Yoshua Bengio, Chris Pal, Pierre-Marc Jodoin, and Hugo Larochelle. Brain tumor segmentation with deep neural networks. *Medical Image Analysis*. 2017, 35, 18–31.

Hazem M Raafat, Andrew K.C Wong. A texture information-directed region growing algorithm for image segmentation and region classification, *Computer Vision, Graphics, and Image Processing*. 1988, 43:1, 1–21.

Havaei M, Davy A, Warde-Farley D, Biard A, Courville A, Bengio Y, Pal C, Jodoin PM, Larochelle H. Brain tumor segmentation with deep neural networks. *Medical Image Analysis,* 2017, 35, 18–31.

He, Haibo, and Edwardo A. Garcia. Learning from imbalanced data. *IEEE Transactions on Knowledge and Data Engineering*. 2009, 21:9, 1263–1284.

Hinton Geoffrey E., Nitish Srivastava, Alex Krizhevsky, Ilya Sutskever, Ruslan R. Salakhutdinov. Improving neural networks by preventing co-adaptation of feature detector. *Neural and Evolutionary Computing*. https://arxiv.org/abs//1207.0580, 2012.

Iftekharuddin K.M. On techniques in fractal analysis and their applications in brian MRI. *Medical Imaging Systems: Technology and Applications, Analysis and Computational Methods*, 2005, 1, 63–86.

Joshi D. M., Rana N. K. and Misra V. M. Classification of Brain Cancer Using Artificial Neural Network, *2nd International Conference on Electronic Computer Technology*, Kuala Lumpur, 2010, 112–116.

Kaus, M. R., Simon K. Warfield, Arya Nabavi, E. Chatzidakis, Peter M. Black, Ferenc A. Jolesz, and Ron Kikinis. Segmentation of meningiomas and low-grade gliomas in MRI. *International Conference on Medical Image Computing and Computer-assisted Intervention,* Springer, Berlin, Heidelberg, 1999, 1–10.

Kavitha A. R, Chellamuthu C. Detection of brain tumour from MRI image using modified region growing and neural network. *The Imaging Science Journal*. 2013, 61:7, 556–567.

Kinahan, P., Muzi, M., Bialecki, B., Herman, B., and Coombs, L. Data from ACRIN-DSC-MR-Brain [Data set]. *The Cancer Imaging Archive*. 2019, DOI: https://doi.org/10.7937/tcia.2019.zr1pjf4i.

Khotanlou Hassan, Olivier Colliot, Jamal Atif, Isabelle Bloch. 3D brain tumor segmentation in MRI using fuzzy classification, symmetry analysis and spatially constrained deformable models, *Fuzzy Sets and Systems*, 2009, 160:10, 1457–1473.

Menze, Bjoern H., Andras Jakab, Stefan Bauer, Jayashree Kalpathy-Cramer, Keyvan Farahani, Justin Kirby, Yuliya Burren. The multimodal brain tumor image segmentation benchmark (BRATS). *IEEE Transactions on Medical Imaging*. 2015, 34:10, 1993–2024.

Marquez, Cristian. Brain tumor extraction from MRI images using MATLAB. *International Journal of Electronics, Communication and Soft Computing Science and Engineering (IJECSCSE)*. 2016, 2:1, 1.

Michael Gonzales. Classification and pathogenesis of brain tumors. *Brain Tumors (Third Edition)*. 2012, 36–58.

Nabizadeh Nooshin and Miroslav Kubat. Brain tumors detection and segmentation in MR images: Gabor wavelet vs. statistical features. *Computers and Electrical Engineering*, 2015, 45, 286–301.

Nair, R.R., David, E. and Rajagopal, S. A robust anisotropic diffusion filter with low arithmetic complexity for images. *EURASIP Journal on Image and Video Processing*, 2019, 1, 48.

Nekovei R. and Ying Sun, Back-propagation network and its configuration for blood vessel detection in angiograms, *IEEE Transactions on Neural Networks*, 1995, 6:1, 64–72.

Prastawa, Marcel, Elizabeth Bullitt, Nathan Moon, Koen Van Leemput, and Guido Gerig. Automatic brain tumor segmentation by subject specific modification of atlas priors 1. *Academic Radiology*. 2003, 10:12, 1341–1348.

Reddick, Wilburn E., John O. Glass, Edwin N. Cook, T. David Elkin, and Russell J. Deaton. Automated segmentation and classification of multispectral magnetic resonance images of brain using artificial neural networks. *IEEE Transactions on Medical Imaging*. 1997, 16:6, 911–918.

Rulaningtyas R. and Ain K. Edge detection for brain tumor pattern recognition, *International Conference on Instrumentation, Communication, Information Technology, and Biomedical Engineering*. 2009, 1–3.

Specht, Donald F. A general regression neural network. *IEEE Transactions on Neural Networks.* 1991, 2:6, 568–576.

Song, Bo, Weinong Chen, Yun Ge, and Tusit Weerasooriya. Dynamic and quasi-static compressive response of porcine muscle. *Journal of Biomechanics.* 2007, 40:13, 2999–3005.

Selvakumar J., Lakshmi A. and Arivoli T. Brain tumor segmentation and its area calculation in brain MR images using K-mean clustering and fuzzy C-mean algorithm. *IEEE-International Conference on Advances in Engineering, Science and Management (ICAESM—2012),* Nagapattinam. 2012, 186–190.

Teuvo Kohonen. The self-organizing map. *Neurocomputing*. 1998, 21:1–3, 1–6.

Tustison N. J. Instrumentation bias in the use and evaluation of scientific software: Recommendations for reproducible practices in the computational sciences, *Front. Neurosci*. 2013, 7, 162.

Zacharaki EI, Wang S, Chawla S, Soo Yoo D, Wolf R, Melhem ER, Davatzikos C. Classification of brain tumor type and grade using MRI texture and shape in a machine learning scheme. *Magnetic Resonance in Medicine*. 2009, 62(6), 1609–1618.

CHAPTER 21

A HYPOTHETICAL STUDY IN BIOMEDICAL BASED ARTIFICIAL INTELLIGENCE SYSTEMS USING MACHINE LANGUAGE (ML) RUDIMENTS

D. RENUKA DEVI* and S. SASIKALA

Department of Computer Science, IDE, University of Madras, Chennai 600005, Tamil Nadu, India

Corresponding author. E-mail: renukadevi.research@gmail.com

ABSTRACT

Artificial intelligence (AI) is the recreation of the anthropological intelligence mechanism by machines and specific intelligence systems. These headways contain erudition, perceptive considerations and self-rectification. Some practices of AI include expert systems, speech recognition and machine vision and learning. Artificial astuteness or the intelligence is advancing dramatically, and it is already transforming our world socially, parsimoniously and politically. There have been paradigms of a shift in the way patients are clinically treated by doctors with AI-assisted excessive limits of data in their gauges. With the degree and dimensions of data raising up at a confounding pace, henceforth the conventional diagnostic approaches have been modernized and there has been a change in the clinical assertion making techniques. The intention of promising technological development in AI, leads to tackle critical health issues with quicker benefits even in shortcoming of the issue. On the same regards and notion machine learning (ML) is also trying to oversee the applications and possibilities of medical theories with assistive symptomatic services, which will drastically progress the accessibility and the accuracy of the medical services for the common man and mankind. In this chapter, we explore the fundamentals and application of ML in biomedical domain. We also discuss the research developments and challenges.

21.1 INTRODUCTION

Artificial intelligence (AI) was framed by John McCarthy, an American computer researcher, in 1956 at the Dartmouth Conference where the persuasion was born. Currently, AI is a parasol terminology that comprehends everything from robotic process automation to medical progress. AI can accomplish errands as identifying patterns and behavior in the data with more precision and efficiency, when compared humankind enabling economic challenge to gain more insight out of their data through industries. With the assistance from AI, enormous amounts of data can be analyzed and processed to map poverty, climatic influence change, automate agricultural production and irrigation along with tailoring health care and learning with predicting consumption patterns. Creating a mutual AI milieu, in which anthropoid and machine work impeccably together takes more than a smart machine. Organizing smart systems in ways we humans find normal and instinctual is both science and the art forms.

Streamlining energy-usage and waste-management and training the behavioral patterns is one of the greatest outcomes of AI in the contemporary technology. AI is gradually changing medical practice. With modern advancements in digitized data procurement, machine learning (ML) and computing substructure the applications of AI are intensifying into areas that were previously thought to be only the province of human expertise. In recent years, there has been an enlarged focus on the practical use of AI in various domains to resolve multifaceted issues. AI is an anthology of numerous technical expertise that impersonate human's cognitive functions. Furthermore, AI is having a greater impact on health care and allied sectors by exploiting ML algorithms (genetic, fuzzy, expert systems, and so on).

It is potentially possible for AI systems to handle both structured and unstructured databases, as the health-care data not only contains formatted inputs, also images, videos, mails, and unformatted data. The technical approach and the technology are extensively used in all varied kinds of health-oriented facets, as well as, considerable application to investigate the health care data in cardiology, neurology, and oncology. Nurses and medical practitioners have already been using this assistive technology in a great way to achieve quick and precise treatment to the patients. An anthology and investigation of medical records, history, a variety of tests, scans, cardiology, and radiology reports is being efficiently managed by AI and digital automation. One of the best examples to portray this would be the Babylon app which uses AI to provide medical consultations and recommends precautionary action using individual medical record with an elevated medical knowledge.

Molly is another classic example of what we have extracted from AI, which is a ML based digital nurse that monitors the patient's health condition and keeps track of the follow ups. This is a remarkable playoff that can save a million health monitoring issues. Conception of drugs is another milestone in the field of AI which saves time, money and sometimes life too. Ebola virus scare, a program powered by AI was used to dissect existing medicine to fight the disease and finally found two medications that may lesson Ebola effectively on a calendar day.

Digital image processing (DIP), a branch of AI applications, has been in use for image analysis and processing. DIP is an evolving area of study by researchers and academicians for the past year. This technique intended for deep analyzes the images from different sources, namely scan images, MRI, CT, and so on. The potential of these techniques is to categorize the features in the image and classify the same into whether the patient is affected by disease or not, and even present the level of risk factor.

ML techniques that were developed for retinal image analysis has exclusive benefits of patient affected by diabetic retinopathy. Some of the features in retina like exudates and micro aneurysm are extracted using AI and ML. The convolutional neural network (CNN)-based AlexNet deep neural networks (DNNs) (Mansour, 2018) is employed for optimal diabetic retinopathy, a computer-aided diagnostic solution which serves as a key contribution to medical history on retinal issues. Google DeepMind has allied on a project with Moorfields Eye Hospital, London to diagnose the causes of diabetic retinopathy and age-related macular degeneration.

Beritelli et al. (2018) recommended system based on neural network (NN) for heartbeat analysis (HBA). HBA is usually done for the patient to uncover the heart diseases. The phonocardiogram records the heart sounds. NN is trained by the known data samples over three thousand in number, followed by feature extraction—Gram polynomial method.

The *Annals of Oncology* recently published the statistics that AI was able to identify the skin cancer patients precisely than doctors. The ML algorithms were applied to the images of patients. The results demonstrate the efficacy of these algorithms in finding the affected patients than human doctors. Billah et al. (2018) recommended an improved feature extraction technique which implemented the polyp detection method for polyp miss rate, thus aiding doctors to pay attention and focus on a specific region. Deep learning NN framework is implemented to address the critical issues of neonatal. Harpreet et al. (2017) proposed the cloud-based integrated Neonatal Intensive Care Unit data analytics framework. This model integrates and tracks the complete assessment sheet of preterm babies.

One of the significant areas where AI is successfully moving to transform, the impact of medicines and providing assistance to pharmacology. It helps in discovering the new combination of drugs. It may become an assistive technology that will empower the medical researchers and practitioners to provide better treatments to serve their patients having some critical diseases. According to Accenture report, the AI market growth is massive in 2021 by ten times compared to previous years.

21.2 MACHINE LEARNING (ML)

AI has its impact in the area of ML where algorithms are developed to learn the similarities in data and develop decision rules from that. Data mining problems, have embedded ML algorithms, constantly combined with statistical methods, to extract knowledge from the given data. The foundation of all the algorithms is a statistical based mathematical model, contribute to various fields NN, Deep Learning (DL), support vector machine (SVM), decision tree, naïve Bayes, random forest, etc., These models were used for analytics and decision-making process. The conceptual model is shown in Figure 21.1.

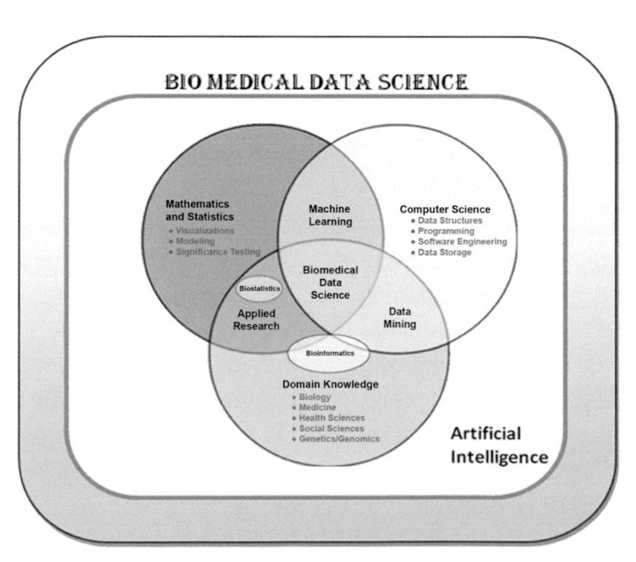

FIGURE 21.1 Biomedical intelligence.

ML also finds its way in numerous health care domains, including diabetes, cancer, cardiology, mental health and so on. Under research, most of the established ML models and tools have explored the potential of prognosis, diagnosis, or differentiation of clinical groups (pathology group and a healthy control group or groups with pathologies), thus signifying capacity towards building computer-based decision support tools. These systems require adequately large datasets and appropriate labels constructed with enough participants and variable inputs provided by clinical experts. The idea is to identify the data structures or variables such as clinical, behavioral, and demographic variables that can be associated with the target outcome, to say whether a person has cancer or not. Hence, useful knowledge can be derived from the accessible medical data after applying ML techniques. This can empower patients to constantly monitor their health status and it also supports health-care professionals in decision making in areas concerning the management, treatment, and follow-up interventions if required.

21.2.1 ML MODELS

ML model is a complete abstraction of the entire system under study. The objective of building a model is to predict the outcome based on the learned inputs or completely optimize the system with the intended objectives. For instance, we build a model for disease detection system, then train them and classify them into two possibilities: Detected or not. This completely involves training, learning, and decision making. This is depicted in Figure 21.2.

Until now, the development in the field of modeling originates from a human perspective. The model reflects upon the knowledge and understanding from this sort of viewpoint. The neurons, molecules, and the immune system are being conceptualized this way. In addition to this type modeling, it can also be extended using computational methods to elaborate data and method. Learning is classified

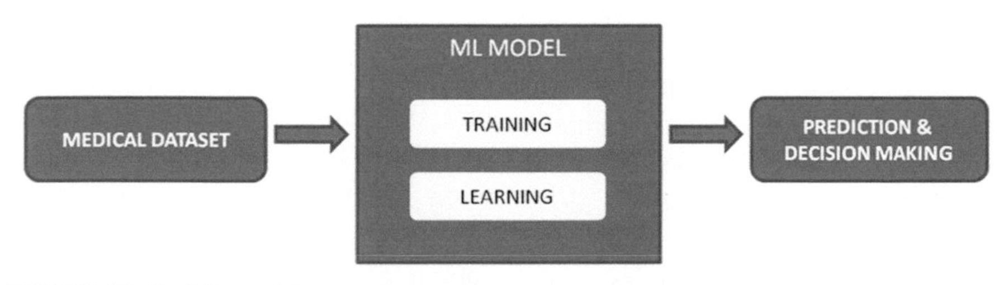

FIGURE 21.2 ML model.

into supervised and unsupervised. Supervised leaning establishes the relationship between the samples of known output. In unsupervised learning, output variable is unknown, it discovers the pattern by its own algorithmic method and devise the predictions. Majority of ML algorithms are classified into either of the above mentioned learning and prediction.

21.2.2 USES OF ML: BIOMEDICAL RESEARCH

In the field of biomedicine, many questions arise, that can be efficiently handled by ML techniques. In certain cases, ML is useful for predictions, answering questions related to drug discovery, whether it is competent to cure the cancer or not. In other cases, it is used to develop a system which overcomes the shortcomings of the manual system. In other instances, the understanding of a system may be enhanced by ML by the revelation of variables that are shared between the system components.

The interpretation of medical images using ML has been an authoritative tool in the diagnosis and assessment of diseases. Legacy systems had discriminative features manually designed for classification such as abnormalities and lesions along with the segmentation of regions of interest such as organs and tissues for different medical applications. Since it was done manually, it required more participation from the side of expert physicians. Nevertheless, the complexity and ambiguity in the images that lead to the limited knowledge for the interpretation of medical images, the automated ML system is used in this domain.

21.2.2.1 MODEL PREDICTION

Basically, ML makes a prediction based on measurable, trained inputs. For instance, studies in psychiatric medicine have used smart phone recordings of a human's day-to-day acts such as their wake-up time, and duration of the exercise time, to predict their mood using ML. In neuroscience, the decoding of brain's neural activity to infer intentions from brain measurements is a common problem. This application is useful to execute the movement of organs when developing interactive prosthetic devices by taking measurements from the brain of a paralyzed subject. Many such problems exist in biomedical research, especially in areas, namely cancer detection, preventive medicine, and medical diagnostics. Such problems, are interested to come up with prediction outputs with high precision. Machine learning methods are built to reach that objective and so to obtain accurate predictions, it is best to rely on such methods.

21.2.2.2 BENCHMARK

Other than describing and predicting data, the other goals is to make the model completely understandable and precise. ML can be exceptionally advantageous by providing a benchmark. The challenge arises when we evaluate the model, the model prediction is prone to errors. As ML is useful to make predictions, it may provide close results to human-produced models. Model evaluation and testing is repeatedly done for achieving the precise outcome, because the model has to be on par with the human generated model with minimal variations. Conversely, if the model completely varies then the basis of the model is heavily misguided.

21.3 MACHINE LEARNING TECHNOLOGIES

In this section, the AI devices or techniques useful in the medical applications are reviewed. In this section, we explore the ML technologies (refer Figue 21.3) that are useful in medical applications. In general, they are categorized into (Jiang et al., 2017),

- conventional ML
- deep learning (DL)
- natural language processing (NLP).

21.3.1 CONVENTIONAL ML

ML algorithms are built on the top of analyzing data. Generally, these kinds of algorithms tend to infer meaningful insights from the medical datasets. The ML technologies are applied to

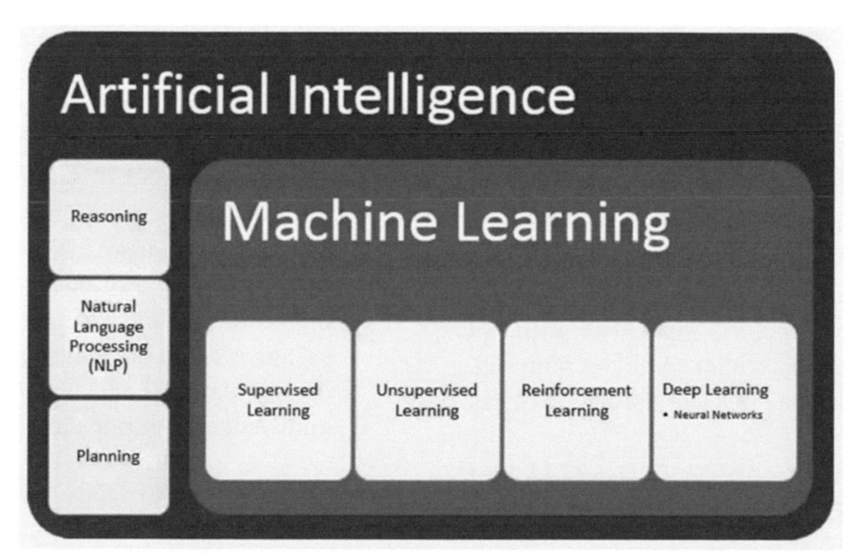

FIGURE 21.3 Machine learning technologies (source: https://www.ibm.com/analytics/machine-learning).

the patient's database, to infer the desired output, that is, condition of health status. The patient's database comprises the basic element of information like name, age, disease, lab report parameters relevant to the disease. It is the combination of transactional and image data like scanned images, CT Scan, MRI, and medication facts collectively.

Furthermore, the outcomes are researched for the level of the disease; this is the output parameter (Y). For instance, in tumor disease prediction, the Y parameter is the size or stage of the disease based on the input parameter (X) of the patient ML algorithms can be divided into two major categories based on whether the outcomes can be incorporated or not. The ML algorithms fall into two categories namely,

- Supervised
- Unsupervised

The unsupervised algorithms are generally used for feature extraction, however supervised algorithm establishes the relationship between X (input) and output (Y) by predicting the output via input. The semisupervised algorithm combines both supervised and unsupervised approaches.

21.3.2 AN EMERGING LEAP IN ML: DEEP LEARNING (DL)

With the leap and bounds of emerging technology, ML can be extended to DL, which abstracts the data model by a deploying number of deep layers (DNNs), thus making the prediction accurate and concise even though if it is a high dimensional data. Here, the process is formulated into two steps,

- Multiple layers process the data. DL extracts the data abstraction through the multilayer learning process. Thus the huge amount of data is processed to extract meaningful insights. As it is a deep consecutive layer, the intention of learning abstraction is done through the layer to layer as is follows hierarchical mechanism, thus lead to the output of each layer is given to the input of the next layer.
- The final output data representation developed by DL algorithms provides constructive information. It is indeed a simpler model working resourcefully on complex data sets. DL also interprets with varied type of data text, image, audio, and video. This system is further extended into derive relational and semantic knowledge from raw data.

The foundations of DNN layers are established on artificial neural networks, which contains multiple deep layers. Having this as a nature of the mechanism, DL can discover more complex nonlinear patterns in the given data. The surge of the volume and complexity of data is one more reason for the mounting call for deep learning. The efficacy of these layers found in handling complex medical data, in contrast to CNN. Big medical database in reality creates a challenge in managing and analyzing complex database. Many researches are intended for medical big data analytics.

21.3.3 NATURAL LANGUAGE PROCESSING (NLP)

The images, the EP and genetic data are machine-understandable and thus ML algorithms can be directly employed after preprocessing or quality control processing. But, when it comes to clinical information, a large proportion of information is in unformatted form related to running text, doctor's note, lab reports, and discharge summaries. To handle this type of unstructured form of data, NLP replaces the conventional ML approaches. NLP processes any form of unstructured data to extract meaningful information.

NLP encompasses two major parts:

- text processing
- classification

21.3.3.1 TEXT PROCESSING

In this step, NLP uncovers the series of disease relevant key words from the unstructured clinical notes. The keywords are identified with the reference form the historical databases.

21.3.3.2 CLASSIFICATION

All the keywords identified in the text processing stage are grouped into a different number of subsets. Then these subsets are analyzed and classified into a two possible outcomes: normal or abnormal. These decision aids the treatment methodology, monitor the health progress and extended clinical support in future. Fiszman et al. (2000) showed that NLP-based system provided antibiotic-assistance system, for an anti-infective therapy alert.

Miller et al. (2017) used NLP laboratory monitoring system for lab adverse effect prediction. The NLP pipelines also helps in disease diagnosis. Castro et al. (2017) developed an NLP-based system for identification of 14 cerebral aneurysms from the unstructured clinical notes. The inferences from this approach, collectively developed key variables for classification of normal patient and affected one with the accuracy of 95 and 86 percentiles on training and validation of the model accordingly. Afzal et al.

(2017) suggested NLP-arterial disease keyword extract system from the narrative clinical notes. The inferred keywords are classified into patients with arterial disease or not, with the accuracy of 90%.

21.4 ADVANCED INTELLIGENCE WITH BIG DATA ANALYTICS

The analysis of health care big data, with ML extends the ascendancy of data science. The emerging health records growing in number of bytes / per second, leads to innovation in handling these data and process it. The electronic health record (EHR) exceeds more than trillion in numbers by the end of this year. This poses diverse challenges such as parallel processing infrastructure and framework, storage management of complex data, highly developed fault tolerant system. The cloud technology based big data frameworks are well suited for complex health care management system(Ngiam et al., 2019). The incorporated health care system, possibly caters the need of all stakeholders of the system. The AI based smart hospital system is one such modern development to monitor the patient at the remote level and give personalized medication.

The fundamental properties of health care big data (Hassan et al., 2019) are tabulated in Table 21.1.

TABLE 21.1 Big Data Characterictics

Characteristics	Explanation
Variety	This is the huge collection of data sets of different kind. For example, image, text, reports, EHR and so on. The sources of data are from wearables, sensor, survey, clinical notes, socio - economic data, pshycosocial data, etc.
Volume	The huge accumulation of data creates dataset ranges from million to trillion KB in size
Velocity	The rate at which the data are collected and analyzed. The real time analytics , require the data analytics tools to process the data as soon as they arrive
Veracity	As the health care decisions are significant steps in data mining, the authenticity of the data is identified by analyzing the different metrics such as originality, origin, and the data collection methods involved
Value	The objective of big data analytics is to derive the meaningful insights and decision making. Interpretation of medical data , deriving decisions are very crucial for diagnosis

The different types of analytics (Abidi et al., 2019) that are applied to health care data are,

- **Decision based**
  - The analytics based decision aids the practitioners for improved medication. The ML decision models are used to generate decision rules, personalized medications, care plans, risk alert, and management.
- **Predictive based**
  - This kind of analytics, process the entire medical records, and predicts the outcome. Based on analysis of past history and prediction is made in the near future.
- **Prescriptive based**
  - This is intensive analytics based on the complex decision rules. Thus require deep knowledge and constraints to analyze big datasets. This is still in the thriving stage of research domain.
- **Comparative based**
  - Comparative analytics assimilate the two or more input methods to derive the conclusion/outcome. The model is generally based on probabilistic methods. In this scenario, both clinical and administrative databases are compared and analyzed for enhanced decisions.
- **Semantics based**
  - This methodology identifies the semantic correlation between the input variables.

Thus, the combination of AI, big data, IoT and ML are ultimately changing the perspective of the conventional analytics to combat the issues of scalability, heterogeneity of data, real-time analytics, single unit/patient level analytics, and complex methodology. Even cloud based system endure the low latency when the data is enormous. This issue is handled by fog computing infrastructure(Anawar et al., 2018), which provides virtualized fog nodes to which IoT devices are connected. The fog infrastructure provides the low latency with a rapid response. To develop a complex health care system, fog computing is always considered as the best options for real time streaming, decision making, and high response time.

21.5 APPLICATIONS

In this section, we explore the applications and proficient ML algorithm in the different facets of biomedical domain. Generally, we uncover the applications that are constructive in many instances such as intense data analytics and management of data accumulation, extract an appropriate element from the huge repository of data (both text and image), reliable treatment: As this technique analyses billions of information very quickly and precisely diagnose with minimal errors, precise system development, reduced cost for the treatment, reduced time to diagnose, achieving the objectivity, keep abreast of technology, simple gadgets and self-monitoring or self assisted guide, in home monitoring, embryo selection of IVF, genome interpretation sick newborns, voice medical coach, potassium blood level prediction, mental health analytics, paramedic diagnosis, promote patient safety, and death prediction in hospital.

21.5.1 NEUROSCIENCE

Neuroscience is the analysis of human nervous system. In neuroscience, decoding and encoding are major study of the brain activity, which is depicted in Figure 21.4. Recent research interests are fascinated towards the cognitive and behavioral neuroscience(Jean-Rémi King et al., 2018).

21.5.1.1 DECODING

The decoding mechanism aims to identify the human intentions based on the brain activity, for instance, to analyze the activity in the brain and to predict the intended movement when they move an exoskeleton with their thoughts. The standard mechanism uses a Wiener filter, where all the brain signals are linearly combined for prediction. However, with the advent of many ML approaches provides a different mechanism in handling the same which gives enhanced performance over the conventional techniques. The ML techniques, linear Wiener filter, nonlinear extension

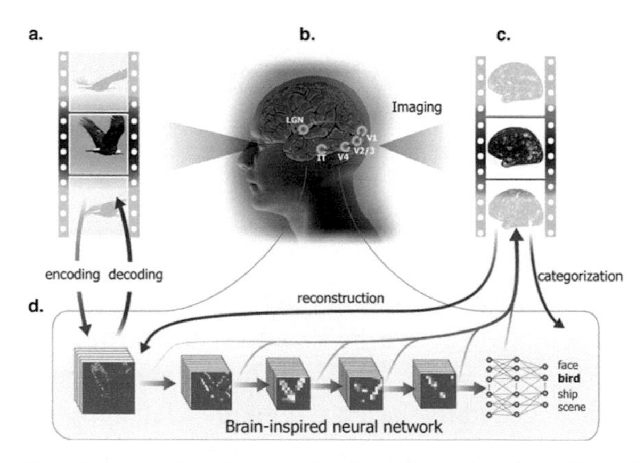

FIGURE 21.4 Encoding and decoding mechanism (Reprinted from Wen, Haiguang, et al. Neural encoding and decoding with deep learning for dynamic natural vision. Cerebral Cortex. 2017, 28(12), 4136-4160. https://arxiv.org/ftp/arxiv/papers/1608/1608.03425.pdf)

(Wiener cascade), Kalman filter, nonlinear SVM , extreme gradient-boosted trees, and NN were used for decoding mechanism.

21.5.1.2 ENCODING

Neural encoding (tuning curve analysis) relate to the study of information representation in a neuron or a brain region to recognize how they transmit with sensory stimuli. Such a study will lead to deriving the insights. The ML model for encoding, plots the signal against the function of external visual stimuli or any brain movements. This simple model leads to decoding mechanism proficient.

21.5.2 DIABETIC EYE RETINOPATHY

Diabetic retinopathy (DR) is on the increase cause of diabetic, which is preventable blindness if treated on time. It is normally found by examining a retinal scan by the trained doctor. In the early stage of this ailment is found, there is a progressive treatment could be given to the patient at the precise time and the issue is effectively handled thus avoiding irreversible blindness condition. Early recognition of this state is significant for better diagnosis. So, the real challenge is predicting the syndrome and act accordingly. This syndrome is prevalent in working age group of adults. ML acts a vital role in the diabetic retinopathy diagnosis, where CNNs are used for diabetic retinopathy staging. The color fundus images are processed by CNN and the outcome is the prediction of the syndrome in a more accurate way.

The absolutely automated system for identifying the syndrome, thus provides an opportunity in preventing the vision loss in our population globally. Many research works are advanced towards combating this issue, by providing leading ML technology one such is CNN algorithms. The reason behind using this architecture is efficacy in processing huge images even learning is made from the raw pixels. This model is not simply for the diabetic retinopathy also used to diagnose other diseases as well.

The Google-Net project developed by Google evaluates the strengths and limitations of the CNN architecture. The 22-layer architecture, with higher accuracy processes the given image with heterogenous sized spatial filters in combination with the low dimensional embedding. The deeper layers learn the system precisely by analyzing the deep features across multiple layers. Every layer is responsible for exploring certain features (i.e.) first layer identifies the edges, second interpret exudate, a classification of the features present. The activation function is applied on the layer (top layer), which maps the input and output variables. This is followed by the normalization procedure of each convolution layer. This is carried out as the batch normalization when the features are elevated.

The scanned images are mixed with both macroscopic and subtle features. Many research methodologies developed for identifying the major features, however the subtle features are crucial for diagnosis. The architectures developed was tested on ImageNet dataset only exploited the macroscopic features. This leads to the new paradigm of model that is capable of identifying even subtle features. The two stage- CNN model uses a pipeline, for feature localization followed by classification. Preprocessing is done to eliminate the nonrelevant features, and network weights are adjusted to deal with the class imbalance. So this model, screens and identifies even the mild disease to multigrade disease detection.

Google developed brain project (refer Figure 21.5), based on DL algorithm that can inspect huge numbers of fundus images and automatically discover DR and diabetic macular edema (DME) with an elevated accuracy. The system is tested with two batches of images (11,711), produced a sensitivity of 96.1% and 97.5% for diabetic retinopathy and a specificity of 93.1% for DME. It is the greatest milestone in Google's research accomplishment to produce high sensitivity and specificity, with a minimal elimination of diseased patient.

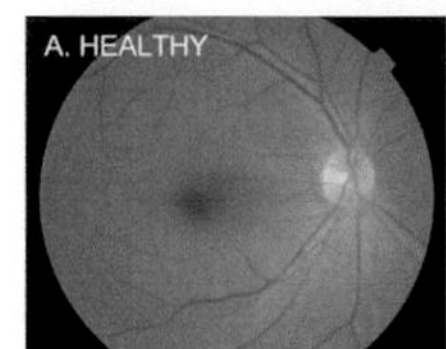

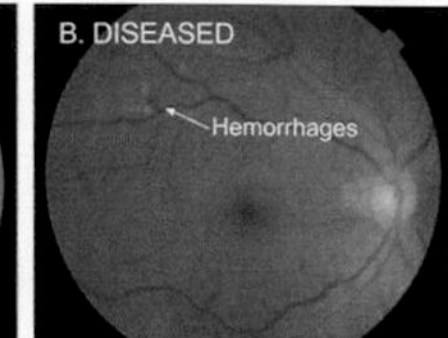

FIGURE 21.5 (A) Healthy retinal fundus image on the left (B) On the right, red spots signifies the affected retina due to DR (source:https://ai.googleblog.com/2016/11/deep-learning-for-detection-of-diabetic.html).

21.5.3 PROSTATE CANCER

Prostate cancer is likely uncommon and non-aggressive in nature. However, the identification this type of cancer poses a challenge in treating the patient, by either surgical method or radiation therapy. So, technology acts a key factor in the measurement of risk factor. In Gleason grade (refer Figure 21.6), the parameter risk stratification is identified leads to further diagnose how cancer cells closely resemble the normal one under the microscopic study. However, this conventional method has a major part of importance in clinical diagnosis, but it is a very complex and subjective technique. This is evident from the studies and report from interpathologist disagreements.

Besides, the trained pathologists are minimal in number when compared to the global call for the prostate cancer treatment across in the globe. From the latest guidelines, the pathologist's report deviates a small percentage by identifying the different Gleason patterns. This depends on the parallel or the supporting technology for the precise identification, treatment, and clinical management. In general, these issues recommend for an improved DL method very much similar to Google metastatic breast cancer detection.

DLS (Nagpal et al., 2019)-based Gleason Scoring of prostate cancer, possibly will improve the accuracy and neutrality of Gleason grading of prostate cancer in prostatectomy specimens. The proposed model categorizes the part of the slide into a Gleason pattern and identifies the tumor region closely resembles the normal cell. It produces and suggest two grades based on Gleason patterns, higher the grade, the greater is the risk factor, thus the patient need utmost care in the treatment. Hence, the DLS competently combats this issue.

21.5.4 *METASTATIC BREAST CANCER*

The microscopic study done by the pathologist is generally considered as the common and effective procedure for the cancer diagnosis and treatment. The challenging factor is in analyzing the cancer spread from the affected region to the nearby lymph nodes. Thus, identification of metastasized part is critical. In TNM cancer staging, identification nodal metastasis are considered.

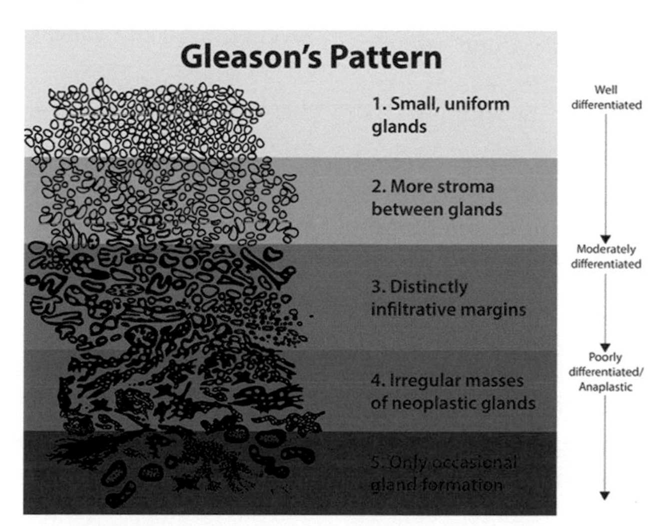

FIGURE 21.6 Gleason pattern (source: https://training.seer.cancer.gov/prostate/abstract-code stage/morphology.html.

In nodal metastasis, the core region of cancer cells where they are created, tend to spread across other parts of the body through the lymph system. The metastasis process majorly influences the breast cancer detection and follow up treatment such as radiation, chemotherapy, and surgical method. The timeline for identifying nodal metastases impart greater significance.

DL-based lymph node assistant (LYNA) provides improved accuracy with the gigapixel slides of lymph nodes from patients. This issue was addressed by the researchers to develop a well optimized algorithm for cancer detection. As these developed methodologies are tested on the real time data, hence the model has to be trained effectively for a random number of test cases and samples before applying the actual data. This demonstrates the strength and weakness of any proposed system. This system should be evaluated by the pathologists, to infer whether the proposed system is beneficial or not.

The research developments in breast cancer detection LYNA algorithm, efficiently supports the breast cancer staging, and assessing its impact in a diagnostic pathology. Still, these methods have limitations when it comes to larger datasets involving multiple slides. This arises the future challenge in DL algorithms.

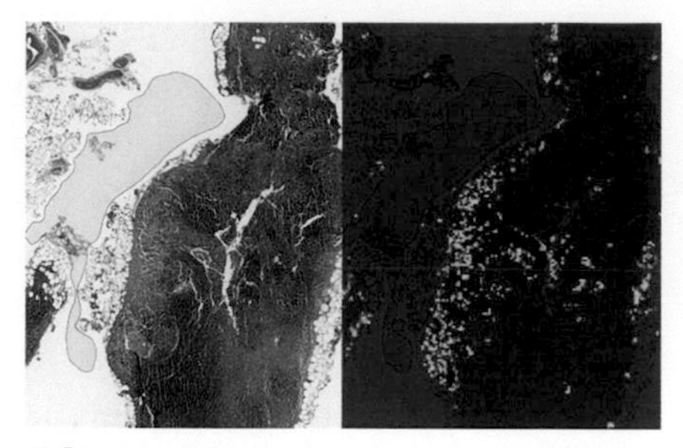

FIGURE 21.7 LYNA.
(*Source*: https://ai.googleblog.com/2018/10/applying-deep-learning-to-metastatic.html)

In Figure 21.7, on the left side is the view of lymph node, right side image where LYNA identifies the tumor region with red color, and blue colored are non- tumor region.

21.5.5 ELECTRONIC HEALTH RECORD (EHR)

EHR maintains the complete history of the patient from the entry into the hospital until their discharge, which is presented in Figure 21.8 EHR is the digitized collection of patient information, not only contains the medical information also the complete treatment plans (Islam et al., 2018), history, and act as a tool for decision making. So many parameters are taken into consideration and predictions made, the date of the patient's discharge from hospital, the recovery time, readmission probability, potential treatment methods, and suggestions. ML techniques aid doctors to accomplish the aforementioned objectives for clinical health care management system. Such a system has the following properties:

- **Scalability:** Any digitized system, should be competent in handling data of minimum to maximum size. As data is escalating for every second, huge amount of information is being accumulated, which should be handled and managed efficiently. The medical records are voluminous in size, as it is the blend of both text and images.
- **Accuracy:** This is the important factor for any proposed system. The prediction accuracy affects the decision making, wrong prediction leads to improper diagnose and treatment. There is much significance is given in gauging how well the system performs in terms of accuracy metrics.

The DL-based EHR system, lessens the burden of manual system. Generally much effort is given for collecting, cleaning and analyzing the clinical data. This system takes up the raw data (raw images in its original form), followed by preprocessing then finally, manipulate relevant records into meaningful insights. The complex EHR data are efficiently handled by the DL system with enhanced accuracy. The decision-making system significantly affects the outcome of the patient data, thus highly developed algorithms are desirable and tested by intensified procedure.

As mentioned in ML model, we did not have to manually select the input parameters. Instead, it establishes the correlation between the input and output parameters by learning the test data. It learns from newest to oldest data, then predicts the outcome. Moreover, this process involves huge database, in such case, recurrent neural networks and

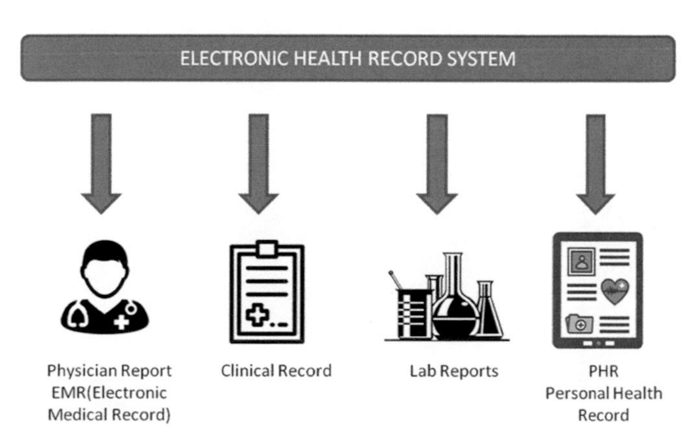

FIGURE 21.8 Electronic health record.

feedforward networks are suggested for decision making.

21.5.6 AUGMENTED REALITY MICROSCOPE FOR CANCER DETECTION

Virtual reality and augmented reality (AR) have been evolved as one of the promising technology for future medical developments. Many research works are inclined towards this area of expertise. In particular, radiological images are important in analyzing the follow up surgical procedure. Izard et al. (2019) proposed the study based on AR technique. The proposed system enhanced the visualization of radiological image from 2D to 3D model. This allows the visualization procedure, effective where multiple dimensional analysis is possible. The images under study is segmented by computer vision and AR technique, providing effective analysis from all dimensions.

The recent development, augmented reality microscope (ARM) (refer Figure 21.8) can possibly help the pathologists in assisting and exploiting the DL based technology. This technology has been more significant in clinical analysis, highly appreciated by pathologists around the globe. The inbuilt light microscope enables the real-time image analysis and provides the results after applying the ML technique into it. The light weight property is essential, because it can be mounted on the existing microscope available in the hospitals. The low cost, readily available components make this system accessible and utilizable across the country.

Modern computational DL built upon TensorFlow, will permit broader technologies to run on this platform. In contrast to the conventional analog microscope, the

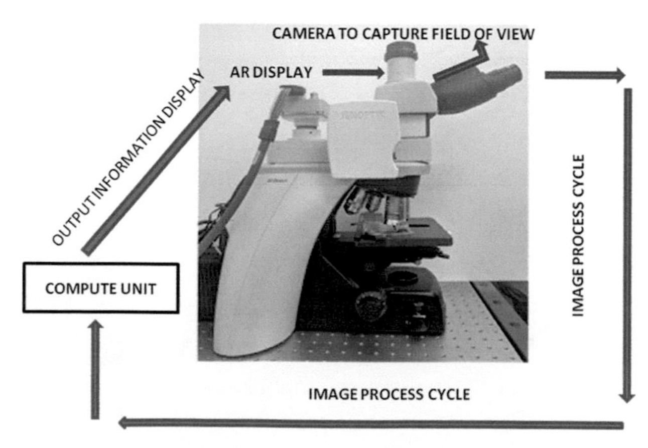

FIGURE 21.9 ARM (*Source*: https://ai.googleblog.com/2018/04/an-augmented-reality-microscope.html)

digitized projection superimposes on the raw image to have the better understanding in quantifying features of interest. Significantly, the system model updates quickly by 10 frames/second as the analyst move the slide. The robust technology makes the analysis in a very quick time and with enhanced prediction accuracy.

21.5.7 DRUG INVENTION

Drug discovery is the process of identifying the new medications after a long run research. Many years have been invested by combing possible methods and tested for new discovery. As the growing population across the globe, the new diseases are also emerging lead into the toughest challenge in treating the same. This identification of compound involves testing millions of drug-like compounds, in the verge of finding new combinations. So, sophisticated tools and technologies are involved in this process.

In recent times, NN (Chen et al., 2018) has been applied in drug identification and screening. The virtual drug screening provides testing for all possible combinations and the computational process is done with high speed with a reduced time factor. The virtual screening is done on the samples of the test experiment under study. The potential increase in data under screening continues to grow, which can be handled with multitask neural networks. The enhanced ML methods in virtual drug screening, increases the impact on this entire process of drug discovery.

21.5.8 SLEEP STAGING

Sleep is imperative for good health of a human being. Sleep staging (refer Figure 21.9) is vital for evaluating sleep disorders. In most of the sleep study, contact sensors are used which may affect the natural sleep and provides biased results. Availability of sleep study research is limited, thus elevates the complexity and demand. The novel approach is required for rapid eye movement (REM), non-REM, and wake staging (macro-sleep stages, MSS) estimation based on sleep sound analysis (Dafna et al., 2018). The different stages of sleep are categorized into,

- awake
- rapid eye movement
- non-rapid eye movement

These are the cycles of different stages occur about ninety minutes. However, these cyclical changes get affected due to modern lifestyle, stressful environment, and poor eating habits that can lead to cardiovascular and cerebrovascular diseases. The assessment and assistive technology are considered essential to monitor the sleep.

The conventional polysomnography (PSG) method measures the spectral analysis of EEG signals. The disadvantage of this method is its bulkiness in size and present us some technical challenges. Thus, collecting the signal recording during sleep is cumbersome. To overcome from this method, many researchers have suggested various other parameters like heart beat rate, inhaling rate, and electrocardiogram (ECG). Alternatively, ECG signals are easier to record and diagnose.

Recent advancements in DNN architecture is used in sleep analysis, which classifies the sleep stages into wake (W), REM, and non-REM. The DNN algorithms competently categorize the sleeping stages and ascertains if any abnormalities present. In comparison with the conventional PSG method, this approach is considered efficient in terms of reduced physiological parameters and reduced factors that affect the sleeping state. This helps to better understand the measurement of usual sleep. This study has great projection in the advancement of monitoring for sleep disorders and respiratory diseases.

21.5.9 WEARABLE TECH (IOT)

In this technically fast and modern era, factor of time is crucial in all aspects. Automation of devices and interconnection is another milestone in the digital technological development. Thus IoT, ML, AI are fabricated into small gadgets, with simple interface likely to enhance the life eminence and expectancy. This is represented in Figure 21.10. Wearables play a vital role nowadays tracking and monitoring the health data in an uncomplicated and user friendly manner. Let us explore how these technologies use the deep analytics and make predictions.

Wearables ranges from arm bands, watches, smart clothing, shoes, smart glasses and so on. Generally, they are used to track pulse rate, blood pressure, and temperature. The recordings and readings from the gadgets were remotely monitored and the real time data is further analyzed by the AI system. Many hospitals incorporated this tracking method after a patient is discharged from the hospital and successfully reduced the readmission rate and an emergency visit to the hospital.

From the report of WHO, it is evident that sixty percent of the health related problems are directly related to individual life cycle. If every individual is keen on tracking health measures regularly, the life expectancy will increase in a greater way. AI aided wearable enables to monitor vital signs, set a reminder for medications, and other health parameters.

21.5.10 CROWDSOURCING—AI

Crowdsourcing is an up-and-coming area of expertise in research. AI researchers are nowadays inclined into crowd opinion to build advanced prediction ML

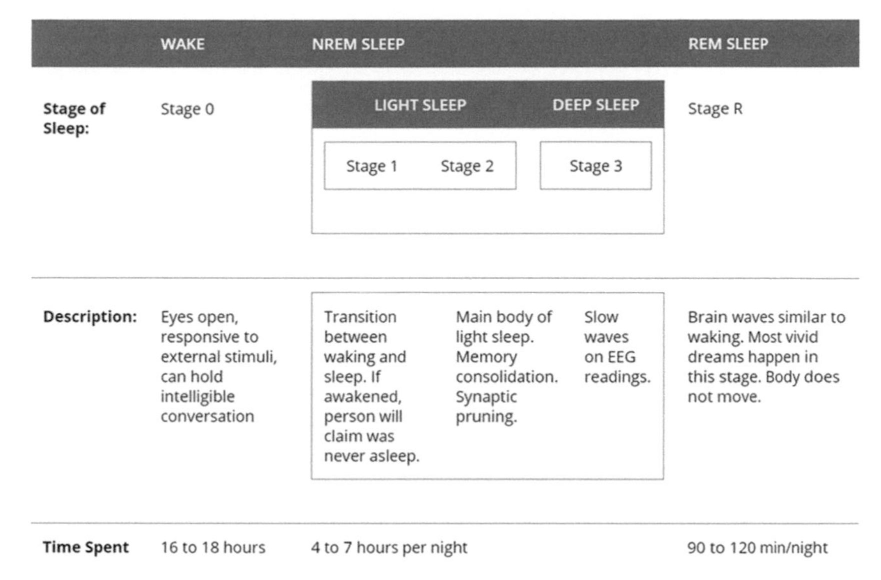

FIGURE 21.10 Sleep stages.
(*Source*: https://www.tuck.com/stages/#how_your_sleep_cycle_changes_with_age)

FIGURE 21.11 Wearable technology.
(*Source*: https://mobisoftinfotech.com/resources/blog/wearable-technology-in-healthcare/)

model based on informed decisions. The crowdsourced-AI tools collaboratively gather information via a common platform. The group members can analyze, share ideas and visualizations. The outcome of such a discussion converges on optimized decisions.

"Speech by Crowd" is one of the NLP-based platforms, to enable the crowdsource approach. But the challenge is how to structure the discussions and extract the significant insights. Swarm Intelligence based Artificial intelligence systems are coupled to attempt these challenges.

21.6 RESEARCH CHALLENGES AND FUTURE DIRECTIONS

Given the scope and application of AI–ML based system in various domains of health care, this area of expertise is in vibrant growth in large scale. The challenges include:

- **Data management**
  Health care data management involves data collection, data storage, processing, and management. Data collection methodologies comprise of different tools and techniques for instance, surveys, medical records, community health care center, lab reports, physician notes, and administrative records. A framework has to be intensified for complex data handling and data management. The advanced health care systems are deployed on the cloud and big data technologies.
- **Reduced complexity with enhanced and simple usage**
  The data analytics tools are generally used for generating reports and visualization of data. To implement AI based health care system, the key challenge is user friendly interface. The complexity of the system implementation is completely hidden from the user/stakeholder perspectives. The user-friendly apps and graphics interface systems are added advantages for predictive analytics.
- **Constructive to the society**
  Every system has a flip side. AI systems are capable of complex decision making at the same time, might be a threat to the human kind if it is used for vicious purposes. The newly developed system should be benefaction to the society.
- **Government AI implementation strategy—Reach to common man**
  An index for artificial readiness compiled by Oxford along with the International Development Research Centre has thrown the light on the Government policies on AI implementation strategy. Two basic queries were raised,
- National government implementing AI-enabled public services
- Beneficial to the community
  The index is measured by taking different metrics of AI systems. It is grouped by four significant metrics like ascendancy, infrastructure, readiness, and services. Recent technologies provide an opportunity to enhance the conventional health care system, hence governments should be prepared to absolutely employ this technology. Governments are instigated to

take up these challenges yet to avoid the associated risks.

- **Not a replacement for health experts, instead as an aiding tool**
  Indeed, the latest AI-enabled treatment enhance the accuracy, but not replace the doctors. It is always considered as the aiding tool, not as the replacement.
- **Reduced cost**
  This assistive technology is only made available to all when it is completely affordable. This factor is considered important as this should be advantageous to the society.

21.7 CONCLUSION

In this chapter, we have effectively analyzed and interpreted various aspects of AI and machine learning (ML) enabled technologies, prodigies and its independent applications. In the current research scenario the developments in AI and ML across various domains along with the applications of these technologies are prominent in biomedical study and the health care industry with proven working methods. A potential machine or deep learning system has the competence to adopt for rapid development in data accumulation and not only the textual data, but it also embraces and examines the image, genetic, and electrophysiology data. There are a need and demand to develop sophisticated algorithms and train the data values rigorously before the system assist in the prediction and suitable treatment suggestions and refinements. It may become an assertive technology that will empower and tutor the medical researchers, practitioners and academicians to provide better study programs along with prescribed treatments to serve their patients having a terminal illness with acute critical diseases. Technology is an ever growing and changing with time as so grows with countless challenges which have to be handled with advanced technology with specific and core AI models. On the same regards and notion ML is also trying to oversee the applications and possibilities of medical theories with assertive symptomatic services, which will drastically improve the accessibility and the accuracy of the medical services for the common man and mankind.

KEYWORDS

- **artificial intelligence**
- **machine learning**
- **biomedical AI**
- **deep learning**
- **big data analytics**

REFERENCES

Abidi, Syed Sibte Raza.; Samina Raza Abidi. Intelligent health data analytics: A convergence of artificial intelligence and big data. Healthcare management forum. Sage CA: Los Angeles, CA: SAGE Publications. 2019.

Afzal, Naveed.; et al. Mining peripheral arterial disease cases from narrative clinical notes using natural language processing. *Journal of Vascular Surgery*. 2017, 65(6), 1753–1761.

Anawar, Muhammad Rizwan.; et al. Fog computing: An overview of big IoT data analytics. *Wireless Communications and Mobile Computing*. 2018.

Beritelli F.; Capizzi G; Sciuto. G. L.; Napoli. C.; Scaglione. F. Automatic heart activity diagnosis based on gram polynomials and probabilistic neural networks. *Biomedical Engineering Letters*. 2018, 8(1), 77–85.

Billah. Mustain.; Sajjad Waheed. Gastrointestinal polyp detection in endoscopic images using an improved feature extraction method. *Biomedical Engineering Letters*. 2018, 8(1), 69–75.

Castro, VM.; Dligach D.; Finan S et al. Large-scale identification of patients with cerebral aneurysms using natural language processing. *Neurology*. 2017, 88, 164–168.

Chen. Hongming.; et al. The rise of deep learning in drug discovery. *Drug Discovery Today*. 2018, 23(6), 241–1250.

Dafna, E.; Tarasiuk. A.; Zigel. Y. Sleep staging using nocturnal sound analysis. *Scientific Reports*. 2018, 8, 1.

Fiszman, M.; Chapman WW.; Aronsky. D.; Automatic detection of acute bacterial pneumonia from chest X-ray reports. *Journal of the American Medical Informatics Association*. 2000, 7, 593–604.

Hassan. Mohammed K.; et al. Big Data Challenges and Opportunities in Healthcare Informatics and Smart Hospitals. Security in Smart Cities: Models, Applications, and Challenges. Springer, Cham. 2019, 3–26.

Islam. Md.; et al. A systematic review on healthcare analytics: Application and theoretical perspective of data mining. Healthcare. Multidisciplinary Digital Publishing Institute. 2018, 6, 2.

Izard. Santiago González.; et al. Applications of virtual and augmented reality in biomedical imaging. *Journal of Medical Systems*. 2019, 43(4), 102.

Jean-Rémi King.; Laura Gwilliams.; Chris Holdgraf.; Jona Sassenhagen.; Alexandre Barachant.; Encoding and Decoding Neuronal Dynamics: Methodological Framework to Uncover the Algorithms of Cognition. 2018, hal-01848442.

Jiang, Fei; Yong Jiang.; Hui Zhi.; Yi Dong.; Hao Li.; Sufeng Ma.;Yilong Wang.; Qiang Dong.; Haipeng Shen, Yongjun Wang. Artificial intelligence in healthcare: past, present and future. *Stroke and Vascular Neurology*. 2017, 2(4), 230–243.

Mansour, Romany F. Deep-learning-based automatic computer-aided diagnosis system for diabetic retinopathy. *Biomedical Engineering Letters*. 2018, 8(1), 41–57.

Miller, TP.; Li Y.; Getz KD.; Using electronic medical record data to report laboratory adverse events. *British Journal of Haematology*. 2017, 177, 283–6.

Nagpal, Kunal.; et al. Development and validation of a deep learning algorithm for improving Gleason scoring of prostate cancer. *npj Digital Medicine*. 2019, 2(1), 48.

Ngiam, Kee Yuan.; Wei Khor. Big data and machine learning algorithms for health-care delivery. *The Lancet Oncology*. 2019, 20(5), e262–e273.

Singh, Harpreet et al. iNICU-Integrated Neonatal Care Unit: Capturing neonatal journey in an intelligent data way. *Journal of Medical Systems*. 2017, 41(8), 132.

CHAPTER 22

NEURAL SOURCE CONNECTIVITY ESTIMATION USING PARTICLE FILTER AND GRANGER CAUSALITY METHODS

SANTHOSH KUMAR VEERAMALLA and
T. V. K. HANUMANTHA RAO*

Department of Electronics and Communication Engineering, National Institute of Technology, Warangal, Telangana 506004, India

Corresponding author. E-mail: tvkhrao75@nitw.ac.in

ABSTRACT

Connectivity is one of the major concerns in human brain mapping. It shows the connections across different brain regions through the nervous system. Until now, the connectivity between the electroencephalogram (EEG) signals has been calculated without taking into the consideration of volume conduction. Even though some of the methods show the flow across the scalp sources, we need a prior assumption about active brain regions. In this chapter, we suggest a new strategy to identify brain sources with their corresponding locations and amplitudes depending on a particle filter. Modeling of the time series (multivariate autoregressive) is used to detect movement and time dependence among the brain sources. Finally, Granger causality techniques have been applied to assess directional causal flow across the sources. We provide a framework to test the analytical pipeline on real EEG information. The results indicate that the suggested strategy is useful for evaluating the directional connections between EEG neural sources.

22.1 INTRODUCTION

Connectivity is considered to be anatomical (structural) functional (symmetrical) or effective (asymmetric) interaction between cerebral subsystems (Bullmore and Sporns, 2009). The analysis of the physical composition of the brain is done at a moment by evaluating the anatomical connections in the brain that is called as anatomical connectivity. Symmetric or functional connectivity is known as moving relationship between areas of the brain during the whole movement of neuronal data. It can be predicted either in a time domain or in a frequency domain by connections between neurons. Efficient or asymmetric connectivity examines the effects of one region of the brain on another. The aim is to detect which regions of the brain can induce other systems during the variation of neuronal data (Nunez and Srinivasan, 2006). The distinction between symmetric and asymmetric connectivities is that asymmetrical connectivity is depicted as a driver's relationship with the recipient, while symmetrical connectivity is defined as a relation between cerebral neurons (Haufe, 2011). Electroencephalogram (EEG) information was used to discover directional movement between neural sources by effective connectivity techniques such as Granger causality (GC) models. Because of the complex relationship between cerebral areas and their practices, it is important to understand the causal connections between cerebral initiation and effective networks among cerebrums neurons (Blinowska, 1992; Haufe and Ewald, 2016; Haufe et al., 2013; Kaminski and Barnett and Seth, 2015; Liao et al., 2011).

Granger casualty constitutes one of the designs that depend on efficient methods for network estimates (Ding et al., 2006). This technique is used for a linear prediction model, such as the multivariate autoregressive (MVAR) model. Directed transfer functions (DTFs) and partial directed coherence (PDC) constitute some of the measures available to this model (Barnett and Seth, 2013). When examining connectivity between cerebral areas, the crucial issue is that associations between scalp EEG signals cannot match the connections between hidden neural sources. It is because EEG signals do not deliver the median activity within the electrode region. This means that each sensor collects a linear superposition of signals from across the brain instead of measuring activity at only one brain site. This superposition of signals creates immediate associations in the data and can detect improper connectivity. It is because the locations of the channel cannot be seen as a physiological location of sources.

Dynamic sources should be found, and the network between them is estimated in the initial section of efficient connectivity assessment. The dipoles can be located using anatomical or physiological information or estimated using various approaches for localization of dipole. At present, there are two main fields of studies in neural generator modeling. The first modeling technique involves the use of imaging models that illustrate the data through a dense group of current dipoles at fixed locations. The second technique is a parametric approach that uses a single equivalent current dipole to replace the dense sets of existing dipoles (Makeig et al., 2002). While imaging-based methods can provide a comprehensive map of neuronal behavior in the brain, a parametric method offers EEG readings directly to a small amount of parameters (Mosher et al., 1992). With the parametric approach, equivalent current dipole models may be more intuitive in their interpretation, explaining the electrical activity of a brain and promoting it in emerging technologies such as brain–computer interface systems. The assessment of places for the corresponding dipole sources in the 3D brain volume using EEG measurements collected from scalp is a significant task of this parametric strategy. In EEG source localization, most of the previous work (Antelis and Minguez, 2009; Galka et al., 2004; Gordon et al., 1992; Mosher and Leahy, 1999; Mohseni et al., 2008; Sorrentino et al., 2007; Van Veen and Buckley, 1988) was based on two hypotheses: (1) the dipole numbers must be fixed and (2) the source locations are static in time. The area and quality assessment of EEG signals electrical current is in any case ill-posed (Miao et al., 2013). However, the number of neural dipoles and their positions varies with time generally. In (Sorrentino et al., 2009), at each step of the measured data, the number of neural dipoles and their positions is dynamically estimated and updated; they are based on random finite sets (RFSs) to handle the unknown number of dipoles. In this chapter, we used a particle filter (PF) for the location by locating the positions and amplitudes of the dipoles, which is equivalent to current dipoles from EEG signals.

The communication between sources should be shown in connectivity evaluations after retrieval of the dipole with its locations. Connectivity is calculated using dipole communication modeling and model-order estimation. The MVAR procedure shows the connection between the sources in the brain, especially in terms of the effect of one variable on the other.

To the best of the authors' knowledge, there is no systemic empirical study on neural source connectivity estimates, relying on both PF and GC techniques. In this chapter, we suggest a new strategy to estimate neural source space connectivity. In this chapter, we propose a framework that the connectivity studies are based on a two-step approach. The first step involves an estimation of the brain sources and its time course using an inverse method, then calculating the connectivity metrics using the estimated time courses of brain source. This approach can be applied to any sort of EEG information, as it can identify the directional connections between dipole sources without knowing the EEG source location beforehand. This chapter mainly aims to identify the area of the brain that can handle statistical dependence in cerebral neurons.

The rest of this chapter is organized as follows. In Section 22.2, we briefly describe the EEG source localization model and introduce the PF with systematic resampling for source localization. Next, we discussed the effective connectivity measures for source connectivity. The results obtained using the proposed technique are demonstrated in Section 22.3. Finally, this chapter is concluded in Section 22.4.

22.2 METHODS

22.2.1 EEG SOURCE LOCALIZATION MODEL

Localization of neural sources based on EEG utilizes scalp potential data to assume the location of underlying neural activity. This methodology involves a forward problem and an inverse problem. The forward problem consists of creating a model of how electrical activity propagates from the source space to the sensor space, and its result is the lead field matrix. Once this model is available, the inverse problem consists of estimating cortical activations given the sensor (scalp) measurements, by imposing some constraints (due to the ill-posed nature of the problem).

The volume conduction model in the head consists of nested concentric spheres (scalp, skull, and brain) with the constant conductivity (Miao, et al., 2013; Federica et al., 2010). To develop a solution for the source localization model in EEG, consider that N dipoles in cerebrum exhibit electrical activity. To evaluate such kind of activity, the multichannel EEG data can be used with $z_t \in \Re$; with n_s number of sensors at a given time t, the forward EEG model is given by

$$z_t = \sum_{i=1}^{N} L_i\left(x_t\left(i\right)\right) s_t\left(i\right) + v_t \qquad (22.1)$$

wherein a 3Dlocalization vector is represented as $x_t(i)$, and the lead field matrix and 3D moment vector (the source signal) for dipole i are denoted by $L_i\left(x_t\left(i\right)\right) = \Re^{ns\times 3}$ and $s_t\left(i\right)$, respectively. The EEG model has the observation noise represented by v_t. To compute a forward model "z_t," the *a priori* knowledge of head geometry, electrode position, and dipole localization is to be known. The representation of N number of dipoles is $x_t = [x_t(1),...,x_t(N)]$, where each single geometric position in 3D is given as $x_t(i) = [x(i)y(i)z(i)]^T, i = 1,2,...,N$.

As there are N number of dipoles, the lead field matrix $L_i(x_t) = \Re^{N \times ns}$ can be written as $L(x_t) = [L(x_t(1)),...,L(x_t(N))]$, and it depends on the location of the dipole $x_t(i)$ at time t. The vector of moments $s_t = \Re^{3N \times 1}$ is $s_t = [s_t(1),...,s_t(N)]^T$, where each single moment or mind source signal in 3D is given as $s_t(i) = [s_x(i)s_y(i)s_z(i)]^T$. Now, (22.1) can be rewritten in a matrix form as

$$Z = L(X)S + V \tag{22.2}$$

This equation can be taken as a measurement equation, and for the state equation, since how the states evolve is unknown, we can take this as a random walk model in the brain source localization problem

$$x_t = x_{t-1} + u_t \tag{22.3}$$

Equations (22.2) and (22.3) can be considered as measurement or observation equation and state equation, respectively. Based on the above two state and observation equation the EEG can be modeled as state-space model. To evaluate the dynamic parameters, that is, location (x, y, and z directions) in the cerebrum, we use the PF by considering the measured signal z_t at time t (Ebinger et al., 2015).

22.2.2 PARTICLE FILTER

The paradigm in state space is well adapted to problems with neural analysis, where an observed signal is influenced by certain unknown factors that change with time. Latent states are referred to as unidentified signals. This approach allows us to solve issues related to the estimation of latent signals and the adaptation of models between latent and observed signals and to statistical testing of their relationship. We must identify a couple of statistical models in order to build a state-space model. In the first model, the state model, the dynamics of the latent states are described. The second model, the observation model, explains how the latent state affects, each time, the probability distribution of the observation process. By considering the measurements, the PF is used to assess the dynamic state variables by approximating the posterior likelihood density function of the unidentified state parameters at each time point (Arulampalam et al., 2002). For such a dynamic system, the state-space model can be described in terms of x_t and z_t as

$$x_t = f\left(x_{t-1}\right) + u_t \quad (22.4)$$

$$z_t = h\left(x_t\right) + v_t \quad (22.5)$$

where x_t consists of N_x number of unidentified parameters at time t, z_t contains N_s number of observations at time t, $f(\cdot)$ and $h(\cdot)$ are considered to be nonlinear functions, where $f(\cdot)$ is related to state transition and $h(\cdot.)$ is related to the state vector along with the observation vector, u_t is the state model error, and v_t is the observation noise. The PF can be used to estimate the joint posterior likelihood density function of x_t at time t using N number of random particles $x^{(i)}_t$ along with weights $w^{(i)}_t, i = 1,2,..., N$, as

$$p\left(x_t / z_t\right) \approx \sum_{i=1}^{N} w(\text{i})\,\text{t}\,\delta\left(x_t - x^{(\text{i})}_t\right) \quad (22.6)$$

where $\delta(\cdot)$ is a Dirac delta function. Using (22.6), the estimated state vector can be written as

$$\hat{x}_t \approx \sum_{i=1}^{N} w^{(\text{i})}_t x^{(i)}_t \quad (22.7)$$

To assign the weights for particles, the importance density plays the vital role; based on the importance density, the sequential importance resampling PF can be used to estimate the states, and it consists of three phases: particle generation, weight calculation, and resampling.

22.2.2.1 PARTICLE GENERATION

The first phase is related to generation of particles denoted by $x^{(i)}_t$. Particles are taken from $q\left(x_t / x^{(\text{i})}_{t-1}, \text{z}_{1:t}\right)$ which is an importance density function, where $z_{1:t} = \left\{z_1, ..., z_t\right\}$.

22.2.2.2 WEIGHT CALCULATION

The particle weights can be computed as follows:

$$w^{(\text{i})}_t \propto w^{(\text{i})}_{t-1} \frac{p\left(z_t / x^{(i)}_t\right) p\left(x^{(i)}_t / x^{(i)}_{t-1}\right)}{q\left(x^{(i)}_t / x^{(i)}_{t-1}, z_{1:t}\right)} \quad (22.8)$$

The sum of all weights should be equal to 1, that is, $\sum_{i=1}^{N} w^{(i)}_t = 1$, provided that an importance density can be taken in such a way that $q\left(x_t / x^{(i)}_{t-1}, z_{1:t}\right) = p\left(x_t / x^{(i)}_{t-1}\right)$. Now, (22.8) becomes $w^{(i)}_t \propto w^{(i)}_{t-1} p\left(z_t / x^{(i)}_t\right)$.

22.2.2.3 RESAMPLING

Resampling is used to avoid rapid degeneracy of particles, that is, most of the particles with high weights will dominate the smaller weight particles. The dominance of these particles degeneracy leads to the poor posterior likelihood density function. To avoid particle degeneration, we used systematic resampling.

Systematic resampling is also called as universal sampling; in this popular technique, we draws only one random number, that is, one direction in the "wheel," for one particle and other particles with the $N - 1$ directions being fixed at $1/N$ increments from that randomly picked direction. Currently, $w_t^{(1)}$ is drained from the regular distribution on $\left(0, \frac{1}{N}\right]$, and whatever is left of the "w_t" information is acquired conclusively (Bolic et al., 2004), that is,

$$w_t^{(n)} \sim \mathrm{U}\left(0, \frac{1}{N}\right),$$
$$w_t^{(n)} = w_t^{(1)} + \frac{n-1}{N}, n = 2,3,\ldots,N \quad (22.9)$$

22.2.3 EFFECTIVE CONNECTIVITY MEASURES

Effective connectivity measures estimate the frequency-domain directional association between cerebrum regions that can be attained from spectral measures by utilizing MVAR models. Subsequently, coordinated communications can be measured by fitting the MVAR model to the time courses of the evaluated sources. These measures were proposed by Granger as, for two signals, if the first signal information can be predicted by using the previous information of the second signal, then it can be stated as casual to the first signal (Granger, 1969). Based on the Granger theory, effective connectivity measures can be categorized as GC, PDC (Baccala et al., 2007), and DTF (Kaminski et al., 2001). The time-series MVAR model equation can be written as

$$s(t) = \sum_{k=1}^{p} r(k) s(t-k) + a(t) \quad (22.10)$$

Here, $s(t)$ represents the $M \times 1$ neural sources time series, where M is the number of sources. $r(k)$ is the $M \times M$ coefficient matrix, which can be attained from the autoregressive (AR) model, and p is the model order of the AR process and can be calculated by the Akaike information criterion (Akaike, 1974) and the Bayesian information criterion (Schwarz, 1978). Rearranging (22.10), we obtain

$$a(t) - \sum_{k=0}^{p} \hat{r}(k) s(t-k) \quad (22.11)$$

where $\hat{r}(k) = -r(k)$ and $\hat{r}(0) = I$. Converting (22.11) to the frequency domain, we obtain

$A(f) = R(f)S(f)$.

Multiplying both sides with $R^{-1}(f)$ gives

$S(f) = Q(f)\,A(f)$, where $R^{-1}(f) - Q(f)$.

Here, $S(f)$ is the matrix of the multivariate process and $Q(f)$ is the transfer function of the system. This transfer function should give the information about the structure of the modeled system.

Now, the GC connectivity measure can be calculated as

$$GC_{mn} = In\left(\frac{\mathrm{var}\left(s_t^m / \hat{s}^m\right)}{\mathrm{var}\left(s_t^m / \hat{s}^m, \hat{s}^n\right)}\right) \quad (22.12)$$

where GC_{mn} represents the relation from $n \rightarrow m$ and $\hat{s}^m$ is the past value of the source m.

PDC is given as

$$PDC_{mn}(\mathrm{f}) = \frac{R_{mn}(\mathrm{f})}{\sqrt{\sum_{i=1}^{p} \left| R_{in}(f) \right|^2}} \tag{22.13}$$

where PDC_{mn} gives the causality from $n \rightarrow m$, PDC_{mn} is in the range from 0 to 1, where 0 represents no connectivity between n and m and 1 represents full connectivity between n and m.

The DTF is given as

$$DTF_{mn}(f) = \frac{Q_{mn}(f)}{\sqrt{\sum_{i=1}^{p} \left| Q_{ml}(f) \right|^2}} \tag{22.14}$$

where DTF_{mn} gives the causality influence from source n to source m at frequency f, and it is in the range of $0 \leq DTF_{mn} \leq 1$, where 0 represents no connectivity between n and m and 1 represents full connectivity between n and m.

22.2.4 *IMPLEMENTATION OF THE PROPOSED APPROACH*

The process of connectivity estimation is as follows:

Calculate the lead field matrix and source space.

The number of dipole sources is unknown and time variant. For this situation, we model these multiple dipole sources as RFSs.

Start analyzing the data at time $t = 1$.

Based on the probabilistic criteria, initialize the particle.

Assign the weights to each particle.

Normalize the weights.

Most of the particles with high weights will dominate the low-weight particles. The dominance of these particles' degeneracy leads to the poor posterior likelihood density function. Use resampling methods to discard the low-weight particles.

Prepare dipole configuration for the next time step.

Extraction of dipoles and their time series using the PF for EEG data.

The MVAR procedure portrays the connection between the sources inside the brain, particularly as far as the effect of one variable on another is concerned. MVAR allows us to derive time- and frequency-domain pictures of causality through the model-order coefficients and through their spectral representation.

The connectivity is calculated from the dipole communication modeling and estimation of model order.

Connectivity measures able to compute causality in the time and frequency domains are GC and PDC/DTF, respectively. Apply these measures and compute the directional connectivity between the sources.

22.3 RESULTS AND DISCUSSIONS

For simulation of the proposed method, the data were obtained from the Brainstorm EEG/Epilepsy dataset (Tadel et al., 2011). The data were recorded at a frequency of 256 Hz using 29 channels (FP1, FP2, F3, F4, C3, C4, P3, P4, O1, O2, F7, F8, T7, T8, P7, P8, Fz, Cz, Pz, T1, T2, FC1, FC2, FC5, FC6, CP1,CP2, CP5, and CP6) as per the 10/20 International framework. The simulation was performed for the EEG data from 10,000 to 20,000 ms over 921,600 samples, as illustrated in Figure 22.1. Note that for real EEG data, we do not have the accurate position of the source dipoles because it is nonlinear. The connectivity procedure contains two phases, that is, source localization and connectivity estimation. In source localization, first, the lead field matrix is computed using the FieldTrip toolbox (Oostenveld et al., 2011). Combination of the lead-field matrix and the EEG data applied to PF for estimating the position and time series of the sources. Figure 22.2a and b demonstrates the source location and amplitudes obtained from the EEG data. It can be observed that, at the given time interval, there are five different sources available, and locations of these sources are given in Table 22.1.

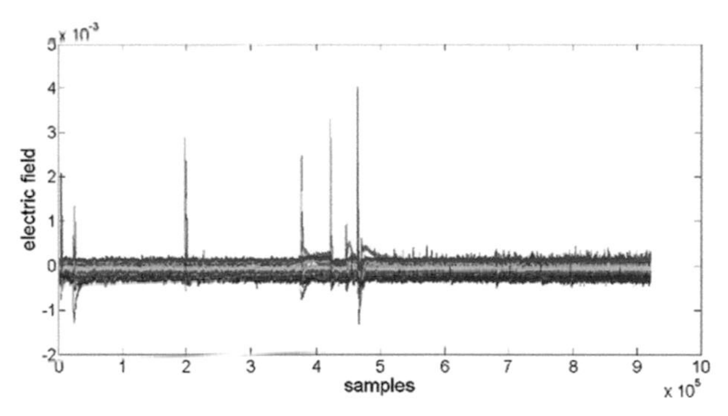

FIGURE 22.1 Real EEG data.

TABLE 22.1 Locations (x, y, and z directions) of the Extracted Sources from the EEG

Source	x	y	z
Source 1	–0.0043	–0.0074	0.0163
Source 2	–0.0160	0.0165	0.0359
Source 3	0.0063	0.0454	0.0280
Source 4	–0.0030	–0.0443	0.0401
Source 5	0.0663	–0.0015	0.0158

After source localization, the source time series had been applied to the MVAR model to find the interactions over time. By using GC methods, effective connectivity measures were obtained; Figure 22.3 shows the connectivity measures from GC. The establishment of the relation between source 1 and source 4, source 4 and source 5, and source 2 and source 3 is demonstrated. Figure 22.4 demonstrates the estimation of PDC with respect to frequency. In Figure 22.4a, the connectivity is

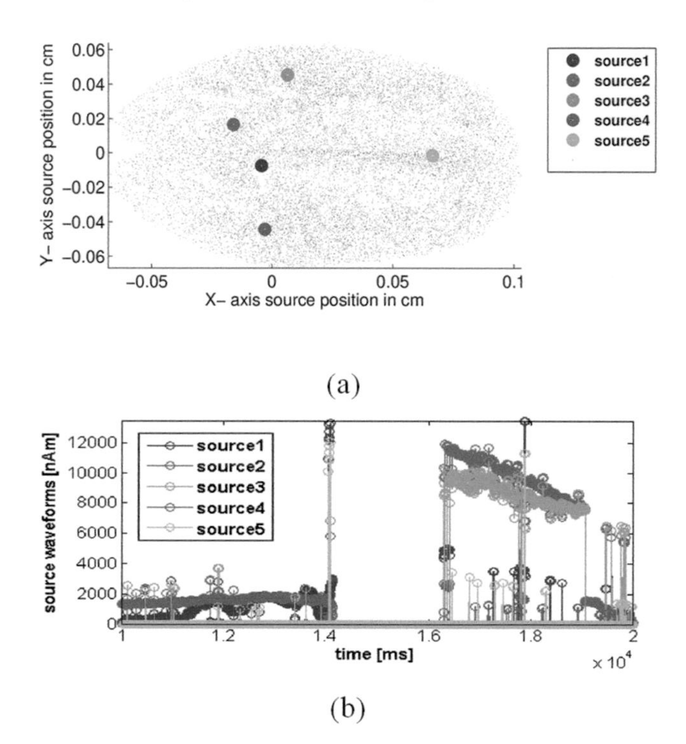

FIGURE 22.2 (a) Source extracted by using the PF. (b) Source amplitudes.

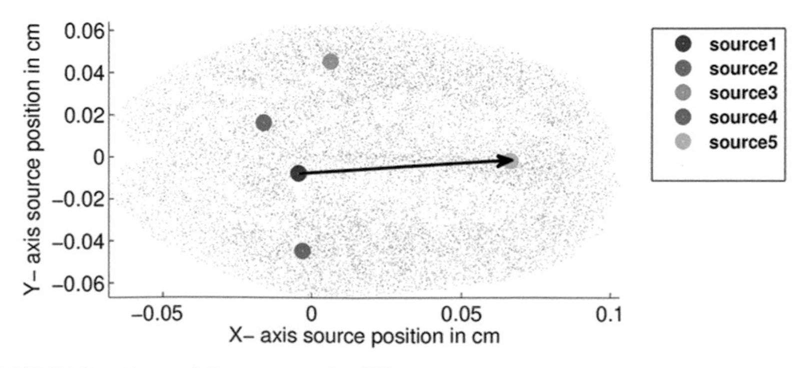

FIGURE 22.3 Connectivity measures by GC.

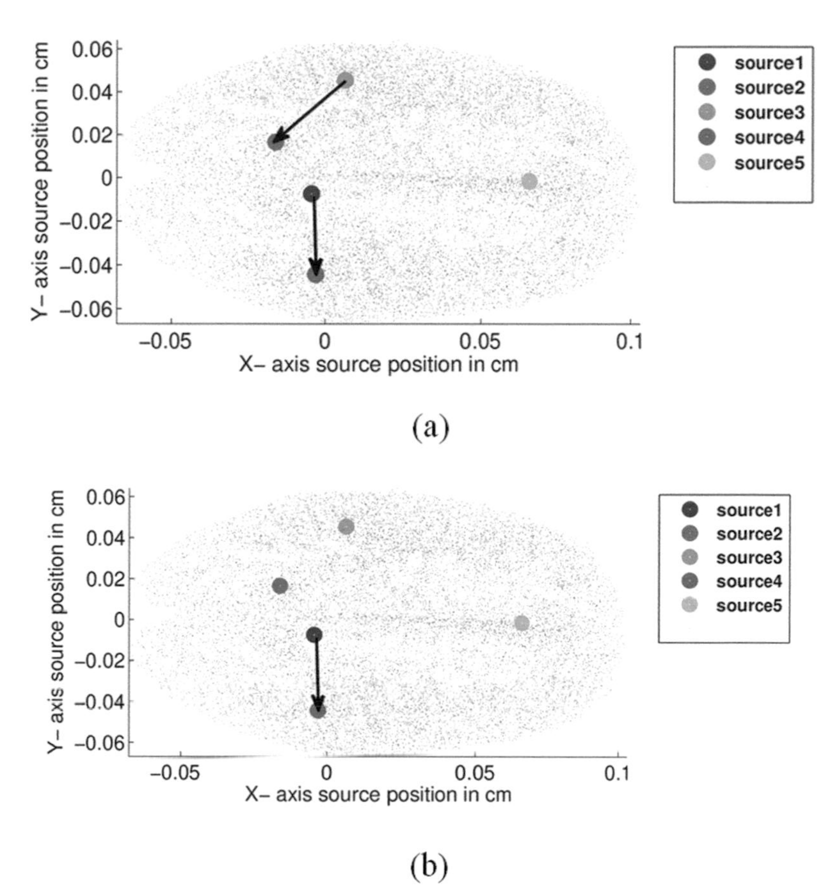

FIGURE 22.4 Connectivity measure by PDC (a) up to 4 Hz and (b) from 4 to 16 Hz.

shown from source 1 to source 4 and from source 3 to source 2 up to 4 Hz. Figure 22.4b shows the connectivity between source 1 and source 4 from 4 to 16 Hz. After 16 Hz, PDC is not showing any connectivity between the sources. In Figure 22.5, DTF connectivity estimation is illustrated. In Figure 22.5a, the connectivity is shown from source 1 to source 2, source 1 to source 3, source 1 to source 4, and source 1 to source 5 up to 4 Hz. Figure 22.5b shows the connectivity between source 1 and source 4 and source 3 and source 2 from 4 to 8 Hz. Figure 22.5c shows the connectivity between source 1 and source 4 from 8 to 16 Hz. After 16 Hz, the DTF does not show any connectivity between the sources. From the effective connectivity estimations, it can be observed that

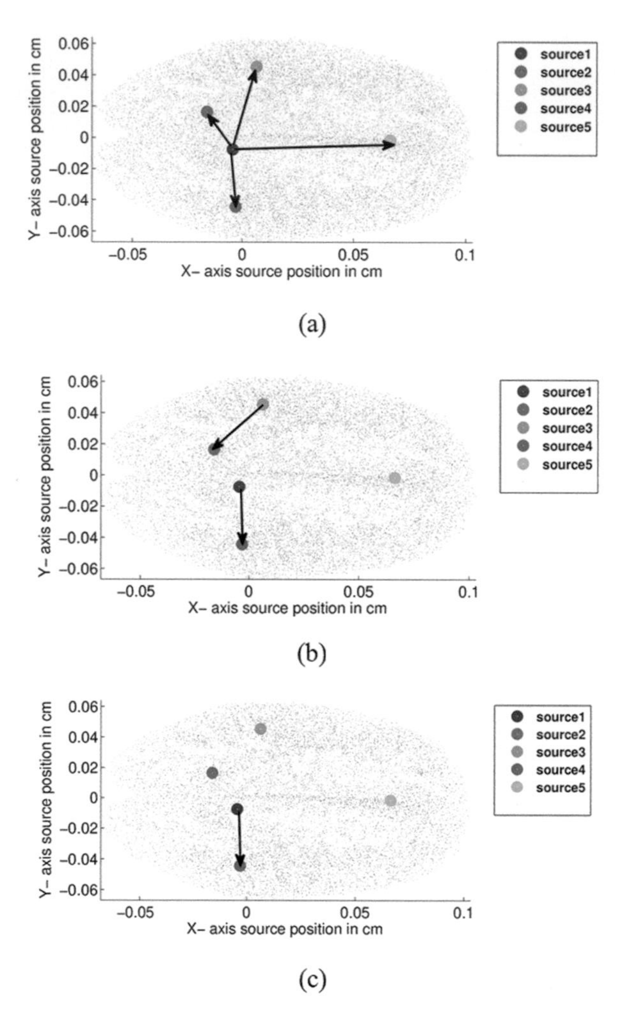

FIGURE 22.5 Connectivity measure by DTF (a) up to 4 Hz, (b) from 4 to 8 Hz, and (c) from 8 to 16 Hz.

GC, PDC, and DTF demonstrate the directional connectivity between source 1 and source 4.

22.4 CONCLUSION

Focusing on the relations and communications among regions of the brain and its useful activities is one of the imperative fields in investigating how brain works. These communications are called brain connectivity. The approach proposed in this chapter was utilized for assessing effective connectivity of cerebrum by using PF and GC methods. In this chapter, the PF was utilized for evaluating the locations of the sources by considering EEG signals as measurement model and MVAR; GC was used for estimating the connectivity between the sources. The proposed strategy is related to dynamic source actuation location. Therefore, it is more precise than different strategies that utilize static source actuation methods. This technique does not require any predefined data (e.g., other model-based strategies). The PF was applied to extract the sources and their amplitudes and applied the multivariate model on estimated sources, using GC techniques, to obtain the effective connectivity measures of the given data. The simulated results show the directed flow among the sources. Multivariate GC-based measures give a suitable framework to building up causal relations between neural populations. GC-based effective connectivity measures deliver the correlation on frequency-specific coupling in neural congregations, and they are vigorous with regard to volume conduction. In addition, they offer potential outcomes to take after dynamical changes of correlation among cerebrum structures. These measures give a significant methodology to investigate the expansive scale neural synchronization and its dynamics.

Financial Disclosure: The authors state no funding involved.

Conflict of Interest: The authors declare no potential conflict of interests.

Ethical Approval: The conducted research is not related to either human or animals use.

KEYWORDS

- **electroencephalography**
- **inverse problems**
- **connectivity**
- **Granger causality**
- **particle filter**

REFERENCES

Akaike H. (1974). A new look at the statistical model identification. *IEEE Trans Autom Control*; 19:716–723.

Antelis J, Minguez J. (2009). Dynamic solution to the EEG source localization problem using Kalman filters and particle filters. In *Int Conf IEEE Eng Med Biol Soc* (pp. 7780).

Arulampalam MS, Maskell S, Gordon N, Clapp T. (2002). A tutorial on particle filters for online nonlinear/non-Gaussian Bayesian tracking. *IEEE Trans Signal Process*; 50:174–188.

Baccala LA, Sameshima K, Takahashi DY. (2007). Generalized partial directed coherence. In: *Proc 15th Int Conf Digital Signal Process* (Cardiff: IEEE). p. 162166.

Barnett L, Seth AK. (2013). The MVGC multivariate Granger causality toolbox: A new approach to Granger-causal inference. *J Neurosci Methods*; 223:5068.

Barnett L, Seth AK. (2015). Granger causality for state space models.

Phys Rev E 91:040101. doi:10.1103/PhysRevE.91.040101

Bolic M, Djuric PM, Hong S. (2004). Resampling algorithms for particle filters: A computational complexity perspective. *EURASIP J Adv Signal Process*; 2004:2267–2277.

Bullmore E, Sporns O. (2009). Complex brain networks: Graph theoretical analysis of structural and functional systems. *Neuroscience*; 10:186–198.

Ding M, Chen Y, Bressler SL. (2006). Granger causality. Basic theory and application to neuroscience. In *Handbook of Time Series Analysis* (B. Schelter, M. Winterhalder, and J. Timmer, Eds.), John Wiley & Sons, Ltd., pp. 437–160.

Ebinger B, Bouaynaya N, Georgieva P, Mihaylova L. (2015). EEG dynamic source localization using marginalized particle filtering. In *2015 IEEE Int Conf Bioinform Biomed*, Washington, DC, USA, pp. 454–457.

Federica V, Fabio M, Fabrizio E, Stefano M, Francesco DS. (2010). Realistic and spherical head modeling for EEG forward problem solution: A comparative cortex-based analysis. *Comput Intel Nerosci*; 2010:972060.

Galka A, Yamashita O, Ozaki T, Biscay R, Valde P. (2004). A solution to the dynamical inverse problem of EEG generation using spatiotemporal Kalman filtering. *NeuroImage*; 23:435453.

Gordon NJ, Salmon DJ, Smith AFM. (1992). Novel approach to nonlinear/non-Gaussian Bayesian state estimation. *IEE Proc F: Radar Signal Process* 140:107113.

Granger C. (1969). Investigating causal relations by econometric models and cross-spectral methods. *Econometrica*; 37:424–438.

Haufe S. (2011). Towards EEG source connectivity analysis. Ph.D., Berlin Institute of Technology, Berlin, Germany.

Haufe S, Ewald A. (2016). A simulation framework for benchmarking EEG-based brain connectivity estimation methodologies. *Brain Topogr* 32:625–642. doi:10.1007/s10548-016-0498-y

Haufe S, Nikulin VV, Muller KR, Nolte G. (2013). A critical assessment of connectivity measures for EEG data: A simulation study. *Neuroimage* 64:120133. doi:10.1016/j.neuroimage.2012. 09.036

Kaminski MJ, Blinowska KJ. (1991). A new method of the description of the information flow in the brain structures. *Biol Cybern* 65:203210. doi:10.1007/BF00198091

Kaminski M, Ding M, Truccolo W, Bressler SL. (2001). Evaluating causal relations in neural systems: Granger causality, directed transfer function and statistical assessment of significance. *Biol Cybern*; 85:145157.

Liao W, Ding J, Marinazzo D et al. (2011). Small-world directed networks in the human brain: Multivariate Granger causality analysis of resting-state fMRI. *Neuroimage* 54:26832694. doi:10.1016/j.neuroimage.2010.11.007

Makeig S, Westerfield M, Jung TP, Enghoff S, Townsend J, Courchesne E, Sejnowski TJ. (2002). Dynamic brain sources of visual evoked responses. *Science*; 295:690694, DOI: 10.1126/science.1066168.

Miao L, Zhang JJ, Chakrabarti C, Papandreou-Suppappola A. (2013). Efficient Bayesian tracking of multiple sources of neural activity: Algorithms and real-time FPGA implementation. *IEEE Trans Signal Process*; 61:633–647.

Mohseni HR, Wilding EL, Sanei S (2008). Sequential Monte Carlo techniques for EEG dipole placing and tracking. In *Sens Array Multichannel Signal Process Workshop* (pp. 95–98).

Mosher JC, Leahy RM. (1999). Source localization using recursively applied and projected (RAP) MUSIC, *IEEE Trans Signal Process*; 47(2):332340.

Mosher JC, Lewis PS, Leahy RM. (1992). Multiple dipole modelling and localization from spatio-temporal MEG data. *IEEE*

Trans Biomed Eng; 39(6):541–557, DOI: 10.1109/10.141192

Nunez PL, Srinivasan R. (2006). *Electric Fields of The Brain: The Neurophysics of EEG.* Oxford University Press: Oxford, UK.

Oostenveld R, Fries P, Maris E, Schoffelen JM. (2011). FieldTrip: Open source software for advanced analysis of MEG, EEG, and invasive electrophysiological Data. *Comput Intel Neurosci*; 2011:156869.

Schwarz G. (1978). Estimating the dimension of a model. *Ann Statist*; 6:461464.

Sorrentino A, Parkkonen L, Pascarella A, Campi C, Piana M. (2009). Dynamical MEG source modeling with multi-target Bayesian filtering. *Human Brain Mapping*; 30:19111921

Sorrentino A, Parkkonen L, Piana M. (2007). Particle filters: A new method for reconstructing multiple current dipoles from MEG data. In *Int Congr Ser*; 1300:173176.

Tadel F, Baillet S, Mosher JC, Pantazis D, Leahy RM. (2011). Brainstorm: A user-friendly application for MEG/ EEG analysis. *Comput Intel Nerosci*; 2011:879716.

Van Veen BD, Buckley K. (1988). Beamforming: A versatile approach to spatial filtering. *IEEE ASSP Mag* 5:424.

CHAPTER 23

EXPLORATION OF LYMPH NODE-NEGATIVE BREAST CANCERS BY SUPPORT VECTOR MACHINES, NAÏVE BAYES, AND DECISION TREES: A COMPARATIVE STUDY

J. SATYA ESWARI[1,*] and PRADEEP SINGH[2]

[1]*Department of Biotechnology, National Institute of Technology Raipur, Raipur, Chhattisgarh 492010, India*

[2]*Department of Computer Science and Engineering, National Institute of Technology Raipur, Raipur, Chhattisgarh 492010, India*

Corresponding author. E-mail: satyaeswarij.bt@nitrr.ac.in

ABSTRACT

Background: In classification, when the distribution class data are uneven, the classification accuracy is biased by the majority class. In the case of the imbalanced data set, the classification methods are likely to perform poorly for minority class examples because they are aimed to optimize the overall accuracy of the class rather than considering the relative distribution of each class.

Procedure: In this chapter, we classify lymph node-negative breast cancer based on their gene expression signatures using a number of classification algorithms, namely, support vector machine (SVM), naïve Bayes, and decision tree. Since the data are imbalanced, we combined these algorithms with the synthetic minority oversampling technique (SMOTE) to handle the problem of classifying imbalanced data and to demonstrate that these techniques provide assistance for classification of imbalanced data sets. The performance of the SMOTE to balance gene expression signature data was examined.

Results: The results demonstrated that the accuracy increased by 8.15%, 15.92%, and 16.78% for SVM, naïve Bayes, and decision tree, respectively, and the area under the receiver operating characteristic curve increased by 32.4% in SVM, 45% in naïve Bayes, and 22.5% in decision tree after balancing the training data.

Conclusion: The results of this study show that the techniques of classification algorithms combined with the SMOTE can provide a significant solution for the class imbalance problem. This work also explains the probable application of classification techniques for the diagnosis of breast cancer and the identification of candidate genes in cancer patients.

23.1 INTRODUCTION

Metastasis is an advanced stage of tumor and can be cured by using belligerent therapy. This therapy is recommended, which led to a substantial decrease in breast cancer death rates. Hence, nowadays, majority of patients with breast cancer do not trust the traditional factors (Alizadeh et al., 2000; Ben-Dor et al., 2000). Even they do not require chemotherapy, early stage and intermediate-stage lymph node-negative patients undergo chemotherapy (Sehgal et al., 2011). Finally, it is needed to identify predictive markers, which are further linked to disease and can calculate the risk of metastasis in individual patients more precisely. There is also a need to accurately classify reappearance of an individual patient's risk of illness to be convinced that patient has received suitable therapy for his illness. In the past few decades, genome-wide expressions profiles with the various numbers of prognostic markers have been recognized (Fisher et al., 1983; Mitra et al., 2006). Microarray is a recent high-throughput technique for the detection of cancer. It helps to analyze large quantities of samples with new or previously generated data to test markers present in tumors. The development of this technique has transformed the cancer research. The cancer can be molecularly characterized by studying the expression profile of mRNA of a large number of genes (Bojarczuka et al., 2011; Perou et al., 2000; Venkateswaru et al., 2012; Yang et al., 2013). This helps in determining alterations of gene expressions between different types of tissue in healthy (control) and cancer patients. Identification of a subset of genes responsible for cancer can be performed by analyzing the vast amount of expression profile of mRNA. The identification of genes, subsequently, leads to their use as a prognostic or diagnostic biomarker for treatment of the cancer. The classification of tumors into different types of cancer has been made possible by the development of molecular classifiers. This method allows separation of tumors into relevant molecular subtypes, which are not feasible by pathological methods. Still, the reliable prediction of a robust gene signature for the detection of cancer becomes a challenge due to the presence of a large number of genes and a small number of patient samples (Eswari et al., 2013, 2015).

Various statistical and computational tools have been used for classification of cancer at molecular level (Eswari et al., 2017, Eswari and Dhagat, 2018). A discrimination of acute lymphoblastic leukemia with acute myeloid leukemia was proposed by Golub et al. (1999). This discrimination was based in a weighted voting scheme, which identified 50 genes. They also predicted the membership of new leukemia cases. The same data set was utilized by Mukherjee et al. to develop a support vector machine (SVM) classifier for classification of samples into acute lymphoblastic leukemia with acute myeloid leukemia (Vuong et al., 2014). Khan et al. classified different forms of small round blue cell tumors using artificial neural network and classified them into 96 genes (Batista et al., 2004). A 70-gene predictor was developed by Veer et al. for the prediction of breast cancer (Sehgal et al., 2011). A tree-based method for classification of leukemia and lymphoma was presented in the works of Zhang et al. (2003). They used tree-based method and constructed random forests for classification of cancers. All of these methods require the usage of statistical techniques for classification of tumors and, hence, pose a hindrance in determination of nonlinear relationships between genes. In contrast, complex models do not provide relationships between genes responsible for tumors. To evaluate gene expression data, naïve Bayes (NB), SVM, and decision trees (DTs) have been rigorously tested for their capability in making a distinction among cancers belonging to various diagnostic categories.

Outdated labor-intensive practices are not competent enough to investigate enormous and multifarious gene classifications encrypting massive volume of evidence. Implementation of new technology classification tools such as NB, SVM, and DTs has driven the expanse of abstraction of biologically noteworthy structures in the gene sequences (Elouedi et al., 2014; Eswari et al., 2013; Hedenfalk et al., 2001; Khan et al., 2001; Liu et al., 2013; Ma et al., 2009). Neural networks and evolutionary algorithms such as differential evolution and genetic algorithms are exploited for optimization and classification of biological systems (Eswari and Venkateswarlu, 2012; Hong and Cho, 2006; Laurikkala, 2001; Quinlan, 2014; Suryawanshi et al., 2020). Some researchers used feature selection methods for microarray analysis (He et al., 2014) and hierarchical clusterings (Tan et al., 2003). NB, SVM, and DTs are efficient methodologies for optimization of systems as a classifier. Microarray technology is used for finding pattern of expression with a large number of genes from all samples of cancer and metastasis in contradiction of expression of the same genes in control gene. Every gene has its own gene expression profile, and the cancer gene also has a unique gene expression profile. Many samples of patients have been taken for appraisal. After comparison, they have placed in groups to find out conjoint genes. Hence, the gene expression profile of microarray genes can have the capability assessing many genes simultaneously. Hence, classification of numerous cancers has been performed. Currently, only a few diagnostic tools are present to recognize patients with cancer risk. In this work, we intended to build up a gene-expression-based methodology and to apply it to provide quantitative predictions on disease outcome for patients with lymph-node-negative breast cancer. Certain investigators such as Taghipour et al. performed studies on Modeling Breast Cancer Progression and Evaluating Screening Policies (Li et al., 2001). However, in this study, classification of breast cancers based on gene expression profiles is carried out by using SVM, NB, and DTs.

23.2 METHODS AND ALGORITHMS

Research has been done elaborately on breast cancer classification by taking the consideration of physical features. These physical features use neural networks, fuzzy logic, NB, and SVM. However, molecular level of classification using gene expression profiles has reported less. Hence, in our study, we rigorously compared with the NB and SVM to classify lymph node-negative breast cancers using MATLAB software.

23.2.1 DATA GENERATION AND MICROARRAY DATABASE CONSTRUCTION

The classification of lymph negative breast cancer requires generation of data and construction of the database. The method for data generation and database preparation was performed using the following steps.

23.2.1.1 DATABASE

A total of 40 gene signatures for estrogen receptor-positive and negative patients, as reported by Wang et al. (Ben-Dor et al., 2000), were selected for the study. The sequences of the gene signatures were taken from the National Center for Biotechnology Information database, where each gene signature has its unique accession number. The obtained data were used for preprocessing and generation of data sets.

23.2.1.2 PREPROCESSING OF DATA AND GENERATION OF DATA SETS

Data encoding is a very important part to improve network performance for the classification of data. The DNA encoding gene signatures of lymph node-negative breast cancer are composed of four nucleotide bases. These four bases are assigned a numerical vector to training SVM, NB, and DT for classification of these sequences. The numerical values of 0, 0.5, 1, and 1.5 were chosen for encoding A, T, G, and C, respectively. In the following, Section 23.2.2 tells the basic idea of SVM, Section 23.2.3 discusses about the NB algorithm, and Section 23.2.4 talks about DT.

23.2.2 SUPPORT VECTOR MACHINE

The SVM algorithm is based on the structured risk maximization theory. This target of this theory is to minimize the generalization error. Sequential minimal optimization version of SVM is used in this experiment. We used complexity parameter $C = 1$ to depict the tolerance degree to errors. Radial basis function kernel, which is an efficient technique for classification, is used in this study. SVM as one of the well-known classifiers can be found in (Eswari and Venkateswarlu, 2016). SVM is a popular machine learning tool for tasks involving novelty detection, regression, or classification.

23.2.2.1 BRIEF ALGORITHM OF SVM

Vapnik proposed the SVM algorithm for density estimation, regression, and classification of data sets (Van't et al., 2002). In SVM, a hyperplane $w.x + b = 0, x_i \in \Re^n$ is found. This hyperplane separates those x_i data points of a given class that lie on the same side of the plane. These data points correspond to the following decision rule: $g(x) = \text{sign}\ (wx + b)$.

To determine the plane, a separating hyperplane is chosen by SVM, which is denoted by $w.x + b = 0$. The plane is chosen such that it is far from the data point x_i and has maximum margin. The hyperplane far away from data points minimizes the chance of wrong decisions during the classification of new data. In other words, the distance of the data points that are closest from the hyperplane is maximized in SVM (Naseriparsa and Kashani, 2014).

23.2.3 NAÏVE BAYES

The NB algorithm is a classification technique, which is based on probability. The advantages of this technique are that it is easier to be applied to different types of data series and provide better results than the existing ones. Suppose *Xip* is fault data without any class label and *H* is a hypothesis such that *X* falls into a class specified as *C*. We aim to establish *Pro (H|Xip)* as the posterior probability representing our confidence in the hypothesis after *Xip*, which is the observation, is given. The Bayesian theorem offers a technique of computing the *Pro(H|Xip)* using probability *Pro(Xip)*, *Pro(H)* and *Prob(Xip|H)* (Ma et al., 2009). The generalized Bayes relation is

$$Pro(H|Xip) = \frac{Pro(Xip|H).Pro(H)}{Pro(Xip)} \quad (23.1)$$

Consider a set of *m* samples *String* = $\{String_1, String_2, \ldots, String_m\}$, where sample String training data set is an *n*-dimensional feature vector $\{Xip_1, Xip_2, \ldots, Xip_n\}$.

Values Xip_i correspond to features $Fe_1, Fe_2, \ldots, Fe_n$.

There are *n* classes $cl_1, cl_2, \ldots, cl_n$, and every sample belongs to one of these classes.

In our model, the value of *n* is 4 as there are four classes.

When an extra data sample *Xip* with an unknown class is provided, class *X* can be predicted by highest conditional probability $Pro\ (Cl_k|Xip)$, where $k = 1, 2, \ldots, n$. The Bayes model is represented as follows:

$$Pro(cl_k|X) = \frac{Pro(Xip|Cl_k).Pro(Cl_k)}{Pro(Xip)} \quad (23.2)$$

Here, *Pro* is kept constant for all *cl*. The product *Pro(Xip|Clk)* needs to be maximized. The former probabilities of the *cl* can be estimated by

$$Pro(cl_k) = \frac{Number\ of\ training\ int\,ances\ of\ Class\ Cl_k}{m} \quad (23.3)$$

where *m* is the total number of training instances.

Taking into account for conditional probability (*C-Pro*) independence assumption between features, the *C-Pro* can be written as follows:

$$Pro(Cl_k|Xip) = P_{t=1}^{n} P(Xip_t|Cl_k) \quad (23.4)$$

where X_t represent features in sample *X* . The probabilities $P(Xt|C_k)$ can be expected from the training data set and are calculated for each attribute columns.

23.2.4 DECISION TREES

DTs predict responses of the data set. This is done by following the decisions in a bottom-up approach, such as a tree, where decisions are from root to leaf node. The responses for classification trees are in the form of "true" or false," whereas the responses are numerical in regression trees. DT adopts a greedy approach to construct DT in a top-down recursive divide & conquer manner. The calculations create a training set (TS) with the *cl* labels. The TS is recursively divided into reduced minisets as the tree is being built. An extension of ID3 (DT algorithm) C4.5 is used in this experiment. The additional features of C4.5 are taken for missing values, continuous attribute value ranges, DT pruning, and rules of derivation. These properties are used as an extension to information gain, which is named as the gain ratio; hence, bias for certain attributes can be considered (Ma et al., 2009).

This technique is represented by three elements, namely, a decision node for a test feature, a branch for one of the feature values, and a leaf that contains objects of the same or similar class. Formation of a DT involves building a tree and classification of the objects in that tree. The tree is built by placing a feature into an appropriate node and assigning a class to each of the leaf. The class of the objects is defined by starting from the root of the tree, to branch, and finally the leaf. The class of the leaf will be assigned as the class of the object (Howland et al., 2013).

23.2.5 TENFOLD CROSS-VALIDATION FOR THE METHODS OF SVM, NB, AND DT

We compare each classification algorithm's performance with imbalance and after removing the imbalance from data using synthetic minority oversampling technique (SMOTE) algorithms and recorded their various performance measures. For the purpose of reliable and stable results in the experiments, a *K*-fold cross-validation strategy is used. A *K*-fold classifier is generally used for classification accuracy measure. In this validation, *K* partitions are made, and out of these, one is used for testing and the rest are used for training. The data set can be shown as follows:

$$\begin{aligned} T_1 &= X_1 \; P_1 = X_2 U X_3 U \ldots U X_k \\ T_2 &= X_2 \; P_2 = X_1 U X_3 U \ldots U X_k \\ T_k &= X_k \; P_k = X_2 U X_3 U \ldots U X_{k-1} \end{aligned} \tag{23.5}$$

Here, T_1, T_2, ..., T_k are the partitions for testing and P_1, P_2, ...P_k are for training. *K* is typically 10 or 30. In this experiment, $k = 10$ has been used. In the 10-fold cross-validation process, the data are split into 10 identical disjoint parts. Nine of the 10 parts are used for training the framework, and one is used for testing. This is performed 10 times, and each time a different part of the data is used for testing. In this work, 40 lymph node-negative gene signatures of metastasis were given for 10-fold cross validation with the help of three techniques such as SVM, NB, and DT. In these 36 gene signatures from DNA, microarrays were taken as training and four were used as validation, and so on. Hence, imbalance in the data sets may be taken into account. We were able to apply methods successfully with the 10-fold cross validation, and the results are presented in the following sections.

23.3 RESULTS

This study presents SVM, NB, and DT to precisely identify the risk of tumor recurrence. This is beneficial to classify the patients as low- and high-risk groups in the case of lymph node-negative breast cancer. Various accuracy measures have been carried out using comparative studies to identify and classify better gene expression profiles of lymph node-negative breast cancer. Various accuracy measures, such as receiver operating characteristic (ROC), precision, accuracy, recall/sensitivity, and root-mean-square error, have been performed and concluded with a better method for identification of gene expression profiles.

23.3.1 ACCURACY MEASURES

Different measures have been proposed for two-class problems, wherein four possible cases (TP, FP, FN, and TN) can be represented in a confusion matrix (see Table 23.1).

23.3.1.1 ACCURACY

The accuracy is measured as

Accuracy (A) = correct classification/total number of classes. (23.6)

While talking about software fault prediction, A does not disclose the discrepancy between FP and FN. Overall, the accuracy that is determined has generally lesser relevance than recall and precision.

TABLE 23.1 Confusion Matrix

		Actual	
Predicted		**Faulty Module**	**Not Faulty Module**
	Faulty Module	TP (True positive)	FP (False positive)
	Not Faulty Module	FN (False negative)	TN (True negative)

23.3.1.2 RECALLING (SENSITIVITY) ABILITY

Recall, also known as the true positive rate or probability of fault detection, is represented as

Recall = correctly predicted faulty modules/total number of actually faulty modules. (23.7)

23.3.1.3 PRECISION ABILITY

Precision that is also referred to as correctness can be given as

Precision = correctly predicted modules/total number of predicted faulty modules. (23.8)

A high-precision value depicts that lesser effort is needed for inspection and testing.

23.3.2 AREA UNDER THE ROC CURVE (AUC)

To compare the results, we have taken the average value of AUC and accuracy in 10-fold cross validation. The AUC represents the most informative and objective indicator of predictive accuracy as known and reported by other researchers (Van't et al., 2002). A better classifier should produce a higher AUC. We have also used the AUC for our study.

23.3.2.1 CRITICAL ANALYSIS FOR CONSIDERING IMBALANCED DATA

In recent years, researchers have observed that an imbalanced data set can prove to be an obstacle for learning and accuracy calculation for learning algorithms (Chawla et al., 2002; Singh et al., 2014; Sweilam et al., 2010; Mahata, 2010). In order to handle the class imbalance problem, SMOTE oversampling is used, and the same procedure as followed in (Gu et al., 2008) is used for implementation. SMOTE (Singh et al., 2014) is a technique for oversampling. This technique is used to enhance the instances of minority class by the nearest neighbor method, which are the synthetic class instances. As this method generated synthetic class instances instead of replicate minority class instances, this overcomes the problem of overfitting.

The first step of the algorithm is the assignment of initial data, named as D. This proceeds with the subset M. Every instance in M is taken as x. Randomization of k from different instances leads to the generation of difference, *Diff*. Various random numbers, ranging from 0 to 1, are generated using *rand fun* => $n = x + Diff * rand$. After addition of n to D, the algorithm is repeated, and finally, the function is ended.

23.3.2.2 THE SMOTE ALGORITHM

SMOTE is a technique to generate an additional data set, wherein the minority samples are oversampled by 192% to obtain a ratio of 1:1. The classical method to determine accuracy cannot be used to determine the performance of the unbalanced data set as the impact of minority class on accuracy is greater than that of the majority class (Gu et al., 2008). Some researchers used the SMOTE for classifying cancer imbalanced data (Blagus, 2012; Ramaswamy et al., 2003; Kothandan, 2015). Thus, alternative methods, such as accuracy, recall, precision, and AUC, should be used for evaluating the performance of the imbalanced data set. From Table 23.2, the original unbalance data set has the accuracy of 62.74% with DT classification, 72.55% with NB classification, and 74.5% with SVM.

TABLE 23.2 Performance Measures of Various Algorithms for Classification of Cancer Data—With Imbalance

	Decision Tree	Naïve Bayes	SVM
ROC	0.59	0.45	0.5
Precision	0.76	0.74	0.745
Accuracy	62.75	72.55	74.51
Recall/ sensitivity	0.73	0.97	1

For all classifiers, the sensitivity value is 73%, 97%, and 100%, respectively. The high accuracy states that the classifier can classify the majority class sample well and can be unsuccessful in classifying the minority set. The data set used in the experiment is balanced by the SMOTE by generating synthetic data of minority class and combining all the majority samples to make equal proportion of each class. This depicted that the cancer classification outcome was better when balanced by the SMOTE than the original unbalance data. With the SVM, NB, and DT classification of microarray data, the sensitivity value was found to be 0.73 even after balancing the data, as shown in Table 23.3.

In this study, the SMOTE was shown to be a good classification technique for all three classifiers: SVM, NB, and DT. From Tables 23.2 and 23.3, it can be observed that accuracy is increased by 15.92% for the DT, 16.78% for NB, and 8.15% in SVM classification results (see Figure 23.1).

TABLE 23.3 Performance Measures of Various Algorithms for Classification of Cancer Data—Without Imbalance

	Decision Tree	Naïve Bayes	SVM
ROC	0.82	0.90	0.82
Precision	0.82	0.85	0.74
Accuracy	78.67	89.33	82.67
Recall/ sensitivity	0.737	1	1

In this study, the major issue in building a good prediction model was found to be not only the ratio of majority and minority samples, but also the requirement for good training samples that can show the properties of data consistent with the corresponding class label assigned to them. In the majority of the cases, the records present in clinical data sets do not truly represent the properties of data consistent with the corresponding outcome label. By the comparison of the results, it can be seen that AUC has increased by 22.5% in DT, 45% in NB, and 32.4% in SVM after balancing the training data (see Figure 23.2).

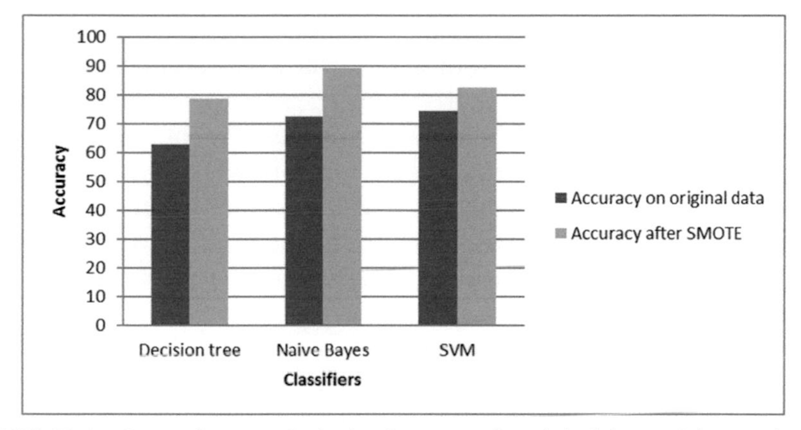

FIGURE 23.1 Comparisons on the basis of accuracy for original data and data used on the basis of SMOTE.

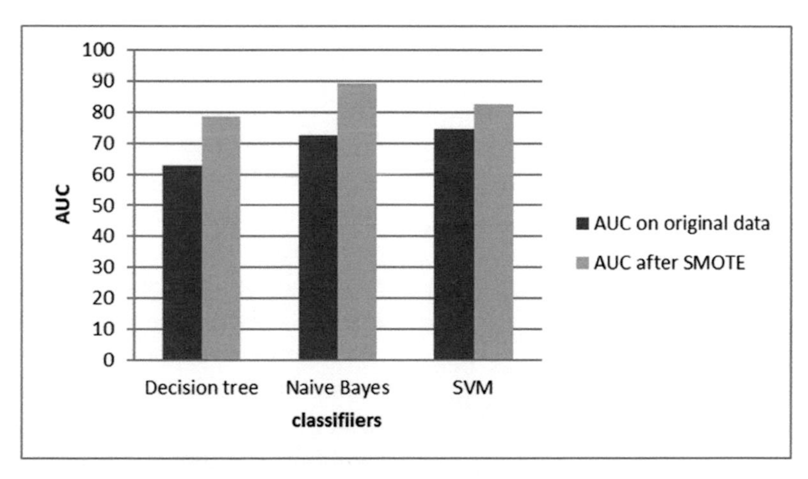

FIGURE 23.2 Comparisons on the basis of AUC for original data and data used on the basis of SMOTE.

23.4 DISCUSSION

In research, molecular classification has been applied to group tumors based on their gene expression, but its practical application in clinics has been hindered. This is because of the fact that a large number of feature genes are needed to construct a discriminative classifier. The need of the hour is the construction of molecular classifiers consisting of a small number of genes. This would be helpful for clinical diagnosis, where a diagnostic assay for the evaluation of several genes in a single test cannot be performed (Dhanasekaran et al., 2001; Eswari and Chimmiri, 2013; Ho et al., 2006; Kharya et al., 2014; Mukherjee et al., 1999; Park et al., 2012; Taghipour et al., 2013). In this study, a molecular classification system using SVM, NB, and DT was developed for lymph node-negative breast cancer.

He et al., in their work, discussed about the possible issues that affect the prognosis of lymph node-negative breast cancer. This includes type of tumor, age and tissue type, the status of hormone receptors, and tumor diameters (Golub et al., 1999). Fisher et al. described that the major predictive feature for breast cancer has been axillary lymph node involvement (ALNI) (Jones et al., 2013). Classification of molecular subtype by computation tools fails to accurately predict ALNI (Ng and Dash, 2006; Vapnik, 2013), but the occurrence of ALNI appears to be more in human epidermal growth factor receptor 2-positive tumors and less in luminal A tumors (Vapnik, 2013; Wang et al., 2005). The identification of predictive factors will help the physicians determine the suitable therapeutic approach. To achieve this, computational tools provide a feasible approach to study gene expression profiles by the molecular level of classification.

In this work, the SMOTE was found to be better classification technique than SVM, NB, and DT, as the accuracy increased by 8.15%, 16.78%, and 15.92% for SVM, NB, and DT, respectively. The AUC was improved by 32%, 45%, and 22.5% in SVM, NB, and DT, respectively, after the training data were balanced.

23.5 CONCLUSION

This chapter presents the techniques to predict breast cancers by lymph node. Generally, the original data are imbalanced, and when the data set is extremely imbalanced, the currently existing classification techniques do not function well on minority class examples. Therefore, the sampling strategy is one of the possible solutions for the imbalanced class problem. In this study, the SMOTE was examined with three machine learning algorithms to classify the test data sets, including DT, NB, and SVM, for lymph node-negative breast cancer. The results of classification were calculated and compared for the class imbalance problem, recall, and AUC. The results obtained from this study indicate that these techniques combined with the SMOTE generally perform better when training is done by imbalanced data.

ACKNOWLEDGMENT

The authors are thankful to the National Institute of Technology Raipur for providing the necessary computational facility to analyze and prepare the manuscript and for permission to publish it.

KEYWORDS

- **DNA microarray-based gene expression profile**
- **classification**
- **breast cancer**
- **machine learning**

REFERENCES

Alizadeh, A.A., Eisen, M.B., Davis, R.E., Ma, C., Lossos, I.S., Rosenwald, A., Boldrick, J.C., Sabet, H., Tran, T., Yu, X., Distinct types of diffuse large B-cell lymphoma identified by gene expression profiling, *Nature*. 403 (2000) p. 503.

Batista, G.E., Prati, R.C., Monard, M.C., A study of the behavior of several methods for balancing machine learning training data, *ACM SIGKDD Explorations Newsletter*. 6 (2004) pp. 20–29.

Ben-Dor, A., Bruhn, L., Friedman, N., Nachman, I., Schummer, M., Yakhini, Z., Tissue classification with gene expression profiles, *Journal of Computational Biology*. 7 (2000) pp. 559–583.

Blagus, R., Lusa, L.. Evaluation of SMOTE for high-dimensional class-imbalanced microarray data. in *11th International Conference on Machine Learning and Applications*. (2012) pp. 89–94.

Bojarczuka, C.C., Lopesb, H.S., Freitasc, A.A., Data mining with constrained-syntax genetic programming: Applications in medical data set. *Algorithms*. 6 (2001) p. 7.

Chawla, N.V., Bowyer, K.W., Hall, L.O., Kegelmeyer, W.P., SMOTE: Synthetic minority over-sampling technique, *Journal of Artificial Intelligence Research*. 16 (2002) pp. 321–357.

Dhanasekaran, S.M., Barrette, T.R., Ghosh, D., Shah, R., Varambally, S., Kurachi, K., Pienta, K.J., Rubin, M.A., Chinnaiyan, A.M., Delineation of prognostic biomarkers in prostate cancer, *Nature*. 412 (2001) p. 822.

Elouedi, H., Meliani, W., Elouedi, Z., Amor, N.B.. A hybrid approach based on decision trees and clustering for breast cancer classification. In *6th International Conference of Soft Computing and Pattern Recognition*. (2014) pp. 226–231.

Eswari, J.S., Anand, M., Venkateswarlu, C., Optimum culture medium composition for rhamnolipid production by pseudomonas aeruginosa AT10 using a novel multi-objective optimization method, *Journal of Chemical Technology and Biotechnology*. 88 (2013) pp. 271–279.

Eswari, J.S., Chimmiri, V., Evaluation of kinetic parameters of an anaerobic biofilm reactor treating pharmaceutical industry wastewater by ant colony optimization, *Environmental Engineering Science.* 30 (2013) p. 527.

Eswari, J.S., Venkateswarlu, C., Optimization of culture conditions for Chinese hamster ovary (CHO) cells production using differential evolution, *International Journal of Pharmacy and Pharmaceutical Sciences.* 4 (2012) pp. 465–470.

Eswari, J.S. and Venkateswarlu, C. Dynamic modelling and metabolic flux analysis for optimized production of rhamnolipids. Chemical Engineering Communications, 203 (2016) pp. 326–338.

Eswari, J.S. and Swasti Dhagat. Surfactin assisted synthesis of silver nanoparticles and drug design aspects of surfactin synthetase. *Advances in Natural Sciences: Nanoscience and Nanotechnology*. accepted. 2018.

Eswari, J.S., Swasti Dhagat, Shubham Kaser and Anoop Tiwari, Molecular docking and homology modelling studies of Bacillomycin and Iturin synthetases for the production of therapeutic lipopeptides, 2017, *Current Drug Discovery Technologies*.

Fisher, Bauer, M., Wickerham, D.L., Redmond, C.K., Fisher, E.R., Cruz, A.B., Foster, R., Gardner, B., Lerner, H., Margolese, R., Relation of number of positive axillary nodes to the prognosis of patients with primary breast cancer. An NSABP update, *Cancer.* 52 (1983) pp. 1551–1557.

Golub, T.R., Slonim, D.K., Tamayo, P., Huard, C., Gaasenbeek, M., Mesirov, J.P., Coller, H., Loh, M.L., Downing, J.R., Caligiuri, M.A., Molecular classification of cancer: Class discovery and class prediction by gene expression monitoring, *Science*. 286 (1999) pp. 531–537.

Gu, Q., Cai, Z., Zhu, L., Huang, B., Data mining on imbalanced data sets, in *International Conference on Advanced Computer Theory and Engineering*. (2008) pp. 1020–1024.

He, J., Wang, H., Ma, F., Feng, F., Lin, C., Qian, H., Prognosis of lymph node-negative breast cancer: Association with clinicopathological factors and tumor associated gene expression, *Oncology Letters*. 8 (2014) pp. 1717–1724.

Hedenfalk, I., Duggan, D., Chen, Y., Radmacher, M., Bittner, M., Simon, R., Meltzer, P., Gusterson, B., Esteller, M., Raffeld, M., Gene-expression profiles in hereditary breast cancer, *New England Journal of Medicine*. 344 (2001) pp. 539–548.

Ho, S.-Y., Hsieh, C.-H., Chen, H.-M., Huang, H.-L. Interpretable gene expression classifier with an accurate and compact fuzzy rule base for microarray data analysis, *Biosystems*. 85 (2006) pp. 165–176.

Hong, J.-H., Cho, S.-B., The classification of cancer based on DNA microarray data that uses diverse ensemble genetic programming, *Artificial Intelligence in Medicine*. 36 (2006) pp. 43–58.

Howland, N.K., Driver, T.D., Sedrak, M.P., Wen, X., Dong, W., Hatch, S., Eltorky, M.A., Chao, C., Lymph node involvement in immunohistochemistry-based molecular classifications of breast cancer, *Journal of Surgical Research*. 185 (2013) pp. 697–703.

Jones, T., Neboori, H., Wu, H., Yang, Q., Haffty, B.G., Evans, S., Higgins, S., Moran, M.S., Are breast cancer subtypes prognostic for nodal involvement and associated with clinicopathologic features at presentation in early-stage breast cancer? *Annals of Surgical Oncology*. 20 (2013) pp. 2866–2872.

Khan, J., Wei, J.S., Ringner, M., Saal, L.H., Ladanyi, M., Westermann, F., Berthold, F., Schwab, M., Antonescu, C.R., Peterson, C., Classification and diagnostic prediction of cancers using gene expression profiling and artificial neural networks, *Nature Medicine*. 7 (2001) p. 673.

Kharya, S., Agrawal, S., Soni, S., Naïve Bayes classifiers: A probabilistic detection model for breast cancer, *International Journal of Computer Applications*. 92 (2014) pp. 26–31.

Kothandan, R., Handling class imbalance problem in miRNA dataset associated with cancer, *Bioinformation*. 11 (2015) p. 6.

Laurikkala, J., Improving identification of difficult small classes by balancing class distribution. in *Conference on Artificial Intelligence in Medicine in Europe*. (2001) pp. 63–66.

Li, L., Weinberg, C.R., Darden, T.A., Pedersen, L.G., Gene selection for sample classification based on gene expression data: Study of sensitivity to choice of parameters of the GA/KNN method, *Bioinformatics*. 17 (2001) pp. 1131–1142.

Liu, H.-C., Peng, P.-C., Hsieh, T.-C., Yeh, T.-C., Lin, C.-J., Chen, C.-Y., Hou, J.-Y., Shih, L.-Y., Liang, D.-C., Comparison of feature selection methods for cross-laboratory microarray analysis, *IEEE/ACM Transactions on Computational Biology and Bioinformatics*. 10 (2013) pp. 593–604.

Ma, J., Nguyen, M.N., Rajapakse, J.C., Gene classification using codon usage and support vector machines, *IEEE/ACM Transactions on Computational Biology and Bioinformatics*. 6 (2009) pp. 134–143.

Mahata, P., Exploratory consensus of hierarchical clusterings for melanoma and breast cancer, *IEEE/ACM Transactions on Computational Biology and Bioinformatics (TCBB)*. 7 (2010) pp. 138–152.

Mitra, A.P., Almal, A.A., George, B., Fry, D.W., Lenehan, P.F., Pagliarulo, V., Cote, R.J., Datar, R.H., Worzel, W.P., The use of genetic programming in the analysis of quantitative gene expression profiles for identification of nodal status in bladder cancer, *BMC Cancer*. 6 (2006) p. 159.

Mukherjee, S., Tamayo, P., Slonim, D., Verri, A., Golub, T., Mesirov, J., Poggio, T., Support vector machine classification of microarray data, (1999).

Naseriparsa, M., Kashani, M.M.R., Combination of PCA with SMOTE resampling to boost the prediction rate in lung cancer dataset, (2014) *arXiv:1403.1949v1*.

Ng, W., Dash, M. An evaluation of progressive sampling for imbalanced data sets, in *6th IEEE International Conference on Data Mining—Workshops*. (2006) pp. 657–661.

Ooi, P. Tan, Genetic algorithms applied to multi-class prediction for the analysis of gene expression data. *Bioinformatics*. 19 (2003) pp. 37–44.

Park, S., Koo, J.S., Kim, M.S., Park, H.S., Lee, J.S., Lee, J.S., Kim, S.I., Park, B.-W., Characteristics and outcomes according to molecular subtypes of breast cancer as classified by a panel of four biomarkers using immunohistochemistry, *The Breast*. 21 (2012) pp. 50–57.

Perou, C.M., Sørlie, T., Eisen, M.B., Van De Rijn, M., Jeffrey, S.S., Rees, C.A., Pollack, J.R., Ross, D.T., Johnsen, H., Akslen, L.A., Molecular portraits of human breast tumours, *Nature*. 406 (2000) p. 747.

Quinlan, J.R., *C4. 5: Programs for Machine Learning*. Amsterdam, The Netherlands: Elsevier. 2014.

Ramaswamy, S., Ross, K.N., Lander, E.S., Golub, T.R., A molecular signature of metastasis in primary solid tumors, *Nature genetics*. 33 (2003) p. 49.

Seema Patel, Shadab Ahmed, Eswari, J.S. Therapeutic cyclic lipopeptides mining from microbes: Strides and hurdles, *World Journal of Microbiology and Biotechnology*. 31 (2015), pp. 1177–1193.

Sehgal, A.K., Das, S., Noto, K., Saier, M., Elkan, C., Identifying relevant data for a biological database: Handcrafted rules versus machine learning, *IEEE/ACM*

Transactions on Computational Biology and Bioinformatics. 8 (2011) pp. 851–857.

Singh, P., Verma, S., Vyas, O., Software fault prediction at design phase, *Journal of Electrical Engineering & Technology*. 9 (2014) pp. 1739–1745.

Suryawanshi N., Sahu J., Moda Y. and Eswari J.S. Optimization of process parameters for improved chitinase activity from thermomyces sp. by using artificial neural network and genetic algorithm. *Preparative Biochem Biotechnol*. (2020), in press.

Sweilam, N.H., Tharwat, A., Moniem, N.A., Support vector machine for diagnosis cancer disease: a comparative study, *Egyptian Informatics Journal*. 11 (2010) pp. 81–92.

Taghipour, S., Banjevic, D., Miller, A., Montgomery, N., Jardine, A., Harvey, B., Parameter estimates for invasive breast cancer progression in the Canadian National Breast Screening Study, *British Journal of Cancer.* 108 (2013) p. 542.

Tan, K.C., Yu, Q., Heng, C., Lee, T.H., Evolutionary computing for knowledge discovery in medical diagnosis, *Artificial Intelligence in Medicine*. 27 (2003) pp. 129–154.

Van't Veer, L.J., Dai, H., Van De Vijver, M.J., He, Y.D., Hart, A.A., Mao, M., Peterse, H.L., Van Der Kooy, K., Marton, M.J., Witteveen, A.T., Gene expression profiling predicts clinical outcome of breast cancer, *Nature*. 415 (2002) p. 530.

Vapnik, V., *The Nature of Statistical Learning Theory*. Springer science & business media: New York, NY, USA, (2013).

Venkateswarlu, K. Kiran, Eswari, J., A hierarchical artificial neural system for genera classification and species identification in mosquitoes. *Applied Artificial Intelligence*. 26 (2012) pp. 903–920.

Vuong, P.T. Simpson, Green, B., Cummings, M.C., Lakhani, S.R., Molecular classification of breast cancer, *Virchows Archiv*. 465 (2014) pp. 1–14.

Wang, Y., Klijn, J.G., Zhang, Y., Sieuwerts, A.M., Look, M.P., Yang, F., Talantov, D., Timmermans, M., Meijer-van Gelder, M.E., Yu, J., Gene-expression profiles to predict distant metastasis of lymph-node-negative primary breast cancer, *The Lancet*. 365 (2005) pp. 671–679.

Yang, C.-H., Lin, Y.-D., Chaung, L.-Y., Chang, H.-W. Evaluation of breast cancer susceptibility using improved genetic algorithms to generate genotype SNP barcodes, *IEEE/ACM Transactions on Computational Biology and Bioinformatics (TCBB)*. 10 (2013) pp. 361–371.

Zhang, H., Yu, C.-Y., Singer, B., Cell and tumor classification using gene expression data: construction of forests, *Proceedings of the National Academy of Sciences*. 100 (2003) pp. 4168–4172.

INDEX

C

D

F

I

R

S

U

V

W

X